Multiscale Cancer Modeling

CHAPMAN & HALL/CRC
Mathematical and Computational Biology Series

Aims and scope:
This series aims to capture new developments and summarize what is known over the entire spectrum of mathematical and computational biology and medicine. It seeks to encourage the integration of mathematical, statistical, and computational methods into biology by publishing a broad range of textbooks, reference works, and handbooks. The titles included in the series are meant to appeal to students, researchers, and professionals in the mathematical, statistical and computational sciences, fundamental biology and bioengineering, as well as interdisciplinary researchers involved in the field. The inclusion of concrete examples and applications, and programming techniques and examples, is highly encouraged.

Series Editors

N. F. Britton
Department of Mathematical Sciences
University of Bath

Xihong Lin
Department of Biostatistics
Harvard University

Hershel M. Safer

Maria Victoria Schneider
European Bioinformatics Institute

Mona Singh
Department of Computer Science
Princeton University

Anna Tramontano
Department of Biochemical Sciences
University of Rome La Sapienza

Proposals for the series should be submitted to one of the series editors above or directly to:
CRC Press, Taylor & Francis Group

Published Titles

Chapman & Hall/CRC Mathematical and Computational Biology Series

Multiscale Cancer Modeling

Edited by

Thomas S. Deisboeck • Georgios S. Stamatakos

CRC Press
Taylor & Francis Group
Boca Raton London New York

CRC Press is an imprint of the
Taylor & Francis Group, an **informa** business

A CHAPMAN & HALL BOOK

First published in paperback 2024

First published 2011 by CRC Press

Published 2019 CRC Press
2385 NW Executive Center Drive, Suite 320, Boca Raton FL 33431

and by CRC Press
4 Park Square, Milton Park, Abingdon, Oxon, OX14 4RN

CRC Press is an imprint of Taylor & Francis Group, LLC

© 2011, 2019, 2024 Taylor & Francis Group, LLC

Library of Congress Cataloging-in-Publication Data

Transatlantic Workshop on Multiscale Cancer Modeling (1st : 2008 : Brussels,
　Belgium) Multiscale cancer modeling / editors, Thomas S. Deisboeck and Georgio S.
　Stamatakos.
　　　p. ; cm. -- (Chapman and Hall/CRC mathematical & computational biology series
　; 34)
　　Includes bibliographical references and index.
　　ISBN 978-1-4398-1440-6 (alk. paper)
　　1. Cancer--Mathematical models--Congresses. I. Deisboeck, Thomas S. II.
Stamatakos, Georgio S. III. Title. IV. Series: Chapman and Hall/CRC mathematical &
computational biology series ; 34.
　　[DNLM: 1. Models, Biological--Congresses. 2.
Neoplasms--physiopathology--Congresses. 3. Neoplasms--pathology--Congresses. 4.
Neoplastic Processes--Congresses. QZ 200 T7723m 2010]
　　RC267.T615 2008
　　362.196'9940011--dc22　　　　　　　　　　　　　　　　　　　　　　　　　2010008845

ISBN: 978-1-4398-1440-6 (hbk)
ISBN: 978-1-03-291924-9 (pbk)
ISBN: 978-0-429-07560-5 (ebk)

DOI: 10.1201/b10407

Visit the Taylor & Francis Web site at
http://www.taylorandfrancis.com

and the CRC Press Web site at
http://www.crcpress.com

*We dedicate this book to
cancer patients and their families.*

Table of Contents

Preface

In 2003, the conclusion of the Human Genome Project identified approximately 25,000 genes. How all these genes and the hundreds of thousands of proteins that they encode interact to form physiological phenotypes and how molecular alterations potentially lead to abnormal patterns, including cancer, is still largely unknown. Genes, cells, and tissues function through many intricate processes that span multiple scales in space and time. Therefore, focusing on a particular level of observation alone may not provide sufficient insight as to the mechanistic relationships across scales. However, the complexity involved is daunting and from an experimental perspective, it is often difficult technically if not prohibitively expensive to alter all parameters involved, reproducibly, in an effort to explore the data space methodically. It is here where *in silico* biology driven by cutting-edge mathematical and computational methods and techniques will have a profound impact. Its translational goals in cancer research range from experimentally-testable hypothesis generation and cross-scale data integration to patient-specific prediction of progression and treatment planning (*in silico* oncology). However, the scientific and technical expertise spectrum necessary to conduct such innovative multiscale modeling research often exceeds the resources of a single research department, institution, or even country. And so this is nothing less than the dawn of a new era of interdisciplinary and multi-institutional collaboration, and a unique opportunity for international exchange to accelerate progress. In recognizing both the considerable challenges and the enormous potential, on October 23–24, 2008, the European Commission and the U.S. National Cancer Institute (NCI) jointly funded the First Transatlantic Workshop on Multiscale Cancer Modeling in Brussels, Belgium. For the first time, this meeting brought together the majority of the top *in silico* modeling groups in the United States and in Europe. This textbook presents the best contributions of this groundbreaking event—the state of the art of multiscale cancer modeling.

Thomas S. Deisboeck, M.D.
Department of Radiology
Massachusetts General Hospital
Harvard Medical School
Boston

Georgios S. Stamatakos, Ph.D.
Institute of Communication and
Computer Systems
National Technical University of Athens
Athens
August 2010

Acknowledgments

Special thanks go to the contributing authors for their hard work that made this textbook a reality. We gratefully acknowledge the support of Dr. Daniel Gallahan (National Institutes of Health) and Dr. Ilias Iakovidis (European Commission) for the First Transatlantic Workshop on Multiscale Cancer Modeling in Brussels in 2008; this workshop's presentations built the basis for the chapters of this textbook. We also thank the excellent Chapman & Hall team for their patience, tenacity, and guidance throughout the publishing process. First and foremost, however, we are indebted to our families for their love, encouragement, and unwavering support to ensure this project goes the distance.

About the Editors

Thomas S. Deisboeck, M.D., is currently Associate Professor of Radiology, Massachusetts General Hospital (MGH), where he directs the Complex Biosystems Modeling Laboratory. He is an affiliated faculty member of the Harvard–MIT Health Sciences and Technology Division (HST) and is a member of the Dana Farber/Harvard Cancer Center. Dr. Deisboeck graduated from Munich's Technical University Medical School and did his internship in the Department of Neurosurgery at the Ludwig–Maximilians University of Munich, Germany. His laboratory's focus is on integrative multiscale cancer modeling and simulation, as well as on related IT infrastructure development (http://biosystems.mit.edu). Dr. Deisboeck is the founding Principal Investigator of the Center for the Development of a Virtual Tumor, CViT (https://www.cvit.org). He currently serves on scientific advisory boards for several large-scale academic consortia in the United States and in Europe. Dr. Deisboeck is coeditor of *Complex Systems Science in Biomedicine* and associate editor of *Biosystems,* and serves on the editorial board of *Molecular Systems Biology* and *Cancer Informatics.* He reviews for many scientific publications and has chaired several international meetings on this topic.

Georgios S. Stamatakos, Ph.D., is Research Professor at the Institute of Communication and Computer Systems, National Technical University of Athens (NTUA) where he has founded and leads the *In Silico* Oncology Group, Laboratory of Microwaves and Fiber Optics. Dr. Stamatakos received the diploma degree in electrical engineering from NTUA, the M.Sc. degree in bioengineering from the University of Strathclyde, Scotland, and Ph.D. degree in physics (biophysics) from NTUA. His research group's focus is on multiscale cancer modeling and simulation with special emphasis on *in silico* oncology (http://www.in-silico-oncology.iccs.ntua.gr). Other interests include bioinformatics and computational bioelectromagnetics and biooptics. Dr. Stamatakos leads the action "Technologies and Tools for *In Silico* Oncology" of the European Commission (EC) and the Japan co-funded, integrated project ACGT: Advancing Clinicogenomic Trials on Cancer (FP6-2005-IST-026996). He also leads the groups Simulation at the Cellular and Higher Levels of Biocomplexity and Integration of the Simulation System of the EC-funded research project ContraCancrum: Clinically Oriented Translational Cancer Multilevel Modelling (FP7-ICT-2007-2-223979). Dr. Stamatakos has co-organized a number of research workshops, including the series of International Advanced Research Workshops on *In Silico* Oncology and the First Transatlantic Workshop on Multiscale Cancer Modelling. He serves as a reviewer for several scientific publications and as coeditor of workshop proceedings. He is a member of the editorial board of Cancer Informatics, the Technical Chamber of Greece, and the Institute of Electrical and Electronics Engineers (IEEE).

Contributors

Ellsworth C. Alvord, Jr.
Department of Pathology
University of Washington
Seattle, Washington

Alexander R.A. Anderson
Integrated Mathematical
   Oncology
H. Lee Moffitt Cancer Center and
   Research Institute
Tampa, Florida

Nicholas Ayache
Asclepios Research Team
Institut National de Recherche en
   Informatique et en Automatique
   (INRIA)
Sophia Antipolis–Méditerranée,
   France

David Basanta
Integrated Mathematical
   Oncology
H. Lee Moffitt Cancer Center and
   Research Institute
Tampa, Florida

Michael Bergdorf
Chair of Computational Science
ETH Zürich
Zürich, Switzerland

Veronika Bordas
Department of Applied
   Mathematics
Harvard University
Cambridge, Massachusetts

Helen M. Byrne
School of Mathematical Sciences
University of Nottingham
Nottingham, United Kingdom

Gargi Chakraborty
Department of Pathology
Department of Applied
   Mathematics
University of Washington
Seattle, Washington

Mark A.J. Chaplain
Division of Mathematics
University of Dundee
Dundee, Scotland

Kwang-Hyun Cho
Department of Bio and Brain
   Engineering
Korea Advanced Institute of Science
   and Technology (KAIST)
Daejon City, Korea

Olivier Clatz
Asclepios Research Team
Institut National de Recherche en
    Informatique et en Automatique
    (INRIA)
Sophia Antipolis–Méditerranée,
    France

Vittorio Cristini
Department of Pathology
University of New Mexico
Albuquerque, New Mexico
Formerly at School of Biomedical
    Informatics
University of Texas Health Science
    Center
Houston, Texas

Thomas S. Deisboeck
Harvard–MIT (HST) Athinoula A.
    Martinos Center for Biomedical
    Imaging
Massachusetts General
    Hospital–East
Charlestown, Massachusetts

Hervé Delingette
Asclepios Research Team
Institut National de Recherche en
    Informatique et en Automatique
    (INRIA)
Sophia Antipolis–Méditerranée,
    France

Pier Poalo Delsanto
Department of Physics
Politecnico di Torino
Torino, Italy

Alexander G. Fletcher
Centre for Mathematical Biology
Mathematical Institute and
    Oxford Centre for Integrative
    Systems Biology
Department of Biochemistry
University of Oxford
Oxford, United Kingdom

Hermann B. Frieboes
School of Biomedical
    Informatics
University of Texas Health Science
    Center
Houston, Texas

Robert A. Gatenby
Departments of Radiology and
    Integrated Mathematical
    Oncology
Moffitt Cancer Center
Tampa, Florida

Philip Gerlee
Department of Molecular and
    Clinical Medicine
Sahlgrenska University Hospital
    Mathematical Sciences
Chalmers and Göteborg
    University
Göteborg, Sweden

Antonio Salvador Gliozzi
Department of Physics
Politecnico di Torino
Torino, Italy

Norbert Graf
Department of Pediatric Oncology
   and Hematology
University of the Saarland
Homburg, Germany

Stanley Gu
Department of Pathology
Department of Bioengineering
University of Washington
Seattle, Washington

Caterina Guiot
Department of Neuroscience
University of Torino
Torino, Italy

Dewi Harjanto
Department of Biomedical
   Engineering
Boston University
Boston, Massachusetts

Sven Hirsch
Department of
   Electrical Engineering
ETH Zürich
Zürich, Switzerland

Larry W. Jean
Program in Computational Biology
Fred Hutchinson Cancer Research
   Center
Seattle, Washington

Jun-Won Kang
Graduate School of
   Medical Science and
   Engineering
Korea Advanced Institute of Science
   and Technology (KAIST)
Daejon City, Korea

Natalia L. Komarova
Department of Mathematics
University of
   California–Irvine
Irvine, California

Ender Konukoglu
Asclepios Research Team
Institut National de Recherche en
   Informatique et en Automatique
   (INRIA)
Sophia Antipolis–Méditerranée,
   France
Now with Microsoft Cambridge

Petros Koumoutsakos
Chair of Computational Science
ETH Zürich
Zürich, Switzerland

Walter Lewis
Department of Biological
   Sciences
Pittsburgh, Pennsylvania

Yingting Liu
Department of Bioengineering
University of Pennsylvania
Philadelphia, Pennsylvania

Bryn Lloyd
Department of Electrical
  Engineering
ETH Zürich
Zürich, Switzerland

John Lowengrub
Department of Mathematics
University of California–Irvine
Irvine, California

E. Georg Luebeck
Program in Computational
  Biology
Fred Hutchinson Cancer
  Research Center
Seattle, Washington

Paul Macklin
Department of Mathematics
University of Dundee
Dundee, Scotland

Philip K. Maini
Centre for Mathematical Biology
Mathematical Institute and
  Oxford Centre for Integrative
  Systems Biology
Department of Biochemistry
University of Oxford
Oxford, United Kingdom

Carlo C. Maley
Molecular and Cellular
  Oncogenesis Program
Systems Biology Division
The Wistar Institute
Philadelphia, Pennsylvania

Susan Massey
Department of Pathology
Department of Applied
  Mathematics
University of Washington
Seattle, Washington

Steven McDougall
Department of Petroleum
  Engineering
Heriot–Watt University
Edinburgh, Scotland

Florian Milde
Chair of Computational Science
ETH Zürich
Zürich, Switzerland

Gary R. Mirams
Department of Physiology
  Anatomy and Genetics
University of Oxford
Oxford, United Kingdom

Philip J. Murray
Centre for Mathematical Biology
Mathematical Institute
University of Oxford
Oxford, United Kingdom

Jeremy E. Purvis
Genomics and Computational
  Biology Graduate Group
University of Pennsylvania
Philadelphia, Pennsylvania

Ravi Radhakrishnan
Department of Bioengineering
University of Pennsylvania
Philadelphia, Pennsylvania

Brian J. Reid
Human Biology and Public Health
  Sciences Divisions
Fred Hutchinson Cancer Research
  Center
Seattle, Washington

Katarzyna A. Rejniak
Integrated Mathematical
  Oncology
H. Lee Moffitt Cancer Center and
  Research Institute
Department of Oncologic
  Sciences
University of South Florida College
  of Medicine
Tampa, Florida

Russell Rockne
Department of Pathology
Department of Applied
  Mathematics
University of Washington
Seattle, Washington

Jonathan Sagotsky
Harvard–MIT (HST) Athinoula A.
  Martinos Center for Biomedical
  Imaging
Massachusetts General Hospital-
  East
Charlestown, Massachusetts

Andrew J. Shih
Department of Bioengineering
University of Pennsylvania
Philadelphia, Pennsylvania

Rita Sodt
Department of Pathology
Department of Computer Science
University of Washington
Seattle, Washington

Ricard V. Solé
ICREA–Complex Systems Lab
Parc de Recerca Biomédica de
  Barcelona
Universitat Pompeu Fabra
Barcelona, Spain

Georgios S. Stamatakos
In Silico Oncology Group
Laboratory of Microwaves and
  Fiber Optics
Institute of Communication and
  Computer Systems
National Technical University of
  Athens
Athens, Greece

Kristin R. Swanson
Department of Pathology
Department of Applied
  Mathematics
University of Washington
Seattle, Washington

Dominik Szczerba
IT'IS Foundation
Zürich, Switzerland

Gabor Székely
Department of Information
  Technology and Electrical
  Engineering
ETH Zürich
Zürich, Switzerland

Alex Walter
Oxford Centre for
    Collaborative Applied
    Mathematics
Mathematical Institute
University of Oxford
Oxford, United Kingdom

Muhammad H. Zaman
Department of Biomedical
    Engineering
Boston University
Boston, Masssachusetts

Zhihui Wang
Harvard–MIT (HST) Athinoula A.
    Martinos Center for Biomedical
    Imaging
Massachusetts General Hospital-
    East
Charlestown, Massachusetts

Evolution, Regulation, and Disruption of Homeostasis and Its Role in Carcinogenesis

Alexander R.A. Anderson, David Basanta, Philip Gerlee, and Katarzyna A. Rejniak*

CONTENTS

* All authors contributed equally to this chapter.

SUMMARY

At its simplest tissue homeostasis is the balance between cell proliferation and apoptosis that preserves the architecture and functionality of a tissue. It is estimated that 50 to 70 billion cells are dying every day in an average adult [1] and approximately the same number of cells need to be born to keep body integrity. This balance is maintained by multiple subcellular, intracellular and extracellular mechanisms including cell genetic management, cell-cell adhesion, paracrine/autocrine signaling, and cell-ECM interactions. The process of carcinogenesis entails the escape from these mechanisms, and the evolution of the tumor cell population toward phenotypes that can exploit or become independent of the normal tissue microenvironmental constraints. In this chapter we consider the mechanisms that regulate normal tissue homeostasis and subsequently homeostatic escape in the development of cancer by using different modeling approaches that examine the role that physical constraints, cell-microenvironment interactions and evolutionary dynamics play.

INTRODUCTION

Homeostasis is a critical property of living beings that involves the ability to self-regulate in response to changes in the environment in order to maintain a certain dynamic balance affecting form and/or function. Homeostasis is of particular importance in multicellular organisms, where it is intertwined with development [2,3]. Organisms have evolved intricate control mechanisms that ensure developmental processes achieve their end points and stabilize (e.g., differentiate) as well as allow for a degree of adaptability to a range of conditions (e.g., stress or damage induced by wounding). This allows for the emergence of a more robust system that can tolerate both external and internal perturbations [4]. However, there are limitations to this tolerance, and often it is the rare events that cause the most disruption [5]; think of the extinction of dinosaurs for an example. From an evolutionary point of view, this is a viable trade-off between the energetic cost of homeostasis versus the fitness benefit it would provide. In practical terms, homeostasis of living multicellular organisms is constrained in terms of the amount of disruption they can cope with and in terms of the amount of time they will remain homeostatic.

In order to understand the transition from normal tissue to invasive cancer, we should first understand how the normal form and function of the tissues under consideration is maintained to achieve a homeostatic balance (emerging from the integration of multiple subcellular, intracellular, extracellular, chemical, and physical signals/constraints). For example, the role of normal epithelial tissue (from which most tumors arise) is to separate the inner body compartments, such as prostate ducts producing prostatic fluid or breast glands secreting milk, from the surrounding environment and to control the exchange of nutrients and waste products between them. Biological homeostasis has to be achieved in a dynamic cellular milieu with a constant cell turnover (a cell lifespan ranges from 3 days for skin cells, and 4 months for red blood cells, to several years for bone cells) and perturbations from various extrinsic factors (e.g., breast duct shrinkage after pregnancy or local tissue damage) and by counteracting induced cellular changes such that homeostasis is restored. Therefore, when a damaged or mutated cell is not functioning as it should, the tissue will try to suppress the damaged cell and prevent further abnormalities. However, if this damaged cell gains a proliferative or migratory advantage over other cells and does propagate, it needs to do so at the expense of other cells and will ultimately defy the constraints imposed by the homeostatic mechanisms employed by the host tissue. In many cases, these constraints are physical, imposing structural constraints on the cells, for example, via cell adhesion, but they may also be chemical, for example, limited metabolite availability. In order to escape homeostasis and overcome these barriers, the mutant cells need to evolve to the point where they can significantly modify their baseline phenotypes and potentially their environment. Therefore, the emergence of an invasive cancer can be viewed as an escape from homeostasis in which the natural synchrony between multiple cellular and microenvironmental variables is perturbed.

The classical approach to oncogenic development views cancer as being a solely genetic disease, whereby genomic mutations are acquired in a stepwise fashion, leading to uncontrolled cell growth, invasion into surrounding tissues, and eventual metastasis (reviewed in Reference [6]). This reductionist view of oncogenesis sees the tumor cells existing in isolation, steadily acquiring mutations, with little interaction with their surrounding environment. In reality, tumor cells are embedded in a matrix of structural extracellular proteins, surrounded by other cells, such as endothelial cells, fibroblasts, and inflammatory and immune cells [7,8]. These multiple cell types make up the tumor microenvironment, and are in continuous dynamic interaction with other stromal and tumor cells. Together, the cells generate a myriad of physical and chemical signals that converge to determine the metabolic, migratory, growth, and survival behavior of the tumor cell. The acquisition of oncogenes alone cannot explain all aspects of tumor development, and there is evidence that escape from normal tissue homeostasis is an essential step in the carcinogenic process. In fact, the gene-centric and microenvironment-centric views of carcinogenesis are to some extent unified under the homeostatic hypothesis, since genetic mutations under microenvironment selection must together define phenotypes that have the potential to escape homeostatic control. By focusing on the cellular phenotype, we can examine what subcellular (e.g., receptor-driven cell processes, cell metabolism), cellular (e.g., cell–cell or cell–ECM adhesion), and environmental perturbations (e.g., nutrient or growth factor distribution, stromal structure) are required for this liberation.

In this chapter, we will consider three different computational models that examine the role of homeostasis in carcinogenesis. Understanding normal tissue formation and maintenance will allow us to better under-stand how cancer can be initiated, and how it develops and progresses. In the first model, we consider a novel approach integrating genetic algorithms and cellular automata to investigate the evolution of homeostatic tissue. In the second, we use an immersed boundary framework to investigate how the disruption of intrinsic cellular responses to extrinsic signals results in a homeostatic imbalance within an epithelial duct. In the final part, we present a model of prostate cancer and examine the importance of both growth factor and stromal interactions in the maintenance of a homeostatic state, even in the presence of cancer.

EVOLVING HOMEOSTATIC TISSUE USING GENETIC ALGORITHMS

During embryogenesis, multicellular organisms follow a developmental program in which tissue architecture results from the interactions between cells and as a result of the processes of mitosis, motility, differentiation, and apoptosis. After reaching maturity, this architecture is maintained through an intricate and finely tuned balance of cell proliferation and loss. Given the consequences of homeostatic disruption (such as aging, psoriasis, or cancer), the organism has to be able to cope with genetic and environmental insults without significant disruption. Recently, a computational model has been proposed by one of the authors to study the evolutionary origin of robust homeostasis [10]. The difficulty of performing experiments to study evolutionary dynamics makes *in silico* approaches particularly useful. In that paper, an evolutionary algorithm (EA) was implemented to evolve the developmental rules of digital organisms, using three-dimensional cellular automata (CA). The developmental rules, shared by all the cells in an organism, match certain external and internal conditions (such as the presence of neighboring cells or the number of divisions the cell has gone through) with cellular actions (motility, division, and apoptosis). At any given time, a cell scans all relevant internal and external conditions and decides upon an action depending on the subset of the 100 rules that constitutes its digital genome. This mechanism has the advantage of making CA more evolvable, and thus their use in conjunction with EAs more efficient than conventional CA [10].

The evolved organisms were selected to grow specific shapes for a number of time steps and to remain homeostatic for the rest of the simulation. During the homeostatic period, a number of different mechanisms were found such that the digital organisms maintained their form. Interestingly, these organisms evolved the capability to recover from severe wounds even though specific evolutionary pressure selecting for wound healing was missing. A study of the digital organisms' evolutionary trajectory showed that organisms that evolved earlier were less capable of coping with environmental insults than those that evolved later (even if they were as fit from a homeostatic point of view). Furthermore, the organisms more capable of wound healing were those that had evolved a tissue-like architecture with a direction flux of cells driving tissue turnover (Figure 1.4A). This mechanism is similar to the stratified architecture that characterizes the human skin or the gut. These results suggest that robustness may be a by-product of the evolution of morphogenetic systems.

Model

Motivated by this work, here we present a model of a tissue developing from a single cell into a homeostatic structure capable of growth control and self-repair. In particular, we focus on the simple case of a monolayer of cells forming a two-dimensional structure, similar to the lining formed by epithelial cells. This problem is studied in the context of an evolutionary hybrid cellular automaton [11–13], an individual-based model in which the behavior of each cell depends on its local environment consisting of a chemical species such as oxygen, which are modeled on a continuous level. Precisely how cells respond to their microenvironment is determined using a feed-forward artificial neural network, which takes extracellular cues as an input, and outputs the phenotype or behavior of the cell. Instead of dictating a given mapping from environment to phenotype (as we have done previously), we will make use of an EA to evolve a cell behavior that gives rise to a homeostatic tissue. In the following, we will first briefly describe the underlying cellular automaton model, and then move on to discuss the implementation of the EA, and the results gathered from it.

Hybrid Cellular Automaton Model (HCA)

The tissue that we simulate is represented by an N × N × M cellular automaton, in which each grid point $\vec{x} = \Delta(i, j, k)$ either contains a cell or is empty. Here, Δ is the lattice constant, which determines the spacing between the grid points, or equivalently, the size of the cells. The cellular automaton is coupled with two concentration fields, one describing the concentration of oxygen $c(\vec{x},t)$ and the other the concentration of a generic growth factor (GF) $g(\vec{x},t)$. The cells on the lattice influence these fields through oxygen and GF consumption, but are also affected by the concentrations, as they serve as inputs to the response network that determines the behavior of the cells.

Response Network

The behavior of each cell is determined by a neural network that takes the number of neighbors on the lattice, the local oxygen concentration and GF concentration as input, and for each possible input calculates a phenotypic response. Phenotypes are limited to four: proliferation, movement, apoptosis and, in the absence of a network response, cellular quiescence. The response network consists of a number of nodes organized into three layers: (1) input, which takes information from the environment; (2) hidden; and (3) output, which determines the action of the cell. The nodes in the different layers are connected with varying weights, determined by

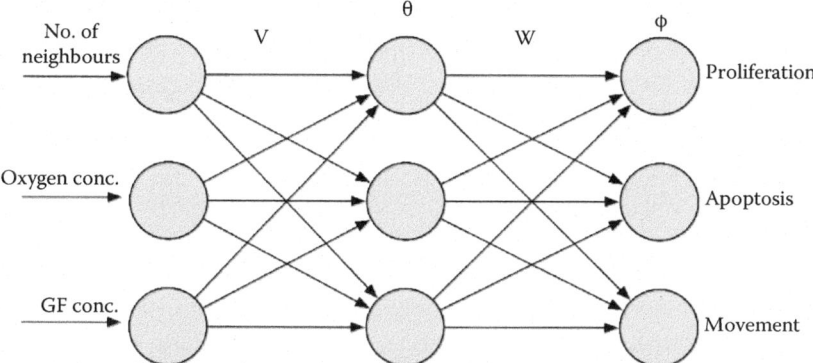

FIGURE 1.1 The layout of the response network that determines the behavior of the cells. The microenvironment of each cell is presented to the input layer of the network. This information is then fed through the network under the influence of the connection matrices V, W and the threshold vectors, and the node with the strongest response in the output layer determines the behavior of the cell. If no node reaches a value above 1/2, the cell becomes quiescent.

two matrices V and W, and the nodes in the hidden and output layer are equipped with internal thresholds θ and ϕ (see Figure 1.1). These parameters fully determine the mapping from environment to phenotype and can therefore be seen as the genotype of the cells. For a more detailed description of the network dynamics, we refer the reader to Reference [14].

Chemical Fields
For the sake of simplicity, we consider only oxygen and a generic GF in our model. All cells are assumed to consume oxygen, although at different rates depending on their phenotype, while only proliferating cells consume GF. Oxygen and growth factor production take place on the domain boundary via diffusion from surrounding tissue or blood vessels (see Reference [14] for a system of diffusion equations that have a similar form). As we are studying a thin slice of tissue ($M<<N$), we will assume that these concentration fields vary insignificantly in the z-direction, and we will thus only solve the equations in two dimensions.

Cellular Automaton
For each time step a cell is in a proliferative state, an internal counter is increased, and when it has reached a certain value t_p corresponding to the time of the cell cycle, the cell divides and a daughter cell is

placed at random in one of the neighboring empty grid points. If no empty space exists, cell division is halted until empty space emerges. If a cell takes on the motile phenotype, it moves at random into one of the empty neighboring grid points. If no empty space exists, the cell remains stationary. Motile cells move with a probability p_m to regulate speed. If apoptosis is the network response, then the cell dies, and its space becomes available in the following time step. A cell can also die from starvation or necrosis due to a lack of oxygen (i.e., $c < c_n$, where c_n is the oxygen level at which cells become necrotic). In order to account for the stochastic nature of cell behavior, we also include a small spontaneous death rate p_d.

The initial conditions of the system are uniform concentrations of oxygen and GF and a single cell with a given genotype (set of network parameters) at the center of the grid. Each time step the chemical concentrations are solved using the discretized equations. The position of each cell is corrected with respect to neighboring cells that have moved or died; that is, a suspended cell drops along the z-axis until it touches another cell or the bottom of the domain. All the cells on the grid are then updated in random order as follows:

1. The microenvironment is sampled, and the response of the network is calculated.

2. The cell consumes oxygen and GF according to the phenotype chosen.

3. The phenotype choice is evaluated, and the grid is updated accordingly.

Evolutionary Algorithm (EA)

Evolutionary approaches have been used to solve various problems in computer science, such as hardware development, image classification, and robot control [15]. What these approaches all have in common is that they try to harness the power of natural selection. Central to this is the notion of a fitness function, which to each candidate solution assigns a value used to rank all the solutions to the problem. Another necessary feature is that the solutions can be randomly modified (mutated) and even mixed with each other.

Here, we want to find a set of network parameters or genotype, which when seeded into a single cell, and given time to grow, give rise to a tissue that is homeostatic. More precisely, we want a single cell to multiply such that the population creates a monolayer of cells that are not stacked on top of each other. This can be formalized as the following fitness function:

$$F = \text{no. of cells in the bottom layer} + \text{no. of empty grid points in all other layers} \qquad (1.1)$$

The goal is then to find a genotype that maximizes this function. One complication is that the fitness function is multiobjective, that is, it contains two distinct parts, and in this case it has been shown that simply summing them is not the best approach [16]. Instead, we employ the sum of weighted ratios, which assigns weights to the different objectives depending on the current minimum and maximum values of the two objectives in the population. If we call the two objectives F_1 and F_2, then the fitness of genotype g is defined as

$$F(g) = \frac{F_1(g) - F_1^{\min}}{F_1^{\max} - F_1^{\min}} + \frac{F_2(g) - F_2^{\min}}{F_2^{\max} - F_2^{\min}} \qquad (1.2)$$

where $F_i^{\min,\max}$ refers to the minimum and maximum value of each objective currently present in the population.

The EA consists of a population of candidate solutions, which are subject to a selection process. We have chosen a tournament-based selection process, which also makes use of a low degree of elitism. This means that a fraction $p_e = 5\%$ of the best solutions are carried unaltered into the next generation, while the rest of the population engage in tournaments. Four solutions are picked at random and are compared in pairs, and this generates two winners and two losers. The winners are carried over to the next generation, while the losers are replaced by the offspring of the winners. The offspring are either generated by single point mutations to the parents' genotypes (occurs with probability 1/2) or by crossing-over their two genotypes (with complementary probability 1/2). This process is repeated until all solutions in the population have engaged in a tournament, and this constitutes one generation in the EA. (See Figure 1.2 for a graphical representation of the selection process.) Point mutations are implemented by changing one of the network parameters (matrix entries or node thresholds) by adding a random member from a normal distribution. The cross-over is implemented by mapping the network parameters of both parents into two vectors, picking two random indices, and swapping the contents above and below these

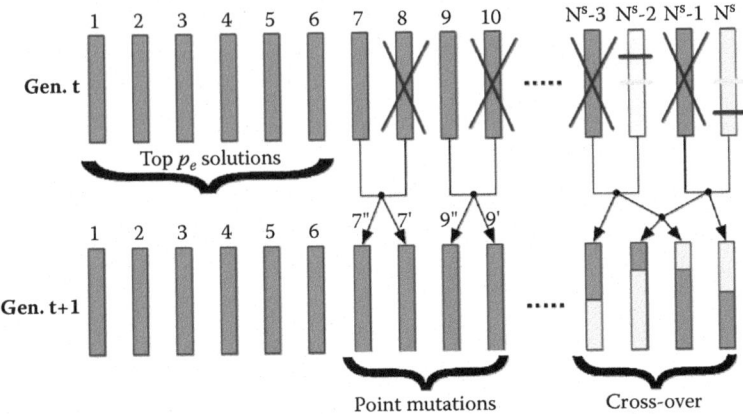

FIGURE 1.2   Schematic of the evolutionary algorithm. In each generation, the top p_e most fit solutions are carried unaltered into the next generation. The remaining genotypes are paired off in tournaments where the winners give rise to offspring either by point mutations or cross-over.

points. In most cases these is disruptive, but in some cases it generates a novel solution with a higher fitness than the parents'.

Each run was started with a population of size N_s = 50, where all genomes are randomly generated. The number of time steps used for each fitness evaluation was t_{max} = 200, which, since the size of the grid is 200 × 200, means that a genotype with proliferative capabilities will be able to fill the entire domain, while it still needs to exhibit homeostatic behavior to receive a high fitness. The total number of generations iterated with the EA was set to T_{max} = 20, which means that we had to run N_s + (1 − p_e) $N_s T_{max}/2$ = 500 realizations of the underlying HCA model.

Results and Discussion

Because the EA is seeded with a random starting population, and only random variation and selection drives the search for a homeostatic genotype, we do not impose any constraint on how the cell population solves the problem of tissue homeostasis. Instead, our aim was to investigate (using our model) which mechanisms emerge for maintaining homeostasis. We achieved this aim by running a number of EAs with different random initial populations of solutions, and analyzing the sequence of solutions generated by them. The different genotypes were then compared to one another by analyzing the phenotypes they give rise to and by investigating their growth dynamics.

The growth patterns generated by the most fit genotypes in one such run are shown in Figure 1.4B. These plots show the distribution of cells in the domain at the end of each fitness evaluation, from a top view, which means that if several cells are stacked on each other only the topmost cell is visible. From this sequence, we can see that the EA has explored several different solutions before settling on a completely proliferating phenotype $T = 11$ to 19. At $T = 1$, the fittest solution consists of a slow-growing phenotype, which forms a tissue with a low cell density. This is then replaced by a phenotype that forms a ring-like structure in which no cells reside in the interior of the domain ($T = 3$). From $T = 5$ to 9, the fittest genotype is one that fills the entire bottom of the domain, and does so by adopting a motile phenotype. This type of solution is then taken over by a fully proliferative phenotype, which remains the fittest solution until the end of the run.

In order to get a better understanding of the tissue architecture generated by the final genotype, we also plotted the cell density in the domain at $t = 200$. This can be seen in the lower part of Figure 1.4B, and shows that most of the tissue consists of a monolayer of cells with isolated cells lying on top (the mean cell density is 1.08 cells/grid point). This means that the genotype meets the requirements of the fitness function fairly well, but we need to understand in more detail how this homeostatic behavior is achieved. A useful way of analyzing the behavior of a genotype in the model is to calculate the behavior of the genotype in all possible microenvironmental conditions. As the input is three dimensional (number of neighbors, oxygen, and GF concentration) and the output is one dimensional and discrete (proliferation, movement, apoptosis, and quiescence), the behavior of the genotype can be visualized as a function from three to one dimensions, in which each point in the input space is associated with a phenotype. This type of plot for the final genotype ($T = 19$) is shown in Figure 1.3a, and from this we can conclude that the possible behavior of the genotype is limited to proliferation and apoptosis. The absence of any quiescence or movement in this apparently homeostatic genotype suggests that apoptosis might play an important role in its behavior.

In order to investigate this further, we measured the number of cell births and cell deaths during a simulation. These rates are shown in Figure 1.3c as a function of time, together with the total number of cells present on the grid. As expected, the birth and death rates are fairly balanced, but they are very high even when the tissue is fully formed, with approximately 83% of the cells in the tissue being replaced every cell cycle.

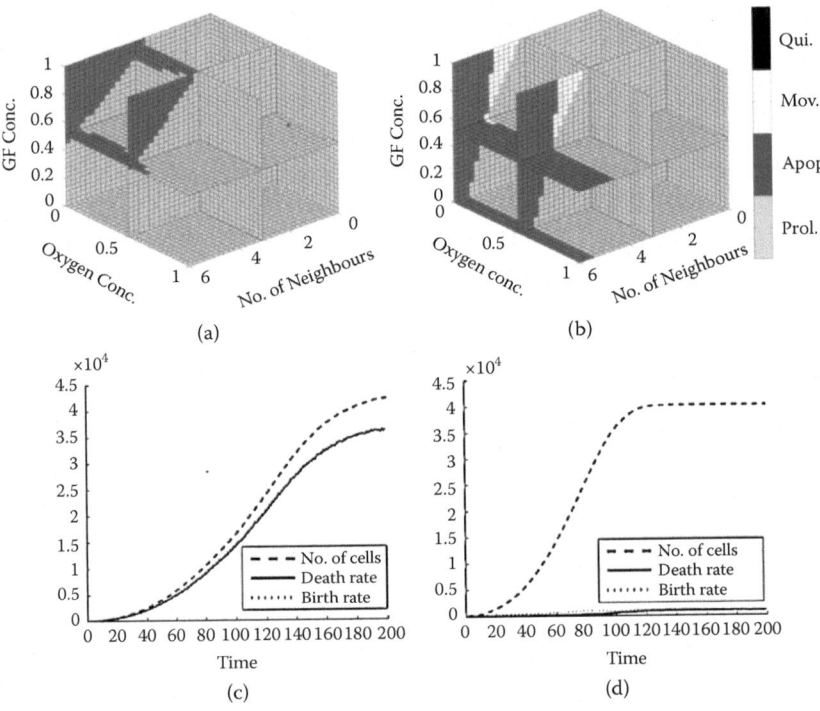

FIGURE 1.3 (a), (b) The behavior of two final genotypes visualized by mapping to each point in input space the associated phenotype determined by the response network. (c), (d) The equivalent time evolution of the total number of cells and the birth and death rates for these two genotypes. Time is scaled according to cell cycles.

This suggests that the final genotype maintains tissue homeostasis by balancing its high proliferation rate with an equally high rate of apoptosis. This type solution is commonly arrived at by the EA, and is in contrast to those solutions that maintain tissue homeostasis by halting cell division when the tissue forms. An example of such a phenotype is shown in Figure 1.3b, which consists mostly of proliferation and apoptosis, but note that there is a small (but obviously significant) subset of the input space that gives rise to motility. Looking at the birth and death rates of this genotype (Figure 1.3d), we see that they are considerably smaller than for the previously analyzed genotype. This genotype has evolved a mechanism to deal with both decreases and increases in local cell density by essentially exploiting the migratory phenotype as a means to become quiescent; that is, if a cell is surrounded by other cells and becomes a motile phenotype, it essentially acts as if it were quiescent.

We have shown that by coupling an HCA model of tissue growth with an EA, we can evolve cell (geno)types that form a homeostatic tissue. The two evolved genotypes analyzed present contrasting mechanisms by which tissue homeostasis can be maintained. One does so by continually shedding cells via apoptosis, and the other by lowering the total proliferation rate of the tissue so that it just balances the spontaneous rate of apoptosis. An important feature that the model does not currently include are somatic mutations, which occur in the tissues of all multicellular organisms. These mutations can transform cell behavior and are the underlying cause of cancer in somatic tissue. In our model, in which no mutations occur (except in the EA), two mechanisms for maintaining a homeostatic tissue emerge from the evolutionary process, but one may be selected over the other by imposing a certain environment. Our conjecture is that a proliferative genotype is more likely to evolve in an environment in which the rate of spontaneous mutations is high, while a conservative genotype is likely to evolve when the mutation rate during cell division is elevated. Clearly, each of the genotypes could produce cancer if sufficiently mutated, but the one that induces massive cellular turnover is more likely to spread damaging mutations than the more conservative one and is therefore more prone to cancer, at least in the short term.

By growing homeostatic tissue structures, evolution highlighted that there are different genetic routes to achieve the same phenotypic outcome. One simplification, however, was to create an idealized single tissue monolayer when in reality most of the epithelial tissues are made up of multiple layers and form ductal-like structures. Next, we consider the development of just such a structure and examine how homeostasis is achieved and what happens when it is perturbed.

HOMEOSTATIC IMBALANCE IN EPITHELIAL DUCTS DRIVEN BY ERRONEOUS CELL RESPONSES

Normal tissue microarchitecture enables individual cells to interact with one another and with the stromal microenvironment either directly via cell membrane receptors or indirectly through a diverse array of soluble factors. Normal cells can respond to changes in their external environment (i.e., chemical or physical stimuli from other cells or from the ECM) by modifying their internal machinery (chemical, physical, genetic) to maintain a stable equilibrium. However, when this homeostatic regulation is disturbed, the intrinsic cell environment may become unstable, resulting in uncontrolled cell behavior, leading to disruption of tissue

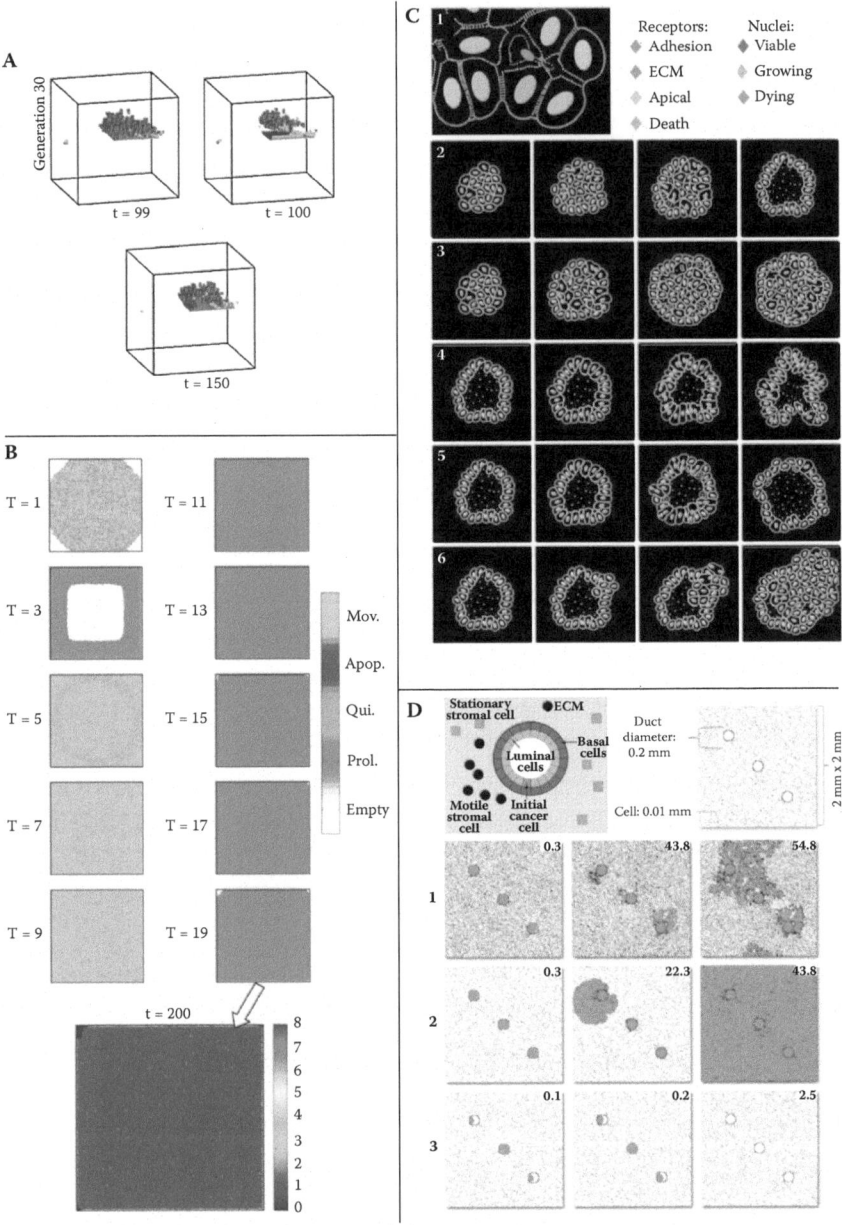

FIGURE 1.4 **(See color insert following page 40)** (A) Example of homeostatic organism with tissue-like architecture and cell flux with upward direction. The digital organism develops for 99 time steps, after which it is wounded by removing a layer of cells and then allowed another 50 time steps to see if it is capable of recovery. (B) Time evolution of the EA showing the most fit genotypes at different

FIGURE 1.4 (**Opposite**) generations in the run, where the process converges on a genotype that predominately proliferates. Arrow: cell density of the final genotype ($T = 19$). (C) Homeostatic imbalance in epithelial ducts driven by erroneous cell responses modeled by the *IBCell*. 1. A cluster of epithelial cells with color-coded receptor and nuclear staining. 2. A sequence of snapshots showing a central cross-section through a normally developing epithelial duct, composed on one layer of tightly packed cells enclosing the hollow lumen. 3. A sequence of snapshots showing that the disruption of cellular responses to the death signals results in lumen repopulation. 4. A sequence of snapshots showing that the disruption of cellular responses to the mitotic signals results in dysfunctional acini manifesting a high degree of abnormally folded epithelial tissue. 5. A sequence of snapshots showing the necessity of switching from symmetrical to asymmetrical cell divisions in order to grow a normal regular epithelial duct. 6. A sequence of snapshots showing that the disruption of cellular responses to the ECM signals results in tumor cell invasion. (D) Schematic of model domain with its key cell types. The basal cells produce TGF-β and help maintain homeostasis. Luminal cells consume TGF-β and can become tumorigenic. The stromal cells occupy locations outside the acini and produce TGF-β in response to TGF-β once it reaches a certain concentration (left). The initial simulation domain configuration is made of three glands, equally spaced and surrounded by stromal cells (right). 1. Simulation in which the tumors break out from the glands and start growing in the mesenchyme. The domain contained 40% of motile and 40% of nonmotile stroma. After about 3 months, each of the glands is entirely occupied by a tumor. After 43 years, two of the tumors have managed to break out from the gland. At the end of the simulation (after about 54 years), the three tumors have merged into a single mass, although its pattern of growth seems to be channeled by the stroma. 2. Simulation in which the tumor takes over the entire prostate. In this simulation, the production of TGF-β is relatively low compared to other simulations, and the proportion of motile stroma is the same as that of nonmotile stroma: 10%. After about 3 months, each one of the glands is occupied by a tumor, and MDE production is already visible. After about 22 years, the tumor in the upper gland breaks out and expands in the mesenchyme. After about 43 years, the tumor has taken over the entire prostate, degraded almost entirely all the membrane, and TGF-β and MDEs can be found everywhere in high quantities. 3. Simulation in which the three tumors are initiated, grow, and die out before they manage to break out from the glands. In this simulation, the proportion of stroma is the same as nonmotilestroma: 10%. The tumor cells produce MDEs at a significantly higher rate than in most of the other simulations. After 60 days, the central gland has almost been taken over by the tumor. After 1 month, the central tumor has produced enough MDEs to degrade the basal membrane, leading to both tumor cells and TGF-β spilling out of the membrane and into the surrounding stroma. After 900 time steps, the simulation shows a situation similar to that at initiation.

microarchitecture and subsequently to malignant changes in the whole organism.

The epithelial glands and cysts are especially interesting examples of tissues maintaining a homeostatic balance due to their finely defined architecture and abundance in many different organs, including the bronchi and alveoli of the lungs, breast ducts and lobules, gastrointestinal crypts, reproductive urinary tracts, and the endocrine glands. In their mature form, these epithelial structures are composed of one layer of tightly packed polarized cells. Critical decisions regulating epithelial tissue integrity, such as cell growth, orientation of cell division, or the induction of cell death, are directed by actions of neighboring cells and interactions with a dynamically evolving ECM milieu. Normal epithelial cells can adapt to certain perturbations in biochemical, genetic, and physical cues sensed from their immediate microenvironment. However, when cell responses are compromised, they may induce a malignant character, filling the hollow lumen of the ducts and breaking through the basement membrane, resulting in loss of tissue homeostasis.

Model

The IBCell was introduced in Reference [17] to model early tumor development, and subsequently applied to investigate the sufficient and necessary conditions for the formation and stability of hollow epithelial acini, three-dimensional cellular culture systems that recapitulate the structure and function of epithelial cysts [18,19]. In this model, the eukaryotic cell is represented as a two-dimensional fully deformable body (Figure 1.4C1), and its structure includes an elastic plasma membrane modeled as a network of linear springs that define cell shape and encloses the viscous incompressible fluid representing the cytoplasm and providing cell mass. These individual cells can interact with other cells and with the environment via a set of discrete membrane receptors located on the cell boundary, and can undergo several life processes, such as cell growth, division, apoptotic death, or epithelial polarization. In particular, each boundary point can be engaged in adhesion either with one of the neighboring cells or with the extracellular matrix, and cell membrane receptors can be used to sense the presence of other cells or extracellular matrix in the local cell vicinity. The host cell can initiate certain cell life processes, such as proliferation, division, apoptotic death, or epithelial polarization, based on its membrane receptors signature (a distribution of growth, death, apical, cell–cell, and cell–ECM adhesion receptors; Figure 1.4C). More precisely, cell growth is modeled by placing point sources and sinks around the cell boundary to model transport of fluid through the cell membrane and, once the cell area is doubled, the contractile ring is formed by introducing springs between opposite points on the cell boundary that upon contraction split the cell into two daughters. Cell epithelial polarity is acquired by developing three distinct cell membrane domains: basal, defined by cell membrane receptors contacting the external media; lateral, defined by cell receptors being in contact with other cells; and apical, facing the hollow lumen. Cell apoptotic death is modeled by placing point sinks and sources along the membrane of the whole cell to release fluid from the cell interior to the extracellular space. The IBCell model is based on the immersed boundary framework and governed by the following set of equations:

$$\rho\left(\frac{\partial u(x,t)}{\partial t}+(u(x,t)\bullet\nabla)u(x,t)\right)=-\nabla p(x,t)+\mu\,\Delta u(x,t)+\frac{\mu}{3\rho}\nabla s(x,t)+f(x,t)$$

(1.3)

$$\rho\,\nabla\bullet u(x,t)=s(x,t),$$

(1.4)

$$f(x,t)=\int_{\Gamma}F(l,t)\,\delta(x-X)(l,t)$$

(1.5)

$$s(x,t)=\sum_{k\in\Xi^{+}}S_{+}(Y_{k},t)\,\delta(x-Y_{k})+\sum_{m\in\Xi^{-}}S_{-}(Z_{m},t)\,\delta(x-Z_{m}),$$

(1.6)

$$\frac{\partial X(l,t)}{\partial t}=u(X(l,t),t)=\int_{\Omega}u(x,t)\,\delta(x-X(l,t))dx,$$

(1.7)

In this system, Equation 1.3 is the Navier–Stokes equation of a viscous incompressible fluid defined on the Cartesian grid $x=(x_1,x_2)$, where p is the fluid pressure, μ is the fluid viscosity, ρ is the fluid density, s is the local fluid expansion, and f is the external force density. Equation 1.4 is the law of mass balance. Interactions between the fluid and the material points $X(l,t)$, on cell boundaries Γ and at point sources Y_k and sinks Z_m placed in the cell local microenvironment are defined in Equations 1.5–1.7. Here, the force density $F(l,t)$ defined on cell boundaries, and the sources $S_+(Y_k,t)$ and sinks $S_-(Z_m,t)$ defined in the cell microenvironment, are applied to the fluid using the two dimensional Dirac delta function δ, while all material boundary points $X(l,t)$ are carried along with the fluid. The boundary forces $F(l,t)$ arise from elastic properties of cell boundaries, from cell-to-cell adhesion, and from contractile forces splitting a cell during its division. The sources $S_+(Y_k,t)$ and sinks $S_-(Z_m,t)$ are chosen such that they balance around each cell separately. More details on the mathematical formulation of IBCell and the implementation of cell life processes can be found in [17,19].

Results

We will illustrate here how changes in cellular responses to microenvironmental cues sensed by host cells via their membrane receptors may result in the disruption of epithelial tissue morphology. Specifically, we will use the biomechanical model, the IBCell, discussed earlier, to develop normal epithelial cysts and examine how these structures change as a result of disrupting cellular responses to three different signals: (1) death, (2) mitotic, and (3) ECM.

Normal Development of Epithelial Cysts

Our model can recapitulate all stages of the development of mammary acini that are the in vitro experimental systems derived from nontumorigenic mammary breast cell line MCF10A and resemble the structure and behavior of breast epithelial cysts [20]. The acinar structure is formed from a single cell that upon consecutive divisions gives rise to an aggregate of randomly oriented cells consisting of two cell populations: outer cells having contact with the extracellular matrix and inner cells surrounded entirely by other cells. Subsequently, cells in the outer layer develop an apico-basal polarity and form a monolayer of epithelial cells. This is followed by apoptotic death of inner cells, resulting in the formation of the hollow lumen and stabilization of the acinar structure (Figure 1.4C2).

Disruption of Cellular Responses to the Death Signals Results in Lumen Repopulation

Normal development of epithelial cysts requires clearance of luminal space via removal of all inner cells. Different processes could be responsible for this inner cell removal; however, it has been shown experimentally that cell apoptotic death is the necessary contributor to the formation and maintenance of the luminal space [22]. This form of cell death is induced on purpose, and often called *programmed cell death*. However, it is not known how this process is triggered. We used the IBCell model to test the hypothesis that cell–cell adhesion regulates cell apoptotic death [18]. We showed that during normal acinar development (Figure 1.4C2), cell apoptotic death is triggered by accumulation of death receptors due to the disassembly of adhesive links with either the emerging polarized cells (along their newly formed apical membrane domains), or with the neighboring inner cells that have started dying. However, when the cell response to death signals sensed from their microenvironment is disrupted, the cells are able to sense free space in their vicinity and reinitiate growth, resulting in the repopulation of the acinar lumen. This escape from tissue homeostatic balance leads to the formation of acinar mutants resembling ductal carcinoma in situ, a noninvasive form of ductal tumors characterized by filled ductal space (Figure 1.4C3).

Disruption of Cellular Responses to the Mitotic Signals Results in Dysfunctional Acini

Structural integrity of epithelial tissues requires both spatial and temporal control of proliferative events to maintain tissue homeostasis and prevent tissue hyperplastic growth. In particular, it also requires a coordination of cell mitotic and cytokinetic events, such as the orientation of mitotic spindle poles that determine the axis of cell division and location of the contractile ring that determines whether or not both daughter cells have equal volumes. It has been observed when examining human breast tissues using electron microscopy [21] that normal epithelial cells acquire two different orientations of cell division: either orthogonal or parallel to the lumen. The orthogonal division results in two luminally positioned daughter cells (symmetric cell division), whereas the parallel division gives rise to one luminally and one basally positioned daughter cell (asymmetric cell division), and culminates in basal cell differentiation or its apoptosis. We used the IBCell to investigate the hypothesis that structural integrity of an expanding epithelial duct requires a switch between symmetric and asymmetric cell divisions [18]. The symmetric cell divisions are necessary to increase the number of epithelial cells in the layer; however, when only symmetric divisions are executed, the created tension will force the duct to bend (Figure 1.4C4) because an excess of epithelial cells is not accompanied by an expansion of duct lumen. The expansion of the inner lumen can be achieved by executing asymmetric divisions, with the basal cells forming the outer epithelium and the inner cells dying by apoptosis (Figure 1.4C5). We have used a simple geometrical rule to switch between symmetric and asymmetric cell divisions, and compared the length of the cell's longest axis with its radius, but other rules, such as the tension from neighboring epithelial cells or the ratio of the lengths of lateral to basal membrane domains, can be also considered.

Disruption of Cellular Responses to the ECM Signals Results in Tumor Cell Invasion

Development of an epithelial acinus from a single cell to a shell of tightly packed cells enclosing the hollow lumen is accompanied by the secretion of various proteins, such as collagens and laminins, that accumulate in the form of a stiff supportive basement membrane surrounding the epithelial structure [22]. Epithelial cells are capable of attaching to the basement membrane through the special cell–ECM transmembrane receptors, such as integrins. We have hypothesized [19,23] that these cell–ECM adhesions, together with cell–cell adhesions, contribute to the growth arrest of a host cell, and showed that the disruption of responses to the ECM adhesion signals may result in the loss of tissue homeostatic balance and initiation of cell hyperplastic growth, leading to luminal filling and invasion of the surrounding tissue (Figure 1.4C6).

Discussion

Maintaining the structural homeostatic balance in tissues is a prerequisite for their proper function. The word *balance* should be stressed here, as certain changes in the tissue architecture do take place even in an adult healthy organism. For instance, it has been determined based on morphological identification of both mitotic and apoptotic events [24], that cell turnover in lobules of the "resting" human breast undergoes significant cyclical changes during the menstrual cycle, with the peak for apoptosis occurring 3 days after the peak for mitosis, at days 25 and 28, respectively. An even more pronounced example of an immense but controlled change in tissue structure is the process of involution: a programmed destruction and removal of the secretory epithelium that developed during pregnancy to enable milk production. This postlactational breast gland regression involves a massive death of epithelial cells, and their replacement by adipocytes and epithelial ducts remodeling to their prepregnant state and function [25].

Loss of such homeostatic balance may lead to nonreversible changes in tissue microarchitecture and subsequently to its malignancy. Using the biomechanical model of epithelial ducts, we showed that the disruption in cell intrinsic responses to extrinsic signals may result in the loss of tissue integrity if such microenvironmental perturbations are exerted in a persistent and prolonged way. In each case of the abnormal acini development considered in this section (Figures 1.4C3, 1.4C4, 1.4C6), the deviations in epithelial duct morphology can be reversed at the early stages. For instance, if the disruption in responses to death signals (Figure 1.4C3) has not been passed from a mother cell to the subsequent generations, the lumen would be cleared eventually as the apoptotic events are continuously executed (red nuclei); however, they are overwhelmed by the proliferative events (green nuclei). The structural integrity of an expanding epithelial duct could be preserved even in the absence of asymmetric division, if proliferative events were limited (round regular duct; Figure 1.4C4, second image); however, increased proliferation accompanied by persistent symmetric cell division results in abnormal tissue geometry. The expansion of an invasive clone in Figure X.C6 can be prevented if the disruption in cell responses to ECM signals was not passed to daughter cells, because normal epithelial cells will die when not attached to the basement membrane (inner cells) and will remain in growth arrest when in contact with the basement membrane (outer cells). In each case, however, the disruption of cellular responses either to the death, growth, or ECM signals that are inherited by all daughter cells result in the interruption of the epithelial structure of the duct and in tumor-like tissue outgrowth.

The epithelial ducts are not only exposed to environmental signals (such as those discussed earlier), but can also actively recruit various stromal cells, and be influenced by the secretion of numerous paracrine factors. These microenvironmental and tissue-wide devices, can potentially, counter the actions of initiated tumor cells reestablishing a homeostatic state. Next, we consider a model of the prostate glandular architecture surrounded by a basal membrane as well as paracrine signals such as TGF-β and investigate the role they could play in constraining tumor progression after initiation.

THE ROLES OF TGF-β AND STROMA IN HOMEOSTATIC ESCAPE IN PROSTATE CANCER

The prostate is a glandular sexual organ composed of ducts lined with luminal secretory epithelium surrounded by a layer of basal epithelial cells. These epithelial acini are encompassed by a stromal compartment. During embryonic development, urogenital epithelial and mesenchymal tissues interact to coordinate the spatial arrangement and eventual differentiation of these tissues into the glandular structure required for prostatic function [26]. In the developed organ, communication between the epithelia and the surrounding stroma maintains homeostasis via paracrine signaling [27,28]. The stroma is separated from the glandular acini by a basement membrane that provides positional information contributing to the maintenance of tissue [29].

The loss of homeostatic interactions between organ tissues in disease has partially been attributed to the loss of the basement membrane [30,31] and an alteration of the type of extracellular matrix [32]. Furthermore, the transformation of prostate epithelial cells by carcinoma-associated fibroblasts was correlated with increased MMP-9 expression [33]. While these results strongly implicate the roles of the basement membrane and the stromal microenvironment in tumor progression, the conflicting data and the vast number of factors involved limit our understanding of the multiple steps by which prostate tumors grow and invade surrounding tissues. TGF-β normally inhibits the proliferation of epithelia through induction of the cell cycle inhibitors p15 and p21 [34]. The determination of whether TGF-β will induce cytostasis or apoptosis in normal epithelia depends on the intensity of their proliferative activity in addition to poorly understood microenvironmental determinants [35,36]. There is therefore a need for further analysis of TGF-β's multiple roles. The TGF-β family of cytokines has many functions, some of which have been accurately modeled computationally, including TGF-β's role in vascular remodeling and hyperplasia and wound repair [37]. Models considering cell–stroma interactions via TGF-β as well as other factors in wound healing and tumor growth were shown to have good qualitative agreement with experimental results [38,39].

Recently [39], we proposed a model of prostate tumorigenesis using a HCA model that integrates five different cellular species (discrete) with three different microenvironmental chemical species (continuous), all of which are thought to play key roles in prostate cancer. Using this HCA model, we will investigate the importance of TGF-β in prostate homeostatic disruption and, in particular, how it regulates tumor–stroma interactions by considering tumors with different degrees of malignancy (in terms of TGF-β and MDE production).

Model

In the HCA model, cells are the discrete entities represented as points in a 2D grid (200 × 200 points representing a 2 × 2 mm slice of prostate; see Figure 1.4D, top right, for the initial conditions). This grid also hosts three microenvironmental variables, which are treated as continuous concentrations: TGF-β, matrix-degrading enzyme (MDE) expression, and Membrane/ECM. Cells can belong to one of five different types: basal epithelial, luminal epithelial, motile stroma, static stroma, and tumor cells. The simulated section of prostate contains three glands arranged along the off-diagonal axis. Each gland has an outside diameter of 19 grid points formed by an inner layer of luminal epithelial cells and an outer one made with basal epithelial cells. The space outside the glands can be occupied by static (muscle or fibroblastic lineages) and motile (monocytes or macrophages) stromal cells (Figure 1.4D, top left).

TGF-β plays a very important role in the model as it is produced, consumed, or utilized in one way or the other by all cell types. Basal cells produce TGF-β and membrane/ECM and also divide to replace basal and luminal epithelial cells that die due to normal attrition [40]. Luminal epithelial cells consume TGF-β and die when surrounded by tumor cells. Static stroma cells consume TGF-β but, over a set threshold, are programmed to produce more, effectively amplifying the TGF-β signal [41]. Motile stroma moves in the direction of the TGF-β gradient and produces extracellular matrix in direct proportion to its concentration. The ability of TGF-β to stimulate myofibroblasts to produce extracellular matrix is well established [42]. Finally, tumor cells appear as mutants of luminal epithelial cells after 10 days in six different positions of the simulated prostate. They require TGF-β to survive, and they proliferate as long as there is sufficient space. They also produce TGF-β and matrix degrading enzymes (MDEs). TGF-β is a well-known promoter of tumor cell proliferation [43] and is elevated in prostate cancer [44]. Alterations in response to TGF-β favor a protumorigenic response, and elevated MDEs have also been observed [45].

The three microenvironmental variables of the model are TGF-β ($T\beta$), MDE (E), and Membrane/ECM (M) concentrations. It is worth noting that the membrane/ECM variable represents both the ECM (a mixture of elements such as collagen, fibronectin, laminin, and vitronectin), which is assumed to be present everywhere outside the glands, and the basement membrane subadjacent to the basal epithelial cells. The dynamics of TGF-β ($T\beta$):

$$\frac{\partial T_\beta(x,y,z)}{\partial t} = \overbrace{\nabla(D(m_0 - M)\nabla T_\beta)}^{\substack{\text{Diffusion Mediated by} \\ \text{Basement Membrane}}} + \overbrace{\alpha_B B_{i,j}}^{\substack{\text{Production by} \\ \text{Basal Cells}}} + \overbrace{\alpha_c C_{i,j}}^{\substack{\text{Production by} \\ \text{Cancer Cells}}} + \overbrace{\alpha_S S_{i,j} T_\beta}^{\substack{\text{Production by Nonmotile} \\ \text{Stromal Cells Scaled by TGF-β}}}$$

$$- \overbrace{\gamma_S M T_\beta}^{\substack{\text{Binding by ECM/BM} \\ \text{Scaled by TGF-β}}} - \overbrace{\gamma_F F_{i,j} T_\beta}^{\substack{\text{Consumption by Motile} \\ \text{Stromal Cells Scaled by TGF-β}}} - \overbrace{\gamma_L L_{i,j}}^{\substack{\text{Consumption by} \\ \text{Luminal Cells}}} - \overbrace{\sigma T_\beta}^{\text{Natural Decay}}$$

(1.8)

which shows that TGF-β diffuses at a rate D modulated by the maximum tissue density, m_0. Production by basal and cancer cells as well as by motile stromal cells is assumed to be in proportion to the local TGF-β concentration at the rates α_B, α_C, and α_S, respectively. It also shows that TGF-β is consumed by luminal and motile stroma cells at the rates γ_L and γ_F. TGF-β also binds to the ECM at a rate γ_S, which depends on the local concentration of TGF-β; also, there is some natural decay of the ligand with rate σ.

MDEs (E) are produced by tumor cells (at rate λ), diffuse (at rate DE), and are depleted as they degrade the ECM and the basement membrane (μ):

$$\frac{\partial E(x,y,t)}{\partial t} = \overbrace{D_E \nabla^2 E}^{\text{Diffusion of Enzyme}} + \overbrace{\lambda C_{i,j}}^{\substack{\text{Production by} \\ \text{Cancer Cells}}} - \overbrace{\mu M E}^{\substack{\text{Used by Degradation of} \\ \text{Basement Membrane}}}$$ (1.9)

Basement membrane/ECM (M) is produced by basal cells (depending on the current local concentration of ECM ensuring the density never exceeds the maximum m_0) and motile stroma (depending on rate α_F, scaled by the local concentration of TGF-β). Finally, the ECM gets degraded by the MDEs at a rate μ:

$$\frac{\partial M(x,y,t)}{\partial t} = \overbrace{\rho B_{i,j}(m_0 - M)}^{\substack{\text{Production by Basal Cells} \\ \text{Provided Membrane is Not} \\ \text{Complete, i.e., } M = m_0}} - \overbrace{\mu M E}^{\substack{\text{Degradation by} \\ \text{Enzyme}}} + \overbrace{\alpha_F F_{i,j} T_\beta}^{\substack{\text{Production by Motile} \\ \text{Stromal Cells Scaled by TGF-β}}}$$ (1.10)

The model is defined such that the initial state of the system is a homeostatic one with birth/death and TGF-β production/consumption being balanced such that no abnormal growth occurs.

Results and Discussion

In order to study the role of stroma in maintaining homeostasis after tumor initiation, we performed a number of simulations with the model, each lasting about 55 years in simulation time (1 time step = 24 h). After 10 time steps, 6 basal epithelial cells, 4 in the central acinus and 1 on each of the two remaining acini, become abnormal epithelial cells initiating tumorigenesis. The simulations tested different configurations in which the proportion of stroma could range from 20% to 40% of the total available space as well as different tumor cell phenotypes (characterized by different rates of TGF-β and MDE production). The configurations considered were a high proportion of motile and nonmotile stroma (40% motile, 40% nonmotile), a high proportion of motile stroma (40% motile, 10% nonmotile), a high proportion of nonmotile stroma (10% motile, 40% nonmotile), and a low proportion of motile and nonmotile stroma (10% motile, 10% nonmotile). The tumor cell phenotypes that were tested produced values of TGF-β and MDE that were lower, equal, or higher to a given nondimensional value by an order of magnitude.

Figure 1.4D1–3 shows an example of simulations that illustrate the three main model outcomes of (1) control, (2) breakout, and (3) dies out. Figure 1.4D1 shows a simulation in which the tumor breaks out from the acini and grows in the surrounding media after a long period of control. Initially, the tumor cells fill the inner space inside the gland ($t = 0.3$). Eventually, the concentration of TGF-β and MDEs is sufficiently high that the basal membrane starts to degrade and TGF-β begins to leak from the gland and attract motile stroma ($t = 43.8$). At the end of the simulation ($t = 54.8$), the tumor has taken a significant portion of the domain, but its growth is constrained by the motile stroma responding to the TGF-β field. Figure 1.4D2 shows a simulation in which tumor cells quickly break from the gland and grow invading, unopposed by the stroma. Finally, Figure 1.4D3 shows tumor cells producing excessive quantities of MDE, which leads to early breakout of the basal membrane and leakage of the TGF-β, without which tumor cells die.

Contrary to our expectations given its centrality in the model, cancer cell TGF-β production does not seem to modify the outcome significantly, whereas stromal configuration has a much clearer effect. These results show that a prostate with a high proportion of stromal cells is more capable of restoring disrupted homeostasis. The role of stroma comes into effect only after the tumor has escaped from the gland, and it is at this point that the role of TGF-β, which mediates the interactions between stroma and tumor, is likely to become dominant.

It is important to remark, once again, that what makes the results particularly relevant is that tumor initiation and progression are performed on a domain that reproduces a prostatic tissue in homeostasis and that the tumors that manage to break from the prostatic gland and progress from PIN are those that manage to decisively disrupt this homeostasis by producing the right amounts of TGF-β and MDEs.

CONCLUSIONS

The models described in this chapter illustrate various aspects of the biological mechanisms that maintain the architecture and function of a homeostatic tissue as well as the consequences of homeostatic disruption. Homeostasis is a crucial feature of living organisms and needs to be underpinned by robust mechanisms capable of coping with genetic and environmental perturbations. Our work shows that homeostasis can evolve using different strategies (e.g., very dynamic characterized by high proliferation balanced by high apoptosis or more static via proliferation induced only by dying cells) and that many of these strategies become increasingly robust as the homeostatic individuals evolve over time. This diversity of homeostatic strategies relates well to different homeostatic tissue types found in multicellular organisms. For example, the epithelial cells lining the colon in humans are shed at a considerable rate. These cells are exposed to a hostile environment where the rate of spontaneous mutation probably is elevated, and in order to avoid harmful mutations accumulating in these cells they are continually removed. A similar tissue architecture is found in the outer layers of the skin, where cells also have a short lifespan. In other tissue types, which are not as exposed, the opposite type of homeostatic mechanism is normally found, that is, only when cells die spontaneously are they replaced.

These homeostatic mechanisms are built using the cells' existing molecular and signaling machinery as well as the architecture of the tissue and other physical constraints. Despite their general robustness against the most common perturbations, some genetic mutations and certain changes in tissue architecture could disrupt homeostasis. There is increasing experimental evidence showing that the restoration of tissue organization is able to repress the malignant phenotype of genetically aberrant cells. For example, when mouse embryonal carcinoma cells (which form malignant tumors upon subcutaneous injection) were fused with normal blastocysts, they were able to give rise to phenotypically normal cancer-free mice [46]. Also, malignant T4-2 cells forming disorganized continuously proliferating colonies can be reverted to near-normal phenotype when grown in the presence of integrin-blocking antibodies [47]. These reverted T4-2 cells formed regular growth-arrested acinar structures with restored apico-basal polarity, reorganized actin cytoskeleton, and were able to remain quiescent for up to 1 month in culture. Chen and co-workers showed that disruption of cytostructure activates the angiogenic switch even in the absence of proliferation and hypoxia, and that restored organization of malignant clusters reduces expression of vascular endothelia growth factor (VEGF) and activation of endothelial cells to levels found in quiescent nonmalignant epithelium [48]. Our own results show that once carcinogenesis has taken place and homeostasis has been disrupted, there is a window of opportunity in which this disruption can be reversed (e.g., in the TGF-β HCA model, an increase in stromal cells could compensate and block tumor progression; in the IBCell model, early disruptive cell outgrowth can be compensated by apoptosis if the erroneous cell response is not passed on to daughter cells) or at least transformed into a different type of homeostasis (e.g., one in which the tumor is not destroyed but permanently contained [49]).

The research we have presented in this chapter, as well as others focused on tissue homeostasis, should lead to a greater knowledge of the mechanisms that underpin it and highlight its limitations. Ultimately, this information will improve our understanding of the types of homeostatic disruptions that could lead to cancer initiation, the likely sequence of phenotypical transformations that would be acquired by tumor cells in those tumors that irreversibly alter the homeostatic balance, and the possible steps that could be taken to reverse them.

ACKNOWLEDGMENTS

We gratefully acknowledge the NIH/National Cancer Institute support from the ICBP (5U54 CA113007) and TMEN (1U54 CA126505) programs.

REFERENCES

1. Reed, J.C., Dysregulation of apoptosis in cancer. *J. Clin. Oncol.*, 1999. 17(9): pp. 2941–53.
2. Wolpert, L., Positional information and pattern formation. *Philos. Trans. Roy. Soc.* Series B, 1981. 295: pp. 441–450.
3. Wolpert, L., *Principles of Development.* 2007: Oxford University Press, USA.

4. Kitano, H., Cancer robustness: Tumour tactics. *Nature*, 2003. 426(6963): p. 125.
5. Carlson, J.M. and J. Doyle, Complexity and robustness. *Proc. Natl. Acad. Sci. USA*, 2002. 99(Suppl. 1): pp. 2538–45.
6. Frisch, S.M. and E. Ruoslahti, Integrins and anoikis. *Curr. Opin. Cell Biol.*, 1997. 9(5): pp. 701–6.
7. Lioni, M. et al., Dysregulation of claudin-7 leads to loss of E-cadherin expression and the increased invasion of esophageal squamous cell carcinoma cells. *Am. J. Pathol.*, 2007. 170(2): pp. 709–21.
8. Lioni, M. et al., Bortezomib induces apoptosis in esophageal squamous cell carcinoma cells through activation of the p38 mitogen-activated protein kinase pathway. *Mol. Cancer Ther.*, 2008. 7(9): pp. 2866–75.
9. In press.
10. Basanta, D. et al., Investigating the evolvability of biologically inspired CA. *Proceedings of the 9th Conference on Artificial Intelligence* (Alife9), Boston, MA, 2004.
11. Gerlee, P. and A. Anderson, An evolutionary hybrid cellular automaton model of solid tumour growth. *J. Theor. Biol.*, 2007. 246(4): pp. 583–603.
12. Gerlee, P. and A. Anderson, A hybrid cellular automaton model of clonal evolution in cancer: the emergence of the glycolytic phenotype. *J. Theor. Biol.*, 2007. 250: pp. 705–722.
13. Gerlee, P. and A.R.A. Anderson, Evolution of cell motility in an individual-based model of tumour growth. *J. Theoretical Biol.*, 2009. 259: pp. 67–83.
14. Gerlee, P. and A.R.A. Anderson, Modelling evolutionary cell behaviour using neural networks: application to tumour growth. *Biosystems*, 2009. 95: pp. 166–174.
15. Banzhaf, W. et al., *Genetic Programming: An Introduction.* 1997: Morgan Kaufmann Publishers.
16. Bentley, P.J., Finding acceptable pareto-optimal solutions using multiobjective genetic algorithms. *Soft Computing in Engineering Design and Manufacturing*, Ed. P. Chawdhry, Roy, R., Pant, R. 1998: Springer-Verlag, London.
17. Rejniak, K.A., An immersed boundary framework for modelling the growth of individual cells: an application to the early tumour development. *J. Theor. Biol.*, 2007. 247(1): pp. 186–204.
18. Rejniak, K.A. and A.R. Anderson, A computational study of the development of epithelial acini: II. Necessary conditions for structure and lumen stability. *Bull. Math. Biol.*, 2008. 70(5): pp. 1450–79.
19. Rejniak, K.A. and A.R. Anderson, A computational study of the development of epithelial acini: I. Sufficient conditions for the formation of a hollow structure. *Bull. Math. Biol.*, 2008. 70(3): pp. 677–712.
20. Debnath, J. et al., The role of apoptosis in creating and maintaining luminal space within normal and oncogene-expressing mammary acini. *Cell*, 2002. 111(1): pp. 29–40.
21. Ferguson, D.J., An ultrastructural study of mitosis and cytokinesis in normal "resting" human breast. *Cell Tissue Res.*, 1988. 252(3): pp. 581–7.

22. Debnath, J., S.K. Muthuswamy, and J.S. Brugge, Morphogenesis and onco-genesis of MCF-10A mammary epithelial acini grown in three-dimensional basement membrane cultures. *Methods*, 2003. 30(3): pp. 256–68.

23. Anderson, A.R.A. et al., Modelling of cancer growth, evolution and invasion: bridging scales and models. *Math. Model. Nat. Phenom.*, 2007. 2(3): pp. 1–29.

24. Ferguson, D.J.P. and T.J. Anderson, Morphological evaluation of cell turn-over in relation to the menstrual cycle in the "resting" human breast. *Br. J. Cancer*, 1981. 44: pp. 177–81.

25. Watson, C.J., Involution: apoptosis and tissue remodelling that convert the mammary gland from milk factory to a quiescent organ. *Breast Cancer Res.*, 2006. 8: pp. 203–207.

26. Hayward, S.W. and G.R. Cunha, The prostate: development and physiology. *Radiol. Clin. North Am.*, 2000. 38(1): pp. 1–14.

27. Cunha, G.R., S.W. Hayward, and Y.Z. Wang, Role of stroma in carcinogen-esis of the prostate. *Differentiation*, 2002. 70(9–10): pp. 473–85.

28. Hayward, S.W., M.A. Rosen, and G.R. Cunha, Stromal-epithelial interactions in the normal and neoplastic prostate. *Br. J. Urol.*, 1997. 79 Suppl 2: pp. 18–26.

29. Bissell, M.J., H.G. Hall, and G. Parry, How does the extracellular matrix direct gene expression? *J. Theor. Biol.*, 1982. 99(1): pp. 31–68.

30. Boudreau, N., Z. Werb, and M.J. Bissell, Suppression of apoptosis by basement membrane requires three-dimensional tissue organization and withdrawal from the cell cycle. *Proc. Natl. Acad. Sci. USA*, 1996. 93(8): pp. 3509–13.

31. Liotta, L., C. Rao, and U. Wewer, Biochemical interactions of tumor cells with the basement membrane. *Annu. Rev. Biochem.*, 1988. 55: pp. 1037–1057.

32. Paszek, M.J. et al., Tensional homeostasis and the malignant phenotype. *Cancer Cell*, 2005. 8(3): pp. 241–54.

33. Phillips, J.L. et al., The consequences of chromosomal aneuploidy on gene expression profiles in a cell line model for prostate carcinogenesis. *Cancer Res.*, 2001. 61(22): pp. 8143–9.

34. Massague, J., S.W. Blain, and R.S. Lo, TGFbeta signaling in growth control, cancer, and heritable disorders. *Cell*, 2000. 103(2): pp. 295–309.

35. Derynck, R., R.J. Akhurst, and A. Balmain, TGF-beta signaling in tumor suppression and cancer progression. *Nat. Genet.*, 2001. 29(2): pp. 117–29.

36. Massague, J., TGF-beta in cancer. *Cell*, 2008. 134(2): pp. 215–30.

37. Budu-Grajdeanu, P. et al., A mathematical model of venous neointimal hyperplasia formation. *Theor. Biol. Med. Model*, 2008. 5(2).

38. Michelson, S. and J. Leith, Autocrine and paracrine growth factors in tumor growth: a mathematical model. *Bull. Mathematical Biol.*, 1991. 53: pp. 639–656.

39. Basanta, D. et al., The role of transforming growth factor-beta-mediated tumor–stroma interactions in prostate cancer progression: an integrative approach. *Cancer Res.*, 2009. 69(17): pp. 7111–20.

40. Wang, Y. et al., A human prostatic epithelial model of hormonal carcinogen-esis. *Cancer Res.*, 2001. 61(16): pp. 6064–72.

41. Peehl, D.M. et al., Loss of response to epidermal growth factor and retinoic acid accompanies the transformation of human prostatic epithelial cells to tumorigenicity with v-Ki-ras. *Carcinogenesis*, 1997. 18(8): pp. 1643–50.

42. Wynn, T.A., Cellular and molecular mechanisms of fibrosis. *J. Pathol.*, 2008. 214(2): pp. 199–210.
43. Wakefield, L.M. and A.B. Roberts, TGF-beta signaling: positive and negative effects on tumorigenesis. *Curr. Opin. Genet. Dev.*, 2002. 12(1): pp. 22–9.
44. Steiner, M.S. and E.R. Barrack, Transforming growth factor-b1 overproduction in prostate cancer: effects on growth in vivo and in vitro. *Mol. Endocrinol.*, 1992. 6: pp. 15–25.
45. Ao, M. et al., Transforming growth factor-beta promotes invasion in tumorigenic but not in nontumorigenic human prostatic epithelial cells. *Cancer Res.*, 2006. 66(16): pp. 8007–16.
46. Mintz, B. and K. Illmensee, Normal genetically mosaic mice produced from malignant teratocarcinoma cells. *Proc. Natl. Acad. Sci. USA*, 1975. 72(9): pp. 3585–9.
47. Weaver, V.M. et al., beta4 integrin-dependent formation of polarized three-dimensional architecture confers resistance to apoptosis in normal and malignant mammary epithelium. *Cancer Cell*, 2002. 2(3): pp. 205–16.
48. Chen, A. et al., Endothelial cell migration and vascular endothelial growth factor expression are the result of loss of breast tissue polarity. *Cancer Res.*, 2009. 69(16): pp. 6721–9.
49. Gatenby, R.A., A change of strategy in the war on cancer. *Nature*, 2009. 459(7246): pp. 508–9.

CHAPTER 2

Cancer Cell

Linking Oncogenic Signaling to Molecular Structure

Jeremy E. Purvis, Andrew J. Shih,
Yingting Liu, and Ravi Radhakrishnan

CONTENTS

INTRODUCTION

The cancer phenotype may be viewed as a pathological dysregulation of cellular signals that control growth, survival, motility, cell–cell connectivity, and DNA synthesis and repair (Hanahan and Weinberg 2000). Generally speaking, the sources of dysregulation are an accumulated set of mutations that produce altered gene products. The interaction of these mutant oncoproteins with normal host signaling mechanisms perturbs proper signaling and confers the oncogenic behavior. For example, mutations affecting the catalytic activity, binding specificity, translation efficiency, or rate of degradation of an enzyme involved in DNA repair can predispose a cell to genomic instability by allowing replication of damaged genomic sequence (Zhivotovsky and Kroemer 2004). The specific identities and combinations of these cancer-causing mutations are known to vary considerably according to tissue and cell type and are strongly interdependent on the cellular/tumor microenvironment (Sjoblom, Jones, et al. 2006). In light of these complexities, mathematical models of cellular signaling networks have become indispensable tools for explaining oncogenic behaviors, predicting resistance mechanisms, and designing molecular therapies to attenuate defective signaling.

How can the altered activities of mutant oncoproteins be represented in a mathematical model? In many cases, wild-type and oncogenic signaling pathways share similar network structures but differ in the kinetic behavior and interaction topologies of only a few oncoproteins (Sharma and Settleman 2007). Unless these differences are resolved quantitatively, one cannot distinguish between normal and cancerous signaling networks in a mathematical model. Moreover, when attempting to resolve such crucial but subtle differences, it may be necessary to switch from a systems perspective of protein interaction networks to a molecular perspective of enzyme activation and protein–protein interaction. This chapter describes an approach for constructing models of intracellular signaling networks in which the oncogenic behavior of the network is encoded through calculations of altered kinetic and structural properties of mutant oncoproteins. Using molecular dynamics and docking simulations, atomistic models are exploited to quantify altered topologies of interactions as well as to provide the missing parameters for network models of both wild-type and oncogenic signaling. The global behavior of these networks may then be compared and functional roles may be assigned to the mutant oncoproteins. An application of this multiscale, multiresolution approach is presented in which structural alterations found in a mutant form of the epidermal growth factor receptor are represented as kinetic perturbations in a model of growth factor signaling. Based on network parameters estimated from molecular-level simulations, simulations at the network level show how small perturbations in molecular structure can lead to a profoundly altered cellular phenotype.

MODEL

A multiscale, multiresolution approach can be devised by encompassing four distinct length/time scales:

- Molecular modeling using Newton's equations: Based on a Hamiltonian $H(r_1,r_2,...) = K + U$; r_i = atomic coordinates, K = kinetic energy, and U = Potential energy, we solve the system of equations (where F_i = Force, m_i = mass, t = time),

$$F_i = -\nabla U = m_i d^2 r_i / dt^2 \qquad (2.1)$$

- Electronic structure using mixed quantum mechanics molecular mechanics simulations: We variationally minimize the energy function E,

$$E = \langle\varphi|H|\varphi\rangle \text{ and } \langle\varphi|\varphi\rangle = 1 \qquad (2.2)$$

where, the bra-ket (Dirac) notation $\langle\varphi|\varphi\rangle$ represents vector dot product and $\langle\varphi|H|\varphi\rangle$ represents the expectation value (Szabo and Ostlund 1996). In Equation 2.2, φ is the electronic wave function satisfying the Pauli's exclusion principle and $H = H_{electronic}$, that is, the electronic Hamiltonian. We then solve Newton's equations given by Equation 2.1 for the nuclear degrees of freedom with some forces derived from $H_{electronic}$.

- Coarse-grained models: Using a coarse-grained Hamiltonian, $H[\lambda(r_1, r_2)] = F[\lambda(r_1, r_2)]$, where F = free energy, λ = generalized/collective coordinate, we solve the generalized Langevin equation (Agrawal, Weinstein, et al. 2008),

$$d\lambda/dt = -M\delta F/\delta\lambda + \xi \qquad (2.3)$$

In Equation 2.3, M represents the mobility term, ξ represents random thermal force satisfying $\langle\xi\rangle = 0$, and $\langle\xi(0)\xi(t)\rangle = 2k_B TM\delta(t)$, where $\langle\rangle$ represents an ensemble average in the equilibrium ensemble, k_B = Boltzmann's constant, and T = temperature.

- Deterministic network models: As an average solution to the Langevin dynamics using F(λ = reaction coordinate) for transitions between reactant and product states of chemical species, we solve deterministic network models for signal transduction using rate laws based on mass-action kinetics.

Based on the aforementioned formulations, the linking of oncogenic signaling to molecular structure in a mathematical model can be achieved in three separate but interconnected modeling steps:

- Deterministic ordinary differential equations (ODE) models are used to represent both wild-type and oncogenic signaling networks that differ in a defined set of molecular species. The pair of networks must both contain the mutant or overexpressed oncoproteins as well as one or more "output" components that can be monitored to evaluate the oncogenic behavior of the system (e.g., a master transcriptional regulator controlling cell survival).

- Molecular docking is used to predict ligand binding in the absence of a ligand-bound crystal structure and functional affinity data. These free energy-based simulations are used to calculate a new set of mutant kinetic parameters based on altered molecular structure.

- Molecular dynamics simulations are used to characterize the structural properties of mutant gene products from an altered polypeptide sequence. These calculations rely on the availability of solved crystal structures and may involve homology modeling.

By integrating these three modeling regimes, the phenotypic differences that define mutant systems at the network level are encoded through fine-scale calculations of the structural and kinetic properties of mutant oncoproteins (Figure 2.1).

The multiscale strategy portrayed in Figure 2.1 is illustrated through a model of dysregulated growth signaling caused by a single amino acid substitution in the epidermal growth factor receptor (EGFR). EGFR is a receptor tyrosine kinase (RTK) that is commonly mutated or overexpressed in human cancers (Mendelsohn and Baselga 2000). A mutant form of the receptor, L834R, exhibits an altered pattern of autophosphorylation caused by differences in its physical structure, binding affinities, and catalytic behavior. These perturbed phosphorylation patterns lead to

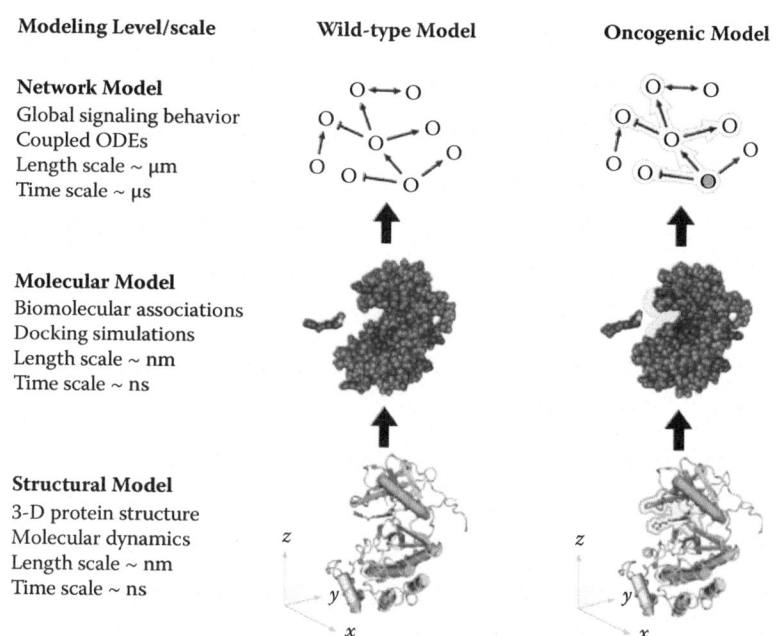

Modeling Level/scale	Wild-type Model	Oncogenic Model
Network Model Global signaling behavior Coupled ODEs Length scale ~ μm Time scale ~ μs		
Molecular Model Biomolecular associations Docking simulations Length scale ~ nm Time scale ~ ns		
Structural Model 3-D protein structure Molecular dynamics Length scale ~ nm Time scale ~ ns		

FIGURE 2.1 Overview of multiscale modeling method for linking molecular structure to oncogenic signaling. The aberrant signaling behavior of oncogenic networks is captured by monitoring the behavior of altered network models, typically ODE reaction networks. Alternate parameterization for the models is provided by docking simulations of mutant oncoproteins, which relies on structural models of the mutant proteins. Structural information is ultimately connected to mutations in genomic sequence. An example of oncogenic signaling behavior caused by altered structural and kinetic properties in a mutant of the epidermal growth factor receptor is presented in the text.

constitutive activation of certain survival pathways that predispose L834R mutants to uncontrolled growth (Choi, Mendrola, et al. 2007). Thus, the goal of this multiscale model will be to track changes in the cellular growth pathways as a result of structural alterations in the mutant receptor. Accomplishing this goal will require a consideration of the receptor's biophysical properties on an atomic scale as well as its interaction with various binding partners and adaptor proteins.

Constructing Mechanistic Models of Oncogenic Signaling

Kinetic models of cellular signaling pathways represent the highest level of modeling in this approach and are used to monitor the global behavior of both wild-type and mutant systems (Figure 2.1). Ordinary differential equations (ODEs) are used to represent coupled kinetic reactions that describe the rates of production and consumption of species in the model (Aldridge, Burke, et al. 2006). For large networks that contain species with posttranslational modifications or multiple binding partners, rules-based modeling (Hlavacek, Faeder, et al. 2006) provides the best method of generating ODEs that encode these kinetic differences.

In the signaling network presented here (Figure 2.2), EGF-induced activation of EGFR occurs through two parallel phosphorylation pathways corresponding to tyrosine 1068 (Y1068) and tyrosine 1173 (Y1173). Phosphorylated Y1068 (pY1068) binds only to the adaptor proteins Gab-1 and Grb2, while phosphorylated Y1173 (pY1173) binds only to the adaptor Shc. The major downstream pathways include EGF-ERK via the Ras-Raf MAP-kinase cascade (Citri and Yarden 2006), and the PI3K-AKT pathway, which results in the activation of the downstream protein—serine/

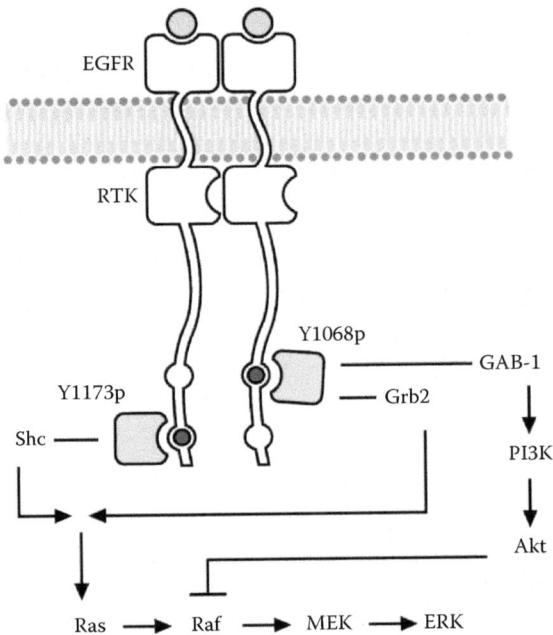

FIGURE 2.2 Network model of EGFR-mediated signaling used in this study. Phosphorylation of the EGFR dimer occurs at either Y1068, which can bind GAB-1 or Grb2, or at Y1173, which binds Shc. Activation of downstream proteins AKT and ERK was used as indicators of cell survival signaling. Multiscale modeling is achieved by calculating changes in dimerization, peptide binding affinity, and phosphorylation in structural mutants of the receptor.

threonine kinase AKT. In choosing the scope of a network, it is important to include one or more species that serve as indicators of the oncogenic potential of the system. Here, we include both AKT and ERK because they are well-studied indicators of EGFR-mediated growth and survival behaviors (Citri and Yarden 2006).

The critical step in distinguishing wild-type and mutant signaling models is defining kinetic differences between the systems. Differences between the wild-type EGFR and L834R models are marked in Table 2.1 and include reactions affecting receptor dimerization, phosphorylation, and peptide binding. Rather than write each reaction equation

TABLE 2.1   Reaction rules for two-site phosphorylation model of the EGF receptor

Event	Reaction rule	Forward	Reverse
Ligand/receptor binding	egfr(l) + egf(r) ↔ egfr(l!1). egf(r!1)	k_1	k_{-1}
Ligand-induced receptor dimerization	egfr(l!1,r) + egfr(l!2,r) ↔ egfr(l!1,r!3).egfr(l!2,r!3)	$k_2{}^*$	k_{-2}
Spontaneous receptor dimerization	egfr(r) + egfr(r) ↔ egfr(r!1). egfr(r!1)	$k_3{}^*$	$k_{-3}{}^\dagger$
Receptor/ATP binding	egfr(r!+,k) + ATP(r) ↔ egfr(r!+,k!1).ATP(r!1)	k_4	$k_{-4}{}^\dagger$
Y1068 entering active site	egfr(y1068~u) ↔ egfr(y1068~b)	k_5	$k_{-5}{}^\dagger$
Y1173 entering active site	egfr(y1173~u) ↔ egfr(y1173~b)	k_6	$k_{-6}{}^\dagger$
Autophosphorylation of Y1068	egfr(r!1,y1068~b).ADP → egfr(r!1,y1068~p) + ADP	$k_7{}^\dagger$	
Autophosphorylation of Y1173	egfr(r!1,y1173~b).ADP → egfr(r!1,y1173~p) + ADP	$k_8{}^\dagger$	
Dephosphorylation of Y1068	egfr(y1068~p) + phos → egfr(y1068~u)	V_9	K_9
Dephosphorylation of Y1173	egfr(y1173~p) + phos → egfr(y1173~u)	V_{10}	K_{10}

Note: The 10 rules generate 328 species and 3324 half-reactions representing all possible molecular intermediates and reaction steps. For simplicity, reaction rules for adaptor protein binding, MAP kinase cascade, and ERK/AKT activation are not shown.

* Denotes $k_2 = k_3$ because the on-rate of dimerization is diffusion limited.

† Denotes rate constants k_{-3}, k_{-4}, k_{-5}, k_{-6}, k_7, and k_8 are each affected by mutation L858R.

! Denotes the site of association for two molecules. For example, egfr(l!1).egf(r!1) are bound through "l" and "r" sites on egfr and egf, respectively.

separately, "reaction rules" (Hlavacek, Faeder, et al. 2006) are used to define general types of interactions between functional domains among the species in the model. For example, the rate constant k_7 in Table 2.1 that describes the catalytic turnover of Y1068 phosphorylation is applied to all forms of the receptor that participate in this phosphorylation reaction (e.g., monomer, dimer, ATP-bound). In this way, rules-based modeling ensures efficient and accurate construction of ODE-based signaling models.

Providing Alternate Parameterization through Docking Simulations

Once the network has been defined and the mutant oncoproteins identified, molecular docking is used to predict ligand binding in the absence of a ligand-bound crystal structure and functional affinity data. Thus, docking simulations provide the missing parameters that characterize the mutant system. Automated docking tools such as AutoDock (Morris, Goodsell, et al. 1998) in combination with more accurate approaches such as free-energy perturbation may be used to predict how small molecules, such as substrates or drug candidates, bind to a receptor of known three-dimensional structure. The binding free energy is calculated based on the intermolecular energy between protein and ligands and changes to the solvation environment. For the EGFR/L834R model, binding modes were determined for ATP as well the C-terminal peptides Y1068 and Y1173 to the catalytic site. A global conformational search was performed using a multiple conformation docking strategy, in which the protein flexibility is taken into account implicitly. Note that rules-based modeling (Hlavacek, Faeder et al. 2006) facilitates the reuse of kinetic parameters calculating through docking simulations.

Resolving the Structure of Mutant Oncoproteins through Molecular Dynamics

In order to perform docking simulations, it is necessary to acquire accurate structural information about the molecules involved. In the EGFR/L834R model, we model the receptor activation characteristics (whether active as a monomer or requires dimerization) of the EGFR receptor tyrosine kinase using molecular dynamics simulations. 10–30 ns trajectories of atomistic and explicitly solvated systems of wild-type and mutant EGFR kinase monomers and dimers are obtained and analyzed for specific stabilizing interactions such as hydrogen bonds and salt bridges, hydrophobic interactions, and conformational changes.

RESULTS

Modeling the EGFR at the network, molecular, and structural levels allows one to determine how point mutations in the EGFR receptor can alter signaling characteristics leading to the onset of oncogenic transformations. The model was constructed in "top-down" fashion, beginning with identical signaling networks that were differentiated by a defined set of mutant oncoproteins. We now examine the effects of these differences in reverse order, beginning with structural alterations in the tyrosine kinase domain and proceeding to observe how these perturbations affect both receptor kinetics and network behavior.

Activation of Wild-Type EGFRTK and L834R Mutant RTK

Crystal structures of the EGFRTK suggest that the conformational switching from an inactive to an active conformation involves a rotation of the αC-helix and the shifting of the activation loop (A-loop) to make way for substrate peptide (harboring the tyrosine residue) and ATP binding. To assess the structural requirements for such a conformational shift, analyses of bond patterns and hydrophobic interactions were performed to identify specific interactions (H-bonds and salt bridges) between residues of the αC-helix, and those of the A-loop needed to reorganize the enzyme and allow conformational switching from inactive to active states. Most of the stabilizing interactions holding the kinase in the inactive conformation are influenced by the dimer-interface residues, supporting an allosteric activation mechanism proposed for the wild-type (Zhang, Gureasko, et al. 2006). Many of these interactions overlap with the residues associated with several clinically relevant mutations, including L834R. The R substitution of L at 834 destabilizes the specific (external H-bonds) interactions associated with A-loop and αC-helix in the inactive but not the active conformations. Thus, our analysis of stabilizing interactions presented in Figures 2.3a and 2.3b serves as a platform for unifying the effects of these mutations at a structural level. An important outcome of these simulations is the notion that the mutant receptor can be active (and thus mediate signaling) even as a monomer, that is, in the absence of any growth factor binding. This establishes a small but crucial variation in network topology between the wild-type and the mutant systems.

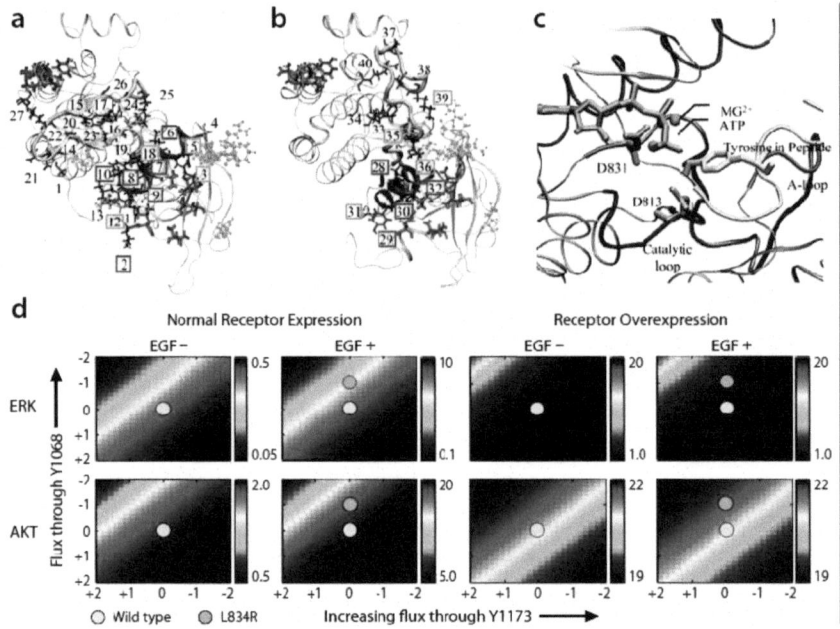

FIGURE 2.3   **(See color insert following page 40)** Structural, kinetic, and network analysis of the effect of L834R mutation in the EGFRTK. Visualization of the stabilizing residues external to A-loop and αC-helix (*blue*), dimer interface residues (*red*), and clinical mutations (*green*) of both the active (*a*) and inactive (*b*) EGFR tyrosine kinases. (*c*) Binding modes for ATP (*cyan*) and the optimal peptide sequence (*yellow*) in the EGFRTK domain. (*d*) Calculated ERK and AKT phosphorylation levels in units of nM (peak-levels over an 1800 s time course) under serum-starved (EGF–) and serum-cultured (EGF+) conditions for cell types with normal EGFR expression and EGFR overexpression. The *x*- and *y*-axes represent log changes in the binding affinity (K_D) of the peptide relative to the wild type.

Ligand and Substrate Binding Affinities for EGFRTK

The structural basis for the context-specific kinetics of the C-terminal tyrosine substrates is provided by our computational docking calculations (Liu, Purvis, et al. 2007). Substrate peptides derived from tyrosine sites of the EGFR C-terminal tail—Y1068 (VPEYINQ) and Y1173 (NAEYLRV)—bound to the wild-type and the L834R mutant EGFR kinase revealed how the structure of the bound peptide–protein complex is altered at the catalytic site due to the arginine substitution of leucine in L834R (Figure 2.3c). By employing this method, we computed the binding affinities for wild-type and L834R mutant RTK binding to two peptide sequences consisting

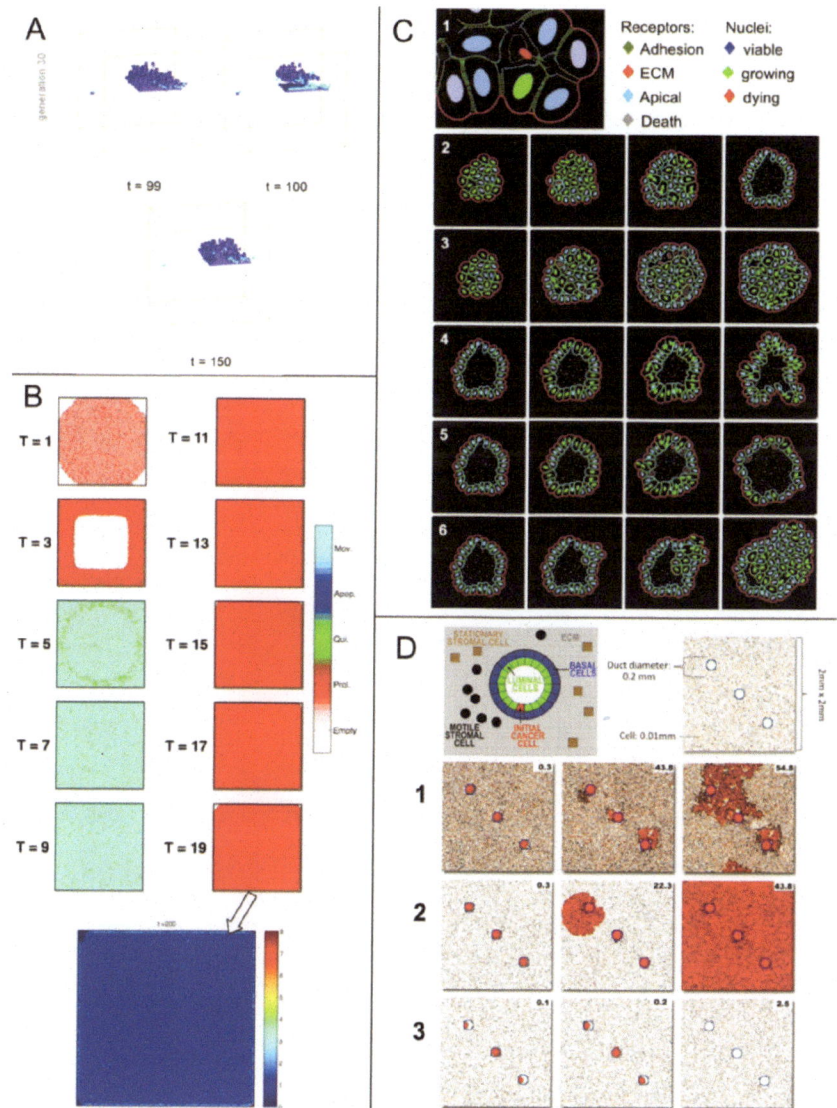

FIGURE 1.4   (**See Figure caption on page 14**)

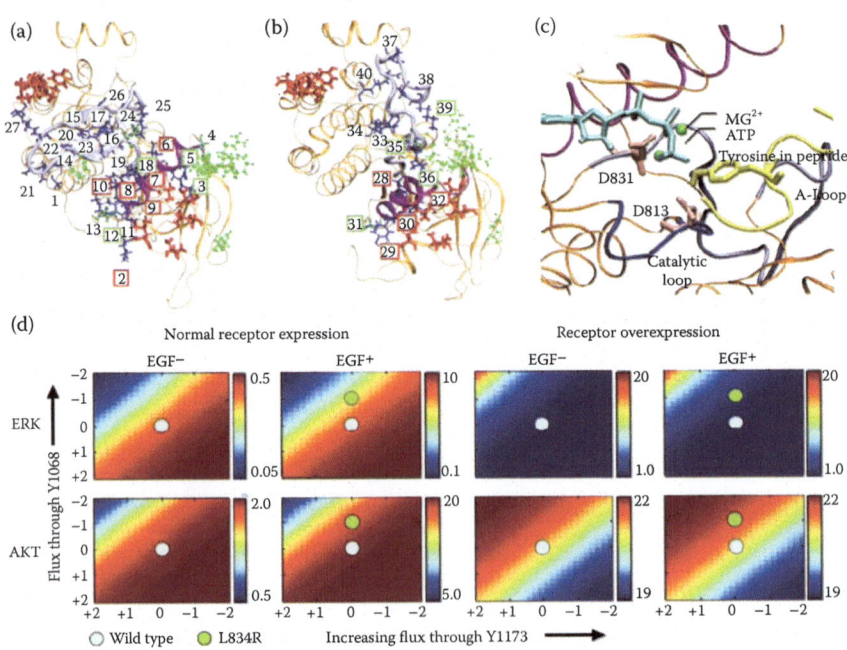

FIGURE 2.3 (**See Figure caption on page 40**)

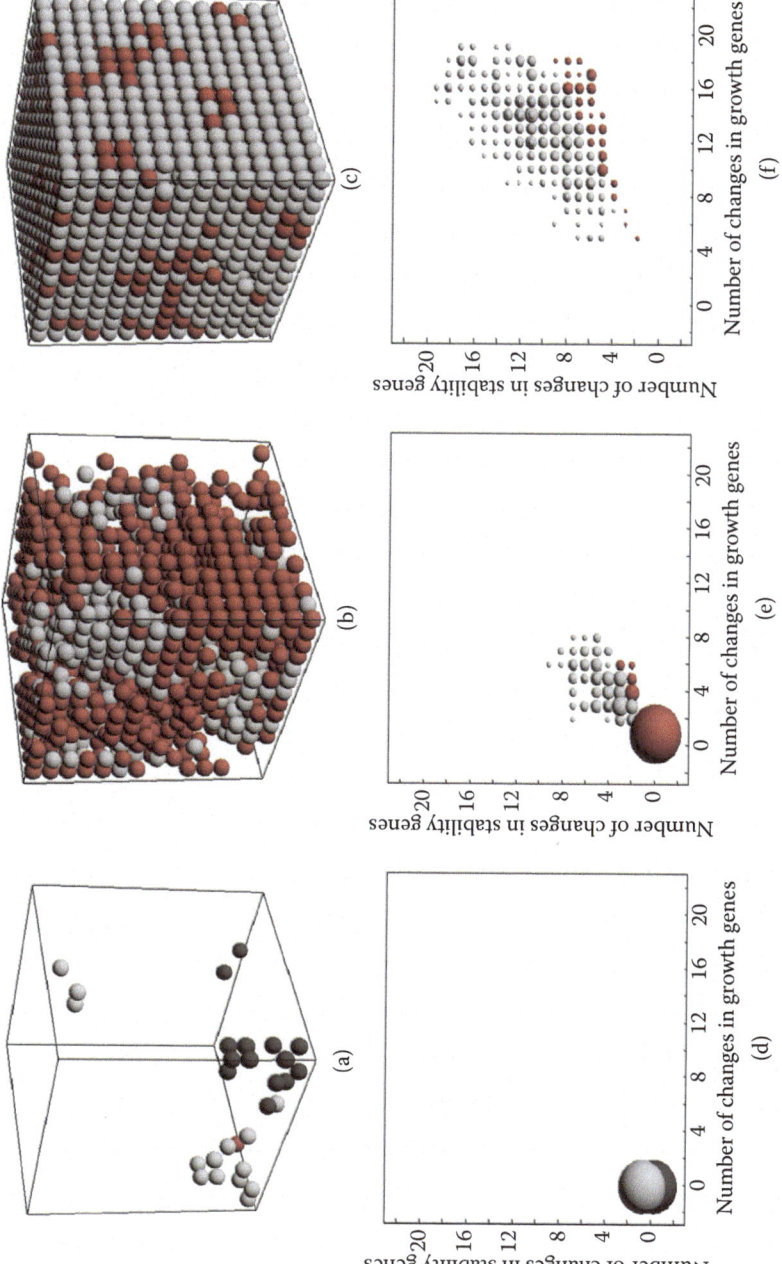

FIGURE 4.5 **(See Figure caption on page 77).**

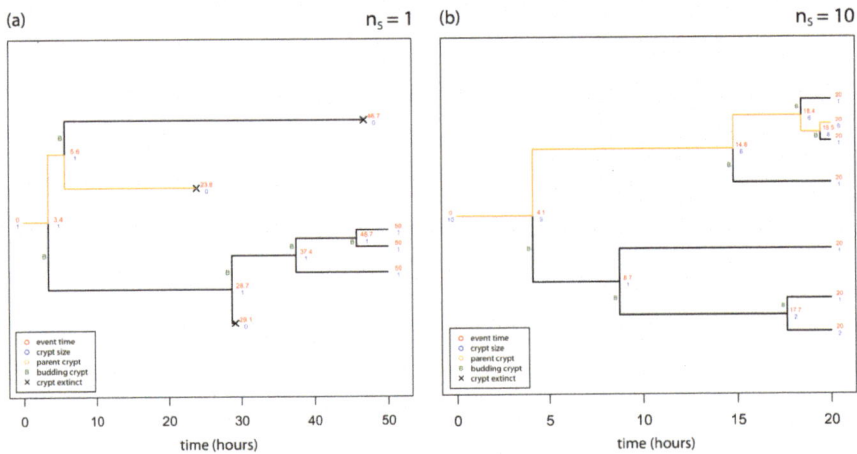

FIGURE 5.4 **(See Figure caption on page 99)**

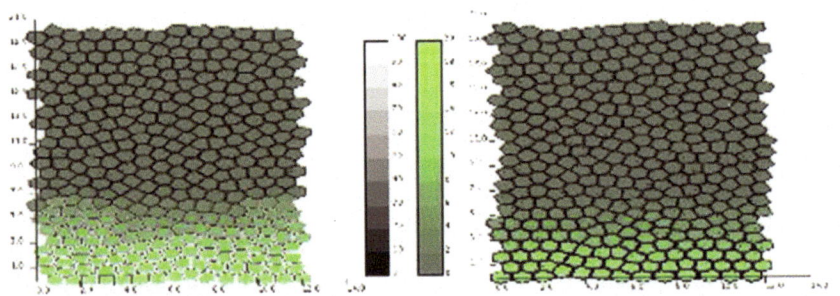

FIGURE 6.5 **(See Figure caption on page 122)**

FIGURE 6.7   (**See Figure caption on page 124**)

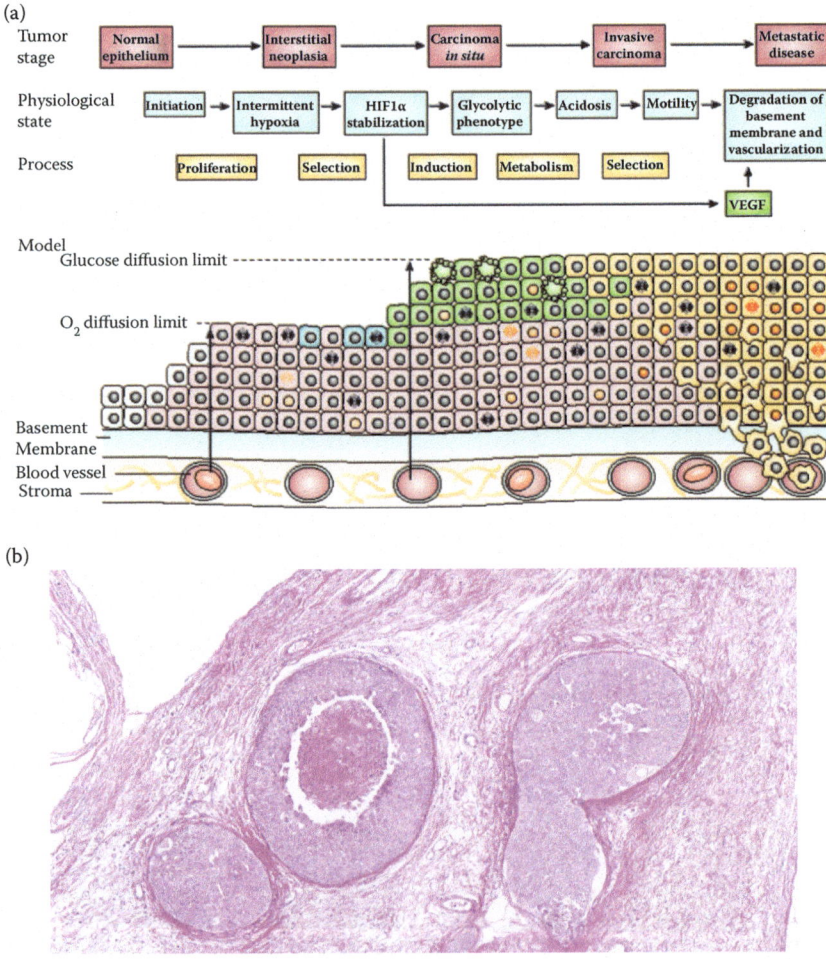

FIGURE 7.1 **(See Figure caption on page 142)**

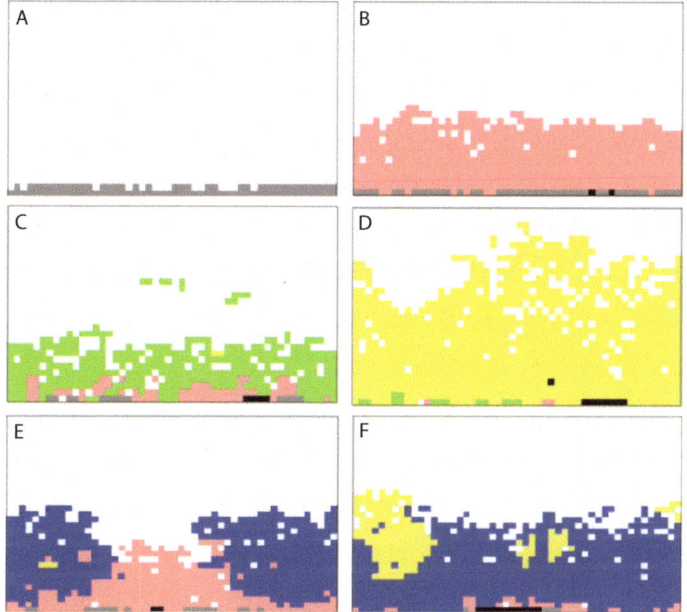

FIGURE 7.2   (See Figure caption on page 144)

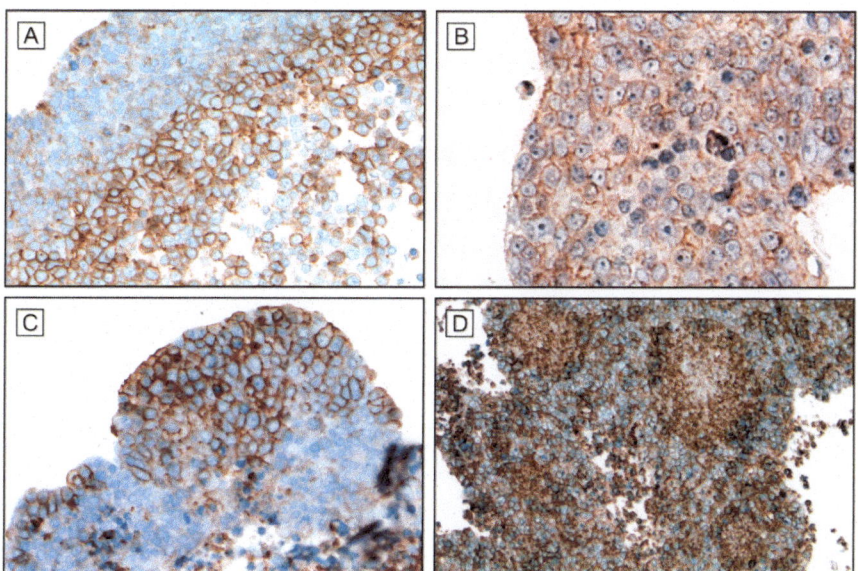

FIGURE 7.3   (See Figure caption on page 146)

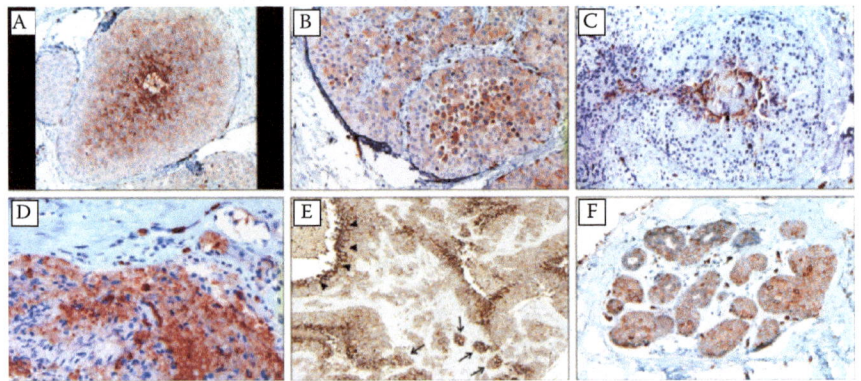

FIGURE 7.4   (**See Figure caption on page 147**)

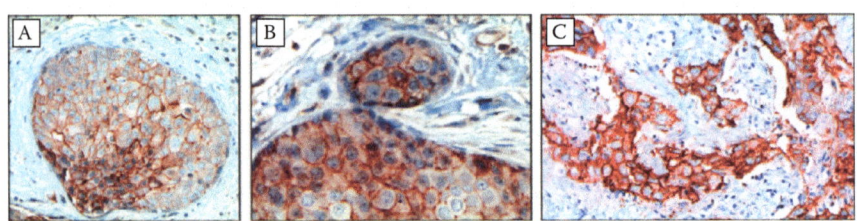

FIGURE 7.5   (**See Figure caption on page 149**)

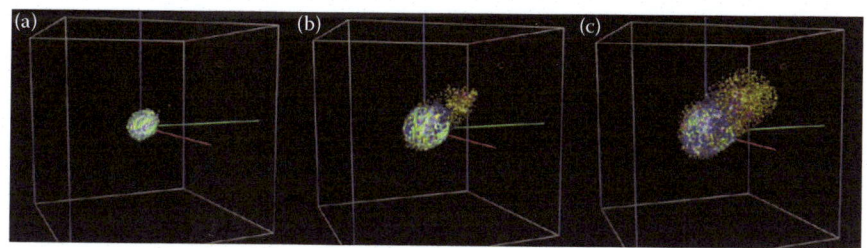

FIGURE 9.2 **(See Figure caption on page 181)**

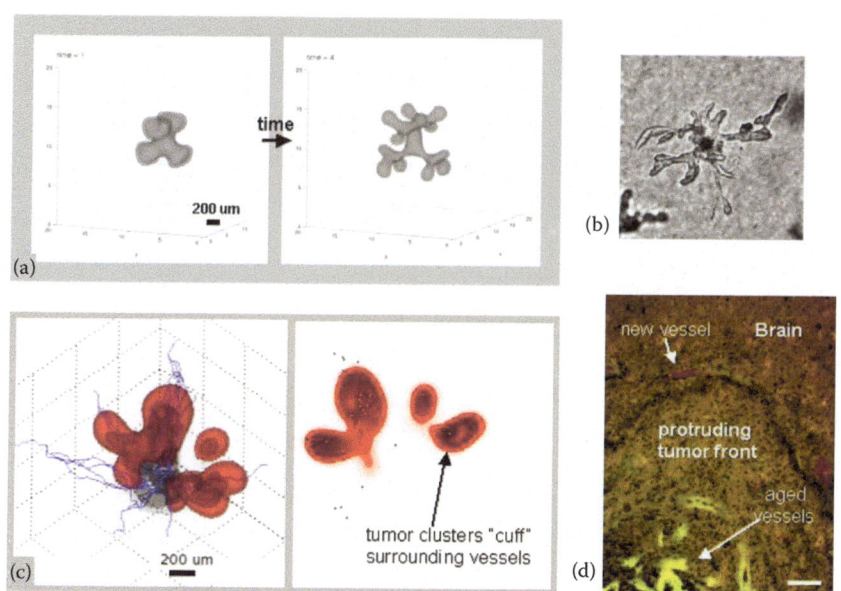

FIGURE 10.3 **(See Figure caption on page 203)**

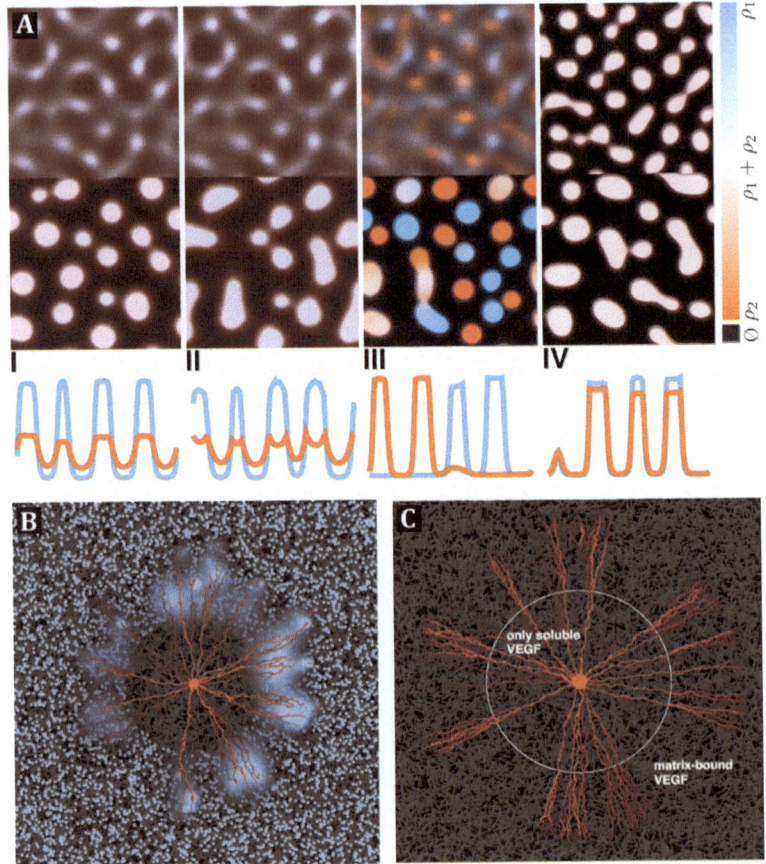

FIGURE 11.4   (See Figure caption on page 226)

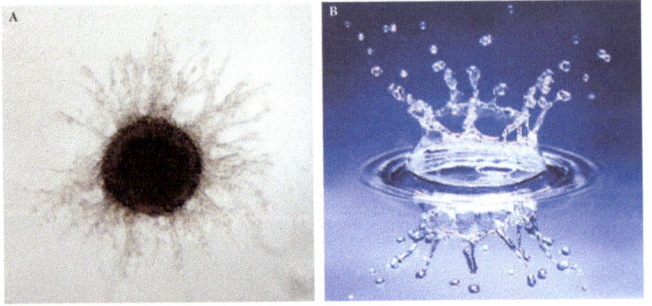

FIGURE 12.5   (See Figure caption on page 246)

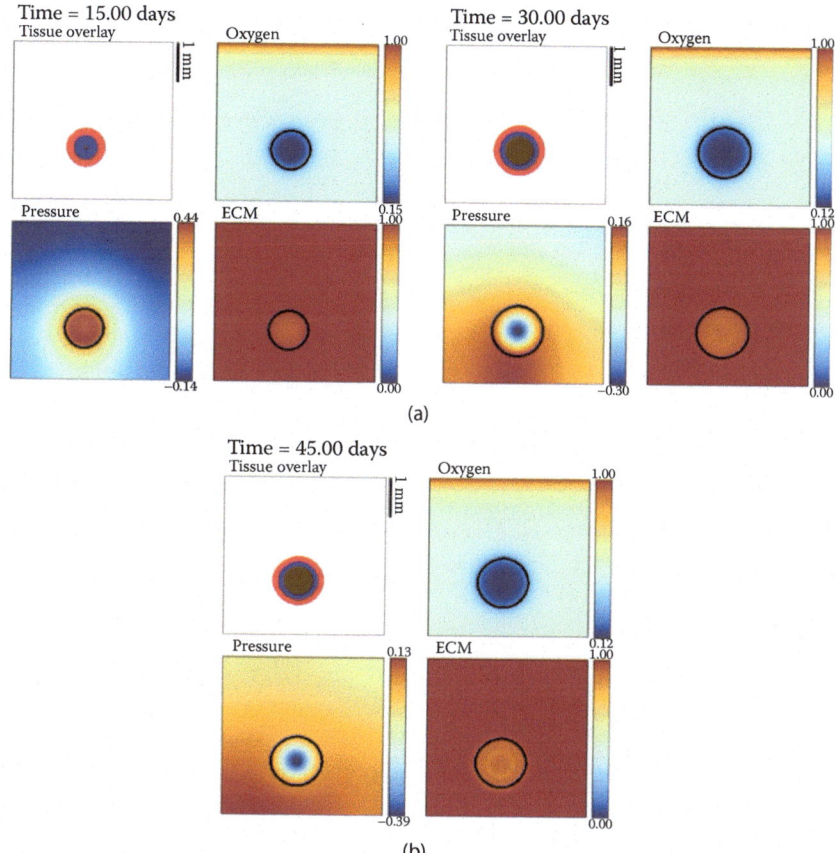

FIGURE 13.4 (See Figure caption on page 285)

FIGURE 14.3   **(See Figure caption on page 318)**

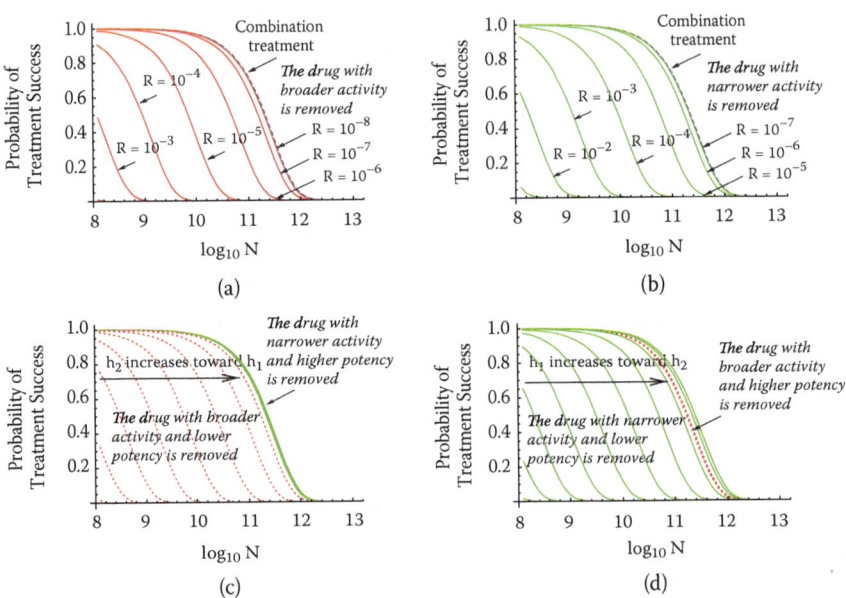

FIGURE 15.4   **(See Figure caption on page 351)**

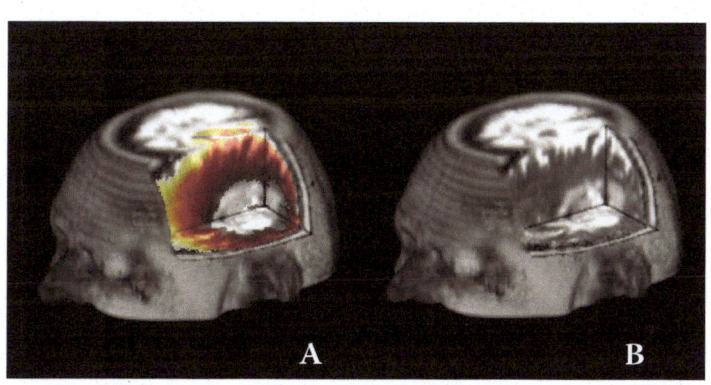

FIGURE 16.3 (**See Figure caption on page 365**)

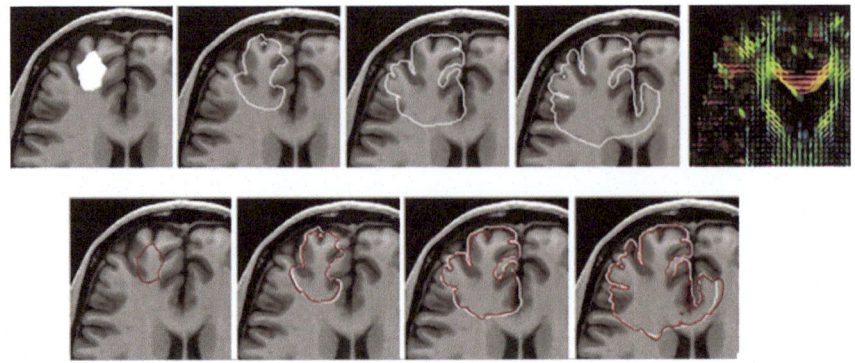

FIGURE 17.4   (See Figure caption on page 396)

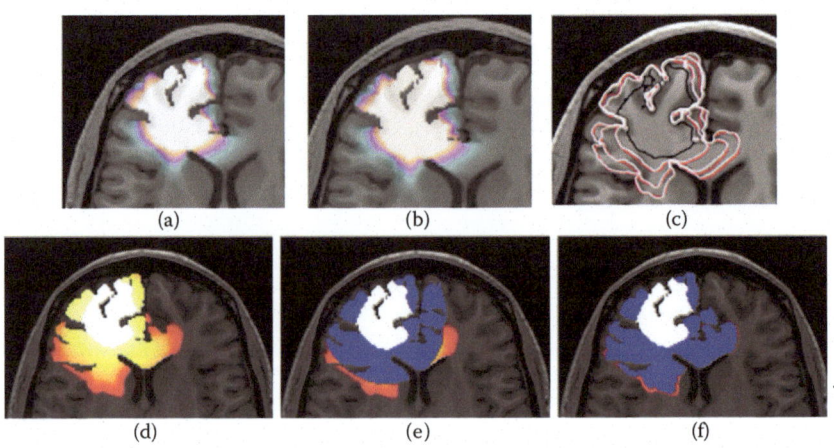

FIGURE 17.5   (See Figure caption on page 398)

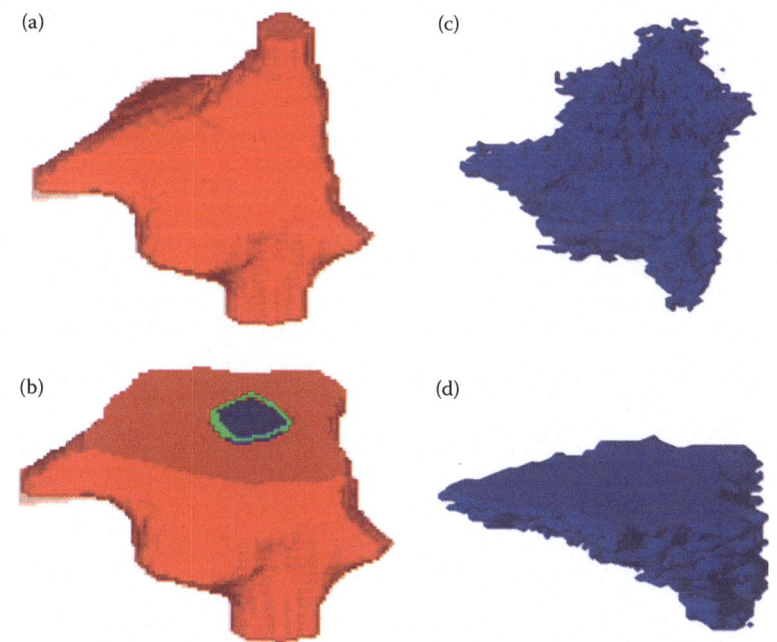

(a)

(c)

(b)

(d)

FIGURE 18.3   (See Figure caption on page 423)

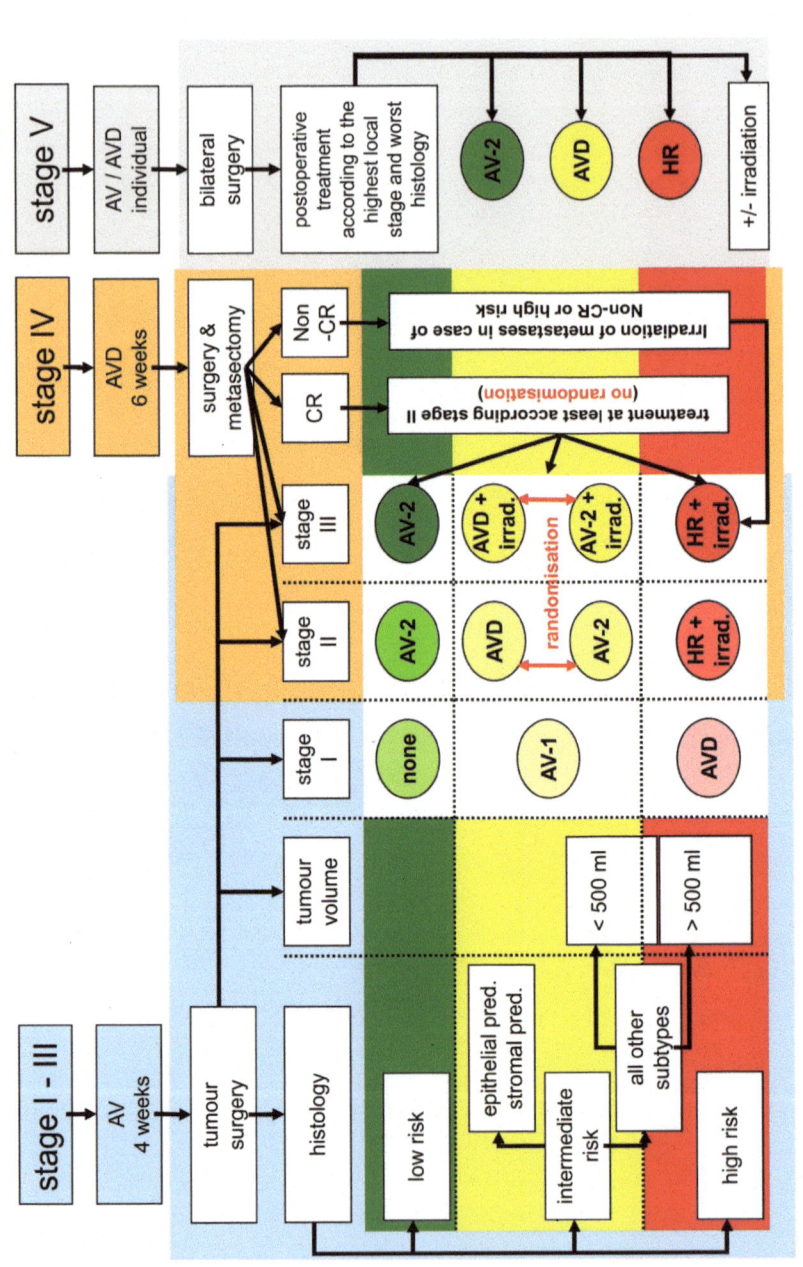

FIGURE 19.2 (See Figure caption on page 440)

of Y1068 and Y1173. These calculations, reported in Table 2.1, are used to parameterize the reactions involving inhibitor binding and substrate phosphorylation in the systems model.

Differential Signaling through EGFRTK

To examine the effects of signaling through Y1068 and Y1173 on the downstream response, a series of 15 min simulations were performed for wild type and mutants under different initial conditions (varying [EGF] and [EGFR]) and monitoring the resulting total phosphorylated ERK and AKT responses (Figure 2.3d). A two-dimensional scan over K_D values associated with Y1068 and Y1173 phosphorylation in which the respective K_D values are allowed to deviate from their default (wild-type) value over a logarithmic range of 5 log units. This was achieved by adjusting k_{-5} and k_{-6} from Table 2.1. The result is a two-dimensional matrix in which each element represents the total ERK or AKT levels from a single simulation involving a unique pair of parameters. In Figures 2.3d and 2.3e each output state is quantified according to the peak level of phosphorylation over the simulated time of 1800 s.

As indicated by the color maps in these scans, the effect of altered affinities of the Y1068 and Y1173 sites to the catalytic domain of the EGFR is that the L834R under normal EGFR expression exhibits differential downstream response, that is, a pronounced decrease in ERK activation (5-fold) and relatively much smaller decrease AKT activation (15% decrease). Our calculated responses for ERK short-term signaling for normal EGFR expression (Figures 2.3d and 2.3e) agree with the experimental observations of Sordella et al. (Sordella, Bell, et al. 2004) and Tracy et al. (Tracy, Mukohara, et al. 2004), who have also reported a pronounced decrease in activated ERK to AKT ratio for the L834R mutant. These results suggest that preferential activation of AKT in L834R could be one of the factors leading to enhanced AKT activation observed in non-small-cell lung cancer cell lines.

DISCUSSION

While the genetic basis of cancer is well appreciated, the resulting complexity in intracellular signaling mechanisms relevant for the conquest of this disease resides at multiple levels of organization, ranging from the subatomic realm involving mutations in individual protein domains to the cellular level of macromolecular assemblies and membrane processes. Relating cancer genotypes to disease phenotypes will be aided by the development of specialized modeling tools to treat the hierarchy of interactions ranging from molecular (nm, ns) to signaling (μm, ms) length and time scales. By introducing increased resolution in phosphorylation kinetics at the receptor level, the network model of EGFR-mediated signaling in wild-type and mutant cells showed how mutant forms of the receptor use an irregular pattern of tyrosine phosphorylation that preferentially activates the survival oncoprotein, AKT.

Recently, this type of multiscale analysis was used to explain why certain networks respond to antitumor tyrosine kinase inhibitors (TKIs) such as erlotinib and gefitinib (Purvis, Ilango, et al. 2008; Shih, Purvis, et al. 2008). Specifically, the branched signaling model was employed to analyze the inhibitory effects of the TKI erlotinib on EGFR phosphorylation and downstream ERK and AKT activation. The results provided a mechanistic basis for the enhanced inhibitor efficacy in mutant cell lines.

Thus, collectively, our results suggest that the clinically identified mutations of the EGFR kinase induce fragility in the stabilizing interactions of the inactive kinase conformation, providing a persistent stimulus for kinase activation even in the absence of any growth factor. At a cellular level, perturbations driving network hypersensitivity through the enhancement of phosphorylated ERK and AKT levels show a striking correlation with observed mutations of specific proteins in oncogenic cell lines as well as the observed mechanisms of drug resistance to EGFR inhibition. Therefore, we suggest that cascading mechanisms of network hypersensitivity/fragility at multiple scales enable molecular-level perturbations (clinical mutations) to induce oncogenic signaling. Moreover, our results describe a possible mechanism for preferential AKT activation in non-small-cell lung cancer lines harboring EGFR activating mutations. This preferential activation of a survival factor makes theses cell lines conducive to pathway addiction, that is, reliance on the L834R EGFR-mediated generation phosphorylated AKT for survival signals. The survival pathway addiction also results in a remarkable sensitivity to TKIs targeting the EGFR kinase.

The computational tools described here are ideal for assessing the likely effect of novel EGFR and HER2 mutations and determining whether the drug-sensitizing mutations implicated in non-small-cell lung cancer also occur in other cancers. Such approaches can also be employed effectively to address the issue of drug resistance to TKI therapy, which in the case of non-small-cell lung cancers is either mediated by point mutations in EGFR kinase (T790M) or the overexpression of HER3 and Met receptors and to investigate other molecular therapeutics targeting for (e.g., VEGF and c-Met). Ultimately, these approaches could be used to optimize the development of small molecule inhibitor therapies.

The multiscale modeling approach illustrated in this chapter enables the incorporation of the molecular context and variability and their impact on intracellular signaling pathways of oncogenic relevance and subsequent cell-fate decisions. This approach also enables the rationalization and prediction of the role and nature of molecular variability in malignant transformed cells as well as drug-sensitive/drug-resistant cells by bridging the gap between molecular resolution/context and intracellular signaling. The approach employed here can be seamlessly integrated with subcellular resolution modeling in agent-based models emphasized in other chapters (see Chapter 9).

ACKNOWLEDGMENTS

We thank Mark Lemmon for valuable input. This work was funded in part by the National Science Foundation and the National Human Genome Research Institute. Computational resources were provided in part by the National Partnership for Advanced Computational Infrastructure.

REFERENCES

Agrawal, N., J. Weinstein, et al. (2008). Landscape of membrane-phase behavior under the influence of curvature-inducing proteins. *Mol. Phys.* 106: 1913–1923.

Aldridge, B. B., J. M. Burke, et al. (2006). Physicochemical modelling of cell signalling pathways. *Nat. Cell Biol.* 8(11): 1195–203.

Choi, S. H., J. M. Mendrola, et al. (2007). EGF-independent activation of cell-surface EGF receptors harboring mutations found in gefitinib-sensitive lung cancer. *Oncogene* 26(11): 1567–76.

Citri, A. and Y. Yarden (2006). EGF-ERBB signalling: towards the systems level. *Nat. Rev. Mol. Cell Biol.* 7(7): 505–16.

Hanahan, D. and R. A. Weinberg (2000). The hallmarks of cancer. *Cell* 100(1): 57–70.

Hlavacek, W. S., J. R. Faeder, et al. (2006). Rules for modeling signal-transduction systems. *Sci. STKE* 2006(344): re6.

Liu, Y., J. Purvis, et al. (2007). A multiscale computational approach to dissect early events in the Erb family receptor mediated activation, differential signaling, and relevance to oncogenic transformations. *Ann. Biomed. Eng.* 35(6): 1012–25.

Mendelsohn, J. and J. Baselga (2000). The EGF receptor family as targets for cancer therapy. [Review] [155 refs]. *Oncogene* 19(56): 6550–65.

Morris, G. M., D. S. Goodsell, et al. (1998). Automated docking using a Lamarckian genetic algorithm and an empirical binding free energy function. *J. Computational Chem.* 19(14): 1639–1662.

Purvis, J., V. Ilango, et al. (2008). Role of network branching in eliciting differential short-term signaling responses in the hypersensitive epidermal growth factor receptor mutants implicated in lung cancer. *Biotechnol. Prog.* 24(3): 540–53.

Sharma, S. V. and J. Settleman (2007). Oncogene addiction: setting the stage for molecularly targeted cancer therapy. *Genes Dev.* 21(24): 3214–31.

Shih, A. J., J. E. Purvis, et al. (2008). Molecular systems biology of ErbB1 signaling: bridging the gap through multiscale modeling and high-performance computing. *Mol. BioSyst.* 4(12): 1151–1159.

Sjoblom, T., S. Jones, et al. (2006). The consensus coding sequences of human breast and colorectal cancers. *Science* 314(5797): 268–274.

Sordella, R., D. W. Bell, et al. (2004). Gefitinib-sensitizing EGFR mutations in lung cancer activate anti-apoptotic pathways. *Science* 305(5687): 1163–7.

Szabo, A. and N. S. Ostlund (1996). *Modern Quantum Chem.* Mineola, New York, Dover Publications.

Tracy, S., T. Mukohara, et al. (2004). Gefitinib induces apoptosis in the EGFRL858R non-small-cell lung cancer cell line H3255. *Cancer Res.* 64(20): 7241–4.

Zhang, X., J. Gureasko, et al. (2006). An allosteric mechanism for activation of the kinase domain of epidermal growth factor receptor. *Cell* 125(6): 1137–49.

Zhivotovsky, B. and G. Kroemer (2004). Apoptosis and genomic instability. *Nat. Rev. Mol. Cell Biol.* 5(9): 752–62.

CHAPTER 3

Has Cancer Sculpted the Genome? Modeling Linkage and the Role of Tetraploidy in Neoplastic Progression

Carlo C. Maley, Walter Lewis, and Brian J. Reid

CONTENTS

INTRODUCTION

The evolutionary transition from unicellular organisms to multicellular organisms with differentiated tissues included a transition in the level of organization at which natural selection acts (Maynard Smith and Szathmary 1995; Buss 1987). Mutations that cause an increase in the reproduction and survival rates in unicellular organisms give those organisms an advantage over their competitors. However, mutations that increase reproduction and survival rates in somatic cells of a multicellular organism can lead to a fatal cancer. Natural selection on multicellular organisms has led to mechanisms for suppressing somatic evolution on multiple levels, including tumor suppressor genes that regulate the growth of cells and the architecture of proliferative epithelia that limits the number of cells that are vulnerable to neoplastic evolution (Cairns 2002, 1975).

It is also possible that selective pressure against cancer may have modified genome architecture. For example, most human cancers appear to arise by chromosomal instability (Cahill et al. 1998) and linkage of tumor suppressor genes (TSGs) with critical genes (CGs) necessary for cell survival could provide an additional mechanism for suppression of somatic cell evolution. Deletion and loss of heterozygosity (LOH) can affect large regions of a chromosome (Nishimura et al. 2002; Lai et al. 2007). Tumor suppressor genes are commonly inactivated by LOH (Meltzer et al. 1991; Cavenee et al. 1983; Pekarsky et al. 2002; Deocampo, Huang, and Tindall 2003) and loss of a single allele of a TSG can lead to a proliferative advantage for the cell and eventually cancer (Wong et al. 2001; Cook and McCaw 2000; Fero et al. 1998). One potential mechanism to suppress such vulnerability would be genetic linkage of a critical gene with the TSG as a result of natural selection against cancer. If the CG is haploinsufficient for cell survival, then LOH in the TSG would be likely to cause LOH in the CG and to lead to cell death rather than progression to cancer. Linkage of CGs with TSGs would then act as a site-specific DNA damage checkpoint.

Tetraploidy is frequently observed in a variety of cancers and precancerous conditions (Haapala et al. 2001; Whang-Peng et al. 1984; Lastowska et al. 1997; Abe et al. 1985; Slaton et al. 1997; Cunningham et al. 1996; Robinson et al. 1996; Shackney et al. 1995; Giaretti 1994; Eskelinen et al. 1992; Dutrillaux et al. 1991; Tachibana et al. 1991). A common pathway to cancer seems to proceed from diploid cells, through a tetraploid intermediate, followed by progression to hypotetraploid aneuploid populations prior to malignancy (Barrett et al. 1999; Giaretti 1994; Merlo et al. 2008).

The mechanism by which the tetraploid population predisposes to cancer is not well understood, although most models postulate that it somehow creates a condition that is permissive for the subsequent evolution of chromosome abnormalities, including LOH (Shackney et al. 1989; Duesberg et al. 1998; Matzke et al. 2003; Li et al. 1997; Galitski et al. 1999; Rajagopalan et al. 2003; Jallepalli and Lengauer 2001). We hypothesized that tetraploidy may be selectively advantageous to neoplastic cells because some tumor suppressor genes are linked in the genome to critical genes necessary for cell survival, and by doubling the genome in a tetraploid cell, precancerous cells are more likely to preserve enough copies of the CG alleles to survive while deleting or inactivating the TSG alleles. Thus, cancers would tend to come from tetraploid cells that happened to duplicate their genomes, either by chance or due to a lesion in a gene involved in mitosis. Here, we consider two types of critical genes. In the first case, inactivation of the CG is recessively lethal for the cell. In the second case, the wild-type CG allele is haploinsufficient for survival, so that a cell with only one active CG allele has some nonzero probability of dying. We call a cell carcinogenic if it has lost all of its TSG alleles but preserved at least one CG allele.

We evaluate the hypothesis that tetraploidy develops during neoplastic progression as an adaptation that allows a neoplastic cell to inactivate a TSG while preserving a linked CG. We also address three questions corresponding to three levels of evolution. First, at the level of the cell, is a tetraploid cell more likely than a diploid cell to inactivate a TSG while preserving a linked CG? Second, at the level of a mosaic population of evolving precancerous cells in a neoplasm, are carcinogenic cells more likely to arise from a tetraploid precursor than a diploid precursor? Third, at the level of the population of organisms, how would selective pressure have sculpted the genomes of multicellular organisms with respect to linkage of critical genes and tumor suppressor genes, assuming that cancer has played an important selective role in the evolution of the genome? A set of mathematical and computational models were used to answer each of these questions.

MODELS

Model of a Single Cell

We first designed a probability model for the evolution of cells in a neoplasm. It is based on four assumptions. (1) Alleles of a TSG or a linked CG can be inactivated individually at some rate s. Sequence mutations, small deletions or insertions, and promoter hypermethylation may all play a part in inactivating an allele of a single gene. (2) One allele of both the TSG and the linked CG may be inactivated by a single event such as a large deletion. We will call this "double-gene inactivation." The rate of double-gene inactivation is encoded as a parameter d. We will initially assume that the single-gene inactivation of the TSG and the CG happen at the same rate s, though this assumption can be easily relaxed. The single-gene inactivation rate s is a combination of the rates of the three mechanisms for inactivating an allele of a gene: point mutations, promoter hypermethylation, and small deletions. Time is updated in the model by allele inactivation events. Because the next inactivation event will either inactivate an allele of one gene, or alleles of both genes at once, $2s + d = 1$. Linkage is thus represented in the magnitude of d, with $d = 0$ being unlinked and d increasing with linkage. (3) If a cell loses all of its CG alleles, it dies. We also introduced a parameter h for the haploinsufficiency of the CG defined as the probability that a cell dies if it only has one intact CG allele. If $h = 1$, then cells with only one CG allele also die. If $h = 0$, then only cells with no active CG alleles die. If $0 < h < 1$, then the cell has a probability h of dying when the first CG allele is inactivated. Thus, h scales the penetrance of a single intact allele of the CG for cellular lethality. (4) Loss of all TSG alleles is a rate limiting step in carcinogenesis. Since malignant tumors generally derive from a single ancestral progenitor cell (Nowell 1976; Sidransky et al. 1992), generation of a cell that has lost its TSG is likely to be a clinically relevant event. For convenience, we will define a cell in this state as *carcinogenic*. The probability model for a single diploid cell is represented in Figure 3.1. The representation of the possible changes in a tetraploid cell is even more complex and has not been shown.

All the possible sequences of inactivation events that may develop in a diploid or tetraploid cell were enumerated computationally. The probability for each sequence was then computed, as

$$p(d,s,h,m,n,k)=\left(\frac{d}{4}\right)^{m}\left(\frac{s}{8}\right)^{n}\left(1-h\right)^{k}$$

for m double-gene inactivation events, n single allele inactivation events, ending in either zero active CG alleles or zero active TSG alleles, and passing through k states with a single active CG allele, which may lead to

Diploid Probability Model

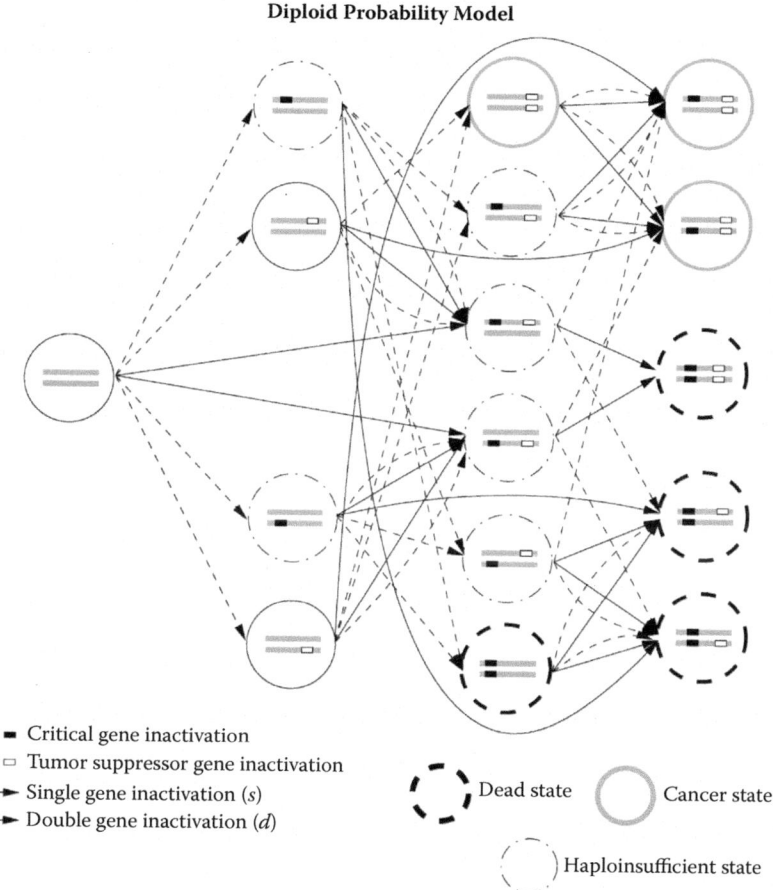

- ∎ Critical gene inactivation
- ▫ Tumor suppressor gene inactivation
- - -► Single gene inactivation (*s*)
- ──► Double gene inactivation (*d*)

Dead state   Cancer state

Haploinsufficient state

FIGURE 3.1 A probability model of the evolution of a diploid cell. Tumor suppressor gene (TSG) lesions (in black) and a critical gene (CG) lesions (in white) as well as double-gene inactivation are illustrated. Over time, transitions caused by single-gene inactivation (point mutations, small deletions, or promoter methylation) are indicated by dashed arrows, and transitions caused by double-gene inactivation (large deletions) are indicated by solid arrows. The four states inside a bold, dashed circle are dead because both of the CG alleles have been lost. The three states inside the thick grey circles are carcinogenic states because both of the TSG alleles have been lost. There is an asymmetry between the carcinogenic and death outcomes because a cell that has lost both alleles of the CG will die regardless of how many TSG alleles are intact. The five states inside the thin dash-dotted circles are states that may die due to haploinsufficiency of the CG gene because they have only one intact allele at that locus. Even for this diploid model, the sequence of possible events is complicated, necessitating the computational enumeration of the tetraploid cell model.

cell death due to haploinsufficiency. There are eight possible single-allele inactivation events and four possible double-gene inactivation events corresponding to the four alleles of the CG and four alleles of the TSG in a tetraploid cell. We calculated the probability of events based on a tetraploid genome. Thus, in order to make diploid and tetraploid cells comparable, half of the events have no effect on a diploid cell because they "happen" in the missing chromosomes. The possible sequences include multiple inactivation events of the same allele, though only the first such event would change the state of the cell.

The model calculated the total probability that a cell inactivates all of its TSG alleles before inactivating all, or all but one, of its CG alleles by summing across all sequences $x_{m,n,k}$, which include m double-gene inactivation events, n single-gene inactivation events, passing through k states with one active CG allele, and end with zero active TSG alleles and at least one active CG:

$$P[\text{inactivation of TSG}| s,d,h,ploidy] = \sum_{x_{m,n,k}} p(d,s,h,m,n,k)$$

Since an event can happen in an already inactivated allele (with probability X), we use the expected number of hits in inactivated alleles before the next event in an active allele $(1/(1-X))$ to calculate the probability of making a state change.

Most cancers require the inactivation of more than one TSG (Renan 1993; Hanahan and Weinberg 2000; Vogelstein and Kinzler 1993). However, our hypothesis applies equally to multiple TSGs and so, for simplicity, we focus on a single TSG. The relative risk of tetraploidy was calculated for each parameter setting by dividing the probability a tetraploid cell becomes carcinogenic by the probability that a diploid cell becomes carcinogenic:

$$RR = P[\text{inactivation of TSG}|s,d,h,4]/P[\text{inactivation of TSG}|s,d,h,2] .$$

Model of a Neoplasm

We incorporated a stochastic simulation of the single-cell model into a simulation of an evolving population of cells in a neoplasm with one additional parameter. Diploid cells duplicated their genomes to become tetraploid cells at some rate t per time interval between mutational events. A population of n ($n = 10^3$, 10^4, or 10^5) cells was represented in a three-dimensional mass. The state of each chromosome in a cell was explicitly represented such that if a double-gene inactivation event occurred in a chromosome that had already incurred inactivation of a single gene, the net effect of the double-gene inactivation was to inactivate the remaining gene on that chromosome. Initially, all cells were diploid with two active alleles in both their TSG and CG loci.

Time was represented in discrete steps denoting the amount of time until the next potential inactivation event in the population. At each time step, the simulation selected a random cell with equal probability. If the cell was diploid, with probability t, the cell's genome was doubled by copying the state of its two chromosomes into two new chromosomes. Next, a random integer between 1 and 4 was generated to specify which copy of the chromosome incurred the inactivation, regardless of the ploidy of the cell. If that number was 3 or 4 in a diploid cell, nothing happened. This guaranteed that every chromosome mutated at the same rate per allele; thus, tetraploid cells incurred twice as many mutations as diploid cells per unit of time. Finally, the simulation inactivated a single allele of the TSG with probability s, a single allele of the CG with probability s, or both the TSG and CG with probability d. If the inactivation reduced the cell to a single CG allele, the cell had probability h of dying. A cell with no intact CG alleles automatically died. The dead cell was then replaced by a neighboring cell that divided to produce two daughter cells with the same chromosomal state of the parental cell. This neighbor was selected through competition between c (usually 4) of the adjacent cells in the three-dimensional grid of cells (Blickle and Thiele 1995). Competition was based on the number of intact TSG alleles (Fodde and Smits 2002). If any of the c neighbors had a single intact TSG allele, one of those neighbors divided to replace the dead cell. Otherwise, a randomly chosen neighbor divided to replace the dead cell. In this way, the TSG was haploinsufficient because the presence of only one intact allele resulted in a phenotype with a competitive advantage over cells with more than one intact allele of the TSG. The parameter c scales the degree of competition. With $c = 1$, there is no competition. As c becomes larger, it becomes easier for a cell with a single TSG allele to spread in the neoplasm.

If a cell became carcinogenic (lost all, 2 or 4, of its TSG alleles), the simulation was stopped and the ploidy of the carcinogenic cell was recorded along with the time (number of inactivation events simulated until a carcinogenic cell was generated). The simulation of the entire neoplastic population was repeated 1000 times to calculate the frequency of carcinogenic cells arising from tetraploid cells rather than diploid cells. This was replicated 50 times for each parameter setting. We ran the model under all combinations of parameters $n = 10^3$, 10^4, and 10^5; $h = 0$, 0.5, and 1; $d = 0.25$, 0.5, and 0.75; and $t = 0.2$, 0.5, and 0.8.

Model of Organismal Evolution

We used a genetic algorithm (Mitchell 1998; Goldberg 1989) to simulate the evolution of a population of organisms under selection from death by cancer. Each organism had "genetic" traits encoding parameters d, t, and h. All traits had minimal values of 0 and maximal values of 1. The traits of the initial population were set randomly from uniform distributions between 0 and 1. There were m organisms in the population, and each organism had a three-dimensional neoplasm of n cells. The population of organisms was simulated for 40 generations, by which time the population genetic traits had generally stabilized. For each generation, we first calculated the fitness of every organism. The fitness of an organism was determined by running the above simulation of a neoplasm to determine the number of mutational events (amount of time) until the organism developed a carcinogenic cell. Thus, organisms with parameters that resulted in a longer time before carcinogenesis had higher fitness scores than those that developed a carcinogenic cell relatively rapidly. After we calculated the fitness scores of the organisms, we generated the population for the next generation by tournament selection, similar to the tournament selection in the competition of cells within the neoplasm. This involved randomly choosing two organisms from the population with uniform probability. The organism with the higher fitness score was chosen to be a parent, and the other organism was returned to the pool of potential parents. This was repeated to select the mate of the first parent from another two random organisms. The result of tournament selection is that the organisms with higher fitness scores produce more offspring than the organisms with lower fitness scores (Blickle and Thiele 1995). The parameters (traits) of the offspring were determined by generating a random number from a Gaussian distribution with mean equal to the average of the parental traits, and standard deviation equal to half the difference between the parental traits. This represents the possible recombinants and mutations in a multigene trait. The standard deviation for the Gaussian distribution had a minimum value of 0.0025 to prevent evolution from stopping altogether. At the end of 40 generations, the average population values for the traits parameters *d*, *t*, and *h* were measured.

RESULTS

Are Tetraploid Cells More Likely to Become Carcinogenic Than Diploid Cells?

Yes, if the TSG and the CG are linked. The relative risk of developing carcinogenic cells in tetraploid cells (Pr[carcinogenesis |4N]/ Pr[carcinogenesis |2N]) is a function of the relative probability (*d*) that a lesion inactivates both the CG and the TSG on a chromosome as well the degree of haploinsufficiency (*h*) of the CG (Figure 3.2). The probability that a tetraploid cell (and a diploid cell) becomes carcinogenic decreases with both increasing linkage (*d*) of the CG to the TSG and increasing haploinsufficiency

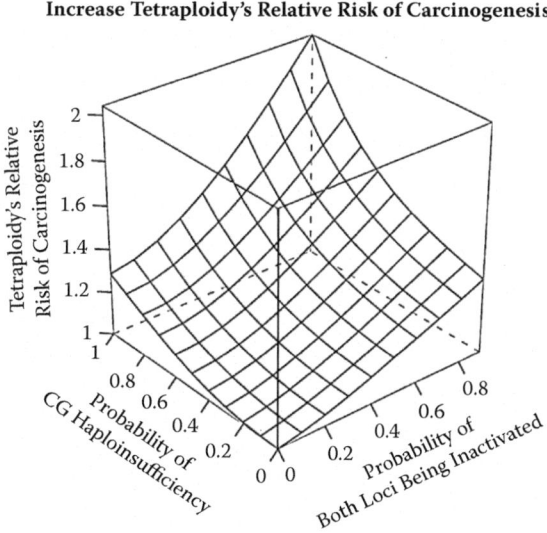

Critical Gene Haploinsufficiency and Close Linkage Increase Tetraploidy's Relative Risk of Carcinogenesis

FIGURE 3.2 The relative risk of carcinogenesis in a single cell due to tetraploidy. Risk is plotted as a function of the relative probability that a lesion deletes both the CG and the TSG (*d*) as well as the probability that a cell with only one functional CG allele dies (*h*). Relative risk is calculated as Pr[carcinogenesis | tetraploidy] / Pr[carcinogenesis | diploidy]. Carcinogenesis is defined as a cell losing all of its TSG alleles before dying due to loss of CG alleles.

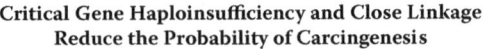

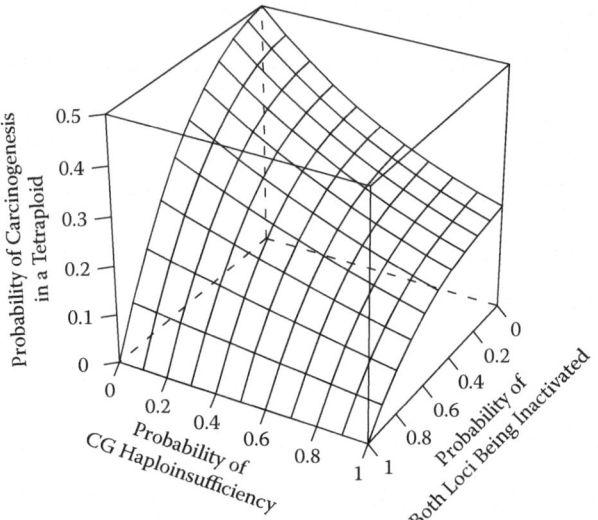

FIGURE 3.3 The probability that a single tetraploid cell becomes carcinogenic, that is, loses all of its TSG alleles before it dies, is due to the loss of CG alleles. The probability of carcinogenesis is plotted as a function of the probability that a lesion inactivates both the CG and the TSG (d) relative to inactivating an allele of a single gene, as well as the probability that a cell with only one functional CG allele dies (h). The direction of the x and y axes has been reversed relative to Figure 3.2 for a better view of the surface.

(h) of the CG (Figure 3.3). However, the relative risk that a tetraploid cell becomes carcinogenic increases with both d and h (Figure 3.2). On average tetraploid cells require more hits per chromosome (Figure 3.4A) than diploid cells (Figure 3.4B) to become carcinogenic, and thus more time before they become cancerous. The time to the emergence of a carcinogenic cell decreases slightly with increasing haploinsufficiency of the CG.

Is a Carcinogenic Cell More Likely to Arise from a Tetraploid Cell Than a Diploid Precursor in a Neoplasm?

Not necessarily. The more closely the TSG and CG are linked, and the higher the degree of haploinsufficiency at the CG locus, the more likely that a carcinogenic cell will derive from a tetraploid cell (Figure 3.5). In addition, the greater the probability that a diploid cell duplicated its genome (t), the longer it takes before a cell evolves that has inactivated all of its TSG alleles (Figure 3.6).

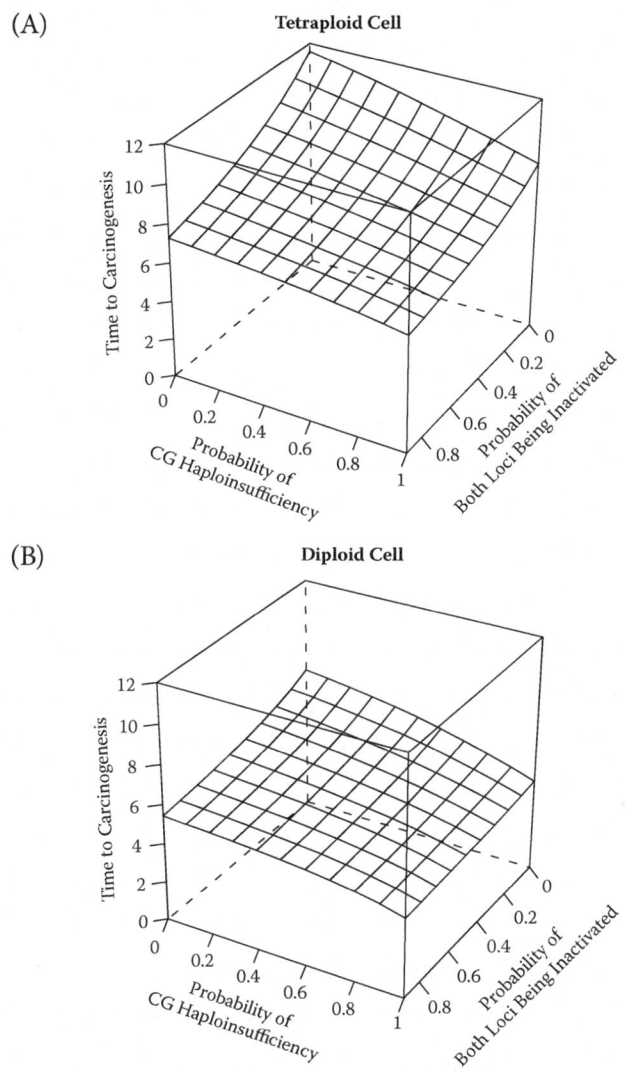

FIGURE 3.4   A tetraploid cell (A) takes longer to become carcinogenic than a diploid cell (B). Tetraploid cells delay carcinogenesis regardless of the linkage between the TSG and CG as represented by the relative probability d that both are lost in a double-gene inactivation event as well as the probability that a cell with only one functional CG allele dies (h). Time is measured by the expected number of inactivation events until a cell has inactivated all of its TSG alleles while preserving some of its CG alleles. Time until carcinogenesis may be greater than the total number of alleles because inactivation events may occur at the same locus multiple times, to no effect.

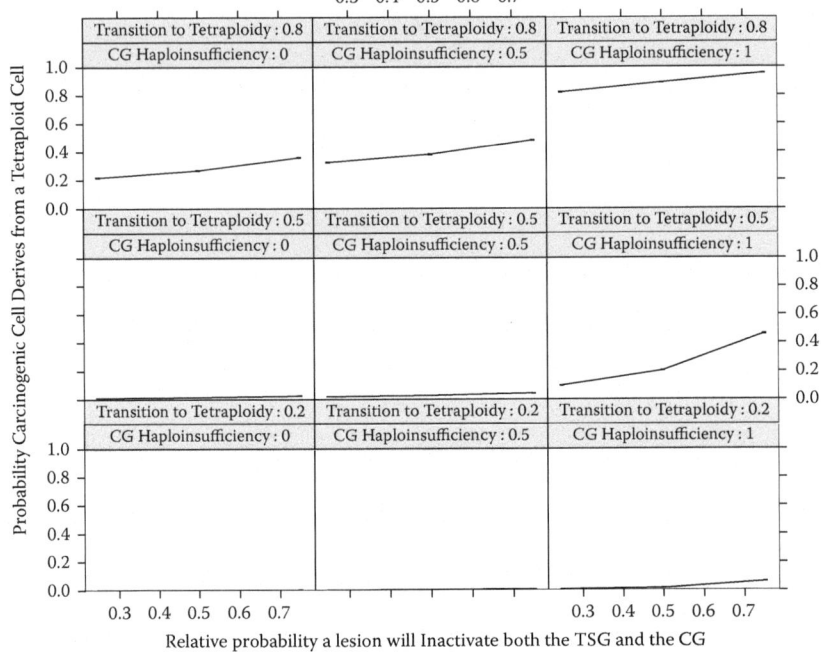

Carcinogenesis in a Tetraploid Clone

FIGURE 3.5 The probability that a carcinogenic cell evolves from a tetraploid cell in a neoplasm. The probability depends on the linkage (d) between the TSG and the CG (x axis) as well as the penetrance of haploinsufficiency in the CG (h) and the relative probability that a diploid cell becomes tetraploid (t). For each parameter setting, 1000 neoplasm simulations were run to calculate the probability, and that was repeated 50 times to estimate statistical error for those probabilities. In all cases, the simulation involved neoplasms of 10^5 cells. Standard error bars are plotted but are < 0.01 in all cases.

How Would Evolution at the Organismal Level Change the Parameters of the System?

The genetic algorithm was run with a population of $m = 1024$ organisms in which each organism had a three-dimensional neoplasm with $n = 1000$ cells. Natural selection was based on the age at which an organism developed a carcinogenic cell in their neoplasm. The longer an organism survived without generating a carcinogenic cell, the higher the probability it would produce more offspring. Over 118 runs of the model, the genetic algorithm maximized linkage between the CG and the TSG

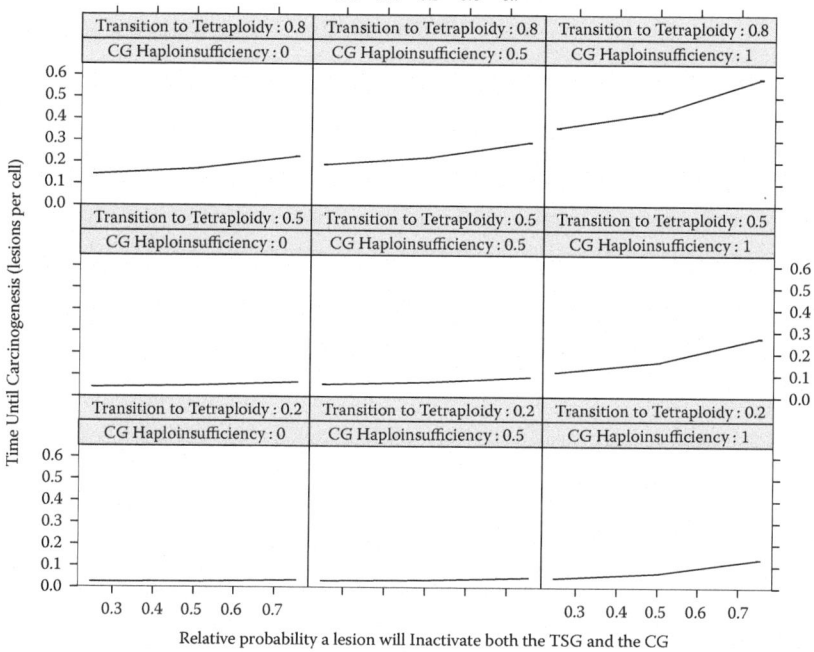

FIGURE 3.6 Time until carcinogenesis in a neoplasm. This shows the average number of mutational events (because time is scaled by mutational events) per cell in a neoplasm before one of the cells evolved with all of its TSG alleles inactivated (carcinogenesis). The emergence of a carcinogenic cell is delayed by increased linkage between the CG and TSG (d), increased haploinsufficiency in the CG (h), and by increased probability that a cell duplicates its genome (t). 50 replicates of 1000 neoplasm simulations were run for each parameter setting. In all cases the simulation involved neoplasms of 10^5 cells. Standard error bars are plotted but are < 0.001 in all cases. The scale of the y axis is much smaller than in Figure 3.5 because time is averaged across all the cells, and many of the 10^5 cells in the neoplasm had not suffered any inactivation events by the time the first carcinogenic cell evolved.

(mean d = 0.9873, standard deviation = 0.0002). It also maximized the penetrance of cellular lethality due to the loss of a single CG allele (h = 0.9903, standard deviation = 0.0019). Furthermore, the chance of a cell making a transition from a diploid state to a tetraploid state was raised, but not maximized (t = 0.774, standard deviation = 0.019). The results were the same for a two-dimensional tumor with 1024 cells.

DISCUSSION

This model addresses three questions regarding the hypothesis that tetraploidy in neoplastic progression is an adaptation that allows a neoplastic cell to inactivate a TSG while preserving a linked CG. Our results suggest: (1) tetraploid cells are more likely than diploid cells to inactivate a TSG while preserving a linked CG; (2) carcinogenic cells are more likely to evolve from a tetraploid precursor when a TSG is linked to a CG than when the TSG and CG are not linked; (3) carcinogenic cells are more likely to evolve from a tetraploid precursor when a TSG is linked to a haploinsufficient CG than when it is linked to a recessively lethal CG; and (4) if cancer has been an important selective pressure in the evolution of multicellular organisms, then TSGs should be tightly linked to CGs that are haploinsufficient for cell survival, and somatic cells should have retained the capacity for genome reduplication.

In our model, tetraploidy increases the chance that a cell will eventually become carcinogenic, but also delays the inactivation of the TSG alleles. The evolution of neoplastic genome duplication is constrained by two opposing forces: an organism would evolve higher fitness if it reduced the chance of genome duplications in its cells because tetraploid cells are more likely to develop cancer than diploid cells, but tetraploid cells take longer than diploid cells to become carcinogenic, and so the organism should evolve higher rates of genome duplication to delay the onset of cancer. The tendency to achieve equilibrium between these two opposing forces results in the evolution of high but not maximal rates of genome duplication in the model. Genome duplication may be an adaptation to buffer the cells against the loss of p53 (TP53) or other lesions that are critical neoplastic events leading to genetic instability.

A priori, we could not be certain whether the results for a single cell would generalize to a neoplasm because a clone with a single intact TSG allele had a competitive advantage over other cells and might expand rapidly throughout the population of neoplastic cells. Does this mean that diploid cells would tend to dominate the neoplasm because they are only a single TSG hit from gaining this competitive advantage? Or would tetraploid cells have an advantage because they are better able to preserve their CGs? Figures 3.5 and 3.6 indicate that carcinogenic cells evolve from tetraploid cells when the CG and TSG are more closely linked (d) and, when the CG is haploinsufficient for cell survival (h), similar to the single-cell model. Tetraploidy also delays the onset of carcinogenic cells in the neoplasm and the loss of all the TSG alleles in a single cell.

Haploinsufficiency of the CG affects time to carcinogenesis differently in the single-cell (Figures 3.4a and 3.4b) and neoplasm (Figure 3.6) models. In a single cell, haploinsufficiency of the CG tends to prevent the longer sequences of events that lead to carcinogenesis, which involve both TSG and CG inactivation events, but does not affect the relatively short sequences of single-gene inactivation of the TSG alone. A high degree of haploinsufficiency tended to reduce the likelihood that a cell became carcinogenic, but if it did, it became carcinogenic through a short sequence of single-gene inactivation events in the TSG, such as point mutations, methylation, or small deletions. Thus, if only the cells that became carcinogenic are considered, a high degree of haploinsufficiency in the CG appears to lead to a short time (sequence of events) until the cell becomes carcinogenic. Within a neoplasm, the total time until a carcinogenic cell emerged in the population of cells, not in a single cell, was measured. Haploinsufficiency of the CG causes the clearance of cells with lesions from the population. This reduces the probability that a cell becomes carcinogenic and delays the time until a carcinogenic cell emerges.

The linkage of a CG to a TSG essentially provides a form of apoptotic response to DNA damage that is specific to the TSG locus. We predict that if cancer has been an appreciable selective force on the evolution of our ancestors, human genomes should have CGs tightly linked to TSGs. Such CGs should be haploinsufficient, unable to maintain the viability of the cell with a single functional allele. The haploinsufficiency of the CGs should only apply to cells in adults, lest they add to the genetic burden of the embryo. Furthermore, evolution should have maintained the capacity for human cells to duplicate their genomes as an adaptation to delay the onset of cancer.

Linkage of TSGs and CGs may explain some of the variability of cancer incidence between tissues. For example, cells in the small intestine may require the expression of a gene linked to a TSG in order to survive, whereas cells in the colon may not require the expression of the same gene. The protective effect of linked CGs might be altered if exposures of our modern lifestyle are sufficiently different from the selective pressures that sculpted the genomes of our multicellular ancestors.

The set of cellular lethal critical genes is unknown and difficult to identify. One potential example of a TSG linked to a CG is the linkage of the TSG $p27$ to the oncogene $KRas2$. $KRas2$ and $p27$ ($CDKN1B$) are within 1 cM on chromosome 12 (Kemp, Kim, and Philipp 2000). $KRas2$ may be a critical gene in that $KRas2$ knockouts are embryonic lethal in mice, though not cellular lethal in all tissues (Koera et al. 1997). Linking a CG that is also an oncogene to a TSG may have the added benefit of creating opposite selective pressures that balance amplification of the oncogene against loss of the TSG. There are a number of TSGs that are closely linked to onco- genes, including *BRCA1* 3Mb from *ERBB2*, *MLH1* 4Mb from *CTNNB1*, *ERCC2* 5Mb from *AKT2*, *MUTYH* 5Mb from *MYCL1*, *CEBPA* 7Mb from *AKT2*, *TCF1* 9Mb from *PTPN11*, and *NF1* 9Mb from *ERBB2*.

Our model only considered a single TSG locus in the genome of a cell. Carcinogenesis is a multistep process that involves more than one TSG in most tissues (Renan 1993; Hanahan and Weinberg 2000; Vogelstein and Kinzler 1993). However, the results of our model should be independent for each locus and so would apply to every TSG locus in a multistep model of cancer progression.

Four different, though not mutually exclusive, hypotheses have been advanced in the literature for the observation of tetraploidy in neoplas- tic progression. (1) Diploid cells duplicate their genomes at some normal rate, but the tetraploid cells are inherently unstable and so more likely than diploid cells to activate oncogenes and inactivate tumor suppressor genes (Shackney et al. 1989; Duesberg et al. 1998; Matzke et al. 2003). (2) Disruption of *TP53* (p53) or some mitotic gene deregulates chromosome segregation and/or mitosis that leads to both tetraploidy and aneuploidy (Lengauer, Kinzler, and Vogelstein 1998; Shackney and Shackney 1997; Cahill et al. 1999; Fodde and Smits 2002; Nowak et al. 2002). Whether chromosomal instability generally comes before or after TSG inactiva- tion has been hotly debated (Rajagopalan et al. 2003; Sieber, Heininmann, and Tomlinson 2003; Moolgavkar and Luebeck 2003). (3) Doubling the genome and then losing large portions of chromosomes may be selected in a neoplasm because it provides a mechanism by which cells can evolve different gene product dosages by changing the number of alleles of the different genes (Li et al. 1997; Galitski et al. 1999; Rajagopalan et al. 2003). (4) Doubling the genome may provide a genetic buffer that allows the cell to survive further chromosome instability (Jallepalli and Lengauer 2001). While our hypothesis is similar to the last two hypotheses, the difference is that we posit linkage between CGs and TSGs that drives selection for tetraploidy.

We did not represent in our model the possibility that tetraploid cells are inherently genetically unstable (Shackney et al. 1989; Duesberg et al. 1998; Matzke et al. 2003), and so might have a higher rate of inactivation than a diploid cell. If this rate is sufficiently high, it might eliminate the delay in the onset of cancer caused by genome duplication. However, changing just the frequency of inactivation should not affect the relative risk of tetraploidy because the elevated inactivation rate would apply equally to both the TSG and the CG. Similarly, an increase in the inactivation rate should not affect the selective pressure to tightly link a CG to a TSG. The relative risk of cancer in tetraploid cells would only decrease if the ratio of single- to double-gene inactivation shifted toward more frequent single-gene inactivation compared to diploid cells. Most hypotheses assume the opposite that tetraploid cells would have increased chromosomal instability relative to diploid cells (Jallepalli and Lengauer 2001; Matzke et al. 2003; Shackney and Shackney 1997) and thus an increased proportion of large deletions that would inactivate alleles of both linked genes.

Barrett's esophagus is a model for human neoplastic progression (Neshat et al. 1994; Reid and Rabinovitch 1988; Maley 2007) and thus a potential test case for hypotheses of neoplastic progression. It is a precancerous condition of the esophagus in which the normal multilayered squamous cells of the esophagus are replaced by hyperproliferative columnar cells. In Barrett's esophagus, flow cytometric tetraploid cell populations predict future progression to aneuploidy (Galipeau et al. 1996). Further, tetraploidy also predicts an increased chance of progression to cancer (RR = 11.7, 95% CI = 6.2 – 22) (Rabinovitch et al. 2001). Tetraploidy is typically observed in cells that have lesions in *TP53* (Galipeau et al. 1996). Both *TP53* and *CDKN2A* (*p16/INK4A*) tumor suppressor genes are commonly inactivated by loss of heterozygosity in Barrett's esophagus (Galipeau et al. 1999; Reid et al. 2001; Wong et al. 2001), and fluorescent in situ hybridization analysis shows that LOH in *TP53* is often, though not always, associated with genome reduplication (Wongsurawat et al. 2006). These observations are all consistent with our model of linkage between TSGs and CGs driving neoplastic progression through a tetraploid intermediate. It is unknown if there are CGs closely linked to *CDKN2A* or *TP53*.

The importance of multiscale modeling is highlighted by the fact that the single-cell model showed that tetraploid cells are more likely to inactivate a TSG than diploid cells, suggesting that tetraploidy should increase the risk of cancer, but the cell–tissue multiscale model showed that the capacity to evolve tetraploidy delays cancer onset, and the cell–tissue–population multiscale model suggests that the capacity to evolve tetraploidy should have been preserved as a tumor suppression mechanism.

Our results suggest that the effects of natural selection on tumor suppressor genes linked to critical genes can explain the observations of somatic genome reduplication during neoplastic progression. This result does not require an assumption of greater chromosomal instability in tetraploid cells relative to diploid cells. Nor does it require an assumption of selective effects of gene dosage alterations due to copy number changes in the genome. This is not to say that our model argues against either chromosomal instability or gene dosage modulation in progression, both of which may be important. We have shown that they are not necessary assumptions to explain the phenomenon of tetraploidy in neoplastic progression.

ACKNOWLEDGMENTS

This work was supported by NIH P01CA91955, NIH R01CA61202, P01 CA91955, P30 CA010815, R01 CA119224, R01 CA140657, NIH K01 CA89267-01, and a CURE supplement to NIH P30 CA016520. Thanks to Patricia Galipeau and Chris Kemp for helpful discussions and suggestions.

REFERENCES

Abe, R., A. Raza, H. D. Preisler, C. K. Tebbi, and A. A. Sandberg. 1985. Chromosomes and causation of human cancer and leukemia. LIV. Near-tetraploidy in acute leukemia. *Cancer Genet Cytogenet* 14 (1–2):45–59.

Barrett, M. T., C. A. Sanchez, L. J. Prevo, D. J. Wong, P. C. Galipeau, T. G. Paulson, P. S. Rabinovitch, and B. J. Reid. 1999. Evolution of neoplastic cell lineages in Barrett oesophagus. *Nat Genetics* 22 (1):106–9.

Blickle, T., and L. Thiele. 1995. A mathematical analysis of tournament selection In *Proceedings of the Sixth International Conference on Genetic Algorithms* edited by L. Eshelman. San Francisco, CA Morgan Kaufmann.

Buss, L. W. 1987. *The Evolution of Individuality*. Princeton, NJ: Princeton University Press.

Cahill, D. P., K. W. Kinzler, B. Vogelstein, and C. Lengauer. 1999. Genetic instability and Darwinian selection in tumors. *Trends Cell Biol* 9 (12):M57–M60.

Cahill, D. P., C. Lengauer, J. Yu, G. J. Riggins, J. K. Willson, S. D. Markowitz, K. W. Kinzler, and B. Vogelstein. 1998. Mutations of mitotic checkpoint genes in human cancers. *Nature* 392 (6673):300–3.

Cairns, J. 1975. Mutation selection and the natural history of cancer. *Nature* 255: 197–200.

———. 2002. Somatic stem cells and the kinetics of mutagenesis and carcinogenesis. *Proc Natl Acad Sci USA* 99 (16):10567–70.

Cavenee, W. K., T. P. Dryja, R. A. Phillips, W. F. Benedict, R. Godbout, B. L. Gallie, A. L. Murphree, L. C. Strong, and R. L. White. 1983. Expression of recessive alleles by chromosomal mechanisms in retinoblastoma. *Nature* 305 (5937):779–84.

Cook, W. D., and B. J. McCaw. 2000. Accommodating haploinsufficient tumor suppressor genes in Knudson's model. *Oncogene* 19 (30):3434–8.

Cunningham, J. M., A. Shan, M. J. Wick, S. K. McDonnell, D. J. Schaid, D. J. Tester, J. Qian, S. Takahashi, R. B. Jenkins, D. G. Bostwick, and S. N. Thibodeau. 1996. Allelic imbalance and microsatellite instability in prostatic adenocarcinoma. *Cancer Res* 56 (19):4475–82.

Deocampo, N. D., H. Huang, and D. J. Tindall. 2003. The role of PTEN in the progression and survival of prostate cancer. *Minerva Endocrinol* 28 (2):145–53.

Duesberg, P., C. Rausch, D. Rasnick, and R. Hehlmann. 1998. Genetic instability of cancer cells is proportional to their degree of aneuploidy. *Proc Natl Acad Sci USA* 95 (23):13692–7.

Dutrillaux, B., M. Gerbault-Seureau, Y. Remvikos, B. Zafrani, and M. Prieur. 1991. Breast cancer genetic evolution: I. Data from cytogenetics and DNA content. *Breast Cancer Res Treat* 19 (3):245–55.

Eskelinen, M., P. Lipponen, S. Marin, H. Haapasalo, K. Makinen, J. Puittinen, E. Alhava, and S. Nordling. 1992. DNA ploidy, S-phase fraction, and G2 fraction as prognostic determinants in human pancreatic cancer. *Scand J Gastroenterol* 27 (1):39–43.

Fero, M. L., E. Randel, K. E. Gurley, J. M. Roberts, and C. J. Kemp. 1998. The murine gene p27Kip1 is haplo-insufficient for tumour suppression. *Nature* 396 (6707):177–80.

Fodde, R., and R. Smits. 2002. A matter of dosage. *Science* 298:761–3.

Galipeau, P. C., D. S. Cowan, C. A. Sanchez, M. T. Barrett, M. J. Emond, D. S. Levine, P. S. Rabinovitch, and B. J. Reid. 1996. 17p (p53) allelic losses, 4N (G2/tetraploid) populations, and progression to aneuploidy in Barrett's esophagus. *Proc Natl Acad Sci USA* 93 (14):7081–4.

Galipeau, P. C., L. J. Prevo, C. A. Sanchez, G. M. Longton, and B. J. Reid. 1999. Clonal expansion and loss of heterozygosity at chromosomes 9p and 17p in premalignant esophageal (Barrett's) tissue. *J Nat Cancer Inst* 91 (24):2087–95.

Galitski, T., A. J. Saldanha, C. A. Styles, E. S. Lander, and G. R. Fink. 1999. Ploidy regulation of gene expression. *Science* 285 (5425):251–4.

Giaretti, W. 1994. A model of DNA aneuploidization and evolution in colorectal cancer. *Lab. Invest.* 71 (6):904–10.

Goldberg, D.E. 1989. *Genetic Algorithms in Search, Optimization, and Machine Learning.* Reading, MA: Addison-Wesley.

Haapala, K., A. Rokman, C. Palmberg, E. R. Hyytinen, M. Laurila, T. L. Tammela, and P. A. Koivisto. 2001. Chromosomal changes in locally recurrent, hormone-refractory prostate carcinomas by karyotyping and comparative genomic hybridization. *Cancer Genet Cytogenet* 131 (1):74–8.

Hanahan, D., and R. A. Weinberg. 2000. The hallmarks of cancer. *Cell* 100 (1):57–70.

Jallepalli, P. V., and C. Lengauer. 2001. Chromosome segregation and cancer: cutting through the mystery. *Nat Rev Cancer* 1:109–117.

Kemp, C. J., K. H. Kim, and J. Philipp. 2000. The murine gene Cdkn1b (p27(Kip1)) maps to distal chromosome 6 and is excluded as Pas1. *Mamm Genome* 11 (5):402–4.

Koera, K., K. Nakamura, K. Nakao, J. Miyoshi, K. Toyoshima, T. Hatta, H. Otani, A. Aiba, and M. Katsuki. 1997. K-ras is essential for the development of the mouse embryo. *Oncogene* 15 (10):1151–9.

Lai, L. A., T. G. Paulson, X. Li, C. A. Sanchez, C. Maley, R. D. Odze, B. J. Reid, and P. S. Rabinovitch. 2007. Increasing genomic instability during premalignant neoplastic progression revealed through high resolution array-CGH. *Gene Chromosomes Cancer* 46 (6):532–42.

Lastowska, M., S. Cotterill, A. D. Pearson, P. Roberts, A. McGuckin, I. Lewis, and N. Bown. 1997. Gain of chromosome arm 17q predicts unfavourable outcome in neuroblastoma patients. U.K. Children's Cancer Study Group and the U.K. Cancer Cytogenetics Group. *Eur J Cancer* 33 (10):1627–33.

Lengauer, C., K. W. Kinzler, and B. Vogelstein. 1998. Genetic instabilities in human cancers. *Nature* 396:643–9.

Li, R., G. Yerganian, P. Duesberg, A. Kraemer, A. Willer, C. Rausch, and R. Hehlmann. 1997. Aneuploidy correlated 100% with chemical transformation of Chinese hamster cells. *Proc Natl Acad Sci USA* 94 (26):14506–11.

Maley, C. C. 2007. Multistage carcinogenesis in Barrett's esophagus. *Cancer Lett* 245 (1–2):22–32.

Matzke, M. A., M. F. Mette, T. Kanno, and A. J. M. Matzke. 2003. Does the intrinsic instability of aneuploid genomes have a causal role in cancer? *Trends Genetics* 19 (5):253–6.

Maynard Smith, J., and E. Szathmary. 1995. *The Major Transitions in Evolution*. Oxford, U.K.: Oxford University Press.

Meltzer, S. J., J. Yin, Y. Huang, T. K. McDaniel, C. Newkirk, O. Iseri, B. Vogelstein, and J. H. Resau. 1991. Reduction to homozygostiy involving p53 in esophageal cancers demonstrated by the polymerase chain reaction. *Biochemistry* 88:4976–4980.

Merlo, L. M., L. Wang, J. W. Pepper, P. S. Rabinovitch, and C.C. Maley. 2008. Polyploidy, aneuploidy and the evolution of cancer. In *Polyploidization and Cancer*, edited by R. Y. Poon. Austin/New York: Landes Bioscience, Inc. and Springer Science+Business Media.

Mitchell, M. 1998. *An Introduction to Genetic Algorithms*. Cambridge, MA: MIT Press.

Moolgavkar, S. H., and E. G. Luebeck. 2003. Multistage carcinogenesis and the incidence of human cancer. *Genes Chromosomes Cancer* 38 (4):302–6.

Neshat, K., C. A. Sanchez, P. C. Galipeau, D. S. Cowan, S. Ramel, D. S. Levine, and B. J. Reid. 1994. Barrett's esophagus: a model of human neoplastic progression. *Cold Spring Harbor Symposia on Quantitative Biol* 59:577–83.

Nishimura, T., N. Nishida, T. Itoh, M. Kuno, M. Minata, T. Komeda, Y. Fukuda, I. Ikai, Y. Yamaoka, and K. Nakao. 2002. Comprehensive allelotyping of well-differentiated human hepatocellular carcinoma with semiquantitative determination of chromosomal gain or loss. *Genes Chromosomes Cancer* 35 (4):329–39.

Nowak, M. A., N. L. Komarova, A. Sengupta, P. V. Jallepalli, I.-M. Shih, B. Vogelstein, and C. Lengauer. 2002. The role of chromosomal instability in tumor initiation. *Proc Natl Acad Sci USA* 99 (25):16226–31.

Nowell, P. C. 1976. The clonal evolution of tumor cell populations. *Science* 194 (4260):23–8.

Pekarsky, Y., A. Palamarchuk, K. Huebner, and C. M. Croce. 2002. FHIT as tumor suppressor: mechanisms and therapeutic opportunities. *Cancer Biol Ther* 1 (3):232–6.

Rabinovitch, P. S., G. Longton, P. L. Blount, D. S. Levine, and B. J. Reid. 2001. Predictors of progression in Barrett's esophagus III: baseline flow cytometric variables. *Am J Gastroenterol* 96 (11):3071–83.

Rajagopalan, H., M. A. Nowak, B. Vogelstein, and C. Lengauer. 2003. The significance of unstable chromosomes in colorectal cancer. *Nat Rev Cancer* 3:695–701.

Reid, B. J., L. J. Prevo, P. C. Galipeau, C. A. Sanchez, G. Longton, D. S. Levine, P. L. Blount, and P. S. Rabinovitch. 2001. Predictors of progression in Barrett's esophagus II: baseline 17p (p53) loss of heterozygosity identifies a patient subset at increased risk for neoplastic progression. *Am J Gastroenterol* 96:2839–2848.

Reid, B.J., and P.S. Rabinovitch. 1988. Barrett's esophagus: a human model of genomic instability in neoplastic progression. In *Accomplishments in Oncology*, edited by J. G. Fortner and B. I. Hirschowitz. Philadelphia: J. B. Lippincott Company.

Renan, M. J. 1993. How many mutations are required for tumorigenesis? Implications from human cancer data. *Mol Carcinogenesis* 7:139–146.

Robinson, J. K., A. W. Rademaker, C. Goolsby, T. N. Traczyk, and C. Zoladz. 1996. DNA ploidy in nonmelanoma skin cancer. *Cancer* 77 (2):284–91.

Shackney, S. E., G. Berg, S. R. Simon, J. Cohen, S. Amina, W. Pommersheim, R. Yakulis, S. Wang, M. Uhl, C. A. Smith et al. 1995. Origins and clinical implications of aneuploidy in early bladder cancer. *Cytometry* 22 (4):307–16.

Shackney, S. E., and T. V. Shackney. 1997. Common patterns of genetic evolution in human solid tumors. *Cytometry* 29:1–27.

Shackney, S. E., C. A. Smith, B. W. Miller, D. R. Burholt, K. Murtha, H. R. Giles, D. M. Ketterer, and A. A. Pollice. 1989. Model for the genetic evolution of human solid tumors. *Cancer Res* 49 (12):3344–54.

Sidransky, D., P. Frost, A. Von Eschenbach, R. Oyasu, A. C. Preisinger, and B. Vogelstein. 1992. Clonal origin of bladder cancer. *N Engl J Med* 326 (11):737–40.

Sieber, O. M., K. Heininmann, and I. P. M. Tomlinson. 2003. Genomic instability—the engine of tumorigenesis? *Nat Rev Cancer* 3:701–8.

Slaton, J. W., C. P. Dinney, R. W. Veltri, C. M. Miller, M. Liebert, G. J. O'Dowd, and H. B. Grossman. 1997. Deoxyribonucleic acid ploidy enhances the cytological prediction of recurrent transitional cell carcinoma of the bladder. *J Urol* 158 (3 Pt 1):806–11.

Tachibana, M., N. Deguchi, S. Baba, S. Jitsukawa, M. Hata, and H. Tazaki. 1991. Multivariate analysis of flow cytometric deoxyribonucleic acid parameters and histological features for prognosis of bladder cancer patients. *J Urol* 146 (6):1530–4.

Vogelstein, B., and K. W. Kinzler. 1993. The multistep nature of cancer. *Trends Genet* 9 (4):138–141.

Whang-Peng, J., T. Knutsen, E. C. Douglass, E. Chu, R. F. Ozols, W. M. Hogan, and R. C. Young. 1984. Cytogenetic studies in ovarian cancer. *Cancer Genet Cytogenet* 11 (1):91–106.

Wong, D. J., T. G. Paulson, L. J. Prevo, P. C. Galipeau, G. Longton, P. L. Blount, and J. B. Reid. 2001. p16 INK4a lesions are common, early abnormalities that undergo clonal expansion in Barrett's metaplastic epithelium. *Cancer Res* 61:8284–9.

Wongsurawat, V. J., J. C. Finley, P. C. Galipeau, C. A. Sanchez, C. C. Maley, X. Li, P. L. Blount, R. D. Odze, P. S. Rabinovitch, and B. J. Reid. 2006. Genetic mechanisms of TP53 loss of heterozygosity in Barrett's esophagus: implications for biomarker validation. *Cancer Epidemiol Biomarkers Prev* 15 (3):509–16.

CHAPTER 4

Catastrophes and Complex Networks in Genomically Unstable Tumorigenesis

Ricard V. Solé

CONTENTS

INTRODUCTION

The unstable character of most cancers, as reflected by the high levels of genomic, cytogenetic, and epigenetic variation, seems an almost universal feature of tumorigenesis. Such cellular disorder is particularly well illustrated by the presence of high levels of aneuploidy: multiple losses and gains of parts or even entire chromosomes can be seen, together with many chromosome rearrangements (Lengauer et al., 1998). It has been properly described as a "gallery of horrors" and such a disorder opens interesting, and I believe largely unanswered questions concerning the nature of cancer itself. The evolutionary dynamics of tumors is thus characterized by selection processes in parallel with unusual levels of genetic variation (Loeb, 1991) more consistent with what we would expect from unicellular systems (Cairns, 1997). Increased genetic instability has been suggested as an adaptive trait of microbial species. When facing high levels of environmental stress, those mechanisms controlling the accurate replication of DNA might be overcome and checkpoints ignored. This is less reasonable in the tissue context, where cooperation among cells and a control of tissue and organ size is an essential part of global homeostasis. Failures in properly replicating the genome face multiple internal controls that force the system to stop dividing or even to die.

The evolutionary nature of cancer progression was earlier highlighted by John Cairns. As he pointed out, although competition and variation is the source of improvement and change in nature, when turning to the competition between individual cells within a complex organism "we see that natural selection has now become a liability" (Cairns, 1975). Looking at cancer as a process where Darwinian evolution plays a major role (Cairns, 1975; Merlo et al., 2006), we can gain real insight into its origins and dynamics. Perhaps not surprisingly, Darwinian evolution has become an important issue within medicine (Greaves, 2007).

Genomic instability seems to be a common trait in many types of cancer (Cahill et al., 1999) and is a key ingredient in the Darwinian exploratory process required to overcome selection barriers. By displaying high levels of mutation, cancer cells can generate a progeny of highly diverse phenotypes able to escape from such barriers (Loeb, 2001; Merlo et al., 2006). In this context, as shown by Maley and co-workers for the premalignant condition known as *Barret's esophagus*, clonal diversity measures (adapted from theoretical works in ecology and evolution) can predict tumor progression to adenocarcinoma. As these authors point out, "progression to cancer through accumulation of clonal diversity, on which natural selection acts, may be a fundamental principle of neoplasia with important clinical implications" (Maley et al., 2006). In this chapter, we consider some recent mathematical and computational models of tumorigenesis involving genetic instability. These models overlap in several ways, incorporating different layers of complexity, spatial context, and ways of introducing the unstable behavior of cancer cells. All of them share a common view of cancer heterogeneity that links it with another type of biological system: RNA viruses. These viruses are known to exhibit high levels of mutation (Domingo et al., 1995). As a consequence of such high levels of mutation, together with high levels of replication, RNA viruses form very diverse populations, which have been named *quasispecies* (Eigen et al., 1987; Schuster, 1994). These clouds of mutants have been shown to behave as the units of selection and are responsible for the rapid adaptation of viruses to their changing environments, particularly, the immune response in vertebrates. One key prediction of the model is that there is a critical mutation rate beyond which viruses are nonviable. Such transition is sharp and, thus, small increases in mutation rate beyond this threshold would effectively eliminate the virus. This prediction has been tested and shown to be correct (Loeb et al., 1991; Coffin, 1995), thus opening a new approach to antiviral therapies.

Some of these ideas are summarized in Figure 4.1. Standard models (a) consider cancer progression as described by a two-population problem (see, for example, Gatenby, 1995, 1996). Here, we show cancer cells as gray spheres occupying some locations within a healthy tissue (*H*, not shown). In the simplest approximation, all cancer cells are considered as having the same kinetic properties, and thus the tumor is described as a single, homogeneous population *C*. A two-dimensional model in the (*H,C*) space can be constructed and compared with predictions from the cellular automaton model (CA). An extension of this simple approach is

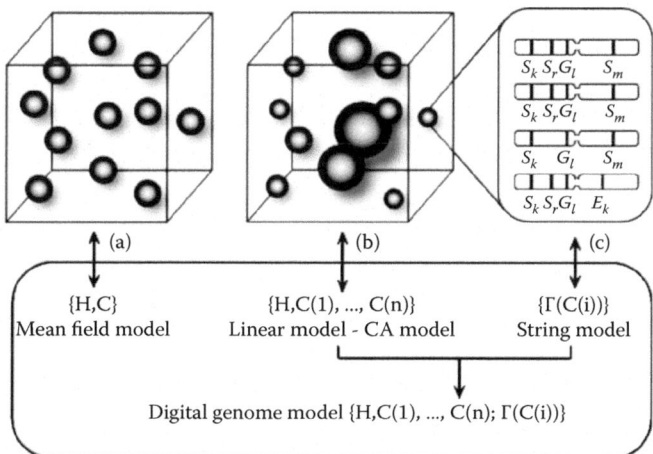

FIGURE 4.1 Different levels of approximation to the dynamics of unstable cancer. These models include homogeneous systems (a) where all cancer cells are equivalent and (b) heterogeneous models where rates of growth and/or death are introduced as continuous numbers and a number of different subpopulations are allowed to exist (here, the size of the spheres indicates the presence of variable traits). A more realistic scenario would consider cells as carrying a genome (c) where genes involved in growth and stability are included, together with a set of essential genes. By incorporating this digital genome description into a spatially explicit model, several key components of tumor progression can be incorporated.

to explicitly introduce the variable character of cells due to genetic heterogeneity (b). In this case, a better population description is achieved, and different effects associated to population noise, bottlenecks, or spatial dynamics are much better represented. This complexity can be made explicit by using a multistate CA model (here, the different radius of cells just indicates different phenotypic continuous traits) and also using a mathematical (mean field) model ignoring spatial interactions but introducing population structure.

Finally, we can also go into a more detailed description of cell organization by considering a string picture (c). Here, each cell and its phenotypic traits are captured by the mapping between mutation affecting genes involved in different locations along the string (Solé, 2003). These digital genomes are a very simplified way of introducing genome organization and can help understanding how growth and instability affect each other. Moreover, by explicitly using the fitness landscape associated with different mutational events, we can also track the exact progression paths followed by the tumor. This approach reveals an unexpected complexity embedded in the presence of a runaway effect, pushing instability levels toward criticality. Several common features shared by RNA viruses and unstable cancer populations can help to better understand some of the counterintuitive patterns displayed by unstable tumors.

MEAN FIELD MODEL

Our first example concerns a very simple characterization of instability based on the assumption that all cancer cells can be considered equal. This is of course in contradiction with the idea that instability generates heterogeneity. In this picture of tumorigenesis, we sacrifice realism in favor of well-defined predictions. The basic model involves two differential equations associated with normal and cancer cell populations. These equations read

$$\frac{dH}{dt} = P_r^0 H - H\Phi(H,C)$$

$$\frac{dC}{dt} = P_r^0 \Gamma(\mu)C - C\Phi(H,C)$$

where P_r^0 indicates the basal cell replication rate associated to normal cells, whereas $\Gamma(\mu)$ indicates the effects of the instability rate μ on the replication of cancer cells. The last term in the right-hand side of both equations introduces selection of master replicating strains. It represents an outflow from the system, and can be easily computed using some additional assumptions. In particular, it can be shown that the previous set of equations is being reduced to a single equation by assuming that the total population size (C + H) is constant. For convenience, we normalize the total population to one. If this assumption is introduced, we have

$$\frac{dH}{dt} + \frac{dC}{dt} = \frac{d(C+H)}{dt} = 0$$

which gives $\Phi(H,C) = P_r^0(H + \Gamma(\mu)C)$. Using this result, we obtain after some algebra:

$$\frac{dC}{dt} = P_r^0(\Gamma(\mu)-1)C(1-C)$$

Such an equation captures the essential dynamics of the model and, in particular, the presence of two possible equilibrium states, namely, a cancer-wins phase ($C = 1$, $H = 0$) and a host-win (i.e., healthy tissue) ($C = 0$, $H = 1$) phase. As it happens with other phase transitions in complex systems (Solé et al., 1996; Solé and Goodwin, 2001), some important lessons can be extracted by understanding the nature and universality of the transition. The critical boundary of a transition is easily obtained from the condition $\Gamma(\mu) = 1$. Now we need to include in the mutation-dependent term some reasonable link between instability, growth, and deleterious effects. This can be done by assuming that instability allows hitting growth-related genes, but also essential genes whose loss or mutation are lethal. One possible choice is

$$\Gamma(\mu) = (1 + \mu n_g \delta G)(1 - \mu)^{n_{HK}}$$

which includes the positive effect associated with growth-related mutations (first term on the right-hand side) and a second component introducing the adverse effects of hitting essential genes. Here, n_g and n_{HK} indicate the number of growth-related and essential (housekeeping) genes, respectively, and δG is the average increase of growth due to the first class of

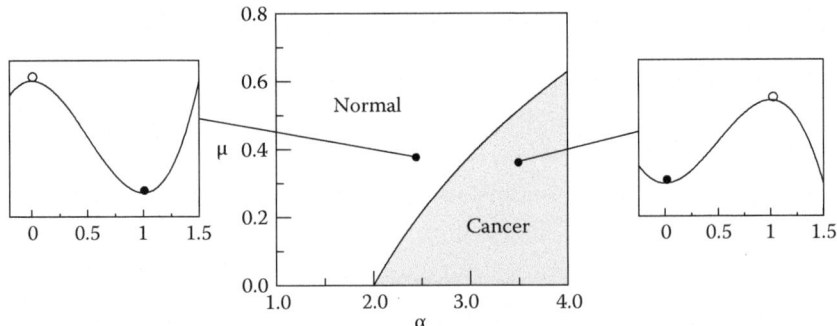

FIGURE 4.2 Mean field model of a two-species interaction system involving cancer and normal cells. Cancer cells are identical in their replication and mutation rates, and a fixed trade-off between replication and mutation is introduced. One prediction from the model is that two possible phases are present: either the tumor fails to propagate or wins and occupies all space. The critical line separating these phases depends both on the efficiency of cancer cells to replicate and the amount of instability. As we can see here, we can shift from the cancer phase (gray) to the healthy tissue phase by increasing instability rates. The two pictures at both sides of the phase diagram are two examples of the associated potential function (whose minima correspond to equilibrium states).

mutations (see next section for a detailed discussion). This choice allows us (for each set of parameters) to find the corresponding critical mutation rate μ_c. An example of the phase diagram associated with this model (for a given set of parameters) is shown in Figure 4.2. The gray area indicates the domain where cancer propagates (here we use $\alpha = n_g \delta G$). We can see that the boundary of this domain is a function of both instability and the selective advantage provided by growth-related mutations. An immediate result from this picture is that small changes in instability (a slight increase) can shift the system from normal to cancer. An additional illustration of this result can be obtained by using the potential function $\phi_\mu(C)$ associated to the previous equation. Specifically, a potential function follows the following property:

$$\frac{dC}{dt} = -\frac{d\phi_\mu(C)}{dC}$$

or, in other words, $\phi_\mu(C) = -\int f(C)dC$, where $f(C)$ is the function describing the growth dynamics of the C population. As defined, this function will be such that the minima of it correspond to the stable equilibrium points associated to our system, whereas the maxima will indicate unstable equilibrium. For our system, we have

$$\phi_\mu(C) = -P_r^0(\Gamma(\mu)-1)\int C(1-C)dC = -P_r^0(\Gamma(\mu)-1)\left[\frac{C^2}{2} - \frac{C^3}{3}\right]$$

The two examples shown in Figure 4.2 illustrate the point.

It can be shown that the critical instability rate scales as the inverse of the number of housekeeping genes. In other words, on a first approximation, we have $\mu_c \approx 1/n_{HK}$. This is an interesting finding, since it predicts a limit (an upper bound) to the maximal amount of genomic instability compatible with viable cancer cells. Another important result is that this type of model (several variations have been explored) seems robust to several relevant modifications, such as the exact functional form of the instability-replication relation (Solé and Deisboeck, 2004). The robustness of our prediction suggests that this type of error catastrophe might play a role in cancer, or might be used in future approaches to cancer treatment, since a small increase in genomic instability close to the boundary can cause the collapse of the population.

This type of model can be extended in several ways. One of them (Solé et al., 2008) included a more accurate description of tissue architecture, in particular, the presence of stem cells and cancer stem cells. In this approach, we expand our picture of cancer organization by adding one important ingredient: cancer stem cells (Reya et al., 2001). In Figure 4.3, we show one example of the type of tissue organization that has been analyzed. Here, a healthy tissue competes with a cancer cell population involving cancer stem cells (Sc) and differentiated cancer cells (C). The CSC population is considered constant (in this way, we can maintain the model complexity under control) and, thus, cancer cells will always be present. Once again, the two-phase scenario is also present, as indicated in the lower picture of Figure 4.5. Here, together with a cancer-wins phase, there is a scenario where both cancer and the healthy tissue coexist. This corresponds to a situation where cancer would be always present, but its relative impact would be controlled by the parameters defining the replicative potential of mutants as well as the limits imposed by instability levels. A chronic state is here defined by the gray area where once again we can keep the system in a safe state of low level, benign configuration provided that instability levels are high enough.

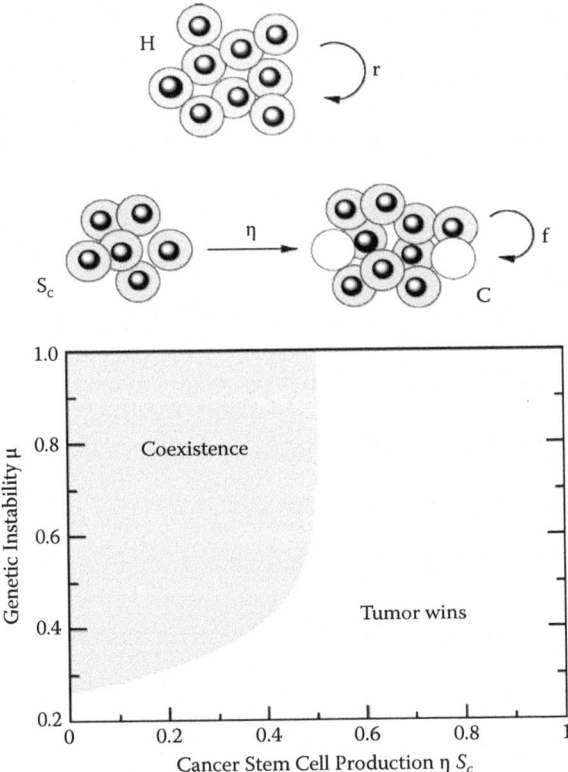

FIGURE 4.3 A model of cancer progression considering the presence of cancer stem cells also displays phase transitions. Now, the existence of a stable, constant compartment of cancer stem cells allows two basic phases to be present. In one the tumor wins, whereas in the second (gray area) coexistence between both can be observed.

An additional extension of the previous models involved a multistep scenario where genetic instability levels could change (increasing) as further mutations accumulate. This was done (Solé et al., 2008) by considering a linear arrangement of cell types involving increasing levels of genomic instability. One given element of the chain displays a given mutation rate, which can affect other genes involved in stability thus further increasing mutation levels. As we move through this linear chain, faster clones allow tumor progression to proceed, but also increase the likelihood of entering inside a dangerous domain of damaging mutation levels. The key result of this model was the finding that the cancer cell population self-organizes around a narrow domain of high instability levels, thus approaching the critical line. Below this domain, a gap is observed, with virtually no cancer cells having intermediate or low levels of instability. This result supports our previous conclusion concerning the possible evolution of unstable tumors toward a fragile organization where cells replicate and adapt but are also likely to experience decline if instability is further increased even by a small amount.

The models outlined earlier are all simplifications of the intrinsic richness associated with genome complexity. Evolution and stochasticity, as well as explicit spatial degrees of freedom, have been ignored or oversimplified. It seems clear that the assumptions implicitly made in our toy models should be explicit. The following section shows how this can be done using a representation of genome architecture that seems to capture some of the evolutionary dynamics that are likely to occur in real tumors.

DIGITAL GENOME MODEL

A different approach that has been taken is to consider the problem of cancer quasispecies in terms of sets of strings of bits (Solé, 2003). In this model approach, each cell is replaced by a string carrying a set of "genes," whose state will be indicated as 1 or 0. Normal tissue defines a particular string ξ that has a given replication rate. Here, we take for convenience the sequence where all bits are 1. On the other hand, mutations can affect any bit in any string in such a way that replication is not accurate and a mutated string appears. In the original formulation of the problem, two populations where considered. One is described by one particular string, whereas the second class included all possible mutants generated from the original sequence. This model did not considered explicit space, and all bits in the string (here representing genes) would only affect replication speed. The model was thus a well-mixed population of strings representing the competition between healthy and tumor cell populations, and confirmed the previous predictions based on the mean field model, in particular, the presence of a well-defined phase transition. An example of these results is summarized in Figure 4.4. Here, we display (a,b) the hypercube of 4-bit strings under two different conditions, corresponding to the two phases of the model. In (a), instability levels are low and efficient mutants have been generated, creating a quasispecies that spreads through part of the landscape. The lower left corner would indicate the population of healthy cells, which is small. If instability rates increase beyond the threshold, the healthy tissue is capable of outcompeting the cancer quasispecies, as shown in (b). The population distribution

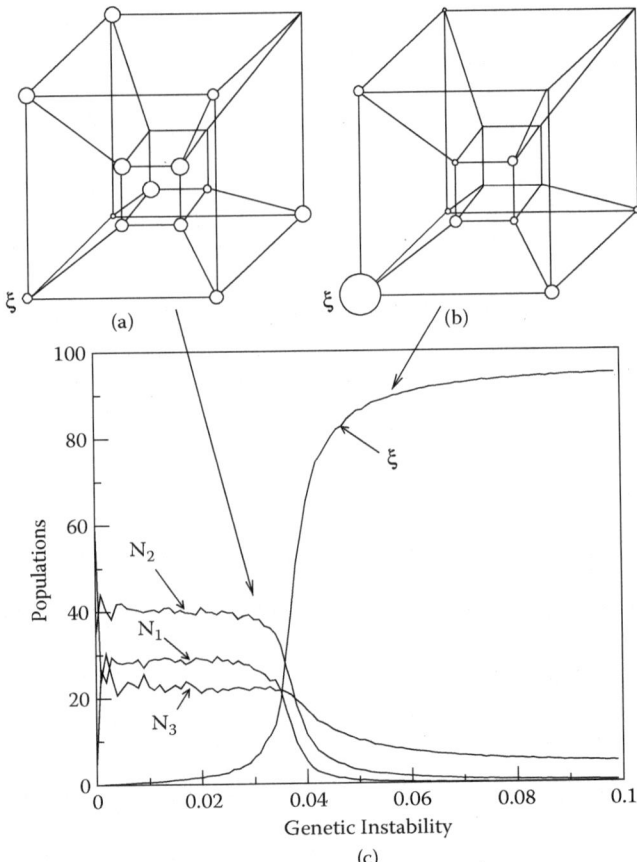

FIGURE 4.4 Phase transitions in the string quasispecies model. Here, a population of mixed strings is used to model a single-peak landscape where the 1111 string has the lower replication rate (the master sequence), whereas all others differ from it at least in one bit and have a smaller replication rate. The upper diagrams (a,b) display the fitness landscape for this small-size example with only $n = 4$ bits. In (a) a moderate level of genetic instability allows generating strings with faster replication rates, whereas for higher levels (b) the deleterious effects associated to increased mutations render mutating strings nonviable. A marked transition is clearly visible.

is displayed in (c), with both the number of strings for the healthy tissue (here indicated as x) and the number of mutant sequences differing in 1, 2, or 3 bits from the master. Although the healthy tissue is unable to win below a critical mutation level, it becomes successful once cancer moves beyond criticality.

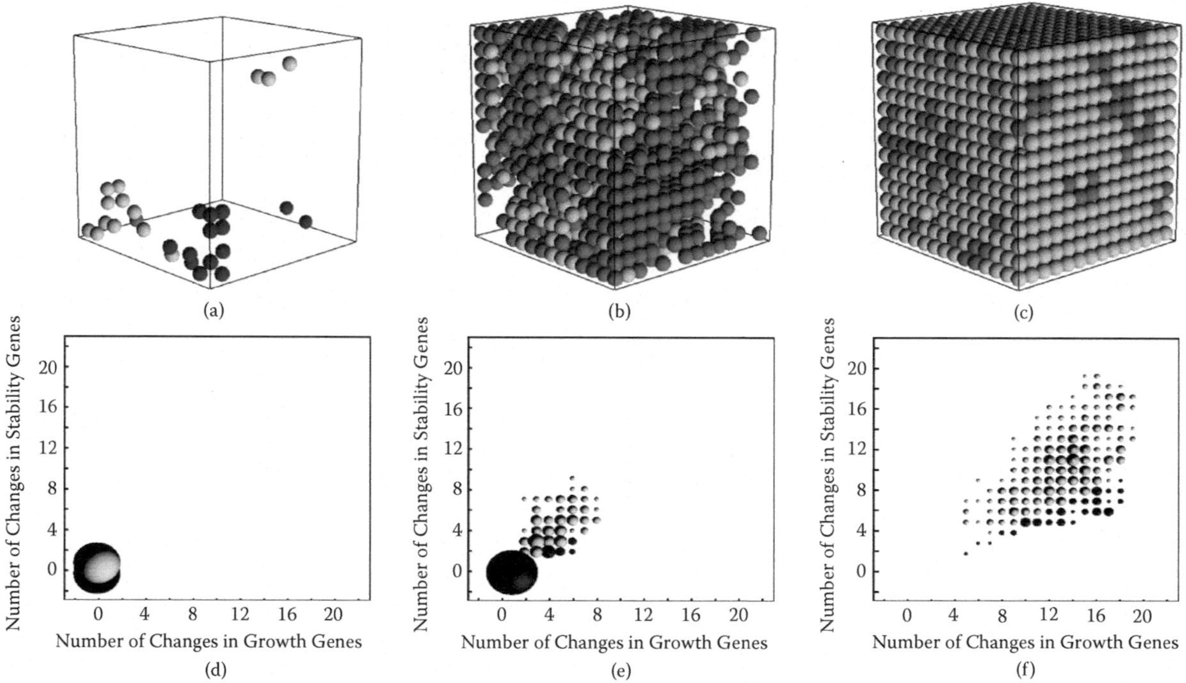

FIGURE 4.5 **(See color insert following page 40)** Digital genome model. The upper pictures (a–c) show an example of the growth of cancer cells (normal cells not shown) through time. This growth takes place together with a constant increase in both instability and replication, as shown in the (d–f) sequence where the formation of a cancer quasispecies can be observed (see text).

A last step in our increasingly detailed modeling efforts will consider the integration of cell-level organization, space, and genome. Each cell is now explicitly located in a given point of a three-dimensional spatial domain. In this way, we take into account the limitations imposed by spatial constrains, in particular, the reduced competition resulting from local interactions. In this context, it has been shown that spatial dynamics in cancer seems strongly influenced by its spatial organization (González-Garcia et al., 2000). In particular, limited dispersal enhances spatial heterogeneity, both at the genetic and phenotypic levels. Now every cell includes an internal description of 3-bit strings, associated each with one class of gene. Specifically, we consider genes linked to proliferation, stability, and those having essential roles in cell survival. Once again, introducing housekeeping genes allows us to set some limits to the levels of instability that can be achieved. The three compartments define this digital genome, and they are

1. A set G of growth-related genes. Here, $G = \{Gj\}$ with $j = 1, ..., n_g$ genes. This set includes genes affecting the rate of replication of a given cell. Their loss or mutation increases the replication rate of cells. This would include both tumor suppressor genes (such as APC or p53) and oncogenes (RAS or SRC). Although they act in different ways (are targeted in opposite ways by genetic alterations) here me make no explicit distinction (Vogelstein and Kinzler, 2006). This assumption simply considers the fact that the impact of both kinds of mutation is an effective increase in the number of cancer cells. This assumption ignores relevant features of tumorigenesis that are not within the scope of our approximation. The exact origins of such driving events (alterations in cell division rates or disruption of checkpoint controls) are not within the scope of our approximation.

2. A set S of stability-associated genes, $S = \{S_j\}$ with $j = 1, ..., n_s$. Mutations in these genes lead to increased levels of mutagenesis. These stability genes (Vogelstein and Kinzler, 2004) are typically genes playing a key role in preserving genome integrity, and their failure can have large effects. In a nutshell, these genes (including BRCA1, BLM, or ATM) keep genetic changes under control. As a consequence, their failure or loss triggers further increases in mutations in other genes. If these mutations affect growth-related genes, the tumor can gain fitness through increased replication. If other stability genes are affected, further instability will be observed.

3. A set H of housekeeping (HK) genes, $H = \{h_j\}$ with $j = 1, ..., n_h$. These genes are associated with essential functions whose failure leads to cell death. In real cells, HK genes are expressed in a constitutive manner in all tissues. Examples would include genes coding for ribosomal proteins actin, GAPDH, and ubiquitin (see Eisenberg and Levanon (2003)).

The genome of the k-th cell, to be indicated as $\Gamma(k)$, is thus defined from the three previous subsets and will be essentially a Boolean string, where a given gene can be in two possible states, namely, 1 and 0, indicating wild-type and mutated loci, respectively. Changes in strings associated to growth or stability will have an impact on cell proliferation although their nature is very different. These strings are, for the k-th cell,

$$G(k) = (G_{1k}, G_{2k}, ..., G_{nk})$$

and for growth-related genes and

$$S(k) = (S_{1k}, S_{2k}, ..., S_{nk})$$

for stability genes. The set of HK genes to be indicated as

$$H(k) = (H_{1k}, H_{2k}, ..., H_{nk})$$

and thus the digital genome is given by:

$$\Gamma(k) = G(k) \cup S(k) \cup H(k)$$

Finally, an additional pair of strings are included, introducing the impact of each mutated gene on the growth or stability (mutation) properties of the cell. These strings are given by:

$$\delta G = (\delta G_1, \delta G_2, ..., \delta G_n)$$

which is position but not cell dependent, for growth and

$$\delta \mu = (\delta \mu_1, \delta \mu_2, ..., \delta \mu_n)$$

for instability effects associated with mutations in each gene. The effects of changes are additive, and thus can be calculated, for the k-th cell as follows:

$$P_r(C(k)) = P_r^0 + \sum_{j=1}^{n} G_{jk} \delta G_k$$

for the growth term and

$$P_\mu(C(k)) = \mu_0 + \sum_{k=1}^{n} S_{jk} \delta \mu_k$$

for the instability one. Since we assume that any damage affecting HK genes is lethal, no probability needs to be introduced for their effect. The two previous expressions give the probabilities of replication and mutation of this cell. For simplicity, we take the same number of genes in each compartment. This approach allows studying tumorigenesis under a multiscale perspective: both cells and cell populations are being taken into account, and the phenotypic traits characterizing cell kinetic parameters are evolvable and implicitly defined by genome structure. We can thus follow the changes taking place within the tumor and what drives them. Two important problems can be addressed here. One is the emergence of unstable clones and genomic heterogeneity under spatial constraints. The second is the patterns displayed by progression paths followed by the tumor cell population. The first is already known to us from the previous model approaches, but now we have little constraints since every property of the cell population is ultimately associated with the microscopic contributions of genome-level changes.

As shown in Figure 4.5, the model displays a tendency to increase both growth rate and instability. These parallel changes seem to result from a coevolution of both instability (which allows to hit growth-related genes) and growth. Clones of cells having one or several mutated proliferation genes will expand, carrying with them those mutations associated with instability. Such mutations then can expand also triggering further increases of instability and accelerating tumorigenesis. Although in some cases instability goes first and in others proliferation, the general trend seems to be a parallel coevolution of both phenomena. The upper row in Figure 4.5 (a–c) shows three snapshots of the model at three different steps in the simulation. Here, a tumor slowly develops from random mutations, which eventually become successful. We can see this evolution under a different perspective by looking at the information provided by how genome composition changes over time. This is illustrated in Figure 4.5d–f, where we display the evolution of the occupancy of the instability-growth space followed by one run of the model. Specifically, we measure how many cancer cells (here displayed as the relative fraction) have a given number of mutated (or lost) genes associated with either replication or mutation. Starting from the lower left corner of this diagram, we can see that a cloud (the cancer quasispecies) develops and expands toward the upper right corner, thus involving mutations in both stability- and replication-related genes. The final outcome of the evolutionary dynamics of these digital genomes is variable. Sometimes, the whole tumor moves to the highest instability-replication levels and remains at that. Sometimes, the initial mutations driving tumorigenesis affect stability genes, afterward followed by replication genes. Others, replication genes come first. However, in general, the typical scenario involves both types of genes, with a slight initial contribution of stability-related mutations. We could say that there is a special type of coevolutionary dynamics here. Mutations in stability genes will be typically neutral, particularly, while far from criticality. However, such mutations will increase the likelihood of hitting growth genes, and the clonal amplification of these will facilitate the expansion of unstable cells, which in turn are likely to produce offspring displaying higher instability and so on.

PROGRESSION PATHS AS COMPLEX NETWORKS

When we analyze the abundance of each string (digital genome) in our unstable tumors, we find that the resulting probability distribution is highly skewed. It is dominated by a few strings having large populations coexisting with many others whose population sizes are rather small. If we plot the frequency $N(m)$ of strings present in m cells in the tumor, it decays as a power law, that is, it follows a distribution $N(m) = Am^{-\beta}$ (here A is a normalization constant). It is interesting to notice that such a shape has been reported from the analysis of chromosome abnormalities in several types of cancer, including breast, colorectal, and renal (Frigyesi et al., 2003). The presence of these power laws has important implications, in particular, in terms of the meaning of taking small samples from the tumor, since the enormous variability associated with these distributions makes statistical averages rather unreliable.

A last message is obtained by watching closely the patterns of network organization associated with the transitions between different genomes. Even at the very small sizes considered in these models (where $n = 20$ genes of each class where used), the potential combinatorics are enormous. Although a heterogeneous model with different levels of growth rate should favor some genes in relation to others, and perhaps lead to a more or less linear chain or gene–gene correlation defining a linear progression, the analysis of the transitions between different genomes (i.e., single-gene mutational events) pictures a rather different, highly nonlinear image (Figure 4.6). Previous work on progression pathways has shown that this is far from a trivial problem, but strong evidence suggests that parallel paths are expected to occur (Subramanian and Axelrod, 2001; Sontag and Axelrod, 2005). However, the general question of what kind of global network organization might be at work has only recently been considered.

Instead of a roughly linear graph, we obtain a complex network of state transitions that describe a scenario where most genomes appearing

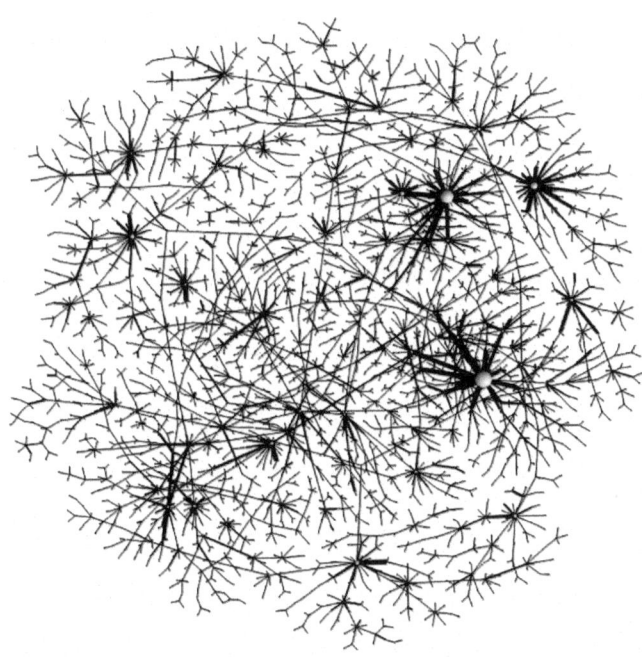

FIGURE 4.6 Complex pathways of tumorigenesis in the digital genome model. Here, we display the complete graph of genome transitions that took place in the process shown in Figure 4.5.

through the progression process have just one or two links with others, whereas a few nodes display a large number of connections. These hubs are typically linked to successful populations of strings from which many other mutants were generated. This network does not follow the standard picture of tumor progression as described by early attempts of understanding the steps required for the process to succeed (Fearon and Vogelstein, 1990). Instead, the pattern of connections follows a rather heterogeneous organization, which can be characterized by means of a scale-free distribution of connections (Albert and Barabási, 2002). More precisely, the probability $P(k)$ of nodes having k links between a given genome $\Gamma(k)$ and another one $\Gamma(k')$ follows a fat-tailed distribution, namely:

$$P(k) = \frac{1}{Z} k^{-\gamma} \exp\left(\frac{k}{K}\right)$$

where Z is a normalization constant and K a given cutoff. This distribution implies (as we can see from Figure 4.6) that most elements have just one or two links, whereas a few of them have many connections. In most complex networks (here too), the exponent γ is bound between two and three. One important consequence of this architecture is that there will be problems in defining the variance associated with the system and thus to properly define statistical significance. It is not difficult to show that, for very large K values, when a power law dominates the distribution, the second-order moment $<k^2>$ diverges, since we have

$$<k^2> \approx \int_1^M k^{2-\gamma} dk = \frac{1}{3-\gamma}\left(M^{3-\gamma} - 1\right)$$

where M indicates the maximal number of links that a node can achieve. As M grows, and given that $2 < \gamma < 3$, this average will rapidly diverge and, as a consequence, the statistical deviations will diverge too. This result gives a rather different picture from a tumor as describable in terms of standard average values and supports the view that better predictors must consider clonal diversity (Maley et al., 2006).

DISCUSSION

The success of unstable tumors in adapting and growing within their hosts creates a paradoxical situation. Aneuploidy is known to be a burden to cell viability and has severe effects on organismal growth and development (Torres et al., 2008). Moreover, aneuploidy provides a source of enormous variation. As David Pellman puts it, "Aneuploid cancers are like Tolstoy's unhappy families: each aneuploid cancer has its own particular abnormal chromosome content, and thus its own abnormal characteristics." Actually, it is well known that there are many more aberrations in solid tumors have been shown to be recurrent (Albertson et al., 2003). The typically reduced fitness caused by aneuploidy, and the great variability associated with cancer progression require an appropriate explanation. Mathematical and computer models explicitly considering the impact of such instability can be useful in order to provide tentative answers. As suggested in early papers (Solé and Deisboeck, 2004), the similarities existing between viral quasispecies and unstable cancer provide a relevant insight. RNA viruses are known to replicate close to the error catastrophe and, thus, an important part of their mutants are nonviable. Living at the error threshold allows these populations to escape from the constant pressure of the immune system (Domingo, 2000). The cost of such elevated mutational load is compensated by the plastic responses that are intrinsic to the quasispecies structure. Information is preserved, and selection forces can act. Once the threshold is crossed, genetic drift dominates the scene and the viruses are no longer able to adapt. Is that the case in unstable tumors? Our models suggest that this might be the case since the presence of an error catastrophe in cancer-normal competition models seems a generic property. On the other hand, the failure of stability-preserving mechanisms that takes place during carcinogenesis, particularly, those associated with balanced segregation of chromosomes, should be expected to trigger multiple cascades of changes. As a consequence, genetic instability should be expected to increase through time since the loss of stability-related genes is irreversible. In other words, if instability can be estimated using some average "mutation" rate, this value should grow over time. The tumor will thus approach the error catastrophe, and how fast this happens will determine how close the evolved cancer population will be to the critical boundary. All this variability seems to articulate rather well with the pattern of pathways emerging from the digital genome model. A scale-free network gives us an appropriate picture of the complex dynamics followed by our *in silico* tumors. Instead of parallel or even linear pathways, we observe the most diverse network of interconnected transitions. As it happens with most complex systems displaying this pattern, robustness and fragility go together. The multiplicity of paths tells us that therapies addressing progression in terms of simple chains of events might fail. The existence of hubs on the other hand is likely to provide new ways of thinking of potential Achilles heels of cancer.

REFERENCES

Albert, R. and Barabási, A.L., 2002. Statistical mechanics of complex networks. *Rev. Mod. Phys.*, 74, 47–97.

Albertson, D. G., Collins, C., McCormick, F., and Gray, J. W. 2003. Chromosome aberrations in solid tumors. *Nat. Genet.* 34, 369–376.

Cahill, D. P., Kinzler, K. W., Vogelstein, B., and Lengauer C. (1999). Genetic instability and Darwinian selection in tumors. *Trends Genet.* 15, M57–M61.

Cairns, J. 1997. *Matters of Life and Death,* Princeton University Press, Princeton, NJ.

Cairns, J. (1975). Mutation, selection and cancer. *Nature* 255, 197–200.

Chow, M. and Rubin, H. (2000). Clonal selection versus genetic instability as the driving force in neoplastic transformation. *Cancer Res.* 60, 6510–6518.

Con, J. M. (1995). HIV population dynamics in vivo: implications for genetic variation, pathogenesis and therapy. *Science* 267, 483–489.

Domingo, E., Holland, J. J., Biebricher, C., and Eigen, M (1995). Quasispecies: the concept and the word, in: *Molecular Evolution of the Viruses* (A. Gibbs, C. Calisher, and F. Garcia-Arenal, Eds.), Cambridge University Press, Cambridge.

Domingo, E. (2000). Viruses at the edge of adaptation. *Virology* 270, 251–253.

Eigen, M. McCaskill, J., and Schuster, P. (1987). The molecular quasispecies. *Adv. Chem. Phys.* 75, 149–263.

Eisenberg, E. and Levanon, E. Y. (2003). Housekeeping genes are compact. *Trends Genet.* 19, 362–366.

Fearon, E.R. and Vogelstein, B. (1990). A genetic model for colorectal tumorigenesis. *Cell* 61, 759–767.

Frigyesi, A., Gisselsson, D., Mitelman, F., Höglund, M. (2003). Power law distribution of chromosome aberrations in cancer. *Cancer Res.* 63, 7094–7097.

Gatenby, R. A. (1995). Models of tumor-host interactions as competing populations. *J. Theor. Biol.* 176, 447–455.

Gatenby, R. A. (1996). Application of competition theory to tumour growth: implications for tumour biology and treatment. *Eur. J. Cancer* 32A, 722–726.

González-García, I., Solé, R. V., and Costa, J. (2000). Metapopulation dynamics and spatial heterogeneity in cancer, *Proc. Natl. Acad. Sci. USA* 99, 12085–12089.

Greaves, M. (2007). Darwinian medicine: a case of cancer. *Nature Rev. Cancer* 7, 213–221.

Lengauer, C., Kinzler, K. W., and Vogelstein, B. (1998). Genetic instabilities in human cancers. *Nature* 396, 643–649.

Loeb, L.A. (1991). Mutator phenotype may be required for multistage carcinogenesis. *Cancer Res.* 51, 3075–3079.

Loeb, L. A., Essigmann, J. M., Kazazi, F., Zhang, J., Rose, K. D., and Mullins, J. I. (1999). Lethal mutagenesis of HIV with mutagenic nucleoside analogs. *Proc. Natl. Acad. Sci. USA* 96, 1492–1497.

Maley, C.C., Galipeau, P.C., Finley, J.C., Wongsurawat, V.J., Li, X., Sanchez, C.A., Paulson, T.G., Blount, P.L., Risques, R., Rabinovitch, P.S., and Reid, B.J. (2006). Genetic clonal diversity predicts progression to esophageal adeno-carcinoma. *Nat. Genetics*, 38, 468–473.

Merlo, L., Pepper, J., Reid, B., and Maley, C 2006. Cancer as an evolutionary and ecological process. Nature *Cancer* 6, 924–935.

Reya, T., (2001). Stem cells, cancer, and cancer stem cells. *Nature* 414, 105–111.

Schuster, P. (1994). How do RNA molecules and viruses explore their worlds? In *Complexity: Metaphors, Models and Reality*, pp. 383–418. G. A. Cowan, D. Pines, and D. Meltzer (Eds.). Addison-Wesley, Reading, MA.

Solé, R. V., Manrubia, S., Luque, B., Delgado, J., and Bascompte, J. (1996). Phase transitions and complex systems. *Complexity* 1(4), 13–26.

Solé, R. V. and Goodwin, B. C. (2001). *Signs of Life: How Complexity Pervades Biology*. Basic Books, Perseus, New York.

Solé, R. V. (2003). Phase transitions in unstable cancer cell populations. *Eur. Phys. J.* B 35, 117–124.

Solé, R. V. and Deisboeck, T. (2004). An error catastrophe in cancer? *J. Theor. Biol.* 228, 47–54.

Solé, R. V., Rodriguez-Caso, C., Deisboeck, T., and Saldanya, J. (2008). Cancer stem cells as the engine of unstable tumor progression. *J. Theor. Biol.* 253, 629–637.

Sontag, L. and Axelrod, D. E. (2005). Progression of heterogeneous breast tumors. *J. Theor. Biol.* 210, 107–119.

Subramanian, B. and Axelrod, D. E. (2001). Progression of heterogeneous breast tumors, *J. Theor. Biol.* 210, 107–119.

Torres, E. M., Williams, B. R., and Amon, A. (2008). *Genetics* 179, 737–746.

Vogelstein, B. and Kinzler, K.W (2004). Cancer genes and the pathways they control, *Nature Med.* 10, 789–799.

CHAPTER 5

A Stochastic Multiscale Model Framework for Colonic Stem Cell Homeostasis

Larry W. Jean and E. Georg Luebeck

CONTENTS

INTRODUCTION

The human colon is lined with a single layer of epithelial cells that undergoes continuous self-renewal by long-lived tissue stem cells compartmentalized into basic proliferative units (*crypts*), each of which is a finger-like invagination into the lamina propria connective tissue of the colon. Significant progress has been made recently in the molecular identification and characterization of intestinal stem cells [1,2], which are located at the base of the crypts, where they give rise to transit-amplifying cells that are committed to different cell lineages (*goblet cells, enterocytes,* and *enteroendocine cells*). The transit-amplifying cells and their differentiated progenies migrate up the crypts toward the intestinal lumen into which they are shed after apoptosis and detachment from the underlying stroma.

It is generally believed that molecular feedback mechanisms among tissue stem cells, their progeny, and interactions with the underlying stroma control the stable maintenance of the intestinal epithelium. Understanding this control, its potential defects, and how they might affect stem cell kinetics during tumorigenesis would clearly provide important input for the development of biologically based cancer models [3–5]. For example, it has been suggested that the mechanisms that control cell cycle checkpoints, DNA repair, and apoptosis are in some ways optimized to delay the onset of neoplastic progression, although experimental evidence for this hypothesis is still lacking [6]. In this chapter, we step away from the biological details of the problem and take a broader view to address the basic question: how do the mechanisms that contribute to the homeostatic control of tissue stem cells manifest themselves in the integration of the cell-level, crypt-level, and tissue-level dynamics? To answer this question, we introduce a stochastic multiscale model for intestinal tissue homeostasis that spans the cellular and tissue scales. The model incorporates explicitly both stem cell and crypt kinetics, including the process of crypt branching. By assuming that crypt branching results from the budding of a new crypt containing one or more stem cells, we identify constraints imposed on the model by the requirement of homeostasis, that is, the overall balance of crypt branching and death while maintaining a constant mean number of tissue stem cells and a stationary number and size distribution for nonextinct crypts. Mathematical expressions such as the crypt survival and the first passage time to crypt branching are derived and used for simulations of crypt phylogenies that facilitate the validation of the derived constraints. We also explore violations of the constraints and their consequences for unconstrained tissue growth. In short, this model represents an attempt to capture effectively the homeostatic control mechanisms by mathematical constraints formulated in terms of the cell-level and crypt-level kinetics.

Colorectal cancer is associated with a number of successive genomic changes [7–13]. Among the earliest changes are (epi)genetic defects that lead to the abrogation of control mechanisms that free mutant stem cells from the crypt constraints that enforce proper cell turnover, allowing them to accumulate in the tissue [14,15]. Quantifying these constraints in terms of the biological parameters describing the crypt dynamics may therefore help us better understand the consequences of defective control mechanisms and their role in tumor initiation and progression.

MODEL OVERVIEW

Stem Cell Divisions and Single Crypt Dynamics

Although they make up only a small subset of the overall cell population in a crypt, stem cells are primarily responsible for maintaining, repairing, and regenerating the single layer of epithelial cells in the intestine. A crypt may be lost due to stem cell death or may bifurcate to produce new crypts. Although the exact mechanisms that trigger crypt bifurcation are unknown, it is commonly assumed that doubling of the number of stem cells in a crypt in response to spontaneous or induced crypt death in its neighborhood is a likely cause (e.g., see Reference [16]). Here, we idealize this view by assuming that "bud-forming" stem cells give rise to distinctly branching crypts and that their formation results from sporadic (asymmetric) stem cell divisions that generate one daughter cell that forms a branching crypt bud and one daughter cell that remains in the parent crypt.

For the mathematical development of the stochastic framework, we focus on the stem cell population within a crypt and ignore transit-amplifying cells and fully differentiated cells as their role is not essential for our arguments. Within the stem cell compartment of the crypt, a *budding crypt* is a population of stem cells derived from progenitors that (in a prespecified time interval) gave rise to crypt bifurcations. In contrast, the *parent crypt* is the lineage excluding the budding crypts. In general, we assume that a parent crypt having n_s stem cells at time s contains $X(t,s)$ stem cells at time t, $s \leq t$, and evolves according to the four fundamental cell division processes shown in Figure 5.1. Incidentally, this assumption is consistent with Potten and Loeffler's concept of *functionally equivalent stem cells* [17–19], which postulates that each stem cell has the same potential to maintain a crypt. Specifically, a stem cell may divide symmetrically to form two stem cells within the parent crypt at a rate of $\alpha(t,s)$ per cell per unit time, it may die or undergo a symmetric cell division that gives rise to two transit-amplifying cells committed to differentiation with a rate of $\beta(t,s)$ per cell per unit time, or it may undergo an asymmetric cell division with rate $\mu(t,s)$ per cell per unit time to form one stem cell and one transit-amplifying cell within the parent crypt, where $Y(t,s)$ is the number of transit-amplifying cells in the parent crypt at time t.

Furthermore, a stem cell may divide with rate $\rho(t,s)$ per cell per unit time to produce one stem cell within the parent crypt and n_b stem cells within a newly formed crypt branch. The reasoning here is that crypt bifurcation (albeit triggered by a single stem cell) can be followed by a short phase of stem cell multiplication. The effective size of a newly born

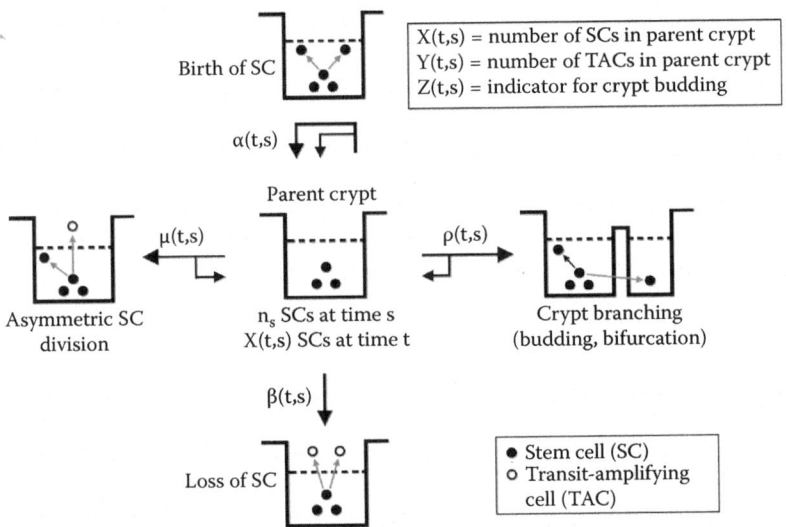

FIGURE 5.1 A multiscale modeling framework for crypt stem cell dynamics. A stem cell undergoes four cell division processes: birth (symmetric division to form two stem cells within the parent crypt), death (apoptosis or symmetric cell division to form two transit-amplifying cells), asymmetric cell division (forms one stem cell and one transit-amplifying cell), and crypt bifurcation (forms $1 + n_b$ stem cells, one in the parent crypt and n_b stem cells initiate a new budding crypt; only $n_b = 1$ is shown). The rates are respectively $\alpha(t, s)$, $\beta(t, s)$, $\mu(t, s)$, and $\rho(t, s)$ per stem cell per unit time.

crypt, $n_b = 1,2,3, ...$, is subsequently also referred to as the *crypt birth size*. Let $Z(t,s)$ be the stochastic indicator variable for a branching by time t from a parent crypt having n_s stem cells at time s:

$$Z(t,s) := \begin{cases} 0, & \text{no crypt branching by time } t \\ 1, & \text{otherwise} \end{cases}. \tag{5.1}$$

Note that both the parent crypt and the budding crypts derived from it are described by the same cell kinetics. Within this framework, a budding crypt may play the role of a parent by giving rise to further budding crypts of its own. We first consider the case where a budding crypt is born with a single stem cell ($n_b = 1$), followed by the general case ($n_b > 1$).

Joint and Conditional Generating Functions of Parent Crypt Sizes

For simplicity, we assume that all cell division parameters are constants, that is, $\alpha(t,s) := \alpha$, $\beta(t,s) := \beta$, $\mu(t,s) := \mu$, and $\rho(t,s) := \rho$. Let $\Psi(x, y, z; t, s, n_s)$ be the joint probability generating function (PGF) of the three processes $X(t, s)$, $Y(t, s)$, and $Z(t, s)$ for a parent crypt having n_s stem cells at time $s \leq t$,

$$\Psi(x,y,z;t,s,n_s) := \sum_{i,j,k} x^i y^j z^k \mathrm{Prob}\{X(t,s)=i, Y(t,s)=j,$$
$$Z(t,s)=k \mid X(s,s)=n_s, Y(s,s)=Z(s,s)=0\} \tag{5.2}$$

Then it satisfies the partial differential equation

$$\partial\Psi(x,y,z;t,s,n_s)/\partial t$$
$$= [\alpha(x^2 - x) + \beta(1-x) + \mu x(y-1) + \rho x(z-1)]\partial\Psi(x,y,z;t,s,n_s)/\partial x, \tag{5.3}$$

with the initial condition $\Psi(x, y, z; s, s, n_s) = x^{n_s}$. A similar expression involving one indicator variable has been applied by Jeon et al. [20]. Since our present goal is mainly to understand the stem cell population dynamics, this formulation makes it convenient to marginalize, and thereby ignore, the asymmetric transit-amplifying cell-generating stem cell division process by setting $y = 1$ in Equation 5.3, which produces the modified PGF

$$\psi(x,z;t,s,n_s):=\Psi(x,y=1,z;t,s,n_s)$$

$$=\sum_{i,k} x^i z^k \text{Prob}\{X(t,s)=i,Z(t,s)=k\,|\,X(s,s)=n_s,Z(s,s)=0\}.$$

$$(5.4)$$

Also, we are interested in the number of stem cells in a parent crypt having n_s stem cells at time $s \leq t$ and that no crypt branching has occurred by time t, whose PGF is obtained by setting $z = 0$ in Equation 5.4, $\psi(x, 0; t, s, n_s) = [\phi(x, 0; t, s)]^{n_s}$, where

$$\phi(x,z;t,s):=\psi(x,z;t,s,n_s=1)$$

$$=\sum_{i,k} x^i z^k \text{Prob}\{X(t,s)=i,Z(t,s)=k\,|\,X(s,s)=1,Z(s,s)=0\}$$

$$(5.5)$$

is the joint PGF for the number of stem cells at time t in a parent crypt having a single stem cell at time $s \leq t$, and crypt branching. $\phi(x, 0; t, s)$ has been shown to satisfy [21,22]

$$\phi(x,0;t,s)=1+\frac{1}{\alpha}\frac{v(w-\eta)e^{-v(t-s)}-w(v-\eta)e^{-w(t-s)}}{((w-\eta)e^{-v(t-s)}-(v-\eta)e^{-w(t-s)})}, \qquad (5.6)$$

where

$$v:=\frac{1}{2}\left[-\alpha+\beta+\rho-\sqrt{(\alpha+\beta+\rho)^2-4\alpha\beta}\,\right], \qquad (5.7)$$

$$w:=\frac{1}{2}\left[-\alpha+\beta+\rho+\sqrt{(\alpha+\beta+\rho)^2-4\alpha\beta}\,\right], \qquad (5.8)$$

$$\eta:=\alpha(x-1). \qquad (5.9)$$

It is also of interest to marginalize the joint PGF of parent crypt size and branching on the parent crypt branching process, which is accomplished by setting $z = 1$ in Equation 5.4. This produces the PGF for the number of stem cells in a parent crypt having n_s stem cells at time $s \leq t$, regardless of whether a budding crypt has formed before time t, $\psi(x, 1; t, s, n_s) = [\phi(x, 1; t, s)]^{n_s}$, where $\phi(x, 1; t, s)$ can be shown to satisfy (see Appendix)

$$\phi(x,1;t,s) = \frac{\beta(x-1)-(\alpha x - \beta)e^{-(\alpha-\beta)(t-s)}}{\alpha(x-1)-(\alpha x - \beta)e^{-(\alpha-\beta)(t-s)}}. \tag{5.10}$$

Equation 5.10 was applied by Luebeck and Moolgavkar [23] to initiation-promotion carcinogenesis models for analyzing the sizes of premalignant lesions. Finally, we are interested in the number of stem cells in a parent crypt having n_s stem cells at time $s \leq t$ *conditioned* on no parent crypt branching by time t, which has the PGF (Appendix)

$$\omega(x;t,s,n_s) := \sum_i x^i \text{Prob}\{X(t,s)=i \mid X(s,s)=n_s, Z(s,s)=Z(t,s)=0\}$$

$$= \left[\frac{\phi(x,0;t,s)}{\phi(1,0;t,s)}\right]^{n_s}. \tag{5.11}$$

Parent Crypt Extinction Time and First Passage Time to Branching

For a parent crypt containing n_s stem cells at time $s \leq t$, let T_e and T_b be the random variables for the time to extinction and the first passage time to branching, respectively. It can be shown (Appendix) that the cumulative density functions (CDFs) for such a crypt satisfy

$$\text{Prob}\{T_e \leq t\}$$
$$= \text{Prob}\{X(t,s)=0 \mid X(s,s)=n_s, Z(s,s)=0\} = [\phi(0,1;t,s)]^{n_s}, \tag{5.12}$$

and

$$\text{Prob}\{T_b \leq t\}$$
$$= \text{Prob}\{Z(t,s)=1 \mid X(s,s)=n_s, Z(s,s)=0\} = 1 - [\phi(1,0;t,s)]^{n_s}, \tag{5.13}$$

where the latter is equivalent to one minus the probability of the crypt having not undergone budding by time t (crypt branching survival).

Similarly, for a parent crypt having n_s stem cells at time $s \leq t$, the crypt extinction time conditioned on no prior occurrences of crypt branching, and the first passage time to crypt branching conditioned on nonextinction, have the respective CDFs

$$\text{Prob}\{T_e \leq t \,|\, T_b > t\}$$

$$= \text{Prob}\{X(t,s) = 0 \,|\, X(s,s) = n_s, Z(s,s) = Z(t,s) = 0\} = \left[\frac{\phi(0,0;t,s)}{\phi(1,0;t,s)}\right]^{n_s} \quad (5.14)$$

and

$$\text{Prob}\{T_b \leq t \,|\, T_e > t\}$$

$$= 1 - \text{Prob}\{Z(t,s) = 0 \,|\, X(t,s) > 0, X(s,s) = n_s, Z(s,s) = 0\}$$

$$= 1 - \frac{\left[\phi(1,0;t,s)\right]^{n_s} - \left[\phi(0,0;t,s)\right]^{n_s}}{1 - \left[\phi(0,1;t,s)\right]^{n_s}}. \quad (5.15)$$

When the cell death rate exceeds the cell birth rate ($\beta > \alpha$), extinction is certain for any parent crypt, regardless of its initial size and the number of prior branchings (Figure 5.2a). Also, depending on the number of stem cells (n_s) in a parent crypt at time s, branching may never occur, since the crypt may suffer extinction before it generates a budding crypt (Figure 5.2b). This occurs more frequently for parent crypts having smaller initial sizes. In general, increasing the initial number of stem cells in a parent crypt shortens its time to first branching and extends its lifetime, since a larger crypt increases the probability of branching and is less susceptible to extinction. Although crypt extinction is certain (assuming $\beta > \alpha$), parent crypts that have not undergone prior branchings are more susceptible to extinction than those that have generated progeny (i.e., given rise to branching crypts) (Figure 5.2c). Again, this can be attributed to the dominance of cell death over cell birth, which drives down the sizes of the crypts, thereby shortening their time to extinction. Finally, conditioned on crypt survival, all parent crypts will eventually generate progeny, regardless of their initial sizes (Figure 5.2d).

As we will show in the next section, Equations 5.14 and 5.15 are especially important for the identification of mathematical constraints on stem cell homeostasis.

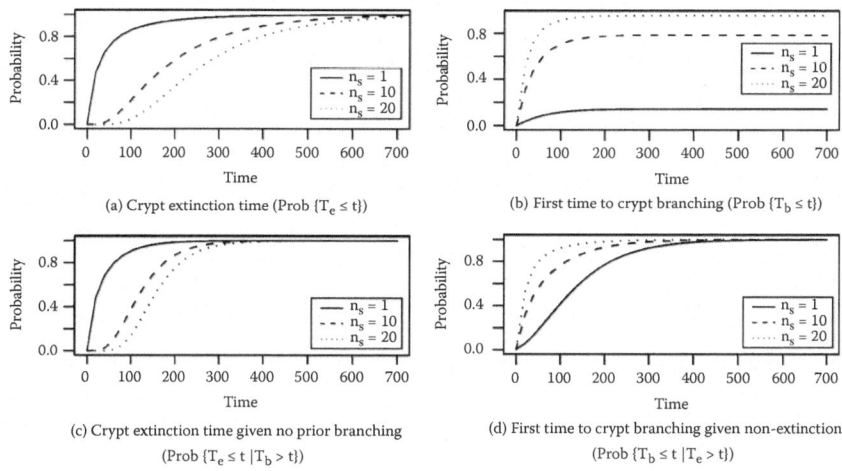

FIGURE 5.2 Cumulative density functions for the time to extinction and the first passage time to branching in a parent crypt having $n_s = 1$, 10, or 20 stem cells at time s, $\alpha = 0.035$, $\beta = 0.042$, $\rho = 0.002$ per stem cell per unit time. (a) Time to crypt extinction. (b) First passage time to crypt branching. (c) Time to crypt extinction given no prior occurrence of crypt branching. (d) First passage time to crypt branching given nonextinction.

TISSUE-LEVEL AND CRYPT-LEVEL CONSTRAINTS FOR HOMEOSTASIS

To derive a constraint for a constant mean number of stem cells in a tissue, regardless of its specific structure, we employ the generalized Luria–Delbrück model introduced by Dewanji, Luebeck, and Moolgavkar [24]. The model was originally developed for quantifying the spontaneity of mutations in bacteria prior to their selection, a phenomenon Luria and Delbrück demonstrated experimentally in their 1943 *Fluctuation Analysis* [25]. In the current context, the generalized Luria–Delbrück model may be applied as follows. Assume that there is a constant number of X stem cells in a tissue, which gives rise to budding crypts according to a Poisson process with rate ρ per stem cell per unit time, and each crypt contains $n_b \geq 1$ stem cells at the time of birth. The stem cells within a budding crypt formed at time $u \leq t$ undergo a birth–death process $\{X(t,u), u \leq t\}$ with $X(u, u) = n_b$ having birth and death rates $\alpha(t, u)$ and $\beta(t, u)$ per stem cell per unit time, respectively. Under constant parameters, that is, $\alpha(t, u) := \alpha$ and $\beta(t, u) := \beta$, it can be shown that the total number of stem cells in the tissue at time t, $X(t)$, satisfies (Appendix)

$$E[X(t)] = \frac{n_b \cdot \rho X}{\beta - \alpha} \left[1 - e^{-(\beta - \alpha)t} \right].$$ (5.16)

Thus, stationarity of the overall stem cell number in the tissue may be obtained by imposing the condition $\lim_{t \to \infty} E[X(t)] = X$ under the assumption $\beta > \alpha$, which produces the tissue-level constraint $\beta = \alpha + n_b \cdot \rho$. The simplicity of this tissue-level constraint is a consequence of the assumed cell-to-cell independence. Although this constraint will guarantee a model tissue with constant mean size (irrespective of the choice of n_b), it does not control its fluctuations over time, nor its possible extinction. Here, we shall not be concerned about this shortcoming as we will consider the tissue to be very large (compared with the crypt stem cell niche) and the overall fluctuations controllable by other means or model extensions that impose additional constraints. For example, the overall stem cell population in a tissue could be controlled effectively by a Prendiville process with reflective boundaries rather than a simple linear birth death process [26,27].

The number of budding crypts in the tissue may be controlled by balancing the loss and the gain of crypts within the tissue. This can be achieved by equating the mean time to crypt extinction conditioned on no prior branching and the mean first passage time to crypt branching conditioned on nonextinction. This is mathematically equivalent to equating the means of these random variables having CDFs given by Equations 5.14 and 5.15. For consistency, we will generally assume that $n_s = n_b$, that is, equality of the arbitrary initial crypt size of the parents at time s with the crypt birth size, unless mentioned otherwise. Upon numerical differentiation of the CDFs and integration for computing the means, we obtain a relationship between α, β, ρ, and n_b defined to be $F_{\text{crypt}}(\alpha, \beta, \rho, n_b) = 0$. For any $n_b \geq 1$, the tissue-level and crypt-level constraints are combined to obtain $F_{\text{crypt}}(\alpha, \alpha + n_b \cdot \rho, \rho, n_b) = 0$, whose solution can be well approximated (via a linear regression) by $\rho = L_b \cdot \alpha$ with a dimensionless constant L_b that renders the composite constraint scale invariant.

The relationships between α and the other parameters (β and ρ) enforced by these constraints are shown in Figure 5.3. For a given α and increasing n_b, ρ decreases and reaches a minimum at $n_b = 4$, while β exhibits a slight increase but remains mostly unchanged up to $n_b = 4$, relative to ρ. This suggests that as budding crypts increase their birth size, they become less vulnerable to extinction, and the tissue responds with a reduction of the branching rate to prevent excessive expansion of the overall tissue. For $n_b > 4$, both the death rate and the crypt branching rate increase, the former

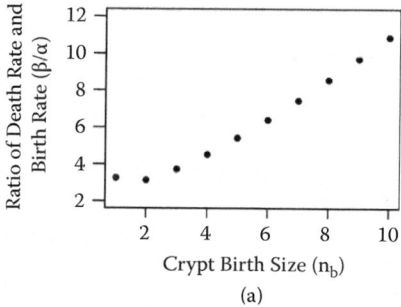
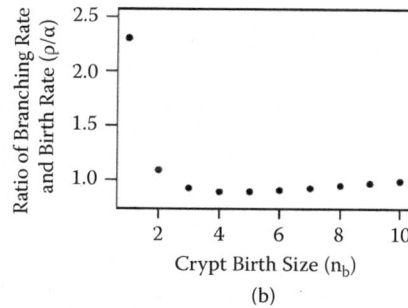

FIGURE 5.3   Crypt birth size versus (a) β/α and (b) ρ/α under the combined crypt-level and tissue-level constraint $Fcrypt(\alpha, a + n_b \cdot \rho, \rho, n_b) = 0$. For each of $n_b = 1,2,3, ..., 10$, a linear regression in the form of $\rho = L_b \cdot \alpha$ is performed on the combined constraint and β/α is obtained via $\beta = \alpha + n_b \cdot \rho = \alpha \cdot (1 + n_b \cdot L_b)$.

TABLE 5.1   Death Rate and Budding Rate Per Stem Cell Per Hour Under the Constraints for $\alpha = 0.009627/(stemcell \cdot h)$ and Crypt Birth Sizes $n_b = 1,2, ..., 10, 15, 20$ Stem Cells

Crypt Birth Size n_b	Death Rate β	Branching Rate ρ
1	0:032685	0:023058
2	0:031468	0:010924
3	0:037398	0:009257
4	0:045290	0:008916
5	0:054257	0:008926
6	0:063972	0:009057
7	0:074416	0:009256
8	0:085498	0:009484
9	0:096912	0:009698
10	0:108763	0:009914
15	0:174840	0:011014
20	0:249341	0:011986

Note: The values are obtained via linear regressions in the form of $\rho = L_b \alpha$ on the combined crypt and tissue level constraint, $F_{crypt}(\alpha, \alpha + n_b \cdot \rho, \rho, n_b) = 0$.

linearly with n_b and the latter approaching a constant close to one, consistent with the observation that β/α has a unit slope as a function of n_b.

In the examples provided here, we assume a symmetric cell cycle time of 72 h (Totafurno et al. [28]). This yields $\alpha = 0.009627/(stemcell \cdot h)$. β and ρ may then be uniquely determined for any n_b 1 by the tissue-level and crypt-level constraints (Table 5.1). To demonstrate that these constraints yield the desired homeostatic (stationary) behavior at the crypt and tissue levels and to obtain tissue-level distributions for the number and sizes of crypts, we simulate crypt phylogenies as described in the following section.

CRYPT PHYLOGENY SIMULATIONS

Define an *event* for a crypt to be either an extinction or a branching. Then, starting with a parent crypt having n_s stem cells at time s, we perform a straightforward Monte Carlo simulation to obtain the occurrence time t_1 of the first event and the corresponding (parental) crypt size, n_{11}, if the event is a branching. If the event is an extinction, there are $c_1 = 0$ nonextinct crypts at time t_1 and further simulation from that branch arrests; otherwise, an additional (budding) branch with $n_{12} = n_b$ stem cells is created, and the sizes of the $c_1 = 2$ nonextinct crypts at t_1, $\{n_{11}, n_{12}\}$, are recorded. For each of $i = 1, 2$, the identical simulation scheme is performed on the ith branch with t_1 and n_{1i} as the initial time and size, respectively, of the crypt. The first event time in each of the two crypts is saved, at which point the crypt lineage loses (via extinction) or gains (via branching) a crypt depending on the nature of the event. The minimum of the two first event times from the two branches, t_2, and the sizes of the c_2 nonextinct crypts at t_2, $\{n_{21}, n_{22}, \ldots, n_{2c_2}\}$, are recorded. The procedure is repeated until the simulation reaches the time of observation, t_K, at which point the sizes of all c_K nonextinct crypts, $\{n_{K1}, n_{K2}, \ldots, n_{Kc_K}\}$, are recorded to yield tissue-level data such as the number and sizes of nonextinct crypts, and the total number of stem cells in the tissue. We use simulated crypt phylogenies to explore in more detail how the mathematical constraints derived earlier affect the tissue at the different levels of organization.

Figure 5.4 shows two illustrations of crypt phylogenies, one starting with a crypt containing a single stem cell (Figure 5.4a), the other with 10 stem cells (Figure 5.4b). Both examples assume $n_b = 1$, $\alpha = 0.009627/$ (*stemcell·h*), $\beta = 0.032685/$(*stemcell·h*), and $\rho = 0.023058/$(*stemcell·h*). In Figure 5.4a, the parent crypt forms six lineages throughout the first 50 h, three of which remain alive at that time. The parent crypt goes extinct after 23.8 h, and the tissue it generates is sustained by the budding crypts, none of which contains more than one stem cell at the time of observation. A tissue generated from a parent crypt staring with 10 stem cells gives rise to six budding crypts over the next 20 h (Figure 5.4b). All seven lineages remain nonextinct at that time, with the largest crypt (the parent) containing six stem cells, while two stem cells occupy the largest budding crypt. In comparison, the tissue with $n_s = 10$ produces more crypts over the same duration than the one with $n_s = 1$; however, we will show that the

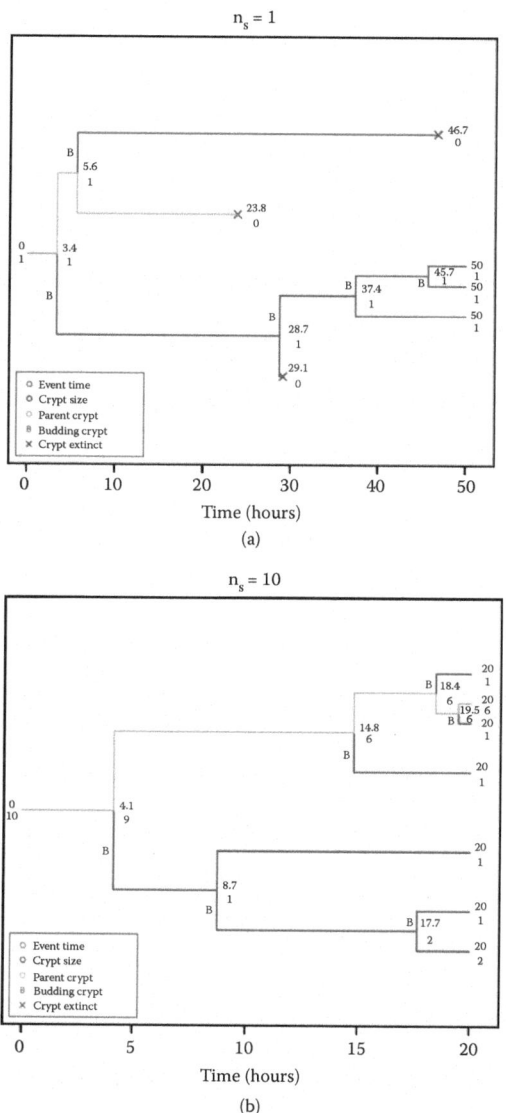

FIGURE 5.4 **(See color insert following page 40)** Phylogenies of tissues generated from a single parent crypt with $\alpha = 0.009627$, $\beta = 0.032685$, and $\rho = 0.023058$
per stem cell per hour. The parent crypt of the sample tissues initially contain
(a) $n_s = 1$ and (b) $n_s = 10$ stem cells, observed after 50 and 20 h, respectively.
For each tissue, the simulation highlights the parent crypt lineage (orange), the
extinct lineages (X), and each budding branch is designated by the green letter
B. The occurrence time of an event (extinction or branching) and the number of
stem cells (blue) at the event time are also shown.

mean numbers of stem cells per crypt converge for the two cases, because they both assume identical values for the crypt birth size n_b.

EXPLORING CRYPT- AND TISSUE-LEVEL BEHAVIOR

Tissues in Homeostasis

First we consider the case where a crypt branch is born with a single stem cell ($n_b = 1$). The crypt-level constraint stabilizes the mean number of nonextinct crypts generated from a parent crypt (Figure 5.5a). For $n_s = 1$ and $n_s = 10$, the stationary mean number of nonextinct crypts are approximately 0.839 and 8.330 per parent crypt, respectively. Stationarity of the overall number of stem cells in the tissue imposed by $\beta = \alpha + n_b \cdot \rho$ and a stationary mean number of crypts in the tissue guarantee that the mean number of stem cells per crypt also reaches stationarity. Figure 5.5b shows that the mean crypt size approaches 1.197 stem cells independent of the choice for the initial crypt size, n_s. We will demonstrate that this independence of the mean number and sizes of crypts on n_s holds also for $n_b > 1$.

The implemented constraints therefore yield model tissues with stationary mean crypt numbers and crypt sizes. To validate the effects of the constraints in tissues whose budding crypts are effectively born with multiple stem cells ($n_b > 1$), we simulate tissues with the same total number of stem cells but with different initial parent crypt sizes. Let N be the number of stem cells in a tissue, then given a parent crypt that initially contains n_s

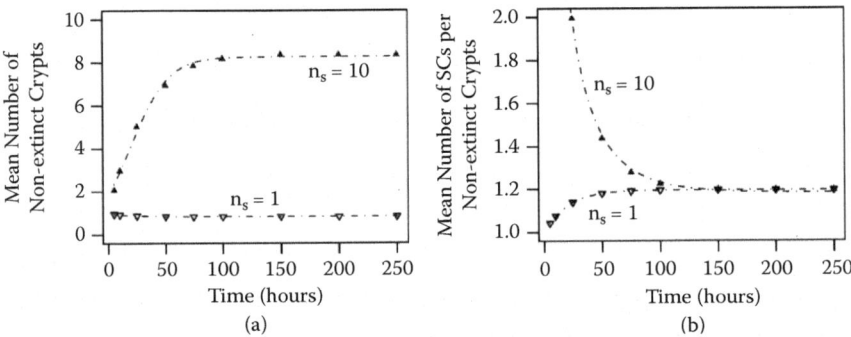

FIGURE 5.5 Mean number and sizes of nonextinct crypts in a tissue generated from a single crypt initially containing $n_s = 1$ or $n_s = 10$ stem cells. (a) Mean number of nonextinct crypts in the tissue. (b) Mean number of stem cells per nonextinct crypt in the tissue. For each case, 1000 samples are generated with rates $\alpha = 0.009627$, $\beta = 0.032685$, $\rho = 0.023058$ per stem cell per hour, and observation times $t = 5, 10, 25, 50, 75, 100, 150, 200,$ and 250 h.

stem cells, this is accomplished by simulating N/n_s such parent crypts per tissue (Figure 5.6). To demonstrate this for $n_s = 2$ and $n_s = 5$, tissues containing 500 stem cells initially (250 and 100 crypts per tissue, respectively) are simulated and analyzed.

Assuming that $n_s = n_b$, the stationary mean number of crypts is approximately 318.880 for $n_b = 2$ and 212.760 for $n_b = 5$ (Figure 5.7a). For $n_s = 10$ on the other hand, the stationary mean numbers for $n_b = 2$ and

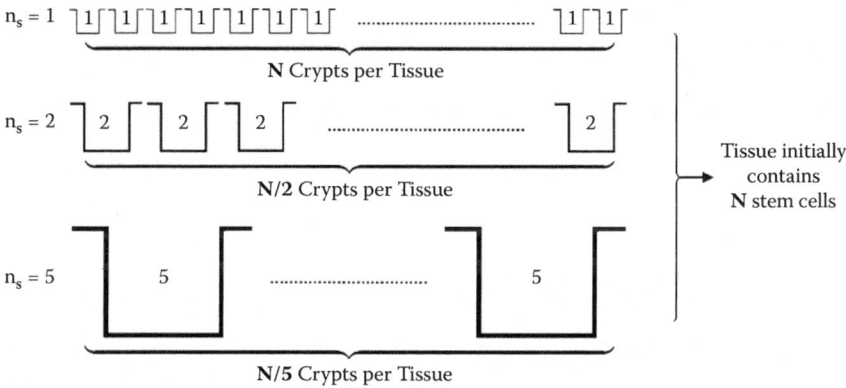

FIGURE 5.6 Constructing tissues having identical total stem cell number but different initial parent crypt sizes. To obtain a tissue containing N stem cells, N/n_s parent crypts each having n_s stem cells initially are simulated.

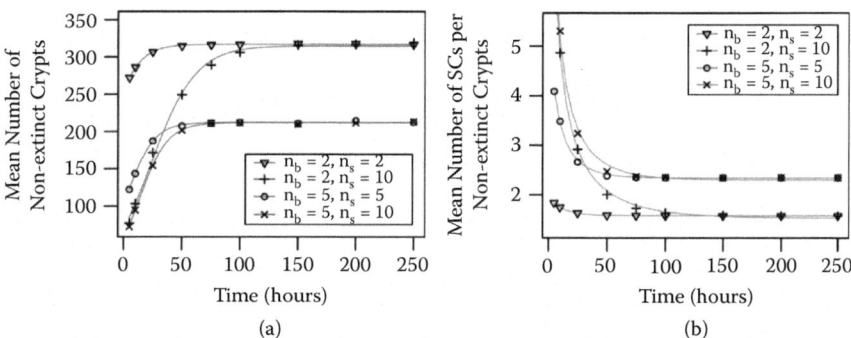

FIGURE 5.7 Mean number and sizes of nonextinct crypts in tissues having crypt birth sizes $nb = 2$ and $nb = 5$ stem cells. (a) Mean number of nonextinct crypts in the tissues. (b) Mean number of stem cells per nonextinct crypt in the tissues. For each case, 1000 samples are generated with rates $\alpha = 0.009627$, $\beta = 0.032685$, and $\rho = 0.023058$ per stem cell per hour, and observation times $t = 5, 10, 25, 50, 75, 100, 150, 200,$ and 250 h.

TABLE 5.2   Stationary Distributions of the Number of Stem Cells in Nonextinct Crypts in a Tissue with $n_b = n_s = 1, 2,$ or 5

$n_b(= n_s)$	Crypt Size (Number of Stem Cells)							
	1	2	3	4	5	6	7	8
1	86.5%	11.6%	1.2%	0.6%	0.1%	0.0%	0.0%	0.0%
2	51.9%	37.4%	8.9%	1.5%	0.3%	0.0%	0.0%	0.0%
5	42.2%	20.2%	14.6%	11.2%	10.3%	1.1%	0.2%	0.2%

Note: In each case, a tissue initially containing 1000 stem cells is simulated with rates $\alpha = 0.009627$, $\alpha = 0.032685$, and $\rho = 0.023058$ per stem cell per hour and observed after $t = 200$ h.

$n_b = 5$ mostly coincide for the first 25 h before splitting off and converging with their respective stationary means. Figure 5.7b indicates that regardless of n_s, $n_b = 2$ generates a tissue having a stationary mean crypt size of 1.56 stem cells, while tissues having $n_b = 5$ are sustained by crypts having 2.35 stem cells on average. Our simulations clearly show that the stationary mean number and size of the crypts in a model tissue of fixed size depend on the crypt birth size parameter n_b, while they are insensitive to the choice of n_s. With the exception of $n_b = 1$, the stationary mean crypt size tends to be smaller than n_b. Furthermore, stationary distributions for crypt sizes within the model tissues show that, regardless of the crypt birth size n_b, crypts containing a single stem cell dominate such tissues (Table 5.2). This may be considered a failure of the model, given that recent experimental evidence suggests that the stem cell number (per crypt) is between four and six in the murine colon [29]. However, the stochastic intercrypt variation of this estimate remains uncertain. In the formulation presented here, only 13.5% of the crypts contain more than one stem cell in a tissue with $n_b = 1$, while such crypts make up 48.1% and 57.8% of tissues with $n_b = 2$ and $n_b = 5$, respectively.

Neoplastic Tissue

Tumor development in crypt-structured tissues, such as the colon or Barrett's esophagus, may be associated with violations of the biological constraints that characterize tissue stem cell homeostasis. Naturally, we expect the number of tissue stem cells to increase over time when the constraints are violated in favor of increased net cell proliferation. This is demonstrated in Figure 5.8, where a tissue with $n_s = n_b = 2$ experiences an increase in the net cell proliferation rate either through a 20% increase in the birth rate α, or a 20% decrease in the death or loss rate β.

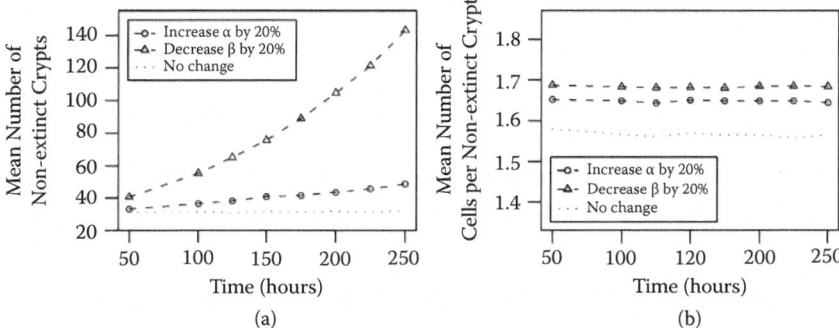

FIGURE 5.8 Mean number and sizes of nonextinct crypts in tissues with violated constraints having crypt birth size of $nb = 2$ stem cells. (a) Mean number of nonextinct crypts in the tissues. (b) Mean number of stem cells per nonextinct crypt in the tissues. For each case, 1000 samples are generated with rates $\alpha = 0.009627$, $\beta = 0.032685$, and $\rho = 0.023058$ per stem cell per hour, and observation times $t = 50$, 100, 125, 150, 175, 200, 225, and 250 h. The constraints are violated by either increasing the birth rate or decreasing the death rate.

In either case, the response is such that the mean number of nonextinct crypts increases over time in a seemingly exponential fashion (Figure 5.8a). However, the case where β is decreased exhibits a much more drastic increase of the crypt number. On the other hand, the mean crypt sizes exhibit only modest increases and interestingly remain mostly constant throughout time for both cases (Figure 5.8b). The former observation can be explained by the relative difference in net cell proliferation, that is, $((1+20\%)\alpha - \beta + \rho)/(\alpha - (1-20\%)\beta + \rho) = (\alpha/\beta) \approx 0.3$, while the latter suggests that crypt sizes are mainly controlled by the crypt branching parameter, ρ, which remains constant in these examples. Thus, our results indicate that during neoplastic progression (and possibly also during fetal development), lesion (or tissue) growth is more likely the result of a downregulation of cell differentiation or apoptosis rather than an upregulation of cell division.

DISCUSSION

Both deterministic [30,31] and stochastic [32,33] models for the dynamics of normal colonic mucosa exist. Most of these models are designed to describe the dynamics at the scale of a single proliferative unit (colonic crypt). A more thorough review is presented by van Leeuwen et al. [34]. Although these crypt-level models provide valuable information at that organizational scale, they offer no insight on the effects of coupling dynamics at the cellular and tissue scales. Here, we introduce a framework that accounts for stem cell dynamics spanning the organizational scales of the stem cells, the proliferative units, and the tissue. Stochastic processes such as cell division, cell death, and crypt branchings are explicitly incorporated into the model framework to describe the relationships among the parameters that yield stable population dynamics at the crypt and tissue levels.

By requiring a constant overall stem cell number in the tissue and balancing the (conditional) times to crypt loss and crypt gain, the model yields constraints sufficient to maintain the mean number of nonextinct crypts, their mean sizes, and thus the mean overall stem cell number in the tissue, although at this level of approximation the fluctuations of the latter remain uncontrolled. This may be remedied by imposing constraints that involve higher moments of the crypt extinction time and the first passage time to crypt branching. We interpret these constraints, which couple the rates of cell division, sporadic or induced cell death, and crypt branching, to capture effectively the complex feedback mechanisms that govern the homeostatic control of stem cells within the normal colonic mucosa.

Our crypt-phylogenic simulations reveal several interesting observations. First, stable numbers of progeny both on the cellular level and on the crypt level are obtained, from which stationarity of the number of nonextinct crypts in the tissue follows. As a function of the crypt birth size (n_b), the constraints decrease the ratio of the crypt branching rate to the symmetric stem cell division rate before approaching one, while they stipulate an increase of the ratio of the death rate to the birth rate with increasing n_b.

Finally, the abrogation of feedback mechanisms that normally maintain the colonic tissue appears to play an important role in colon tumorigenesis. Initiating mutations, that is, mutations that lead to clonal expansions by uncompensated increases in crypt bifurcation and/or uncontrolled increases in crypt size likely violate (at least locally) the constraints identified here. We explore the effect of such violations by varying α and β independently, in one case increasing α by a fraction, in the other decreasing β by the same fraction. Both scenarios effectively increase the net cell proliferation rate, $\alpha + \rho - \beta$. However, because the tissue constraint stipulates $\beta > \alpha$, decreasing the rate of stem cell loss, β, results in a higher net cell proliferation rate compared with increasing α (by the same fraction). Thus, our framework predicts that tissue expansions are accompanied by crypt proliferation in the tissue while the crypt sizes remain relatively unaffected. Alternatively, the spreading of neoplasms via crypt bifurcations may also be driven by direct increases in the crypt branching rate, ρ.

The consequences of such a violation, in comparison to the perturbations in stem cell kinetics so far tested, remain to be explored.

APPENDIX

Derivation of Equation 5.10

Let $g(t, s)$ and $G(t, s)$ be defined by

$$g(t,s):=\exp\left\{-\int_s^t [\alpha(u,s)-\beta(u,s)]du\right\} \tag{5.17}$$

and

$$G(t,s):=\int_s^t \alpha(u,s)g(u,s)du, \tag{5.18}$$

respectively; then for $\alpha(t, s) = \alpha$ and $\beta(t, s) = \beta$, it is easy to see that $g(t, s) = e^{-(\alpha-\beta)(t-s)}$ and $G(t,s)=\frac{\alpha}{\alpha-\beta}\left[1-e^{-(\alpha-\beta)(t-s)}\right]$. According to Dewanji et. al. [24], the probability generating function (PGF) is

$$\phi(x,1;t,s)=1-\frac{x-1}{(x-1)G(t,s)-g(t,s)}, \tag{5.19}$$

which can be shown to satisfy

$$\phi(x,1;t,s)=\sum_i x^i \text{Prob}\{X(t,s)=i\,|\,X(s,s)=1,Z(s,s)=0\}$$

$$=1-\frac{x-1}{(x-1)G(t,s)-g(t,s)}$$

$$=1-\frac{x-1}{(x-1)\dfrac{\alpha}{\alpha-\beta}\left[1-e^{-(\alpha-\beta)(t-s)}\right]-e^{-(\alpha-\beta)(t-s)}}$$

$$=1-\frac{(\alpha-\beta)(x-1)}{\alpha(x-1)-\left[\alpha(x-1)+(\alpha-\beta)\right]e^{-(\alpha-\beta)(t-s)}}$$

$$=\frac{\alpha(x-1)-(\alpha x-\beta)e^{-(\alpha-\beta)(t-s)}-(\alpha-\beta)(x-1)}{\alpha(x-1)-(\alpha x-\beta)e^{-(\alpha-\beta)(t-s)}}$$

$$=\frac{\beta(x-1)-(\alpha x-\beta)e^{-(\alpha-\beta)(t-s)}}{\alpha(x-1)-(\alpha x-\beta)e^{-(\alpha-\beta)(t-s)}}. \tag{5.20}$$

Derivation of Equation 5.11

$$\omega(x;t,s,n_s) = \sum_i x^i \text{Prob}\{X(t,s)=i \mid X(s,s)=n_s, Z(s,s)=Z(t,s)=0\}$$

$$= \frac{\sum_i x^i \text{Prob}\{X(t,s)=i, Z(t,s)=0 \mid X(s,s)=n_s, Z(s,s)=0\}}{\text{Prob}\{Z(t,s)=0 \mid X(s,s)=n_s, Z(s,s)=0\}} \quad (5.21)$$

$$= \psi(x,0;t,s,n_s) / \psi(1,0;t,s,n_s) = \left[\phi(x,0;t,s) / \phi(1,0;t,s)\right]^{n_s}.$$

Derivation of Equation 5.14

$$\text{Prob}\{T_e \le t \mid T_b > t\}$$

$$= \text{Prob}\{X(t,s)=0 \mid X(s,s)=n_s, Z(s,s)=Z(t,s)=0\}$$

$$= \frac{\text{Prob}\{X(t,s)=0, Z(t,s)=0 \mid X(s,s)=n_s, Z(s,s)=0\}}{\text{Prob}\{Z(t,s)=0 \mid X(s,s)=n_s, Z(s,s)=0\}} \quad (5.22)$$

$$= \psi(0,0;t,s,n_s) / \psi(1,0;t,s,n_s) = \left[\phi(0,0;t,s) / \phi(1,0;t,s)\right]^{n_s}.$$

Derivation of Equation 5.15

$$\text{Prob}\{T_b \le t \mid T_e > t\}$$

$$= 1 - \text{Prob}\{Z(t,s)=0 \mid X(t,s)>0, X(s,s)=n_s, Z(s,s)=0\}$$

$$= 1 - \text{Prob}\{Z(t,s)=0, X(t,s)>0 \mid X(s,s)=n_s, Z(s,s)=0\} /$$

$$\quad \text{Prob}\{X(t,s)>0 \mid X(s,s)=n_s, Z(s,s)=0\}$$

$$= 1 - \left[\text{Prob}\{Z(t,s)=0 \mid X(s,s)=n_s, Z(s,s)=0\} /\right.$$

$$\quad \left(1 - \text{Prob}\{X(t,s)=0 \mid X(s,s)=n_s, Z(s,s)=0\}\right) \quad (5.23)$$

$$\quad - \text{Prob}\{Z(t,s)=0, X(t,s)=0 \mid X(s,s)=n_s, Z(s,s)=0\} /$$

$$\quad \left.\left(1 - \text{Prob}\{X(t,s)=0 \mid X(s,s)=n_s, Z(s,s)=0\}\right)\right]$$

$$= 1 - \left(\psi(1,0;t,s,n_s) - \psi(0,0;t,s,n_s)\right) / \left(1 - \psi(0,1;t,s,n_s)\right)$$

$$= 1 - \frac{\left[\phi(1,0;t,s)\right]^{n_s} - \left[\phi(0,0;t,s)\right]^{n_s}}{1 - \left[\phi(0,1;t,s)\right]^{n_s}}.$$

Derivation of Equation 5.16

Let $\theta(z; t, u, n_b)$ be the PGF for the number of stem cells at time t in a budding crypt that is born with n_b stem cells at time $u \le t$. It is easy to see that $\theta(z; t, u, n_b = 1) \equiv \phi(z, 1; t, u) = 1 - (z-1)/((z-1)G(t,u) - g(t,u))$ and satisfies

$$\partial\theta(z; t, u, 1)/\partial z \big|_{z=1} = \frac{-(z-1)G(t,u) - (z-1)}{\{(z-1)[G(t,u) - g(t,u)]\}^2} \tag{5.24}$$

$$= 1/g(t,u) = e^{-(\beta-\alpha)(t-u)},$$

for $g(t, u)$ and $G(t, u)$ given by Equations 5.17 and 5.18, respectively, under the constant-parameters assumption; that is, $\alpha(t, u) = \alpha$, $\beta(t, u) = \beta$. Now, let $\Theta(z; t)$ be the PGF for the overall stem cell number in a tissue having a stationary number of X stem cells, where any budding crypt is born with $n_b \ge 1$ stem cells. Then, according to Parzen [35], it satisfies

$$\Theta(z; t) = \exp\left\{\int_0^t \rho X[\theta(z; t, u, n_b) - 1] du\right\}, \tag{5.25}$$

where $\theta(z; t, u, u_b) = \theta(z; t, u, 1)^{nb}$. From this and the trivial fact that $\theta(1; t, u, 1) \equiv \Theta(1; t) = 1$, the mean of $X(t)$ can be derived as

$$E[X(t)] = \partial\Theta(z; t)/\partial z \big|_{z=1}$$

$$= \Theta(1; t) \cdot \int_0^t \rho X \cdot \left\{\frac{\partial}{\partial z}[\theta(z; t, u, 1)^{n_b} - 1]\right\}\Big|_{z=1} du$$

$$= n_b \cdot \rho X \int_0^t \theta(1; t, u, 1)^{n_b - 1} \cdot \{\partial\theta(z; t, u, 1)/\partial z\}\big|_{z=1} du \tag{5.26}$$

$$= n_b \cdot \rho X \int_0^t e^{-(\beta-\alpha)(t-u)} du$$

$$= \frac{n_b \cdot \rho X}{\beta - \alpha}[1 - e^{-(\beta-\alpha)t}].$$

REFERENCES

1. N. Barker, J.H. van Es, J. Kuipers et al., Identification of stem cells in small intestine and colon by marker gene Lgr5, *Nature*, 119 (2007), pp. 1003–1008.
2. N. Barker, H. Clevers, Tracking down the stem cells of the intestine: strategies to identify adult stem cells, *Gastroenterology*, 113 (2007), pp. 1755–1760.

3. T. Reya, S.J. Morrison, M.F. Clarke, I.L. Weissman, Stem cells, cancer, and cancer stem cells, *Nature*, 414 (2001), pp. 105–111.

4. M.S. Wicha, S. Liu, G. Dontu, Cancer stem cells: an old idea—a paradigm shift, *Cancer Res.*, 66, (2006), pp. 1883–1890.

5. C.T. Jordan, M.L. Guzman, M. Noble, Cancer stem cells, *N. Engl. J. Med.*, 335 (2006), pp. 1253–1261.

6. J. Cairns, Mutation selection and the natural history of cancer, *Nature*, 255[5505] (1975), pp. 197–200.

7. E.R. Fearon, B. Volgelstein, A genetic model for colorectal tumorigenesis, *Cell*, 61[5] (1990), pp. 759–767.

8. L.A. Aaltonen, P. Peltomäki, F.S. Leach et al., Clues to the pathogenesis of familial colorectal cancer, *Science*, 260[5109] (1993), pp. 812–816.

9. P. Peltomäki, R.A. Lothe, L.A. Aaltonen et al., Microsatellite instability is associated with tumors that characterize the hereditary non-polyposis colorectal carcinoma syndrome, *Cancer Res.*, 53[24] (1993), pp. 5853–5855.

10. K.W. Kinzler, B. Vogelstein, Cancer-susceptibility genes. gatekeepers and caretakers, *Nature*, 386[6627] (1997), pp. 761–763.

11. D. Hanahan, R.A. Weinberg, The hallmarks of cancer, *Cell*, 100[1] (2000), pp. 57–70.

12. L.D. Wood, D.W. Parsons, S. Jones, J. Lin et al., The genomic landscapes of human breast and colorectal cancers, *Science*, 318[5853] (2007), pp. 1108–1113.

13. S. Jones, W.D. Chen, G. Parmigiani, F. Diehl, N. Beerenwinkel et al., Comparative lesion sequencing provides insights into tumor evolution, *Proc. Natl. Acad. Sci. USA*, 105[11] (2008), p. 4283.

14. N. Barker, R.A. Ridgway, J.H. van Es, M. van de Wetering, H. Begthel, M. van den Born, E. Danenberg, A.R. Clarke, O.J. Sansom, H. Clevers, Crypt stem cells as the cells-of-origin of intestinal cancer, *Nature*, 457[7229] (2009), pp. 608–611.

15. R.A. Phelps, S. Chidester, S. Dehghanizadeh, J. Phelps, I.T. Sandoval et al., A two-step model for colon adenoma initiation and progression caused by APC loss, *Cell*, 137[4] (2009), pp. 623–634.

16. D.L. Chao, J.T. Eck, D.E. Brash, C.C. Maley, E.G. Luebeck, Preneoplastic lesion growth driven by the death of adjacent normal stem cells, *Proc. Natl. Acad. Sci. USA*, 105[39] (2008), pp. 15034–15039.

17. M. Loeffler, R. Stein, H.E. Wichmann, C.S. Potten, P. Kaur, S. Chwalinski, Intestinal crypt proliferation. I. A comprehensive model of steady-state proliferation in the crypt, *Cell Tissue Kinet.*, 19 (1986), p. 627.

18. C.S. Potten, M. Loeffler, A comprehensive model of the crypts of the small intestine of the mouse provides insights into the mechanisms of cell migration and the proliferation hierarchy, *J. Theor. Biol.*, 127 (1987), pp. 381–391.

19. M. Loeffler, B. Grossmann, A stochastic branching model with formation of subunits applied to the growth of intestinal crypts, *J. Theor. Biol.*, 150 (1991), pp.175–191.

20. J. Jeon, R. Meza, S.H. Moolgavkar, E.G. Luebeck, Evaluation of screening strategies for pre-malignant lesions using a biomathematical approach, *Math. Biosci.*, 213 (2008), pp. 56–70.
21. W.F. Heidenreich, E.G. Luebeck, S.H. Moolgavkar, Some properties of the hazard function of the two-mutation clonal expansion model, *Risk Anal.*, 17[3] (1997), p. 391.
22. W.F. Heidenreich, Heterogeneity of cancer risk due to stochastic effects, *Risk Anal.*, 25[6] (2005), p. 1589.
23. E.G. Luebeck, S.H. Moolgavkar, Stochastic analysis of intermediate lesions in carcinogenesis experiments, *Risk Anal.*, 11[1] (1991), pp. 149–157.
24. A. Dewanji, E.G. Luebeck, S.H. Moolgavkar, A generalized Luria-Delbrück model, *Math. Biosci.*, 197 (2005), pp. 140–152.
25. S.E. Luria, M. Delbrück, Mutations of bacteria from virus sensitivity to virus resistance, *Genetics*, 28 (1943), p. 491.
26. B.J. Prendiville, Discussion: symposium on stochastic processes, *J. Roy. Statist. Soci.* B, 11 (1949), p. 273.
27. Q. Zheng, Note on the non-homogeneous Prendiville process, *Math. Biosci.*, 148 (1998), p. 1–5.
28. J. Totafurno, M. Bjerknes, J. Cheng, Crypt and villus production in the adult intestinal epithelium, *Biophys. J.*, 52 (1987), pp. 279–294.
29. N. Barker, The canonical Wnt/beta-catenin signalling pathway, *Methods Mol. Biol.*, 468 (2008), pp. 5–15.
30. C.S. Potten, C. Booth, D. Hargreaves, The small intestine as a model for evaluating adult tissue stem cell drug targets, *Cell Prolif.*, 36 (2003), p. 115.
31. F.A. Meineke, C.S. Potten, M. Loeffler, Cell migration and organization in the intestinal crypt using a lattice-free model, *Cell Prolif.*, 34 (2001), p. 253.
32. M. Loeffler, T. Bratke, U. Paulus, Y.Q. Li, C.S. Potten, Clonality and life cycles of intestinal crypts explained by a state dependent stochastic model of epithelial stem cell organization, *J. Theor. Biol.*, 186 (1997), pp. 41–54.
33. M. Loeffler, A. Birke, D. Winton, C.S. Potten, Somatic mutation, monoclonality and stochastic models of stem cell organization in the intestinal crypt, *J. Theor. Biol.*, 160 (1993), p. 471.
34. I.M.M. van Leeuwen, H.M. Byrne, O.E. Jensen, J.R. King, Crypt dynamics and colorectal cancer: advances in mathematical modelling, *Cell Prolif.*, 39 (2006), pp. 157–181.
35. E. Parzen, Stochastic processes, *Holden-Day Series in Probability and Statistics*, Holden-Day Inc., San Francisco, CA (1962).

CHAPTER 6

Multiscale Modeling of Colonic Crypts and Early Colorectal Cancer

Alexander G. Fletcher, Gary R. Mirams,
Philip J. Murray, Alex Walter,
Jun-Won Kang, Kwang-Hyun Cho,
Philip K. Maini, and Helen M. Byrne

CONTENTS

INTRODUCTION

Colorectal cancer accounts for 13% of all cancers in the United Kingdom, with around 35,300 new diagnoses and 16,000 deaths occurring each year (http://info.cancerresearchuk.org). Colorectal cancer is predominantly a disease associated with old age, with 80% of diagnoses being made in patients over the age of 60. As a result of longer life expectancy and declining fertility rates, the proportion of people in this age group is growing faster than any other. In the future, colorectal cancer is therefore sure to rise in prevalence (http://www.who.int/topics/ageing/en).

Colorectal cancers originate from the epithelium that covers the luminal surface of the intestinal tract. This epithelium renews itself more rapidly than any other tissue, being completely replaced every 2–3 days in mice [1] and 5–6 days in humans [2]. The renewal process requires a coordinated program of cell proliferation, migration, and differentiation, which begins in the crypts of Lieberkühn that descend from the epithelium into the underlying connective tissue (see Figure 6.1). At the base of each crypt, a small number of stem cells proliferate continuously, producing transit amplifying cells, which migrate up the crypt axis and divide several times before differentiating into the various cell types that constitute the epithelium (enterocytes, goblet cells, and enteroendocrine cells). Upon reaching the crypt orifice, cells undergo apoptosis and are shed into the lumen.

Under normal conditions, the foregoing cellular processes are tightly regulated by biochemical and biomechanical signals. It is believed that

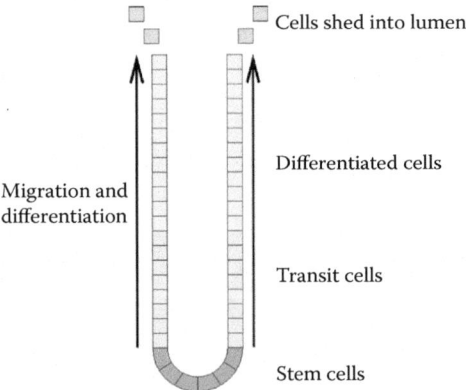

FIGURE 6.1 Schematic of a colonic crypt. Stem cells at the crypt base proliferate continuously, producing transit amplifying cells that migrate up the crypt and differentiate. Cells at the top of the crypt undergo apoptosis and are shed into the lumen.

the first stage of colorectal cancer is caused by the accumulation of genetic alterations that disrupt normal crypt dynamics and cause cells to increase their net proliferation rates. The associated proliferative excess generates biomechanical stress within the crypt, which may deform in order to accommodate the additional cells. The dysplastic cell population may expand further by invading neighboring crypts and/or inducing crypt fission, leading to the formation of an adenoma. Identifying the mechanisms that govern the cellular dynamics of normal crypts is therefore fundamental to understanding the origins of colorectal cancer.

The Wnt pathway is known to play a key role in stem cell maintenance [3,4], cell–cell adhesion [5], cell-fate specification (cell differentiation) [6], central nervous system patterning [7], and tissue development [8,9]. Wnt is an extracellular factor that, when detected by receptors on the outer cell membrane, triggers a cascade of events, culminating in upregulation of intracellular β-catenin levels [10]. A cell's response to Wnt signaling is believed to be mediated predominantly through the concentration and subcellular localization of β-catenin [11]. At the base of the crypt, high levels of Wnt are believed to encourage "stemness" (lack of differentiation), proliferation, and high cell–cell adhesion. By contrast, the low-Wnt environment at the top of the crypt stimulates cells to stop proliferating, differentiate, and weaken their bonds of cell–cell adhesion, preparing them for apoptosis and sloughing into the lumen at the top of the crypt [12].

Most cancers can be initiated by a wide number of different mutations, but almost all colorectal cancers carry activating mutations in a single pathway, the Wnt pathway, with over 80% carrying a double truncation mutation in the gene that encodes the protein APC [13,14]. Thus, the Wnt pathway plays a crucial role in the initiation of colorectal cancer.

As in many cases in biology, colorectal cancer emerges from the interaction of processes that span many different spatial scales. At the genetic level, mutations occur that cause intracellular processes to respond inappropriately to homeostatic cues. This, in turn, affects behavior at the tissue level due to abnormal apoptotic and mitotic responses. Multiscale mathematical modeling can provide insight into how such a complex, highly regulated system operates, both normally and pathologically. A multiscale model cannot account for everything, and in order for a model to be computationally tractable, we must simplify processes at each level. For example, we can exploit different timescales, or use Boolean approaches to simplify the biochemical/metabolic pathways that operate within individual cells. At the tissue level, we need to consider different ways of modeling a collection of cells, ranging from individual cell-based models right through to the continuum limit. When constructing a multiscale model, which simplifications are appropriate and how processes at each level should be combined remain open questions.

In this chapter, we illustrate the challenges inherent in multiscale modeling by taking colorectal cancer as an example. In the next section, we describe a multiscale model that incorporates simple subcellular models of the Wnt signaling pathway and the cell cycle into a discrete, mechanical model of cell movement in a colonic crypt. This model has been used to investigate several aspects of crypt behavior and to explore different ways of coupling these effects within a fully integrated tissue-level model. The results of these investigations are discussed in the following section. We then conclude with a discussion of alternative modeling approaches and avenues for further work.

STRUCTURE OF THE MULTISCALE MODEL

Mathematical modeling of Wnt regulation of cell activity within intestinal crypts presents a formidable challenge as the Wnt pathway plays an important role in determining a range of cell-level behaviors (e.g., adhesion, proliferation, cell–cell interaction) via mechanisms that are not yet fully understood. In order to investigate how mutations in the Wnt pathway affect crypt dynamics, we therefore require a multiscale framework that takes into account these cell-level behaviors. We now describe a multiscale model in which simple subcellular models of the Wnt signaling pathway and the cell cycle are embedded within a discrete, mechanical model of cell movement.

Wnt Signaling Model

Various mathematical models of Wnt signaling have been proposed. Lee et al. (2003) [15] model the Wnt pathway by a system of nonlinear ordinary differential equations (ODEs), which describe the evolution through time of key cytoplasmic protein concentrations, including β-catenin. This model is analyzed by Mirams et al. (2009) [16], who exploit the different timescales involved to reduce the system to a single ODE, which determines how β-catenin evolves in response to a Wnt stimulus. In addition to providing biological insight into the roles of different proteins on different timescales, this type of systematic model reduction is extremely useful in order to achieve tractable computation times for multiscale simulations.

The localization of subcellular β-catenin has been modeled, also as a system of nonlinear ODEs in [17]. This model is used to examine various hypotheses about underlying biochemical mechanisms; for example, whether β-catenin undergoes a conformational change that favors its involvement in cell–cell adhesion rather than transcription, or whether its fate is determined solely by competition for binding partners.

Wnt-Dependent Cell-Cycle Model

The cell cycle is the orderly sequence of events in which a cell duplicates its contents before dividing into two cells. Since cancer is a disease associated with uncontrolled cell proliferation, the cell cycle constitutes a major target for anti-cancer drug development. This has stimulated extensive experimental research and the formulation of detailed mathematical models designed to enhance understanding of the regulatory networks involved and to explore potential therapeutic interventions. Such models are typically formulated as systems of coupled nonlinear ODEs that characterize changes in the levels of key cell-cycle proteins [18].

We employ the model for the Wnt pathway developed by van Leeuwen et al. (2007) [17] to calculate the associated position-dependent levels of gene expression and use these to link the outcome of the Wnt model to the cell-cycle model developed by Swat et al. (2004) [18], as shown in Figure 6.2. As a result, near the bottom of the crypt, where cells are exposed to high levels of Wnt, the production of Wnt-dependent cell-cycle control proteins is enhanced and cells progress through the cell cycle. In contrast, near the crypt orifice where Wnt levels are low, little or no cell division takes place. Full details of the subcellular models of Wnt signaling and the cell cycle are given in [19].

Mechanical Model

A variety of discrete model frameworks can be used to describe the mechanical behavior of tissue, ranging from lattice-based models, cell-center ("point mass") models, and vertex-based ("non-point-mass") models [20]. We use a tessellation-based, cell-center approach, in which the centers of adjacent cells are connected by linear springs [21] and a Delaunay triangulation is performed at each time step, in order to determine cell–cell connectivity.

Following [21], we determine cell movement by balancing the forces exerted on an individual cell by its neighbors with a drag force.

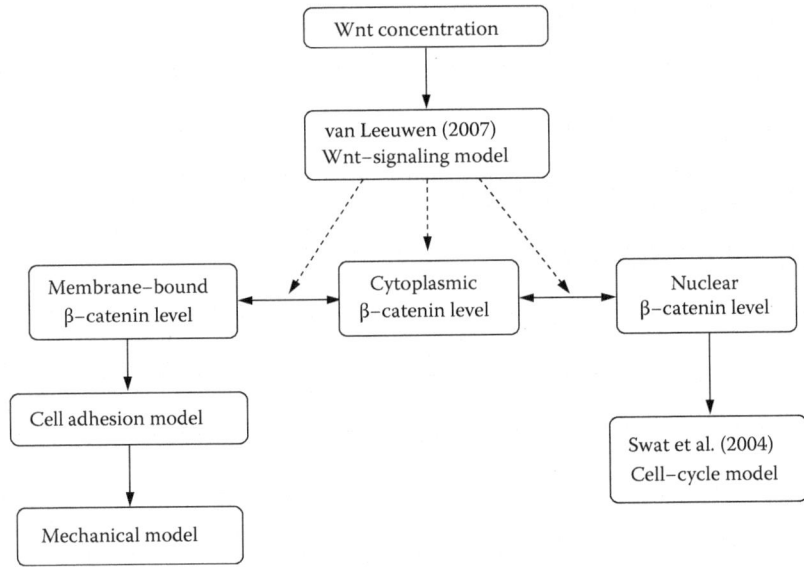

FIGURE 6.2   Influences of the Wnt-signaling model inside a single cell. Note that the Wnt concentration that is experienced depends on the position of the cell within the crypt. The cell-adhesion model influences the motion of the cell, and the cell-cycle model influences the proliferation (and hence again the dynamics) of the cell; thus, the output influences the cell position and changes the input to the Wnt-signaling model. Each cell in a multiscale simulation carries its own Wnt-signaling model.

Specifically, let $\mathbf{r}_i$ be the position of the center of cell i, and define $\mathbf{r}_{ij} = \mathbf{r}_j - \mathbf{r}_i$, and $\hat{\mathbf{r}}_{ij} = \mathbf{r}_{ij}/|\mathbf{r}_{ij}|$. The force exerted on cell i by an adjacent cell j is defined to be

$$\mathbf{F}_{ij} = \mu_{ij}\,\hat{\mathbf{r}}_{ij}(|\mathbf{r}_{ij}| - s_{ij}), \tag{6.1}$$

where μ_{ij} is the spring constant and s_{ij} is the prescribed rest length between cells i and j (i.e., the distance between them for which the force of interaction vanishes). In order to investigate the effect of variable cell–cell adhesion, in the section titled, "Variable Cell–Cell and Cell–Matrix Adhesion" we will consider three choices for the spring constant μ_{ij}. In the first case, $\mu_{ij} \equiv \mu$ takes the same constant value for all neighboring cells i, j. In the second case, to avoid an unrealistically strong attraction between distant neighboring cells, we suppose that μ_{ij} increases with the cell–cell contact length. In this case we take

$$\mu_{ij}(t) = \mu e_{ij}(t)\sqrt{3} / L \tag{6.2}$$

where $e_{ij}(t)$ is the length of the edge between cells i, j and L is the distance between neighboring cell centers in an equilibrium, hexagonal lattice (in such a regular lattice, $e_{ij} \equiv L/\sqrt{3}$, so the first case is recovered). In the third case, we assume that the spring constant depends on the concentration of β-catenin–E-cadherin complexes on the cell membrane, these being determined from the Wnt signaling model (see section titled "Wnt Signaling Model"). In particular following [17], we use the following expression to determine the spring constant connecting cells i and j:

$$\mu_{ij}(t) = \mu e_{ij}(t)\min\left\{ B_i(t)C_{Ai}(t) / E_i(t), B_j(t)C_{Aj}(t) / E_j(t) \right\} / Q_A. \tag{6.3}$$

Here, C_{Ai} denotes the Wnt-dependent concentration of adhesion complexes on the surface of cell i; E_i and B_i denote its perimeter and surface area, respectively; and Q_A is a scaling factor that ensures that under equilibrium conditions, the first case is recovered (for details see [19]).

The total force exerted on cell i by its neighboring cells is

$$\mathbf{F}_i = \sum_j \mathbf{F}_{ij}, \tag{6.4}$$

where the sum is over all cells j that are connected to cell i. An overdamped limit is assumed, for which inertial effects are negligible compared to dissipative terms, so that the equation of motion of cell i is

$$\nu_i \frac{d\mathbf{r}_i}{dt} = \mathbf{F}_i, \tag{6.5}$$

where ν_i is the drag coefficient of cell i. In order to investigate the effect of variable cell–substrate adhesion, in the section titled "Variable Cell–Cell and Cell–Matrix Adhesion" we will consider two different cases for the drag coefficient. In the first case, $\nu_i \equiv \nu$ takes the same constant value for all cells i. In the second case, we suppose that the drag coefficient is proportional to the surface area of contact between a cell and the underlying basement membrane, since a larger cell has more focal adhesions. In this case, we prescribe

$$v_i(t) = (d_0 + d_1 B_i(t))v, \qquad (6.6)$$

where the parameters d_0, d_1 satisfy $d_1 = 2(1-d_0)/(\sqrt{3}L^2)$ so that for an equilibrium, hexagonal lattice we recover the first case.

The equation of motion is discretized numerically using a forward Euler approach, from which it is straightforward to deduce that the position of the cell at time $t + \Delta t$ is related to its position at time t via

$$\mathbf{r}_i(t+\Delta t) = \mathbf{r}_i(t) + \frac{\Delta t}{v_i}\mathbf{F}_i(t). \qquad (6.7)$$

The rest length s_{ij} between cells is assumed to be the typical diameter of a crypt cell. When a cell divides, as determined by its internal cell-cycle model, a new cell is placed at a smaller fixed distance in a random direction. The rest length s_{ij} between the two daughter cells increases linearly over the course of an hour to the mature cell rest length (to emulate the mitosis phase of the cell cycle). Thus, the nuclear β-catenin influences the cell-cycle model (and so indirectly the mechanics as extra cells are added), and membrane-bound β-catenin influences the mechanical model. Intracellular β-catenin is influenced by cell position due to the imposed Wnt gradient along the crypt axis, which feeds back and influences the cell cycle and mechanical models.

Methodology and Implementation Using Chaste

For simplicity we focus on an individual crypt, treating the three-dimensional tubular crypt as a monolayer of cells lying on a cylindrical surface. We take a discrete approach, modeling each cell individually. For simulation purposes, it is convenient to roll the crypt out onto a flat planar domain and impose periodic boundary conditions on the left and right sides. The structure of the multiscale model is depicted in Figure 6.3. It comprises the three interlinked modules discussed earlier: a model of the Wnt signaling pathway [17]; a model of the cell cycle [18], which together with the Wnt model determines each cell's proliferative behavior; and a mechanical model of cell movement [21].

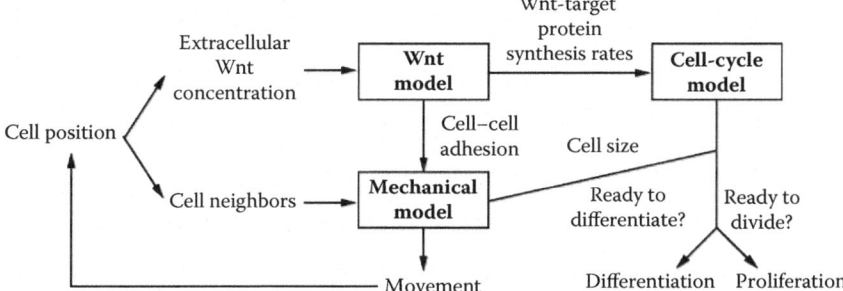

FIGURE 6.3   Diagram illustrating the modular nature of our multiscale crypt model. The occurrence of cellular events (proliferation, differentiation, migration) is monitored at discrete time steps t_n. By coupling Wnt signaling, cell cycle, and mechanical models, we are able to predict the spatiotemporal behavior of every cell at time t_{n+1}, given the state of the system (e.g., intracellular protein levels, cell position, Wnt stimulus, location of neighboring cells) at time t_n and the model parameters.

Chaste (Cancer, Heart and Soft Tissue Environment) is a collaborative software development project that is designed to act as a high-quality multi purpose library supporting computational simulations for a wide range of biological problems. In this context, "high-quality" means that the software is extensible, robust, fast, accurate, and maintainable and uses state-of-the-art numerical techniques. It is also open-source, and so can be adapted by other developers. Chaste has been developed by a multidisciplinary team including mathematicians and software engineers. This ensures that the code is well structured as a piece of software, while at the same time practical and useful as a computational modeling tool. While it is a generic extensible library, to date attention has focused on the fields of cardiac electrophysiology and tumor growth [22].

Chaste is written using an agile method adapted from a technique known as "eXtreme Programming" [23]. This programming methodology is characterized by test-driven development, in which a test is written to cover any new functionality in the code before it is implemented [24]. This enables developers rapidly to discover, diagnose, and fix bugs in the code. The main Chaste code has been written in object-oriented C++, which leads naturally to more modular code: software that is easier to abstract, to modify, and to document. This is especially advantageous for multiscale models of the type considered in this chapter, as it allows different simulations to be generated in a straightforward manner, by using the appropriate components, and preventing unnecessary repetition of code. Further details on Chaste, including visualization movies and user support, are available at http://web.comlab.ox.ac.uk/chaste/.

RESULTS

The multiscale model described earlier has been used to study several aspects of normal crypt behavior and to investigate coupling of processes occurring across a number of spatial scales. We now summarize our results to date.

Wnt Signaling in the Crypt

It has been postulated that a Wnt gradient exists in the crypt, stimulating proliferation at the base and promoting differentiation toward the top. Use of the multiscale model in [19] led us to predict that a Wnt gradient along the entire crypt axis is not necessary to provide a β-catenin (and hence proliferation) gradient. Indeed, Wnt expression in a neighborhood of (approximately) the three cells at the base of the crypt is sufficient to establish a proliferation pattern that extends throughout the crypt; this is because cells move up the crypt more quickly than their Wnt signaling pathways can adapt to the reduction in the local Wnt stimulus. These results are illustrated in Figure 6.4, where the height at which a cell divides, and the corresponding cell-cycle duration are recorded in a scatter graph, for a crypt containing stationary cells and another containing cells that move.

Van Leeuwen et al. (2009) [19] perform simulations of the multiscale model in order to compare the distribution of β-catenin inside each cell in the crypt, under the two hypotheses stated earlier (the simpler hypothesis states that β-catenin fate is determined by competition for binding partners, whereas the second hypothesis proposes that β-catenin can undergo a conformational change that favors binding to E-cadherin at the cell membrane). The results of such simulations are shown in Figure 6.5. The different patterns of β-catenin associated with each hypothesis suggest that it should be possible to discriminate between them by measuring the distribution of β-catenin within the epithelial cells that line a crypt.

Mitotic Labeling

Mitotic labeling experiments are often used to characterize the proliferation and cellular dynamics of intestinal crypts (e.g., [25]). These

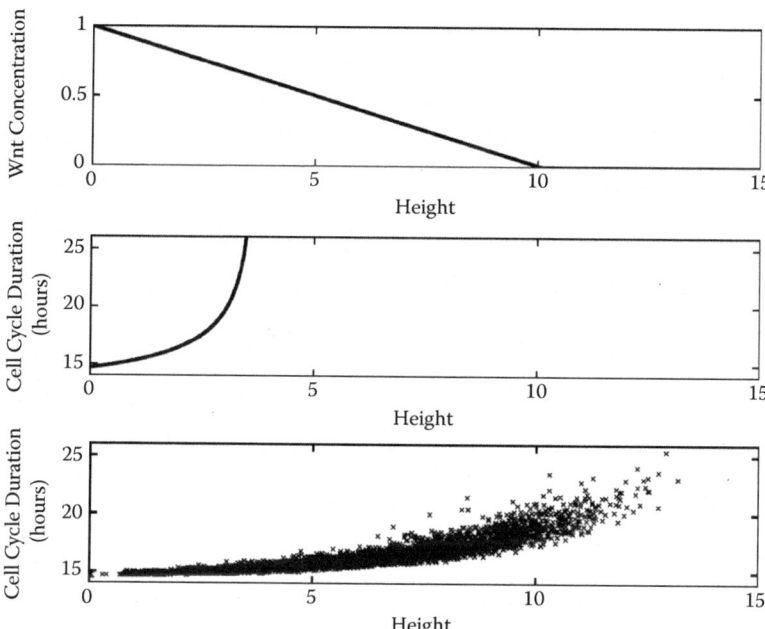

FIGURE 6.4 The cell-cycle duration response of a coupled Wnt signaling and cell-cycle model to varying Wnt stimuli. Simulation performed in a crypt that is 23 cells high. Top: The Wnt gradient imposed upon the crypt. Middle: Cell-cycle durations if cells are held in fixed positions; the predicted Wnt threshold for cell division is about 0.66. Bottom: Cell-cycle durations in a dynamic crypt simulation; for each cell in the simulation, the cell-cycle time is plotted as a function of the cell's position at the time of division.

experiments involve injecting laboratory rodents with an agent that is incorporated into cells during the S phase of the cell cycle and is passed on to their progeny. The distribution of clonal populations can be monitored over time by dissecting the crypts longitudinally and recording the positions of labeled cells along the two dissection lines. Given a sample containing several crypts, the outcome of the experiment is summarized in the form of a labeling-index (LI) curve, which shows the percentage of labeled cells per cell position at the time of sacrifice. We have used our multiscale model to perform similar in silico LI experiments. At time $t = 0$, we label all cells that are in the S phase. The simulation proceeds under the assumption that labeled cells behave in the same manner as their unlabeled counterparts, except that they transmit labels to their daughters. After a fixed time, we stop the simulation and perform a virtual crypt dissection. The LI curves obtained from the virtual dissections descend

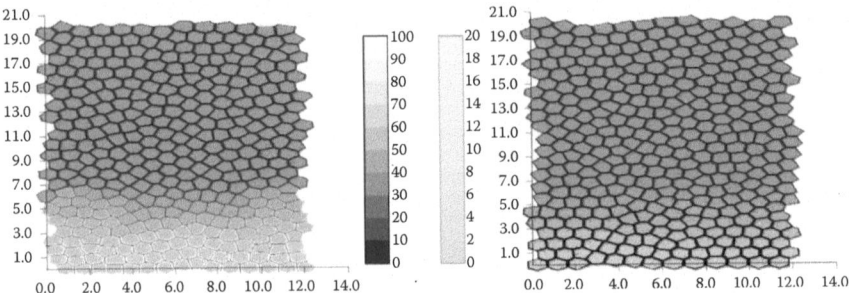

FIGURE 6.5 **(See color insert following page 40)** A quasi-steady simulation of cells stained for β-catenin, under two different hypotheses, as discussed [17] and as implemented in van Leeuwen et al. (2009). The Chaste visualizer displays the concentrations of nuclear and cytoplasmic levels of β-catenin on the green scale and membrane-bound β-catenin on the grey scale, facilitating a qualitative comparison with crypt staining or GFP-labeling experiments.

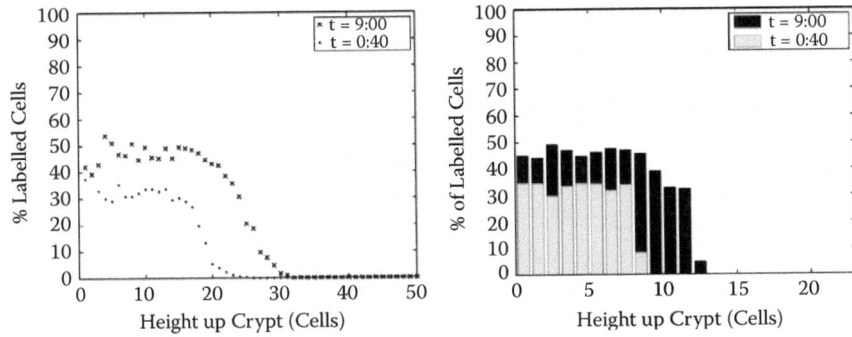

FIGURE 6.6 Results of virtual labeling-index experiments. Data obtained from 250 crypt simulations. Height up the crypt is expressed in units of length L. (left) Percentage of labeled cells per position along the dissection lines. Bullet points and crosses correspond to results obtained 40 min and 9 h after labeling, respectively. (right) True average percentage of labeled cells as a function of distance from the crypt base. Grey and black bars represent the results obtained 40 min and 9 h after labeling, respectively. (Reproduced with permission from van Leeuwen et al. *Cell Prolif.* 42 doi:10.1111/j.1365–2184.2009.00627.x. 2009.)

gradually, suggesting a smooth decrease in the percentage of labeled cells with increasing distance from the crypt base (Figure 6.6a). However, our model shows clearly segregated proliferative and differentiated populations, with an abrupt boundary between labeled and unlabeled cells in the averaged data (Figure 6.6b). This discrepancy is due to dissection and suggests that data from standard LI experiments may tend to overestimate the true position of the labeled cells.

Clonal Expansion and Niche Succession

Over time, the progeny of a single stem cell may dominate an entire crypt via a process termed *monoclonal conversion*, since the resulting crypt consists of a single clonal population [26]. Since mutations occur in single cells, the process of monoclonal conversion is important in the context of carcinogenesis as a mutant clone descended from this single cell has to persist in a crypt, by proliferating and eventually dominating it, in order for a mutant clone to gain a foothold in the colonic epithelium. Once a crypt has become mutant monoclonal, the mutant population can spread to neighboring crypts, either by top-down invasion, or through a process called *crypt fission* whereby a crypt divides into two.

Our multiscale model is ideally suited to study expansion of a clonal population *in silico*, and to predict conditions under which a crypt may become monoclonal. The main advantage is the ability to follow a clone's progress in real time, something that is impossible with current experimental techniques. We simulate the experiments of Taylor et al. (2003) [27], in which the progeny of cells with mitochondrial DNA (mtDNA) mutations that are functionally neutral are tracked. Such cells express a phenotype, for example, cytochrome-c oxidase (CcO) deficiency, which appear blue in histochemical stainings. In addition to wild-type crypts, Taylor et al. (2003) [27] observed crypts either partially or wholly filled with blue cells. In the former, "there is a ribbon of CcO-deficient cells within an otherwise normal crypt that is entirely compatible with the view that there are multiple stem cells in some crypts."

We investigated clonal expansion for two alternative model assumptions: first, following [21], the stem cells were fixed at the crypt base and assumed to divide asymmetrically; and second, following [19], the stem cells were unpinned and their proliferative behavior determined by the local Wnt stimulus. The results presented in columns I and II of Figure 6.7 reveal that if the stem cells are fixed in position, then an initial blue-stained stem cell invariably generates a thin, blue trail that moves upward toward the crypt orifice. Discontinuities in the clone can occur, due to waiting times between consecutive cell divisions. Importantly, although the trail's pattern can change in time, it does not expand laterally. Thus, under the original model assumptions, we are unable to capture the broad, wavy blue ribbons observed by Taylor et al. (2003) [27]. In contrast, as columns III and IV of Figure 6.7 show,

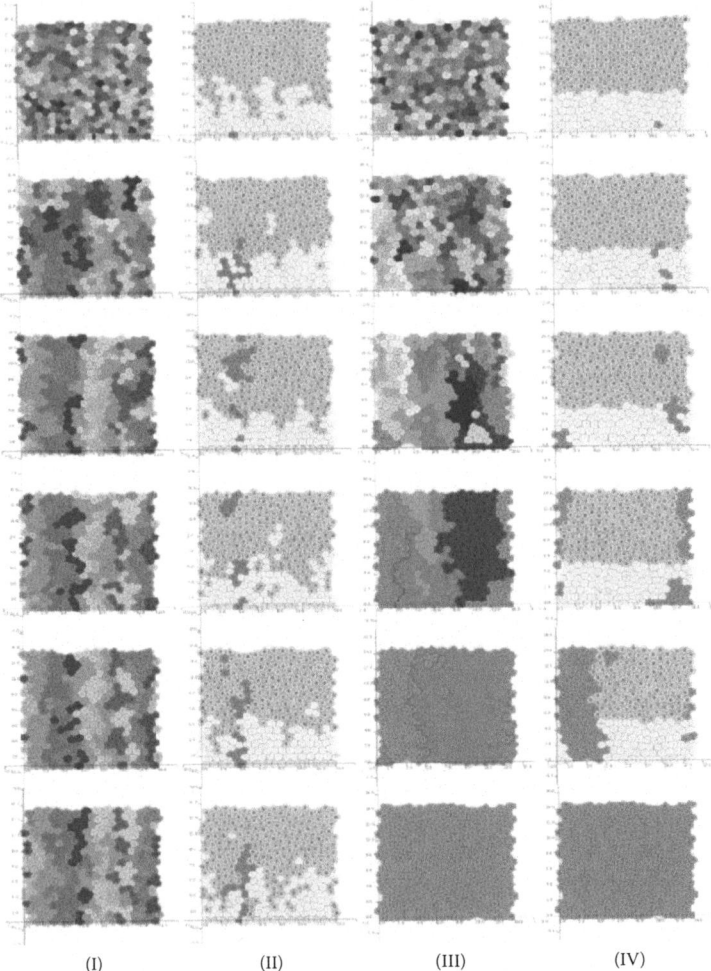

(I) (II) (III) (IV)

FIGURE 6.7 **(See color insert following page 40)** Clonal expansion in the crypt. Each column shows six snapshots from two independent *in silico* experiments performed with the model in [21] (columns I and II) and standard ($\mu_{ij} \equiv \mu$ and $v_i \equiv v$) model in [19] (columns III and VI), respectively. At time $t = 0$, a single cell is stained with a blue dye. This label is transmitted from generation to generation, giving rise to a clonal population of labeled cells. Columns II and IV highlight how the labeled populations evolve in time, whereas columns I and II show the clonal composition of the crypt. In column II, the stem cells, which are pinned to the base of the crypt, are highlighted in green. In the DMC simulation (columns III and IV), the population of labeled cells eventually takes over the crypt. (Reproduced with permission from van Leeuwen et al. *Cell Prolif.* 42 doi:10.1111/j.1365–2184.2009.00627.x. 2009.)

if the stem cells are free to move and cell fate is determined by local environmental conditions then, over time, clonal populations either expand in size or become extinct. In particular, the progeny of a single cell will eventually populate the entire crypt, and further, this cell will always eventually leave the crypt. These results suggest that cell "stem-ness" may depend on local biochemical cues rather than being an intrinsic property of a cell.

Variable Cell–Cell and Cell–Matrix Adhesion

As discussed in the section titled "Mechanical Model," we have considered a number of different cases regarding the dependence of cell–cell and cell–matrix adhesion on cell shape and Wnt signaling. In order to compare the impact of these different model assumptions on cell kinetics, we followed the dynamics of a standard crypt simulation (in which $\mu_{ij} \equiv \mu$ and $v_i \equiv v$; denoted NN) for 800 h and then repeated this for three other cases: the case of area-dependent cell–matrix adhesion only (denoted YN); the case of contact-edge-dependent cell–cell adhesion only (denoted NY); and the case of both contact-edge-dependent cell–cell adhesion and area-dependent cell–matrix adhesion (denoted YY). Results are shown in Figure 6.8. We find that YN cells located near the crypt base are larger than their NN counterparts. This is because in the YN case, if two cells of different sizes are attached by a compressed spring, the smaller cell moves apart more rapidly than the larger one. Consequently, small newborn cells leave the crypt base quicker than in the NN case. We also find that in the NY case, cells are more hexagonal in shape, and the crypt is densely populated with closely packed cells. In this case, the dependence of spring forces on cell size could eventually lead to a critical situation in which migration ceases completely; this can be prevented in the YY case, where variable cell–matrix and cell–cell adhesion are considered.

Hypotheses for Crypt Invasion

It is a matter of great debate how a single, mutant cell establishes a mutant epithelium within the crypt [28]. Two mechanisms have been suggested: top-down and bottom-up morphogenesis. Under top-down morphogenesis, a mutant cell at the top of a crypt expands not only laterally and downward but also invades (adjacent) crypts containing normal epithelium [29]. Under bottom-up morphogenesis, the mutant cell originates at the base of the crypt and increases in number through proliferation, until its progeny populate the entire crypt [30].

The model has been used to investigate the behavior of cells with APC or β-catenin mutations, the most common in colorectal cancer [31], within

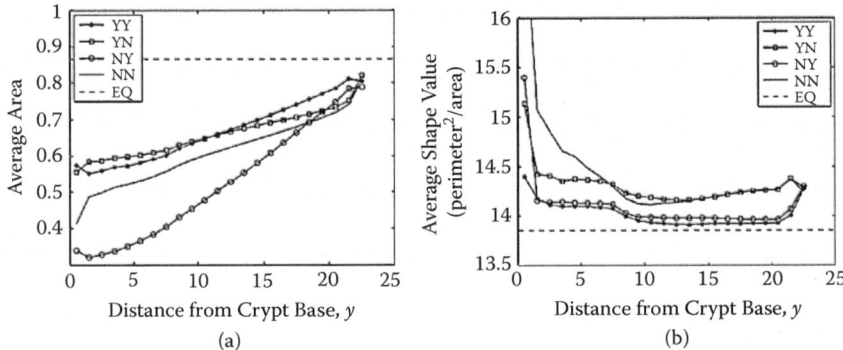

FIGURE 6.8 Dependence of cell size and geometry on cell adhesion. Results from four crypt simulations in dynamic equilibrium with different mechanical assumptions: NN = standard model; YN = area-dependent cell-matrix adhesion only; NY = contact-edge-dependent cell–cell adhesion only; YY = contact-edge-dependent cell–cell adhesion and cell-size-dependent cell–matrix adhesion; and EQ = values for hexagonal equilibrium lattice. Height up the crypt is expressed in units of length L. (a) Average cell area as a function of cell position. (b) Average shape value (perimeter^2/area) as a function of cell position. (Reproduced with permission from van Leeuwen et al. *Cell Prolif.* 42 doi:10.1111/j.1365–2184.2009.00627.x. 2009.)

the crypt. Mutations in these proteins enable cells to proliferate independently of Wnt [10]. Such mutant cells have also been shown to have a more rigid cytoskeleton [32], higher levels of cell–stroma [13] and stronger cell–cell adhesion [33]. We model these changes by allowing the damping constant to depend on whether the cell is mutant or not. The model was then used to establish the properties a mutant cell would require to allow top-down and bottom-up morphogenesis to occur. Numerical simulations reveal that mutant cells, which do not proliferate in a Wnt-dependent manner, can establish themselves within the crypt if they have higher levels of cell–substrate adhesion and a more rigid cytoskeleton. Top-down morphogenesis requires higher levels of cell–substrate adhesion and cytoskeleton rigidity than bottom-up morphogenesis.

DISCUSSION

In this chapter, we have presented a computational framework that allows us to integrate biological processes that act across a broad range of spatial scales. We have considered the model in the context of colorectal cancer and used it to address issues such as the role of Wnt signaling in the crypt, the process of monoclonal conversion, and the effects of model assumptions regarding cell–cell and cell–stroma adhesion.

In modeling the dependence of cell proliferation on Wnt signaling, we have neglected other pathways that are known to play an important role in regulating crypt structure.

Bone morphogenetic protein (BMP) signaling, which converges with the Wnt pathway to regulate β-catenin, is thought to control the process of stem cell self-renewal [34]. Dysregulation of BMP signaling can result in crypt fission and excessive quantities of crypt-like structures [35], as observed in humans with juvenile polyposis syndrome. The control of the Eph/ephrin signaling pathway may also be highly relevant in ensuring the proper crypt structure, as demonstrated by the fact that loss of expression of EphB receptors is correlated with the onset of invasive behavior [36]. Lastly, all proliferating cells in the crypt largely depend not only on Wnt but also Notch signaling; neither pathway is sufficient on its own to maintain proliferation [37]. Future work will involve the construction of mathematical models to investigate how these different pathways interact to control the proliferation of cells within the crypt, and incorporation of these models within the multiscale framework described in this chapter.

Many of the results presented in the section titled "Results" are consistent with independent experimental observations of colonic crypts. However, to have confidence in the model, we should account for the model assumptions that are implicit in our cell-center model by contrasting our model with other discrete model frameworks. In particular, it remains to be established which discrete model is best suited to a given biological problem.

Cell-center models, such as that presented in the section titled "Mechanical Model," can efficiently simulate cell proliferation, growth, and migration in the crypt. Moreover, it is straightforward to incorporate differential cell–cell adhesion [38–41] and to vary cell–substrate adhesion by varying the cellular drag coefficients. However, a disadvantage of such cell-center models is their reliance on the Delaunay triangulation, meaning that the number of vertices and the shapes of the cells do not change smoothly [42]. An alternative approach is cell-vertex modeling, in which cells are treated as polygons in 2D or polyhedra in 3D [43]. In cell-vertex models, the dynamics of each cell is governed by the movement of its vertices, these being determined by explicitly calculating the resultant forces or minimizing a global energy function. Cell-vertex models can describe changes in cell shape more realistically than cell-center models.

This is particularly important in the context of crypt modeling as we may wish to couple cell shape and surface areas to subcellular control models, as described in the section titled "Mechanical Model." Cell-vertex models are particularly suitable for modeling differential cell–cell adhesion, an important feature of cell dynamics in the crypt, as common mutations in colorectal epithelial cells are thought to affect cell–cell adhesion. However, the inclusion of differential cell–substrate adhesion is not so straightforward, as the drag terms include contributions from cells surrounding a given vertex. While cell-vertex models do not require the computation of a Delaunay triangulation at each time step, the higher spatial resolution considered in cell-vertex models results in a larger system dimension than that of a cell-center model. Osborne et al. (2010) [44] have developed a cell-vertex model of the crypt and, using numerical simulations, have found that it exhibits qualitatively similar behavior to our cell-center model.

A major problem with discrete models, especially those incorporating stochastic behavior, is their computational intensity. For example, in the case of our multiscale model, a large number of simulations are needed to determine how a proliferative advantage bestowed on mutant cells translates into an increase in their probability of becoming the dominant clonal population within a crypt, and how this increased probability varies with the location of the initial mutation within the crypt. Moreover, as the molecular details of subcellular pathways become increasingly more complex, systematic and rational model reduction becomes a critically important tool, as a modeling approach that simply includes all known molecular details quickly becomes intractable. One resolution of this problem is to develop a continuum model that replicates the qualitative features of the original discrete model. We can then apply mathematical techniques to analyze the coarse-grained model and, for instance, establish quickly the necessary phenotypic traits for mutant cells to take over a crypt via the top-down and/or bottom-up morphogenesis. Such continuum models can be derived either formally [45] or phenomenologically [44].

By viewing the epithelial cells that line a crypt as a one-dimensional chain of connected linear springs, Murray et al. (2009) [45] have formally derived a continuum model for cell number density. This model comprises a reaction-diffusion equation with a spatially non-uniform proliferation term and a nonlinear diffusive flux term, with diffusion coefficient

$$D(q) = \frac{\mu}{vq^2}, \tag{6.8}$$

where q denotes the cell number density and μ and v denote the spring constant (assumed the same for all cell–cell interactions) and damping constant (assumed the same for all cells). As Figure 6.9 shows, there is generally good qualitative agreement between the cell velocities obtained with this coarse-grained model and those obtained from our 2D multiscale model. A discrepancy between the two models for smaller values of μ arises from the assumption that the crypt is one-dimensional.

Using a phenomenological approach, Osborne et al. (2010) [44] have developed a 2D continuum model for a crypt in which cells are treated as an incompressible viscous fluid obeying Darcy's law. As Figure 6.10 shows, model simulations compare reasonably well with the multiscale model, as well as with a cell-vertex model of the crypt. However, the continuum

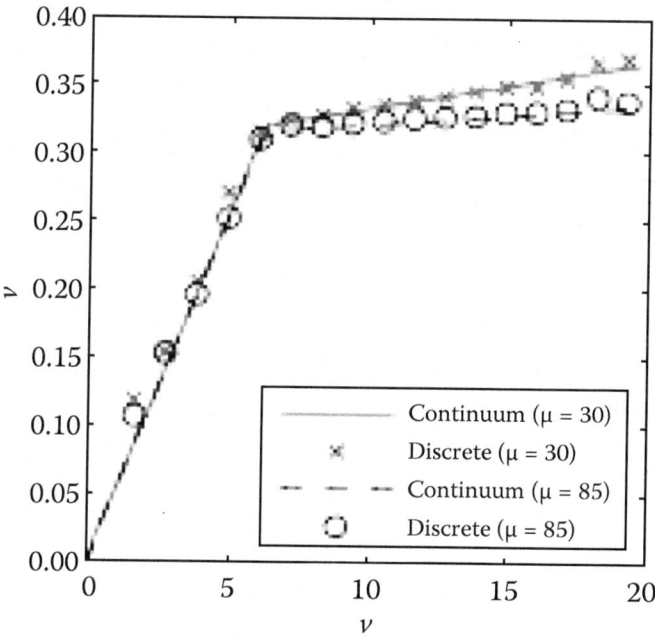

FIGURE 6.9 Steady-state crypt velocities, v, plotted against crypt height, y. Thousand simulations of a 2D periodic crypt were run, and the average cell velocities (markers) were compared with the velocities predicted by the corresponding continuum model (lines). In this plot $y_C = 0.3$, $T_C = 14$, and $L = 20.1$. The circles and solid line correspond to $\mu = 40$, whereas the crosses and dashed line correspond to $\mu = 80$.

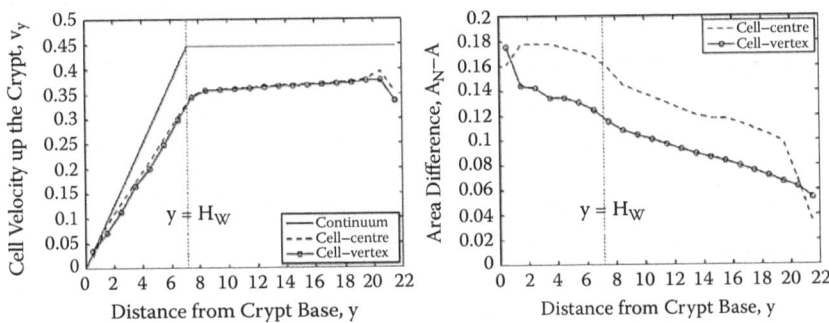

FIGURE 6.10   Comparison of cell-center, cell-vertex, and continuum crypt models. In each graph, the dotted vertical line $y = H_W$ delineates the upper boundary of the Wnt-stimulated region: cells proliferate for $y \le H_W$, and for $y > H_W$ they do not. Left: Dependence of average speed up the crypt axis on distance from the crypt base. Right: Dependence of average cell compression (natural cell area minus actual cell area) on distance from the crypt base.

model does slightly overestimate cell velocities within the crypt, as a result of the assumption of cell incompressibility, which in 1D corresponds to the limit $\mu \to \infty$ in the Murray et al. model [45].

There are now a multitude of such integrative models in the literature (see, for example [46–48]). Similar to these, the modeling approach discussed in this chapter suffers from the problem that we have made simplifications at each scale and, while we can investigate the errors induced at each level, we have not developed a theory for how to do this across scales. This remains an open question. Therefore, an important future challenge for the modeling community is to develop a systematic way of constructing such models. As described earlier, one possible way to approach this might be in the recent research that aims to develop continuum models of individual-based computational schemes (see, for example, [45,49,50]). This allows us not only to see precisely where the different modeling assumptions at the cell-level affect tissue-level behavior, but may also allow us to, in the future, use the well-developed mathematical machinery for partial differential equations to address key problems in multiscale modeling.

ACKNOWLEDGMENTS

The authors gratefully acknowledge funding from the EPRSC as part of the Integrative Biology program (GR/572023/01). PJM and PKM were partially supported by NIH Grant U56CA113004 from the National

Cancer Institute. PJM, KHC, PKM, and HMB were partially supported by PMI2 (British Council). PKM was partially supported by a Royal Society Wolfson Merit Award. JWK and KHC acknowledge the support received from the National Research Foundation of Korea grant funded by the Korea Ministry of Education, Science & Technology, through the Systems Biology grant (20090065567), the Nuclear Research grant (M20708000001-07B0800-00110), the 21C Frontier Microbial Genomics and Application Center Program (Grant 11-2008-10-004-00), the World Class University grant (R32-2008-000-10218-0), and the BRL (Basic Research Laboratory) grant (2009-0086964).

REFERENCES

1. Okamoto, R. and Watanabe, M. 2004. Molecular and clinical basis for the regeneration of human gastrointestinal epithelia. *J. Gastroenterol.* 39:1–6.
2. Ross, M.H., Kaye, G.I., and Pawlina W. 2003. *Histology: A Text and Atlas.* Lippincott Williams & Wilkins.
3. Reya, T. and Clevers, H. 2005. Wnt signalling in stem cells and cancer. *Nature* 434:843–850.
4. Pinto, D. and Clevers, H. 2005. Wnt control of stem cells and differentiation in the intestinal epithelium. *Exp. Cell Res.* 306:357–363.
5. Hinck, L., Nelson, W.J., and Papkoff, J. 1994. Wnt-1 modulates cell-cell adhesion in mammalian cells by stabilizing β-catenin binding to the cell adhesion protein cadherin. *J. Cell Biol.* 124:729–741.
6. Logan, C.Y. and Nusse, R. 2004. The Wnt signaling pathway in development and disease. *Annu. Rev. Cell Dev. Biol.* 20:781–810.
7. Hall, A.C., Lucas, F.R., and Salinas, P.C. 2000. Axonal remodeling and synaptic differentiation in the cerebellum is regulated by WNT-7a signaling. *Cell* 100:525–535.
8. Church, V.L. and Francis-West, P. 2002. Wnt signalling during limb development. *Int. J. Dev. Biol.* 46:927–936.
9. Nelson, W.J. and Nusse, R. 2004. Convergence of Wnt, β-catenin, and cadherin pathways. *Science* 303:1483–1487.
10. Ilyas, M. 2005. Wnt signalling and the mechanistic basis of tumour development. *J. Pathol.* 205:130–144.
11. Fodde, R. and Brabletz, T. 2007. Wnt/β-catenin signaling in cancer stemness and malignant behavior. *Curr. Opin. Cell Biol.* 19:150–158.
12. Gaspar, C. and Fodde, R. 2004. Wnt signalling/APC dosage effects in tumorigenesis and stem cell differentiation. *Int. J. Dev. Biol.* 48:377–386.
13. Sansom, O.J., Reed, K.R., Hayes, A.J. et al. 2004. Loss of Apc in vivo immediately perturbs Wnt signaling, differentiation, and migration. *Genes Dev.* 18:1385–1390.
14. Sansom, O.J., Meniel, V.S., Muncan, V. et al. 2007. Myc deletion rescues Apc deficiency in the small intestine. *Nature* 446:676–679.

15. Lee, E., Salic, A., Kruger, R., Heinrich, R., and Kirschner, M.W. 2003. The roles of APC and Axin derived from experimental and theoretical analysis of the Wnt pathway. *PLoS Biol.* 1:116–132.
16. Mirams, G.R., Byrne, H.M., and King, J.R. 2010. A multiple timescale analysis of a mathematical model of the Wnt/ β-catenin signalling pathway. *J. Math. Biol.* 60, 131–160.
17. van Leeuwen, I.M.M., Byrne, H.M., Jensen, O.E., and King, J.R. 2007. Elucidating the interactions between the adhesive and transcriptional functions of β-catenin in normal and cancerous cells. *J. Theor. Biol.* 247:77–102.
18. Swat, M., Kel, A., Herzel, H. 2004. Bifurcation analysis of the regulatory modules of the mammalian G1/S transition. *Bioinformatics* 20:1506–1511.
19. van Leeuwen, I.M.M., Mirams, G.R., Walter, A. et al. 2009. An integrative computational model for intestinal tissue renewal. *Cell Prolif.* 42 doi:10.1111/j.1365–2184.2009.00627.x.
20. Weliky, M., Minsuk, S., Keller, R., and Oster, G. 1991. Notochord morphogenesis in Xenopus laevis: simulation of cell behavior underlying tissue convergence and extension. *Development* 113:1231–1244.
21. Meineke, F.A., Potten, C.S., and Loeffler, M. 2001. Cell migration and organization in the intestinal crypt using a lattice-free model. *Cell Prolif.* 34:253–266.
22. Pitt-Francis, J., Pathmanathan, P., Bernabeu, M.O. et al. 2009. Chaste: a test-driven approach to software development for biological modelling. *Computer Physics Communications* 180, 2452–2471.
23. Beck, K. and Andres, C. 2004. *Extreme Programming Explained: Embrace Change.* Addison-Wesley, Boston.
24. Pitt-Francis, J., Bernabeu, M.O., Cooper, J. et al. 2008. Chaste: using agile programming techniques to develop computational biology software. *Phil. Trans. Roy. Soc. A* 366:3111–3136.
25. Sunter, J.P., Appleton, D.R., De Rodriguez, M.S.B., Wright, N.A., and Watson, A.J. 1979. A comparison of cell proliferation at different sites within the large bowel of the mouse. *J. Anat.* 129, 833–842.
26. McDonald, S.A.C., Preston, S.L., Greaves, L.C. et al. 2006. Clonal expansion in the human gut: mitochondrial DNA mutations show us the way. *Cell cycle* 5:808–811.
27. Taylor, R.W., Barron, M.J., Borthwick, G.M. et al. 2003. Mitochondrial DNA mutations in human colonic crypt stem cells. *J. Clin. Invest.* 112:1351–1360.
28. Wright, N.A. 2000. Epithelial stem cell repertoire in the gut: clues to the origin of cell lineages, proliferative units and cancer. *Int. J. Exp. Pathol.* 81:117–143.
29. Shih, I.M., Wang, T.L., Traverso, G. et al. 2001. Top-down morphogenesis of colorectal tumors. *Proc. Natl. Acad. Sci. USA* 98:2640–2645.
30. Preston, S.L., Wong, W.M., Chan, A.O. et al. 2003. Bottom-up histogenesis of colorectal adenomas: origin in the monocryptal adenoma and initial expansion by crypt fission. *Cancer Res.* 63:3819–3825.

31. Giles, R.H., van Es, J.H., and Clevers, H. 2003. Caught up in a Wnt storm: Wnt signaling in cancer. *Biochim. Biophys. Acta.* 1653:1–24.

32. Näthke, I. 2006. Cytoskeleton out of the cupboard: colon cancer and cytoskeletal changes induced by loss of APC. *Nat. Rev. Cancer* 66:967–974.

33. Bienz, M. 2005. β-catenin: a pivot between cell adhesion and Wnt signalling. *Curr. Biol.* 15:64–67.

34. He, X.C., Zhang, J., Tong, W.-G. et al. 2004. BMP signaling inhibits intestinal stem cell self-renewal through suppression of Wnt-β-catenin signaling. *Nat. Genet.* 36:1117–1121.

35. Haramis, A.P.G., Begthel, H., van den Born, M. et al. 2004. *De novo* crypt formation and juvenile polyposis on BMP inhibition in mouse intestine. *Science* 303:1684–1686.

36. Battle, E., Bacani, J., Begthel, H. et al. 2005. EphB receptor activity suppresses colorectal cancer progression. *Nature* 435:1126–1130.

37. van Es, J.H., van Gijn, M.E., Riccio, O. et al. 2005. Notch/γ-secretase inhibition turns proliferative cells in intestinal crypts and adenomas into goblet cells. *Nature* 435:959–963.

38. Galle, J., Loeffler, M., and Drasdo, D. 2005. Modeling the effect of deregulated proliferation and apoptosis on the growth dynamics of epithelial cell populations in vitro. *Biophys. J.* 88:62–75.

39. Ramis-Conde, I., Drasdo, D., Anderson, A.R.A., and Chaplain, M.A.J. 2008. Modelling the influence of the E-cadherin-β-catenin pathway in cancer cell invasion: a multi-scale approach. *Biophys. J.* 95:155–165.

40. Schaller, G. and Meyer-Hermann, M. 2005. Multicellular tumor spheroid in an off-lattice Voronoi-Delaunay cell model. *Phys. Rev. E* 71:51910.

41. Walker, D.C., Southgate, J., Hill, G. et al. 2004. The epitheliome: agent-based modelling of the social behaviour of cells. *Biosystems* 76:89–100.

42. Brodland, G.W. 2004. Computational modeling of cell sorting, tissue engulfment, and related phenomena: a review. *ASME Applied Mechanics Rev.* 57:1–30.

43. Honda, H., Tanemura, M., and Nagai, T. 2004. A three-dimensional vertex dynamics cell model of space-filling polyhedra simulating cell behavior in a cell aggregate. *J. Theor. Biol.* 226:439–453.

44. Osborne, J.M., Walter, A., Kershaw, S.K. et al. 2010. A hybrid approach to multiscale modeling of cancer. *Phil. Trans. Roy. Soc. A* (in press).

45. Murray, P.J., Edwards, C.E., Tindall, M.J., and Maini, P.K. 2009. From a discrete to a continuum model of cell dynamics in 1D. In preparation.

46. Alarcón, T., Byrne, H.M., and Maini, P.K. 2005. A multiple scale model for tumor growth. *Multiscale Model Simul.* 3:440–475.

47. Ribba, B., Saut, O., Colin, T., Bresch, D., Grenier, E., and Boissel, J.P. 2006. A multiscale mathematical model of avascular tumor growth to investigate the therapeutic benefit of anti-inasive agents. *J. Theor. Biol.* 243:532–541.

48. Macklin, P., McDougall, S., Anderson, A.R.A., Chaplain, M.A.J., Cristini, V., and Lowengrub, J. 2009. Multiscale modelling and nonlinear simulation of vascular tumour growth. *J. Math. Biol.* 58:765–798.

49. Alber, M., Chen, N., Lushnikov, P., and Newman, S. 2007. Continuous macroscopic limit of a discrete stochastic model for interaction of living cells. *Phys. Rev. Lett.* 99:168102.

50. Fozard, J.A., Jensen, O.E., Byrne, H.M., and King, J.R. 2009. Continuum approximations of individual-based models for epithelial monolayers. *Math. Med. Biol.* doi:10.1093/imammb/dqp015.

CHAPTER 7

The Physical Microenvironment in Somatic Evolution of Cancer

Robert A. Gatenby

CONTENTS

INTRODUCTION

The transition from normal tissue to invasive cancer is a multistep process in which increasingly malignant cellular populations emerge over time (1–3), generally coincident with accumulating genomic mutations. This is often described as "somatic evolution" (4–5) because it appears formally analogous to Darwinian evolution in nature. While this conceptual model is well accepted, the interactions with phenotypic properties and environmental selection forces that determine individual fitness remain ill defined. Furthermore, the language of evolution is often employed in carcinogenesis without full explanation. For example, it is often stated that, during carcinogenesis, some random mutations "confer a selective growth advantage" resulting in clonal expansion and subsequent tumor growth.

However, precisely how a genomic change alters the phenotype and how a phenotypic trait interacts with environmental growth constraints and selection factors remains vague. Thus, while the conceptual model is appealing and well accepted, the dynamics governing the Darwinian interactions of altered cellular genotypes with changing microenvironments often remain unclear. Theoretical models of tumor development typically include a sequence of genomic mutations and epigenetic changes synchronous with progressive drift of cellular populations from normal through premalignant lesions to invasive cancer (6). Line drawings ("Vogelgrams") (6) have been developed to correlate alterations in specific oncogenes and tumor suppressor genes with a linear progression from normal tissues through premalignant lesions (large and small polyps) to invasive colorectal cancer. This approach, although useful conceptually and pedagogically, is overly simplified, ignoring, for example, the stochastic nature of mutations, mitigating intracellular processes such as the chaperone function of heat shock proteins, and the critical role of microenvironmental selection factors that determine the fitness of any given phenotype. The role of the mutation rate in driving somatic evolution remains the subject of debate. Loeb and others (7) hypothesize an increased mutation rate due to defects in chromosomal stability, or DNA repair pathways is necessary as a forcing function to produce the number of genomic changes required for evolution of invasive cancer. This assumes the background mutation rate is insufficient to allow the necessary carcinogenic mutations to accumulate in the human life span. The role of the mutator phenotype is supported by observation of large numbers of mutations in most cancer cells (8) and increased mutation rates in early colon and esophageal cancers (9,10).

On the other hand, Tomlinson and others cite (11,12) empirical evidence and mathematical models to demonstrate that normal mutation rates are sufficient for tumor evolution in microenvironments generating strong clonal selection. Bissell and colleagues (13–16) have published a number of studies showing that microenvironmental factors such as the extracellular matrix (ECM) and admixed normal cell populations alter tumor cell proliferation independent of permanent genomic change and find that, in some stages of the somatic evolution of the malignant phenotype, the environment plays a greater role than mutagenesis. Finally, the mutator hypothesis does not typically incorporate epigenetic phenomenon such as DNA methylation and acetylation or intracellular factors such as heat shock proteins that can maintain phenotypic robustness in the face of genomic heterogeneity. In fact, reversible changes in phenotype are observed in Bissell's studies and are clearly dependent on environmental factors. This phenomenon likely plays an important role in carcinogenesis and development of metastases [17]. As with any nonlinear process, the complex multistep transformation of normal cells to invasive cancer will not be fully understood without formal mathematical models (18,19). To this end, a number of quantitative models of carcinogenesis have been developed based on methods adapted from information theory, cellular automaton models, and evolutionary game theory. Insights from these models, in conjunction with experimental observations, have yielded a number of insights into the Darwinian dynamics of somatic evolution.

MODEL: EVOLUTIONARY GAME THEORY

Evolutionary Game Theory allows the concept of somatic evolution to be formalized and framed mathematically to examine the cellular and intracellular dynamics (20) that lead to the evolution of specific properties of the malignant phenotype (17,20). In general, a volume of tissue contains ns distinct cellular populations designated x_i, $i = 1, ..., n_s$ and described by a phenotype vector $\mathbf{u}i$ composed of multiple scalar components. Population and mean phenotype vectors are

$$\boldsymbol{x} = [x_1 \ ... \ x_{n_s}]\tag{7.1}$$

$$\mathbf{u} = [\mathbf{u}_1 \ ... \ \mathbf{u}_{n_s}]\tag{7.2}$$

where x_i is the number of individuals in population i and $\mathbf{u}$ is adaptive phenotypic properties of each cell population. This could be linked to quantitative data by designating each orthogonal axis in $\mathbf{u}$ to be the genes in a microarray. Alternatively, they could be specific measurable phenotypic properties such as proliferation rate, glucose uptake, etc. Note that "mean phenotype" assumes some phenotypic diversity within each cellular population as observed in clonal populations of both normal and transformed cells (13), and this diversity is typically represented through suitable distribution functions. In this somatic ecosystem, cellular fitness, defined by proliferative capacity, may be determined through a fitness-generating function (G-function) (21–26) with a virtual variable, v. Setting the virtual variable equal to the phenotype of a population produces its fitness, which is a function of $\mathbf{x}$, $\mathbf{u}$, and substrate concentration R. The relationship between fitness H_i and the G-function is

$$G(v, \mathbf{u}, \mathbf{x}, R)_{v = u_i} = H_i(\mathbf{u}, \mathbf{x}, R) \quad i = 1, \dots, n_s. \tag{7.3}$$

The population dynamics may be written as n_s fitness functions or one fitness-generating function.

$$\dot{x}_i = x_i H_i(\mathbf{u}, \mathbf{x}, R) = x_i G(V, \mathbf{u}, \mathbf{x}, R)\big|_{v=u_i} \tag{7.4}$$

The G-function simplifies writing the equations of motion and provides a conceptual advantage for understanding system evolution as a plot of G versus v for fixed $\mathbf{u}$, $\mathbf{x}$, and R is a geometric representation of the adaptive landscape upon which evolution takes place. We present later some results based on a multiple G-function model describing tumor growth and development within somatic ecosystems. For now, we use a single G-function model—a simpler approach that, nevertheless, yields identical qualitative results to the more comprehensive model.

$$G(v, \mathbf{u}, \mathbf{x}, R) = B_n \left(1 - \frac{\sum_{i=1}^{n_s} a(v, \mathbf{u}) x_i}{K(v)} \right) \left(\frac{E(v) R^2}{R_0^2 + R^2} - m \right). \tag{7.5}$$

It is immediately apparent that the first right-hand term in parentheses is the Lotka–Voltera equation. K is the carrying capacity, and α is the quantitative effects of one population on another. The second term represents substrate dynamics, where R is the concentration of a critical substrate, substrate uptake obeys Michelis–Menten kinetics (hence, the E and R_0 terms), and m represents the substrate utilization to maintain basic cell function. The overall value of this term must be positive (i.e., substrate uptake must exceed basal demand) for proliferation to occur. Thus, in Equation 7.5, cell populations in vivo are subject to several growth constraints: (**1**) "Tissue organizational" controls are included in the first right-hand term these include: (**a**) intracellular factor, that determine population density [$K(v)$] through growth promoters such as oncogenes and growth inhibitors, including tumor suppressor genes, senescence, and apoptosis pathways, and (**b**) extracellular controls generated within the environment [defined by $a(v, \mathbf{u})$] through cell–cell interactions or products of other cell phenotypes, such as the ECM, soluble growth promoters, etc., consistent with studies demonstrating that environmental factors exert significant control in normal tissue development (31–33). (*Note that both $K(v)$ and $a(v, \mathbf{u})$ are lumped phenomenological terms.*) (**2**) Substrate availability (second right-hand term), that is, cells must obtain substrate in excess of basal metabolic demand m to supply energy and macromolecules for proliferation. $B_n = d_n c_n$, where c_n is a constant converting excess substrate into new cells, and d_n is maximum proliferation rate. We assume normal cells under physiologic conditions are not subject to substrate limitations, so their proliferation is controlled solely by tissue controls. Pathological exceptions include acute or chronic ischemia such as stroke, myocardial infarction, or diabetic ulcers.

RESULTS: THE PHYSICAL MICROENVIRONMENT AND EVOLUTION OF THE MALIGNANT PHENOTYPE

Evolutionary Game Theory

When these evolutionary models are applied to carcinogenesis, several interesting conclusions are reached:

1. Initial tumor cell growth is controlled by normal tissue constraints generated by cellular interactions with other cell populations, the extracellular matrix, and soluble or insoluble growth factors. Thus, cellular adaptation in early carcinogenesis will favor phenotypic alterations that reduce these constraints, such as loss-of-function mutations in tumor suppressor genes and gain-of-function mutations in oncogenes. That is, since proliferation of normal cells under physiologic conditions is controlled by the social constraints in Equation 7.5 (the first right-hand term) and not substrate limitations (the second right-hand term), evolutionary pressures favor mutations that reduce cellular sensitivity to normal growth constraints. Thus, the initial evolution of tumor cells requires loss-of-function

mutation in tumor suppressor genes and gain-of-function mutations in oncogenes similar to the conventional view of carcinogenesis as expressed in, for example, the "Vogelgram."

2. Global evolutionary dynamics that determine the time course of this process are governed by:

$$\dot{\mathbf{u}}_i = \sigma_i^2 \left.\frac{\partial G(v,\mathbf{u},\mathbf{x},R)}{\partial v}\right|_{v=u_i} \tag{7.6}$$

where σ_i is the variance of the phenotypic distribution around the mean and $\partial G/\partial v$ is the slope of the fitness landscape representing the change in fitness for a given change in phenotype.

There are two points in Equation 7.6 that may provide insight into somatic evolution. First, the evolutionary rate is dependent on phenotypic and not genotypic diversity or, more broadly, evolution selects phenotypes not genotypes. Nevertheless, the dependence of evolutionary dynamics on σ_i reflects the increased rate with which phenotypically diverse populations explore the fitness parameter space and does support the hypothesis that increased mutation rates, by generating multiple phenotypic variants, promotes carcinogenesis (7). Second, Equation 7.6 demonstrates that evolution is not solely dependent on phenotypic diversity (and, therefore, the mutation rate), because cellular populations may evolve even with limited phenotypic diversity (i.e., low mutation rates) if microenvironmental conditions generate strong clonal selection pressures increasing $\partial G/\partial v$. This is similar to modeling results by Tomlinson (12) and consistent with observations by Bissell et al. (13–17) that tumorigenesis of genetically stable populations may be promoted or suppressed by wounding, peritumoral stromal cells, ECM alterations, changes in growth factor concentrations, etc. (31,32). (3) Even multiple mutations in oncogenes and tumor suppressor genes only led to self-limited growth (Figure 7.1). This is because growth of tumor cells is eventually limited by substrate availability resulting from cellular proliferation. This predicted a previously unknown era in carcinogenesis in which somatic evolution was dominated by microenvironmental hypoxia and acidosis.

Evolutionary-model-predicted carcinogenesis proceeds through two distinct phases (33–36). The second of these phases, in which cellular growth in premalignant lesions is limited by substrate limitation, had not been previously identified in traditional theoretical models of carcinogenesis. This led us to reexamine in detail the cellular and environmental dynamics that might result in substrate-limited evolution and the role of these interactions in emergence of the glycolytic phenotype during carcinogenesis (34–47). How could this occur in typical premalignant lesions such as colon polyps or breast ductal carcinoma in situ (DCIS)? This requirement for substrate limitation led to the realization that, while premalignant lesions are often characterized as highly vascularized, this is true only in a macroscopic sense. That is, while a premalignant lesion such as a polyp or carcinoma in situ may have a vascular stroma, the hyperplastic epithelia are physically separated from their blood supply by a basement membrane. This is illustrated in Figure 7.1 as the hyperplastic epithelium of a carcinoma in situ is clearly delimited from the stroma by a thin basement membrane. Blood vessels are confined to the stromal compartment and, hence, early carcinogenesis and development of the malignant phenotype actually occur in an avascular environment. As a result, substrates, such as oxygen and glucose, must diffuse from the vessels across the basement membrane and through layers of tumor cells, where they are metabolized. The diffusion and consumption of substrate was modeled by Krogh (43) as early as 1919 through reaction-diffusion equations that demonstrated oxygen concentrations will decrease with distance from a capillary such that oxygenated cells were limited to a distance of less than 150 μm from a blood vessel (38). In the 1950s, empirical studies by Thomlinson and Gray showed that viable tumor cells were not observed at distances greater than 160 μm from blood vessels, consistent with Krogh's calculations. Subsequent experimental studies in window chambers in animal models have demonstrated that near-zero partial pressure of oxygen (pO_2) is observed at distances of only 100 μm from a vessel (39,40).

Thus, premalignant lesions, provided their basement membranes remain intact, will inevitably develop hypoxic regions near the oxygen diffusion limit, as persistent proliferation leads to a thickening of the epithelial layer, pushing cells ever more distant from their blood supply, which remains on the other side of the basement membrane (Figure 7.1). At this penumbral layer, microenvironmental selection forces will favor phenotypes that adapt to harsh environments (through resistance to hypoxia and acid-induced cell toxicity) and successfully compete for scarce resources, such as oxygen and glucose.

This is consistent with the model predictions of an era in carcinogenesis dominated by substrate limitation. Low oxygen concentrations appear to be the first substrate limitation confronting neoplastic cell populations as reaction-diffusion models, and empirical studies have shown that pO$_2$ decline more rapidly with distance from blood vessels than do glucose levels. Although the presence of hypoxia in premalignant in situ lesions has not been measured directly, it can be inferred

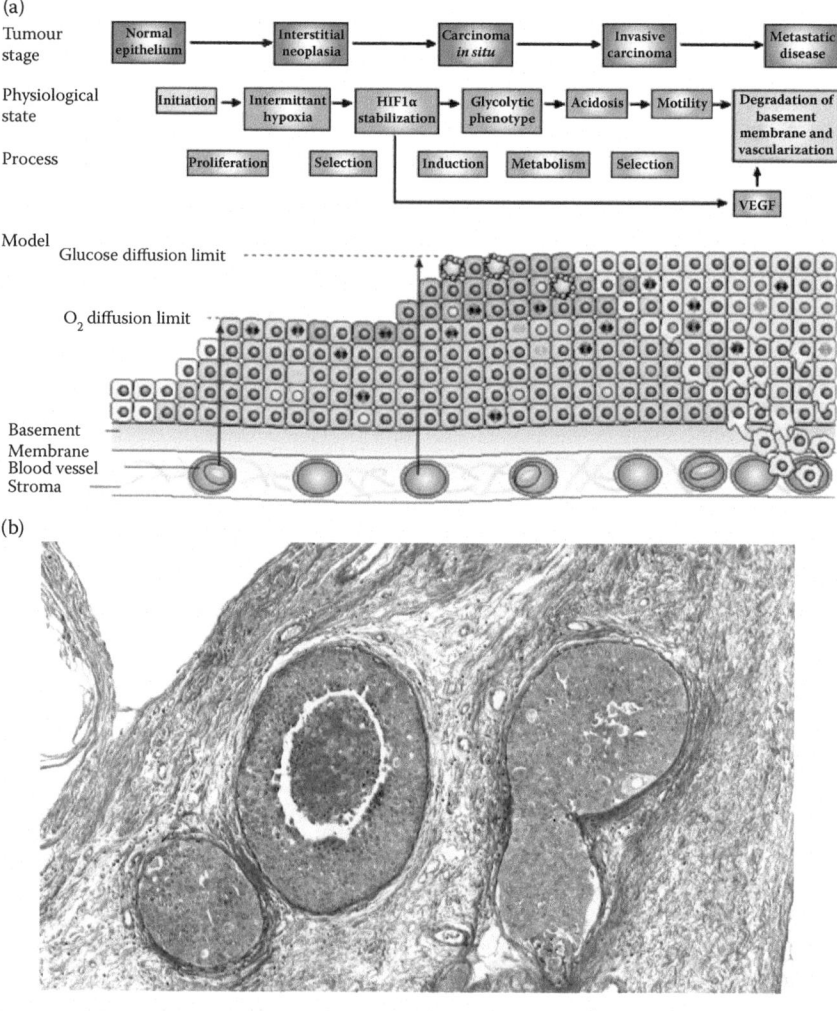

FIGURE 7.1   **(See color insert following page 40)**

from the frequent observation of necrosis in these lesions and by demonstration of hypoxia-inducible enzymes such as carbonic anhydrases IX and XII in late stage DCIS, particularly adjacent to areas of necrosis (41). While the upregulation of glycolysis is a successful adaptation to hypoxia/anoxia, it also has significant negative consequences due to increased acid production, which causes significant decreases in local extracellular pH. Prolonged exposure of normal cells to an acidic microenvironment typically results in necrosis or apoptosis through p53 and caspase-3-dependent mechanisms (42). The physiological trigger for apoptosis may be the collapse of the transmembrane H^+ gradient that occurs with intracellular acidosis, but other factors may also play a role. Thus, constitutive upregulation of glycolysis requires adaptation to the negative effects of extracellular acidosis through resistance to apoptosis or upregulation of membrane transporters to maintain normal intracellular pH. Intracellular pH is maintained by multiple families of H^+ transporters, which are coexpressed and redundant. Na^+/H^+ exchange and vacuolar H^--ATPases have both been observed to be upregulated in cancers, and vacuolar H^+-ATPase may confer resistance to apoptosis (43–45). Additional adaptations may also be required as the increased glucose consumption rates further decrease glucose concentrations. Cellular competition for this increasingly limited resource will therefore increase and favor phenotypes with greater numbers of either high V_{max} (e.g., GLUT-1) or low K_m (e.g., GLUT-3) glucose transporters. Such upreg-

FIGURE 7.1 **(Opposite)** From reference 42. The prediction of a substrate-dominated era of carcinogenesis from mathematical models led to recognition of the role of the anatomy and physiology of epithelial surfaces in somatic evolution The hypothesized substrate and metabolite diffusion-reaction kinetics and their effects on tumor cell evolution are shown in (a). As proliferation carries tumor cells farther and farther from the basement membrane (and the underlying blood vessels), they initially become hypoxic resulting in upregulation of glycolysis. This, in turn, results in acidification of the environment creating new environmental selection forces that promote phenotypic changes that reduce acid-mediated cytotoxicity. This cellular population has a profound proliferative advantage because it creates an acidic environment that is toxic to other populations. In (b) H&E micrograph of DCIS demonstrates tumor cells proliferating into the duct remaining separated from the underlying vessels by the intact basement membrane.

ulation of glucose transporters has been observed during carcinogenesis in esophageal, gastric, breast, and colon cancers (46–50).

Modified Cellular Automata Models

The potential boundary conditions imposed by the surface anatomy of ducts were tested by Smallbone et al. (51,52) *in silico*, using a modified cellular automaton approach that followed the history of individual cells to examine phenotypic evolution as well as the mutual interactions of cells and the changing microenvironment within a duct (Figure 7.2). The two-dimensional model is composed of an $M \times N$ array of automaton elements with a specific rule set governing their evolution, as well as glucose (g), oxygen (c), and H$^+$ (h) fields, each satisfying reaction-diffusion equations. A two-dimensional automaton was used to focus on growth away from the basement membrane, rather than along the duct. The avascular geometry

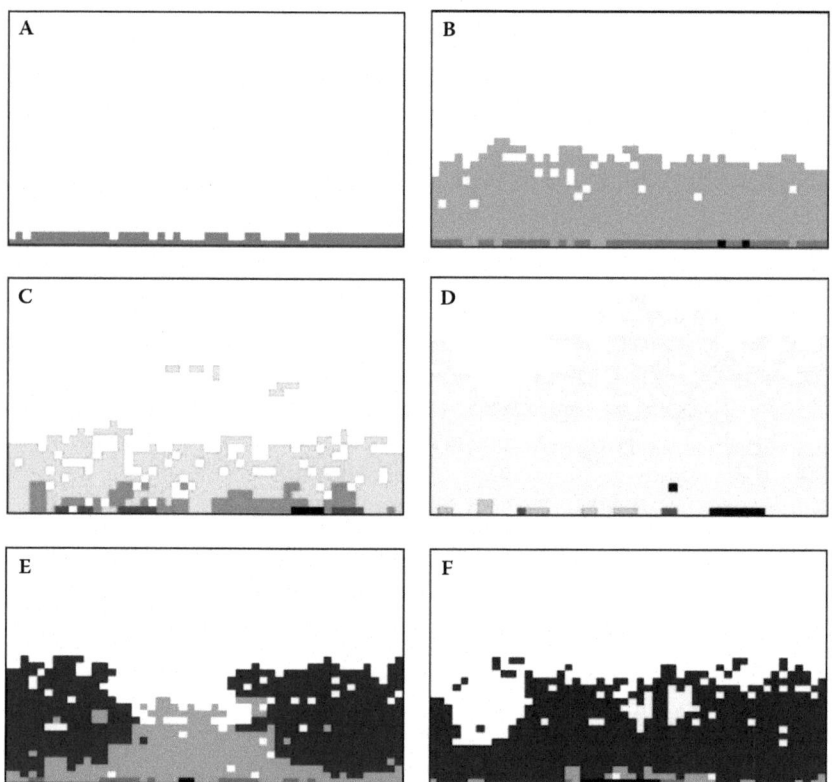

FIGURE 7.2   **(See color insert following page 40)**

of premalignant epithelia by assuming that one edge of the array represents the basement membrane. Initially, the automaton consists of a layer of a normal epithelial tissue, so the initial array consists of normal cells at the basement membrane and is vacant elsewhere. As well as proliferation and death, the cells may randomly undergo three possible heritable changes, either through mutations or epigenetic changes such as alterations in the methylation patterns of promoters. The cells may become (1) hyperplastic, allowing growth away from the basement membrane; (2) glycolytic, increasing their rate of glucose uptake and utilization; or (3) acid resistant, requiring a lower extracellular pH to induce toxicity. These three changes give rise to eight different phenotype combinations and, thus, eight competing cellular populations. Simulations from these models are shown in Figure 7.2, which demonstrates the temporal evolution of a typical cellular automaton model of tumor arising on the surface of a duct.

FIGURE 7.2 **(Opposite)** From References 51 and 52 showing simulations of intraductal evolution from the mathematical model described in the text showing potential pathways in ductal carcinoma in situ. Simulations start with a single layer of normal epithelial cells (grey cells) on a basement membrane (A). All simulations found that initial growth occurred only when mutations produced a hyperproliferative phenotype (pink cells) (B) through mutations in oncogenes, tumor suppressor genes, etc. Growth into the lumen eventually ceased, however, due to hypoxia and acidosis (B). Without additional cellular evolution, this population remains limited. Additional growth occurred following two possible sequences: (1) heritable changes that upregulate glycolysis. This population with constitutive upregulation (green cells) (C) allow this new population to replace the hyperplastic cells and to extend further into the lumen. However, clonal expansion is eventually limited by acid-mediated toxicity. This promotes evolution of a glycolytic, acid-resistant phenotype (yellow cells) which rapidly replaces all other extant populations in a highly aggressive, infiltrative pattern extending to the basement membrane and farther into the lumen (D). (2) A second pathway begins with development of an acid-resistant population (blue cells). This population expands and replaces many of the hyperplastic population (E) but growth remains limited by hypoxia promoting emergence of a phenotype with upregulated glycolysis and acid resistance (yellow cells) identical to the population in (C). However, unlike in (C), this phenotype initially grows into the normoxic region forming nodules of varying size (F). These eventually coalesce into a pattern essentially identical to the appearance in (D).

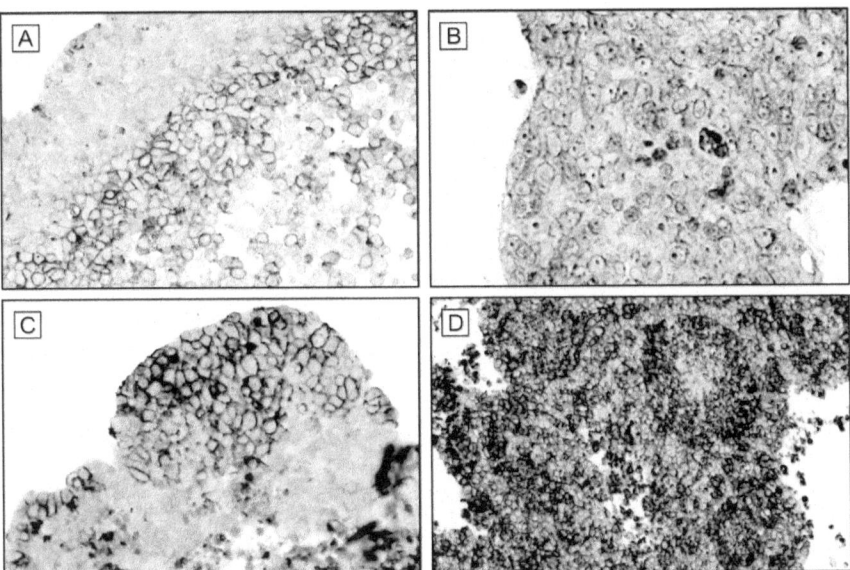

FIGURE 7.3 **(See color insert following page 40)** Multiple immunohistochemistry images from MCF7 cells grown in spheroids at 1 and 15 days following initial seeding. (A) Demonstrates GLUT-1 antibody distribution on day 1 showing upregulation only in the hypoxic core of the spheroid adjacent to areas of necrosis. (B) Shows NHE-1 staining diffusely consistent with constitutive upregulation. (C) Shows and (D) are two different spheroids each showing cluster of cells with upregulation of GLUT-1 in the normoxic regions of the spheroids. This growth pattern is identical to the nodular morphology predicted in Figure 7.2.

The top-left image consists of normal epithelial (gray) cells aligned along the basement membrane. The top-right image represents 100 generations of the model and shows pink hyperplastic cells (i.e., those with mutations in oncogenes and tumor suppressor genes) extending into the lumen of the duct and away from the basement membrane. The lower-left image is after 250 generations and shows replacement of many of the hyperplastic cells by glycolytic cells (green) in hypoxic regions of the premalignant lesion. In the lower-right image, following 300 generations, the glycolytic cells, hyperplastic cells, and normal cells have been largely replaced by cells that are both glycolytic and acid resistant (yellow). This represents an evolutionary sequence driven by microenvironmental hypoxia and acidosis that produce sufficient toxicity to force cellular evolution to more adapted phenotypes. Note that significant hypoxia and acidosis can be expected in the duct only a few cell layers (values shown on the x-axis) from the basement membrane. Figure 7.4 shows an increased expression

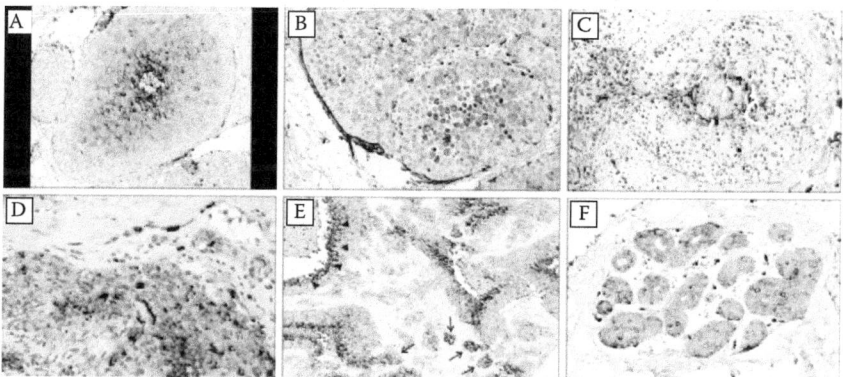

FIGURE 7.4 **(See color insert following page 40)** Immunonistochemistry demonstrating glucose transporter 1 (GLUT-1) distribution in Ductal Carcinoma In-Situ (DCIS) and invasive breast cancer. (A) Shows central distribution of upregulated GLUT-1 with a gradient of intensity that parallels the transition from normoxia to hypoxia as predicted in Figure 5B and similar to the gradient observed in spheroids at day 1 (Figure 6A). (B) Demonstrates a nodule of cells that predominantly demonstrate upregulation of GLUT-1 in the periphery of DCIS similar to the nodules seen in spheroids (Figure 5C and D). (C) Demonstrates extension of cells with upregulated GLUT-1 from the periluminal regions directly into a focus of invasion. (D) Show populations of cells with increased GLUT-1 expression in the periphery of DCIS adjacent to foci of microinvasion in which the cells also have increased GLUT-1 expression. Note the diffuse intracellular staining (i.e. membrane, cytoplasmic, and nuclear). (E) Demonstrates a region of DCIS in the upper left increased GLUT-1 expression only in the luminal, hypoxic cells (arrowheads). In the lower right are foci of microinvasion with increased GLUT-1 expression (arrows). Note the cells with increased GLUT-1 expression adjacent to the basement membrane in the adjacent tumor filled duct. (F) Demonstrates GLUT-1-positive cells in an invasive cancer.

of GLUT 1, presumably in response to hypoxia 3 or 4 cell layers from the basement membrane as predicted by the math models.

In Vitro Studies

Tumor spheroids grown in microgravity are unusually large, reaching about 1 cm in diameter and have diffusion-reaction kinetics dominating their in vivo physical characteristics, similar to those of intraductal tumors (53,54). As a result, they develop hypoxia, necrosis, apoptosis, and regional variations in cellular phenotypes. These spheroids have been used as an experimental model to reproduce variations in tumor microenvironmental and phenotypic adaptation predicted to occur by mathematical models. Specific attempts to test model predictions of nodular morphology of some evolving populations led to experiments with MCF-7, which exhibit much less aerobic glycolysis than the more aggressive, metastatic cell lines such as MDA-231 or MDA-438 but are resistant to acid-mediated toxicity. For this reason, it was anticipated that evolution of the MCF-7 cells to a more glycolytic and aggressive phenotype under hypoxic, acidic environmental conditions could be observed and that the subsequent growth dynamics would at least initially lead to nodular morphology. Initial experiments (55) showed that the MCF-7 spheroid were similar in size (up to 10 mm in diameter) to DCIS observed clinically. Environmental and cellular heterogeneity similar to those seen in DCIS was reproduced in the nodules, which developed zones of hypoxia, apoptosis, and necrosis (Figure 7.3). Despite the internal dynamics with environmental heterogeneity and cellular proliferation and death, the nodules have been maintained in a steady state of size over a period of 30 days.

All three spheroids harvested on day 1 exhibited evidence of hypoxia in the core with necrosis. Immunohistochemistry (IHXC) stains demonstrated upregulation of GLUT-1 in cells about 100–50 mm from the edge, representing an adaptation to hypoxia (Figure 7.5). IHC staining for NHE-1 was observed throughout all of the cells of the spheroid at all time points so that there was no evidence of regional or temporal variations in NHE-1 (Figure 7.5). In two (of three) spheroids harvested on day 15 following initiation and two (of three) spheroids harvested on day 30, cells exhibiting increased GLUT-1 were observed in the periphery of the spheroid (Figure 7.4), which is normoxic based on both the simulations and the absence of an increased GLUT 1 expression on the spheroids on day 1. In each spheroid, the nodules varied in size—a pattern remarkably similar to the model simulations. GLUT-1 and HIF-1a were coexpressed (not shown), indicating that increased glycolysis was regulated by stabilization of HIF-1a.

Clinical Observations

Gatenby et al. reviewed (55) 20 clinical specimens with DCIS for evidence of cellular adaptations predicted by the mathematical model and

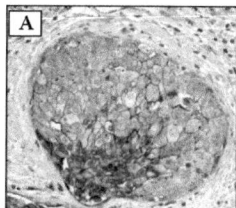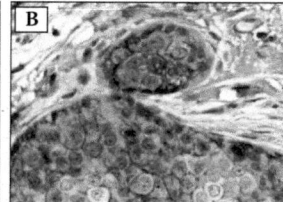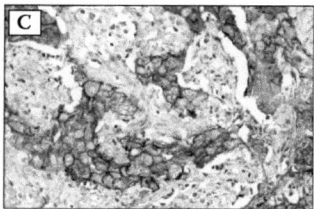

FIGURE 7.5 **(See color insert following page 40)** Immunohistochemistry showing sodium-hydrogen exchanger 1 (NHE-1) distribution in DCIS. In (A) NHE-1 expression is increased in a region of the DCIS extending into the normoxic region where there is a focal bulge into the basement membrane. In (B), there is a population of cells exhibiting increased NHE-1 expression in the periphery of DCIS adjacent to a focus of microinvasion which also exhibits increased expression of NHE-1. (C) Demonstrates upregulated NHE-1 in cells within an invasive breast cancer.

observed in spheroids. In all but one of the samples, tumor cells with upregulated GLUT-1 were observed in at least some regions of DCIS. In all 19 samples, an increased GLUT-1 expression was present in the central regions of intraductal tumors (Figure 7.4). In 17 of 19 specimens, cells with upregulated GLUT-1 were also observed in the peripheral (presumably normoxic regions) regions of some of the intraductal tumors. In all of these cases, the cells formed clusters similar in pattern to the model simulations and spheroids, suggesting that adaptation to acidosis typically precedes constitutive upregulation of glycolysis. This differs from the modeling results in which upregulated glycolysis preceded the development of acid resistance in most simulations. The reason for this will be the subject of further study.

In four specimens, a focus of microinvasive tumor was observed adjacent to a tumor-filled duct. In each of these cases, the cells in the invasive tumor demonstrated upregulated GLUT-1, as did the cells in the periphery of the DCIS immediately adjacent to the focus of microinvasion. Upregulation of GLUT-1 was observed in cells within four of five invasive cancers (Figure 7.4). The cellular expression of GLUT-1 was often both membranous and cytoplasmic—a pattern previously observed by Brown et al. (56). Regions of cells with an increased expression of NHE-1 were observed in all eight specimens examined (Figure 7.5). In DCIS, the distribution typically showed areas of an increased expression both centrally and in the periphery. Distinctive nodules, such as those seen in GLUT-1 distribution, were not observed. In the four cases of microinvasion, an upregulated NHE1 expression was observed both in the invasive cells and in the DCIS cells immediately adjacent to the foci of microinvasion. In all three cases of invasive cancer examined, NHE-1 was upregulated diffusely in the tumor cells.

Similar results were recently reported by Lee et al. (57), who demonstrated a hypoxia–glycolysis–acidosis sequence in cervical cancer and by Pinheiro et al., who demonstrated upregulation of monocarboxylate transporters 1 and 4 (a presumed adaptation to acidosis) in the transition from in situ to invasive gynecological cancers (58).

DISCUSSION

Can Perturbations of the Physical Microenvironment Inhibit Carcinogenesis?

Ultimately, the goal of understanding the role of the physical microenvironment in carcinogenesis is the translation into potential clinical prevention strategies. There have been to date no explicit experimental or clinical attempts to perturb the physical microenvironment during carcinogenesis to determine the effects on transition from in situ to invasive cancer. However, it is possible that such experiments have been inadvertently performed by individuals who self-induce brief episodes of systemic acidosis through exercise. Indeed, there is accumulating evidence that regular physical activity is a potent cancer prevention strategy. Friedenreich and Orenstein (59) recently reviewed over 170 observations for epidemiological studies and concluded that evidence for decreased cancer risk with increased physical activity was convincing for breast and colon cancer, probably for prostate cancer, possible for lung cancer, and unknown for all other sites. The mechanism by which regular physical activity reduces cancer risk is unknown. It is widely speculated that the beneficial effects in breast cancer arise from exercise-induced alteration of hormone levels, although no specific mechanism has been developed experimentally. Furthermore, the benefits of exercise have been demonstrated in pre- and postmenopausal women and in a number of other cancer sites as noted earlier. Smallbone et al. have proposed the hypothesis that the observed protective effects may be mediated by the transient systemic acidosis associated with exercise. Specifically, studies have shown that even moderate exercise may reduce arterial pH to less than 7.3 for as much as an hour. This would briefly alter the diffusion-reaction kinetics within an in situ cancer, perhaps causing tumor cell necrosis and a delay in further somatic evolution. Preliminary *in silico* studies have suggested that brief, transient episodes of systemic acidosis may, in fact, delay the transition from in situ to invasive cancers by many years. These interesting results suggest a need to further explore the potential clinical effects of perturbation of the physical microenvironment in both cancer prevention and treatment.

CONCLUSION

The role of the physical microenvironment in carcinogenesis has not been well investigated. However, an iterative research approach in which mathematical modeling is integrated into in vitro, in vivo, and clinical observations has suggested that oxygen, acid, and glucose concentrations within premalignant tumors strongly affect their subsequent evolution to an invasive cancer. From this work, novel insights have been gained into the roles of the unique anatomy and physiology of epithelial surfaces in carcinogenesis. Because epithelial tumors grow on an intact basement membrane, they remain separated from their blood supply, and substrate and metabolites must reach tumor cells through diffusion-reaction kinetics. As tumor cells proliferate further and further from the basement membrane, they will develop regions of hypoxia (which is often cyclical) requiring upregulation of glycolysis. The consequent increase in acid production and absence of blood vessels to remove excess H _results in the development of a potentially toxic, acidic microenvironment. This, in turn, requires adaptations to reduce acid-induced cytotoxicity. This sequence appears critical for subsequent evolution of invasive cancer because it confers a profound adaptive advantage. Specifically, when tumor cells adapted to the hypoxic, acidic regions of in situ cancers move into normoxic regions, they create an acidic environment (because of constitutively upregulated glycolysis) that is toxic to the local cells but not to themselves. That is, the hypoxia–glycolysis–acidosis sequence produces a mechanism that allows for transition to an invasive phenotype (see Figure 7.5 panel C). If the critical role of the physical microenvironment in evolution of the malignant phenotype is confirmed, this work will provide opportunities for new cancer prevention strategies.

REFERENCES

1. Garcia, S.B., Novelli, M., and Wright, N.A. The clonal origin and clonal evolution of epithelial tumors. *Int. J. Exp. Path.* 81:89–116. 2000.
2. Ilyas, M., Straub, J., Tomlinson, I.P.M., and Bodmer, W.F. Genetic pathways in colorectal and other cancers. *Eur. J. Cancer* 35:335–351, 1999.
3. Gray, J.W. and Collins, C. Genome changes and gene expression in human sold tumors. *Carcinogenesis.* 21:443–452, 2000.
4. Nowell, P.C. The clonal evolution of tumor cell populations. *Science* 194:23–28, 1976.
5. Clarke, R., Dickson, R.B., and Brunner, N. The process of malignant progression in human breast cancer. *Ann. Oncol.* 1:401–407, 1990.
6. Fearon, E.R. and Vogelstein, B. A genetic model for colorectal tumorigenesis. *Cell* 61:759–767, 1990.
7. Loeb, L.A. Mutator phenotype may be required for multistage carcinogenesis. *Cancer Res.* 51:3075–3079, 1991.
8. Peinado, M.A., Malkhosyan, S., Velazquez, A., and Perucho, M. Isolation and characterization of allelic losses and gains in colorectal tumors by arbitrarily primed polymerase chain reaction. *Proc. Natl. Acad. Sci. USA* 89:10065–10069, 1992.
9. Shih, I.S., Zhou, W., Goodman, S.N., Lengauer, C., Kinzler, K., and Vogelstein, B. Evidence that genetic instability occurs at an early stage of colorectal tumorigenesis. *Cancer Res.* 61:818–822, 2001.
10. Rabinovitch, P.S., Reid, B.I., Haggitt, R.C., Norwood, T.H., and Rubin, C.E. Progression to cancer in Barrett's esophagus is associated with genomic instability. *Lab. Invest.* 60:65–71, 1988.
11. Chow, M. and Rubin, H. Clonal selection versus genetic instability as the driving force in neoplastic transformation. *Cancer Res.* 60:6510–6518, 2000.
12. Tomlinson, I.P., Novelli, M.R., and Bodmer, W.F. The mutation rate and cancer. *Proc. Natl. Acad. Sci. USA* 93:14800–14803, 1996.
13. Boudreau, N., Sympson, C.J., Werb, Z., and Bissell, M.J. Suppression of ICE and apoptosis in mammary epithelial cells by extracellular matrix. *Science* 267:891–893, 1995.
14. Wang, F., Hansen, R.K., Radisky, D., Yoneda, T., Barcellos-Hoff, M.H., Petersen, O.W., Turley, E.A., and Bissell, M.J. Phenotypic reversion or death of cancer cells by altering signaling pathways in three-dimensional contexts. *J. Natl. Cancer Inst.* 94:1494–1503, 2002.
15. Weaver, V.M., Petersen, O.W., Wang, F., Larabell, C.A., Briand, P., Damsky, C., and Bissell, M.J. Reversion of the malignant phenotype of human breast cells in three-dimensional culture and in vivo by integrin blocking antibodies. *J. Cell Biol.* 137:231–245, 1997.
16. Park, C.C., Bissell, M.J., and Barcellos-Hoff, M.H. The influence of the microenvironment on the development of the malignant phenotype. *Mol. Med.* 6:324–329, 2000.
17. Gatenby, R.A. and Vincent, T.L. An evolutionary model of carcinogenesis. *Cancer Res.* 63:6212–6220, 2003.

18. Gatenby, R.A. and Maini, P. Modelling a new angle on understanding cancer. *Nature,* 420:462, December 5, 2002.
19. Gatenby, R.A. and Maini, P. Mathematical oncology—Cancer summed up. *Nature,* 421:321, January 23, 2003.
20. Buerger, H., Otterbach, F., Simon, R., Poremba, C., Diallo, R., Decker, T., Riethdorf, L., Brinkschmidt, C., Dockhorn-Dworniczak, B., and Boecker, W. Comparative genomic hybridization of ductal carcinoma in situ of the breast—evidence of multiple genetic pathways. *J. Pathol.* 187:396–402, 1999.
21. Jiang, F.J., Desper, R., Papadimitriou, C.H., Schaffer, A.A., Kallioniemi, O.P., Richter, J., Schraml, P., Sauter, G., Mihatsch, M.J., and Moch, H. Construction of evolutionary tree models for renal cell carcinoma from comparative genomic hybridization data. *Cancer Res.* 60:6503–6509, 2000.
22. Kerangueven, F., Noguchi, T., Coulier, F., Allione, F., Wargniez, V., Simony-Lafontaine, J., Longy, M., Jacquemier, J., Sobol, H., Eisinger, F., and Birnbaum, D. Genome-wide search for loss of heterozygosity shows extensive genetic diversity of human breast carcinomas. *Cancer Res.* 57:5469–5474, 1997.
23. Gatenby, R.A. and Vincent, T. Application of quantitative models from population biology and evolutionary game theory to tumor therapeutic strategies. *Mol. Cancer Ther.* 2:919–927, 2003.
24. Vincent, T. and Gatenby, R.A., Somatic evolution of cancer. International Game Theory Review Dynamic Games Applications and Methods. 8:331–346, 2006.
25. Vincent, T.L. and Van, M.V., and Goh, B.S. Ecological stability, evolutionary stability, and the ESS maximum principle. *Evol. Ecol.* 10:567–591, 1996.
26. Vincent, T.L., Cohen, Y., and Brown, J.S. Evolution via strategy dynamics. *Theor. Pop. Biol.* 44:149–176, 1993.
27. Brown, J.S. and Vincent, T.L. A theory for the evolutionary game. *Theor. Pop. Biol.* 31:140–166, 1987.
28. Cohen, Y., Vincent, T.L., and Brown, J.S. Does the G-function deserve an F? *Evol. Ecol. Res.* 3:375–377, 2001.
29. Vincent, T.L. and Brown, J.S. Stability in an evolutionary game. *Theor. Pop. Biol.* 26:408–427, 1984.
30. Vincent, T.L. and Brown, J.S. Evolution under nonequilibrium dynamics. *Math Modelling* 8:766–771, 1987.
31. Grobstein, C. Mechanism of organogenetic tissue interaction. *Natl. Cancer Inst. Monogr.* 26:279–299, 1967.
32. Kratchwil, K. Organ specificity in mesenchymal induction demonstrated in the embryonic development of the mammary gland of the mouse. *Dev. Biol.* 20:46–71, 1969.
33. Petersen, O.W., Ronnov-Jessen, L., Howlett, A.R., and Bissell, M.J. Interaction with basement membrane serves to rapidly distinguish growth and differentiation pattern of normal and malignant human breast epithelial cells. *Proc. Natl. Acad. Sci. USA* 89:9064–9068, 1992.

34. Racker, E. History of the Pasteur effect and its pathobiology. *Mol. Cell. Biochem.* 5:17–23, 1974.

35. Semenza, G.L. et al. The metabolism of tumours: 70 years later. Novartis Foundation Symposium 240, 251–260, 2001.

36. Warburg, O. *Ueber den Stoffwechsel der Tumoren.* Constable, London, 1930.

37. Hawkins, R.A., Phelps, ME. PET in clinical oncology. *Cancer Metastasis Rev.* 7:119–142, 1988.

38. Weber, W.A., Avril, N., and Schwaiger, M. Relevance of positron emission tomography (PET) in oncology. *Strahlentherapie und Onkologie* 175:356–373, 1999.

39. Gambhir, S.S. Molecular imaging of cancer with positron emission tomography. *Nat. Rev. Cancer.* 2:683–693, 2002.

40. Bos, R. et al. Biologic correlates of (18)fluorodeoxyglucose uptake in human breast cancer measured by positron emission tomography. *J. Clin. Oncol.* 20:379–387, 2002.

41. Burt, B.M. et al. Using positron emission tomography with [(18)F]FDG to predict tumor behavior in experimental colorectal cancer. *Neoplasia* (New York) 3:189–195, 2001.

42. Gatenby, R.A. and Gillies, R.J. Why do cancers have high levels or aerobic glycolysis? *Nat. Rev.* 4:891–898, 2004.

43. Krogh, A. The number and distribution of capillaries in muscles with calculations of the oxygen pressure head necessary for supplying the tissue. *J. Physiol.* 52:409–415, 1919.

44. Thomlinson, R.H. and Gray, L.H. The histological structure of some human lung cancers and the possible implications for radiotherapy. *Br. J. Cancer* 9:539–549, 1955.

45. Dewhirst, M.W., Secomb, T.W., Ong, E.T., Hsu, R., and Gross, J.F. Determination of local oxygen consumption rates in tumors. *Cancer Res.* 54:3333–3336, 1994.

46. Helmlinger, G., Yuan, F., Dellian, M., and Jain, R.K. Interstitial pH and pO_2 gradients in solid tumors in vivo: high-resolution measurements reveal a lack of correlation. *Nat. Med.* 3:177–182, 1997.

47. Wykoff, C.C. et al. Expression of the hypoxia-inducible and tumor-associated carbonic anhydrases in ductal carcinoma in situ of the breast. *Am. J. Pathol.* 158:1011–1019, 2001.

48. Gatenby, R.A. and Gawlinski, E.T. A reaction-diffusion model of cancer invasion. *Cancer Res.* 56:5745–5753, 1996.

49. Ober, S.S. and Pardee, A.B. Intracellular pH is increased after transformation of Chinese hamster embryo fibroblasts. *Proc. Natl. Acad. Sci. USA* 84:2766–2770, 1987.

50. McLean, L.A., Roscoe, J., Jorgensen, N.K., Gorin, F.A., and Cala, P.M. Malignant gliomas display altered pH regulation by NHE1 compared with nontransformed astrocytes. *Am. J. Physiol.* 278:C676–C688, 2000.

51. Smallbone, K., Gavaghan, D.J., Gatenby, R.A., and Maini, P.K. The role of acidity in solid tumour growth and invasion. *J. Theor. Biol.* 235:476–484, 2005.

52. Smallbone, K., Gatenby, R.A., Gillies, R., Maini, P., and Gavaghan, D. Metabolic changes during carcinogenesis: potential impact on invasiveness. *J. Theor. Biol.* 244:703–713, 2007.

53. Dabos, K.J., Nelson, L.J., Bradnock, T.J., Parkinson, J.A., Sadler, I.H., Hayes, P.C., and Plevris, J.N. The simulated microgravity environment maintains key metabolic functions and promotes aggregation of primary porcine hepatocytes. *Biochim. Biophys. Acta.* 1526:119–130, 2001.

54. Doolin, E.J., Geldziler, B., Strande, L., Kain, M., and Hewitt, C. Effects of microgravity on growing cultured skin constructs. *Tissue Eng.* 5:573–581, 1999.

55. Gatenby, R.A., Smallbone, K., Maini, P.K., Rose, F., Averill, J., Nagle, R.N., and Gillies, R.J. Cellular adaptations to hypoxia and acidosis during somatic evolution of breast cancer. *Br. J. Cancer.* 3:97(5):646–653, 2007.

56. Brown, R.S., Goodman, T.M., Zasadny, K.E., Greenson, J.K., and Wahl, R.L. Expression of hexokinase II and Glut-1 in untreated human breast cancer. *Nuclear Med. Biol.* 29:443–453, 2002.

57. Lee, W.Y., Huang, S.C., Hsu, K.F., Tzeng, C.C., and Shen, W.L. Roles for hypoxia-regulated genes during cervical carcinogenesis: somatic evolution during the hypoxia-glycolysis-acidosis sequence. *Gynecol Oncol.* 108(2):-377–384, 2008.

58. Pinheiro, C., Longatto-Filho, A., Ferreira, L., Pereira, S.M., Etlinger, D., Moreira, M.A., Jubé, L.F., Queiroz, G.S., Schmitt, F., and Baltazar, F. Increasing expression of monocarboxylate transporters 1 and 4 along progression to invasive cervical carcinoma. *Int J Gynecol Pathol.* 27(4):568–574, October 2008.

59. Friedenreicih, C.M. and Orenstein, M.R. Physical activity and cancer prevention: etiologic evidence and biological mechanisms. *J. Nutr.* 132: 3456S–3464S, 2002.

CHAPTER 8

Multiscale Modeling of Cell Motion in Three-Dimensional Environments

Dewi Harjanto and Muhammad H. Zaman

CONTENTS

INTRODUCTION

Both individual and collective cell migration are hallmarks of cancer invasion preceding metastasis and the fatal outcome of the disease.[1] To date, basic paradigms of cell migration, including the migration of cancer cells, have been extensively studied in reductionist 2D systems that insufficiently reflect the complexity of cancer invasion in vivo.[2] These reductionist approaches have been extremely powerful in identifying the genetic and epigenetic factors involved in various stages of tumor development and progression, but are also limited in their capacity to capture the systems-level behavior seen in vivo. Additionally, the artificiality of the substrate, dimensionality, and unrealistic geometric constraints can affect the observed signaling pathways quantified.[3] Migration experiments in 2D environments are also blind to biochemical and biomechanical effects of the surrounding matrix, resulting in an incomplete understanding of the tumor cell migration process.

Quantitative imaging and modeling of cellular motion in native-like three-dimensional matrices offer a far superior alternative that can capture a more realistic picture of in vivo cell migration in a controlled environment.[4-7] Not surprisingly, recent studies have shown marked differences between molecular, macromolecular, and cellular events in 2D and 3D environments.[3] These initial results have also shown improved agreement with in vivo observations, highlighting the need and necessity of measuring tumor cell migration in native-like 3D environments.

The traditional paradigm for migration, developed primarily from research on fibroblasts and keratinocytes moving on 2D substrates, consists of a multistep process.[8-10] First, the cell polarizes, adopting an elongated morphology via changes in the actin cytoskeleton. The cell then forms pseudopodia by extending out actin filaments at the leading edge to explore the matrix. Next, the cell interacts with the matrix, with integrins binding to extracellular matrix (ECM) components and clustering for signaling. The cell forms focal contacts in response to the integrin clustering, as the actin-binding proteins vinculin, paxilin, and alpha-actinin colocalize, a process regulated by kinases, including Rho-GTPases. The cell then contracts its cytoskeleton via actomyosin, which is regulated through the Rho-Rho kinase (ROCK) signaling pathway, pushing the cell forward. Finally, the cell detaches at the trailing edge, with the rear focal contacts cleaved via proteases such as calpain and sheddases. The integrins at the trailing edge surface are either released into the environment or absorbed into the cell endocytotically for reuse as new focal contacts are formed at the cell's leading edge, and the cycle repeats. In 3D, a proteolytic step is added after the adhesion step, where proteases such as matrix metalloproteinases (MMPs) are recruited to the focal contact sites at the cell surface to locally degrade the ECM and to cleave latent MMPs, activating more proteolytic enzymes for the removal of sterically hindering matrix obstacles.[11-13] An illustration comparing 2D and 3D migration is presented in Figure 8.1.

2D cell migration

3D cell migration

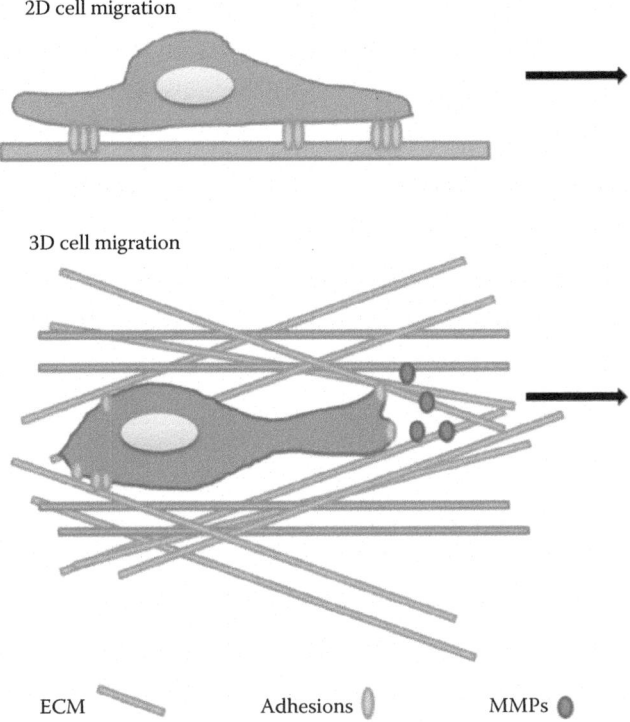

ECM   Adhesions   MMPs

FIGURE 8.1   On 2D substrates, cells form stable focal adhesions with the planar ECM. In 3D matrices, cells are capable of forming adhesions in all three dimensions. Proteolysis of the ECM barriers also becomes more crucial.

MODELS

Mathematical modeling has greatly contributed to our understanding of cell migration.[8,14] A major strength of computational approaches is their ability to isolate parameters that experimentally may be very difficult to extract.[15] Simulations also have the added advantages of affordability and efficiency over experimental work. Models have been developed that not only quantify a wide spectrum of experimental results, but are also being used to focus and predict the outcome of future experiments. This predictive power of models has allowed for the creation of new knowledge and a deeper understanding in 2D regarding the modes of motility, the balance between traction and adhesion, the role of matrix mechanics, and signaling pathways regulating motility. While the bulk of modeling efforts still continue to focus on migration in 2D due to the vast amount of literature available to calibrate and validate the models, the last few years have seen a significant surge in modeling efforts to study cell migration in 3D at both cellular and subcellular levels. It remains a challenge to create a unified model that accounts for the multiple levels of detail—cellular and molecular—as there is a dearth of good quantitative experimental and computational data to draw on. Nonetheless, the power of such a model would be tremendous in not only mathematical and computational fields, but also in fundamental and applied clinical sciences, ranging from predicting the outcome of a single mutation in a given receptor or identifying a specific drug target. At the moment, we are far from such an idealistic situation. Yet, there have been significant improvements in development and integration of mathematical and computational models that have allowed for an improved understanding of the migration process and predicting the outcomes of new experiments. In the following sections, we present an overview of a number of modeling strategies that are currently employed to probe cell migration at various length and timescales.

Force-Based Models

Force-based models account for cellular generation of traction forces to calculate motility tracks based on parameters such as receptor–ligand interactions. There have been 2D force-based models, including the work of DiMilla et al.[16] in which the researchers used receptor-ligand binding kinetics and implemented a viscoelastic-solid model to account for the cell mechanics.[1] DiMilla et al. found that cell velocity is biphasically dependent on cell–substrate adhesiveness, intracellular contractile force, and cell rheology. The model also predicted the importance of asymmetry in the distribution and adhesiveness of receptors. This work is an example of computational modeling guiding experimental efforts.

Zaman et al. present a 3D model that calculates traction forces, F_{traction}, at the front and rear end of the cell based on the receptor adhesiveness, characterized by dimensionless parameters $\beta_f(t)$ and $\beta_b(t)$ that are functions of the number of receptors, ligand density, and their binding constant, and the force per ligand–receptor complex, $F_{\text{R-L}}$, a function of matrix stiffness.[17] The tractions forces $F_{\text{trac-f}}$ and $F_{\text{trac-b}}$ in the forward and backward directions, respectively, are given by the equations

$$F_{\text{trac-f}} = F_{\text{R-L}} \times \beta_f(t). \qquad (8.1a)$$

$$F_{\text{trac-b}} = F_{\text{R-L}} \times \beta_b(t). \tag{8.1b}$$

The model also includes the forces due to the cell protruding into the matrix, $F_{\text{protrusion}}$, modeled with an experimentally defined magnitude and randomized direction, and the viscous drag, F_{drag}, a function of cell velocity and geometry and matrix viscosity, due to the resistance of the viscoelastic ECM against cell movement. For simplicity, the model assumes a spatially uniform distribution of ligands and equal binding constants for the integrins at the front and rear of the cell. Cell velocity is then obtained from setting the net force acting on the cell to zero (Equation 8.2).

$$F_{\text{total}} = F_{\text{drag}} + F_{\text{traction}} + F_{\text{protrusion}} = 0. \tag{8.2}$$

The time step of the model is 600 s, which is too large to account for cell dynamics at the edges such as actin waves and lamellipodial contractions.

The model predicts a biphasic response in cell velocity to adhesiveness, cell detachment force, matrix stiffness, and ligand concentration, with peak velocity achieved at intermediate levels of each parameter. Asymmetry in receptor concentrations between the front and rear ends increases cell velocity. The results qualitatively agree with what has been found in 2D experimental and computational work, but with the addition of the matrix properties the model provides a more 3D, physiological perspective.

However, the model, similar to other force-based algorithms, only predicts the movement of a single cell while cells in vivo generally migrate as a population. The model also fails to account for changing cell morphology, and the effect of proteolysis and ECM remodeling on matrix mechanics is completely neglected. While the simulation outputs velocity values, the numbers are not necessarily accurate, and the model is more useful when considered qualitatively. The model nonetheless is useful for identifying key cell and matrix parameters that are responsible for cell motility in vivo.

Stochastic Random Walk Models

Stochastic models of persistent random walk in 3D matrices are extensions of 2D migration models such as that of Tranquillo.[18–20] The path taken by each cell is determined by solving the Langevin equation numerically. The model selects a cubic volume element, and cells are distributed uniformly within this volume. A random velocity vector is assigned to each cell where each component of the velocity is selected randomly from a Gaussian distribution. This Gaussian distribution is directly proportional to the size of the time step. Parkhurst et al. used this kind of model to predict neutrophil motility in a 3D environment using a time step of 0.1 s.[21] After each time step, the velocity and the location of each cell are updated. The model defines the root mean square displacement $\{D^2\}t$ of the cell as a function of root mean square speed S_n and persistence P as follows:

$$\{D^2\}t = 2\,(S_n)^2 P(t - P + Pe\text{-}t/P) \tag{8.3}$$

(This formula is often used to extract speed and persistence from time-lapse images in experimental studies of cell migration.) Experimentally derived random motility coefficient and persistence parameter values are available in the literature and can be used in the Parkhurst model. After simulation over a period of time, the path taken by each cell in three dimensions is generated. By comparing their computer simulations to experimental results, Parkhurst et al.[21] determined that the 3D paths generated by the simulation are similar to the path taken by particles in a Brownian motion. By fitting the mean square displacement values for a different population of cells to calculate random motility coefficiency and persistence, they determined that with small cell populations (around 10 cells), the variation in random motility coefficiency and persistence is high, whereas with larger populations (greater than 50 cells), the estimates of random motility coefficiency and persistence approach experimental values. Hence, they conclude that a cell population of as few as 100 cells is enough to predict population behavior.

The strength of this model lies in the fact that population behavior can be predicted. Even though it is the paths of individual cells that are being predicted, population effects are still visible. The downside to this model is that dynamic effects such as traction and drag are not incorporated into the model. Also, the effect of matrix stiffness and porosity are not apparent in the model. Even though the population as a whole is evaluated in this model, it still fails to account for the fact that aggregation of cells is a possibility. However, this model is quite useful for validating the experimental results of a small population of cells in a 3D environment.

Another generalized stochastic model of 3D cell migration is presented by Dickinson.[22] Dickinson proposes a generalized random walk model in 3D that accounts for migration guided by haptoxtaxis, chemotaxis, or contact guidance in an anisotropic environment. The model presents migration on different timescales, accounting for both the short-timescale locomotive fluctuations that are well represented with a random walk along with the longer-timescale, more diffusive migration observed. However, the model only predicts the movement of a single cell.

Reaction-Diffusion-Based Multiple Cell Spheroid Models

Multiple cell spheroid models simulate the movement of cell masses by accounting for cell proliferation and necrosis, based on the diffusion of nutrients and oxygen. Differential rates of cell proliferation and cell death lead to pressure gradients that induce cell locomotion. Growth factors and chemokines[23] may also be included in such models.

Frieboes et al. propose a model of 3D tumor invasion using diffusion of nutrients, oxygen, and growth factors.[24,25] Necrosis is induced when nutrient and oxygen levels fall below a defined threshold level. Cell velocity is modeled to be proportional to local pressure, calculated using Darcy's law, and the buildup of tumor mass through cell division. Cell adhesion forces are represented with surface tension. Growth is calculated by using the conservation of mass. The results of the model suggest that there is more mitosis occurring at the edges of the tumor mass, where there is more perfusion of nutrients and oxygen; in contrast, in the middle of the spheroid, there are inadequate levels of nutrients, oxygen, and growth factors, so cells there die rather than proliferate. The authors conclude that diffusion gradients drive invasion by changing the tumor morphology, with high proliferation rates at the tumor rim resulting in subtumor formation, increasing the surface area exposed to free nutrients (i.e., less has to diffuse into the core to support the tumor). The breaking off of clusters is thus more conducive to metastasis. The computational results agreed well with experimental data.

Stein et al. present a model of a tumor cell population $u_i(r,t)$ based on a simple reaction-diffusion equation (Equation 8.4) in which cells proliferating at a given rate g are considered to be diffusing with a diffusion constant D that correlates with motility.[26,27]

$$\frac{\partial u_i}{\partial t} = D\nabla^2 u_i - v_i \nabla_r \cdot u_i + s\delta(r - R(t)) + gu_i\left(1 - \frac{u_i}{u_{max}}\right). \tag{8.4}$$

The spheroid is assumed to be radially symmetric. The authors distinguish the cells in the core of the tumor from the invasive cells on the rim of the spheroid, with invasive cells further defined by a shedding rate from the core surface, s, and radial velocity, v_i. r is the radial coordinate from the tumor core, and $R(t)$ is the core radius at time t. Parameter values are selected based on data from experiments that were also conducted by the group. They proposed that cells are more directed in their motility due to loss of cell–cell adhesion, chemotaxis away from waste or toward oxygen and nutrients, or haptotaxis because of MMP-mediated ECM breakdown. The model results suggest that the loss of cell–cell adhesion (increased shedding) could account for the more directed migration observed in the wild-type U87 cells compared to U87ΔEGFR cells, a glioblastoma model cell line that has a mutation in EGFR that increases invasiveness.

A major drawback of reaction-diffusion based models is that they neglect matrix properties, such as density, porosity, and stiffness, which are significant factors in 3D migration. They also fail to account for subcellular events such as lamellipodial formation and retraction. However, reaction-diffusion models are useful for simulating the migration of cell populations with chemokines, albeit only qualitatively at this point.

Monte Carlo Models

Monte Carlo methods can also be applied to model 3D migration. They are useful for explicitly accounting for the ECM and changing many parameters qualitatively.

Rubenstein and Kauffman model a multicellular spheroid (glioma) invasion by extending a cellular Q-Potts simulation to assign states to discrete lattice sites that represent cells as well as ECM.[28] Initially, a spheroid of cells is positioned in the middle of an ECM lattice. The cells are in one of three states: proliferative, quiescent, or necrotic, with cells changing between the three depending on waste and nutrient levels, and necrosis being a terminal state. Proliferating cells result in more cells replacing ECM sites. The ECM assumes either collagen fiber or nonfibrous matrix states. The authors applied the differential adhesion hypothesis, a thermodynamic principle that states that cells move about until they reach an arrangement that gives the lowest adhesion energy. Different adhesion energies are given between two adjacent sites (cell–cell or cell–ECM), with values selected based on experimental data. The lattice evolves to adopt the lowest energy state. While the model is actually in 2D, the results may be interpreted as cross sections of 3D cell masses, analogous to images obtained from confocal microscopy, as the simulation sought to account for migration phenomena observed in 3D ECM.

By running the simulation with increasing collagen concentration, the authors found a biphasic response in invasion radius and spheroid size with peak values observed at intermediate concentrations and stiffness. At high collagen concentrations, an excess of ECM ligands result in increased sterics and higher adhesion forces that have to be overcome, whereas at low collagen concentrations, there is inadequate adhesion for traction force generation and movement. The model is useful for qualitatively determining how modifying ECM parameters such as fiber density, diameter, length, and matrix porosity affects migration. Shortcomings of the model include the fact that proteolysis is not implemented as the ECM is ultimately conserved (once cells move from a site, it reverts to ECM) and diffusion of nutrients and waste is not represented.

Zaman et al. pursue a different approach to modeling migration using Monte Carlo methods.[29,30] They apply a 3D lattice, representing ECM, with the lattice spaces occupied by ECM ligands to varying extents. The pore size of each lattice space is then a function of the number of ligands, the area of the ligand, and the lattice space (ALS) cross section as given by the following equation

Pore Size

$$= (\text{ALS cross section}) \times \left[1 - (\text{Number of ligands}) \times \frac{\text{Area per ligand}}{\text{ALS cross section}} \right]$$

$$(8.5)$$

The pore size is varied such that it ranges from 10% to 90% of the lattice space. Cells are allowed to move from one lattice space to another if the ratio ψ of the cell cross section and the new lattice space's pore size falls within a certain regime (0.36–1, such that there are adequate ligands for adhesion and traction force generation). The model also allows for cell deformability so that cells may migrate even to spaces that have pore sizes smaller than the cellular cross section, as well as proteolysis when the deformation of cells still does not provide enough space for migration. So, three types of migration behavior are allowed for in this model: simple migration, when the pore size exceeds the cell size; cell-deformability-assisted migration, when cells must squeeze through smaller pores; and proteolysis-assisted migration, where sterically hindering ECM obstacles are degraded to increase the pore size. The Monte Carlo routine is offered in Figure 8.2. The model outputs cellular

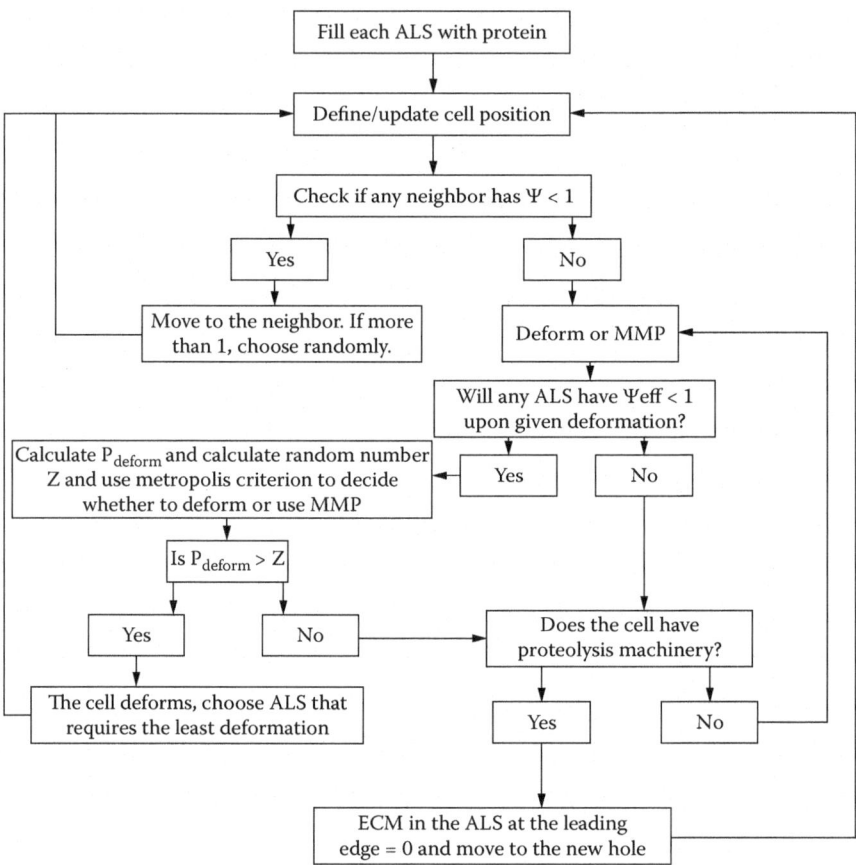

FIGURE 8.2 A schematic of the Monte Carlo routine for the discussed model. (Image reprinted with permission from Zaman, M.H. et al. *Ann. Biomed. Eng.* 35, 91–100, 2007.)

persistence and velocity. The simulation found a bimodal response in velocity and resistance to ligand concentration and pore size, as even when cells are capable of deformation and proteolysis, they are unable to migrate through very small pore sizes. Deformability and proteolysis contribute most at moderate ligand concentrations and intermediate pore sizes, with MMPs being more important in achieving persistent migration. The results correlate well with experimental results. However, this model fails to account for the deposition of ECM and phenomena on the subcellular level such as individual receptor–ligand interactions and the formation of pseudopodia.

Integrating Length and Timescales at a Single-Cell Level

As mentioned in the introduction, the process of migration is inherently multiscale. Small changes in structure or sequence of adhesion receptors, at the molecular level, have significant consequences for the ability of cells to migrate, both at a single cell and population level. Similarly, micro-level changes in the extracellular matrix influence processes at the macro level.

Despite the clear connection between length and timescales from nano to macro (and microsecond to hours in the case of time), to date true multiscale models of cell motion, in 2D or 3D, do not exist. There are two major reasons for the dearth of robust multiscale models. The first problem is the lack of biological data that are necessary to provide high-quality input data. For example, efforts to connect length scales through molecular dynamics simulations require high-quality three-dimensional structures of proteins and macromolecular complexes. Most of the proteins involved at the cell–matrix interface are membrane proteins and, hence, notoriously hard to crystallize. Coarse-grained models, which assume certain structural motifs, end up being largely qualitative or nonspecific, providing little predictability. Similarly, large-scale systems-level data providing kinetic binding constants for subcellular proteins are also scarce. Ab initio structure prediction is also not a viable option due to large sizes of the proteins involved. In addition, first principle structure prediction is often unreliable for transmembrane proteins.

The second problem is rooted in lack of models that are able to bridge the timescales. This is, in part, connected to the first problem. However, independent of input data, there are very few models that are able to connect events at the molecular level to the macromolecular and cellular level. Coarse-graining is often ad hoc and unable to predict anything beyond what can be easily predicted through scaling arguments. This problem is further amplified due to enormous computational costs associated with modeling processes that can scale from microseconds, or picoseconds, to minutes or hours. Computational efforts made in this area have largely been on single proteins or protein fragments, which provide little information at the cellular level. Performing simulations on mutations or various homologous domains is computationally prohibitive.

Despite these major challenges, there have been a few efforts in the recent past that have shown significant promise. A number of these multiscale efforts have tried to connect cellular motion with either cell–matrix or cell–cell interactions. In the majority of these efforts, cellular adhesion has been used as a stepping stone for modeling migration, intravasation, and extravasation. Chaplain and co-workers[31,32] have used cadherin junctions and cell–cell adhesion in their multiscale tumor models to predict tumor cell intravasation. The authors have used continuum equations with cadherin adhesion pathways to predict transendothelial migration. The model developed shows good agreement with experimental findings and is able to connect molecular adhesion to cellular outcomes. However, the model is blind to cadherin structure and mutations, as well as interactions of cells with extracellular matrix through integrins. In another multiscale adhesion-migration model, Zaman and co-workers have studied the role of integrin in migration.[33,34] Their approach is rooted in equilibrium statistical mechanics and is focused on adhesion of integrins to extracellular ligands. The focus of these set of studies is more on the molecular structure, adhesion, and the free energy landscape to capture the role of cell–matrix adhesions. The study of Zaman and co-workers, however, does not take into account external nonequilibrium forces, such as shear, and does not provide information on cell–cell connectivity.

DISCUSSION

The field of modeling cell migration in 3D is relatively young as compared to migration models of 2D cell cultures. The field of multiscale models of single-cell motion is even younger. Nonetheless, the merger of these two intellectually rich fields is exciting for not only mathematicians and modelers but also for cell biologists, cancer biologists, and clinical oncologists for a wide variety of reasons ranging from fundamental questions to drug design and delivery.[35-39]

Despite recent developments in modeling single-cell processes in 3D, a number of key challenges remain before the multiscale migration models can be useful in making quantitative predictions. Among these, incorporation of matrix mechanical, structural, and chemical components ranks at the top. As mentioned earlier, cellular processes do not occur in isolation, but are informed and influenced by changes in the surrounding matrix. While there are a number of studies focused on developing multiscale models of collagen fibers and the extracellular network, they ignore cellular motion, matrix remodeling, and general cell–matrix interactions. Similarly, multiscale models of single-cell migration also lack the molecular details necessary to make systems-level predictions. While coarse-graining is necessary for computational efficiency, it does not have to assume the matrix as a continuum of ligands with uniform mechanical properties. Just as the structure, orientation, and organization of cellular receptors plays a role in cellular motion and invasion, the organization, porosity, structural integrity, and compliance of the matrix also controls tumor cell motion. To date, no effort has been made to integrate various multiscale models of matrix with those of cell migration. While it may not be completely straightforward, a dedicated attempt to develop integrated models is definitely worthwhile and may lead to new insights into cellular migration and adhesion.

The other key challenge that remains largely unresolved is the integration of mechanochemical factors regulating migration with the signaling cascades. This is particularly challenging since not only is the kinetic information about some of the most important signaling pathways, at best, sketchy, but also the spatiotemporal resolution necessary to model is completely absent. In the absence of these two critical components, a truly multiscale model may be quite challenging. However, the very idea of a multiscale model of migration has to involve mechanosensation, inside-out signaling, and information flow. In this regard, a starting point can be integrin signaling, which is fairly well studied and incorporates both mechanical and chemical components for its function.

In the end, perhaps the biggest and most immediate challenge is to develop and foster close connections with experiments that not only validate the predictions, but also provide the necessary input needed at various levels and length scales of the model. Experiments in 3D cultures are improving rapidly and are able to provide an unprecedented level of simultaneous information about the matrix and the cell that was previously unavailable. Modeling a complex and dynamic process of cell motion in 3D would require a new level of connection between experiments and modeling, not only for validation, but also for building the bridges that connect the flow of information across orders of magnitude of length and timescale. Ultimately, our ability to fully understand tumor invasion and migration rests upon building these bridges, which will provide quantitative information linking events that connect events at the genetic level with outcomes at the cellular and multicellular level.

ACKNOWLEDGMENTS

We would like to thank the National Institutes of Health for its generous support (1R01CA132633).

REFERENCES

1. Hanahan, D. and Weinberg, R.A. The hallmarks of cancer. *Cell* 100, 57–70, 2000.
2. Friedl, P. and Brocker, E.B. The biology of cell locomotion within three-dimensional extracellular matrix. *Cell. Mol. Life. Sci.* 57, 41–64, 2000.
3. Zaman, M.H., Trapani, L.M., Sieminski, A.L., MacKellar, D., Gong, H., Kamm, R.D., Wells, A., Lauffenburger, D.A., and Matsudaira, P. Migration of tumor cells in 3D matrices is governed by matrix stiffness along with cell-matrix adhesion and proteolysis. *Proc. Natl. Acad. Sci. USA* 103, 10889–10894, 2006.
4. Friedl, P. and Bröcker, E.-B. (Eds. H. Heine and M. Rimpler) 7–18 *Extra cellular matrix and aging*, (Gustav Fischer Publ., Stuttgart, Jena, Lübeck, Ulm; 1997).
5. Friedl, P. and Brocker, E.B. (Ed. D.P. Hader) 9–21 *Image Analysis: Methods and Applications*, (CRC Press, Boca Raton, London, New York; 2001).
6. Friedl, P. and Gilmour, D. Collective cell migration in morhogenesis and cancer. *Nat. Rev. Cell. Mol. Biol.* (invited), 2009.
7. Friedl, P., Zanker, K.S., and Brocker, E.B. Cell migration strategies in 3-D extracellular matrix: differences in morphology, cell matrix interactions, and integrin function. *Microsc. Res. Tech.* 43, 369–378, 1998.
8. Lauffenburger, D.A. and Horwitz, A.F. Cell migration: a physically integrated molecular process. *Cell* 84, 359–369, 1996.
9. Ridley, A.J., Schwartz, M.A., Burridge, K., Firtel, R.A., Ginsberg, M.H., Borisy, G., Parsons, J.T., and Horwitz, A.R. Cell migration: integrating signals from front to back. *Science* 302, 1704–1709, 2003.
10. Webb, D.J. and Horwitz, A.F. New dimensions in cell migration. *Nat. Cell Biol.* 5, 690–692, 2003.
11. Friedl, P. and Wolf, K Tumour-cell invasion and migration: diversity and escape mechanisms. *Nat. Rev. Cancer* 3, 362–374, 2003.
12. Wolf, K., Mazo, I., Leung, H., Engelke, K., von Andrian, U.H., Deryugina, E.I., Strongin, A.Y., Brocker, E.B., and Friedl, P. Compensation mechanism in tumor cell migration: mesenchymal-amoeboid transition after blocking of pericellular proteolysis. *J. Cell Biol.* 160, 267–277, 2003.
13. Wolf, K., Wu, Y.I., Liu, Y., Geiger, J., Tam, E., Overall, C., Stack, M.S., and Friedl, P. Multi-step pericellular proteolysis controls the transition from individual to collective cancer cell invasion. *Nat. Cell Biol.* 9, 893–904, 2007.
14. Mogilner, A. Mathematics of cell motility: have we got its number? *J. Math. Biol.* 58, 105–134, 2009.
15. Rangarajan, R. and Zaman, M.H. Modeling cell migration in 3D: status and challenges. *Cell Adh. Migr.* 2, 106–109, 2008.
16. DiMilla, P.A., Barbee, K., and Lauffenburger, D.A. Mathematical model for the effects of adhesion and mechanics on cell migration speed. *Biophys. J.* 60, 15–37, 1991.

17. Zaman, M.H., Kamm, R.D., Matsudaira, P., and Lauffenburger, D.A. Computational model for cell migration in three-dimensional matrices. *Biophys. J.* 89, 1389–1397, 2005.

18. Tranquillo, R.T. and Lauffenburger, D.A. Stochastic model of leukocyte chemosensory movement. *J. Math. Biol.* 25, 229–262, 1987.

19. Tranquillo, R.T., Lauffenburger, D.A., and Zigmond, S.H. A stochastic model for leukocyte random motility and chemotaxis based on receptor binding fluctuations. *J. Cell. Biol.* 106, 303–309, 1988.

20. Shreiber, D.I., Barocas, V.H., and Tranquillo, R.T. Temporal variations in cell migration and traction during fibroblast-mediated gel compaction. *Biophys. J.* 84, 4102–4114, 2003.

21. Parkhurst, M.R. and Saltzman, W.M. Quantification of human neutrophil motility in three-dimensional collagen gels: effect of collagen concentration. *Biophys. J.* 61, 306–315, 1992.

22. Dickinson, R.B. A generalized transport model for biased cell migration in an anisotropic environment. *J. Math. Biol.* 40, 97–135, 2000.

23. McElwain, D.L. and Pettet, G.J. Cell migration in multicell spheroids: swimming against the tide. *Bull. Math. Biol.* 55, 655–674, 1993.

24. Frieboes, H.B., Zheng, X., Sun, C.H., Tromberg, B., Gatenby, R., and Cristini, V. An integrated computational/experimental model of tumor invasion. *Cancer Res.* 66, 1597–1604, 2006.

25. Frieboes, H.B., Lowengrub, J.S., Wise, S., Zheng, X., Macklin, P., Bearer, E., and Cristini, V. Computer simulation of glioma growth and morphology. *Neuroimage.* 37(Suppl. 1), S59–S70, 2007.

26. Stein, A.M., Nowicki, M.O., Demuth, T., Berens, M.E., Lawler, S.E., Chiocca, E.A., and Sander, L.M. Estimating the cell density and invasive radius of three-dimensional glioblastoma tumor spheroids grown in vitro. *Appl. Opt.* 46, 5110–5118, 2007.

27. Stein, A.M., Demuth, T., Mobley, D., Berens, M., and Sander, L.M. A mathematical model of glioblastoma tumor spheroid invasion in a three-dimensional in vitro experiment. *Biophys. J.* 92, 356–365, 2007.

28. Rubenstein, B.M. and Kaufman, L.J. The role of extracellular matrix in glioma invasion: a cellular Potts model approach. *Biophys. J.* 95, 5661–5680, 2008.

29. Zaman, M.H., Matsudaira, P., and Lauffenburger, D.A. Understanding effects of matrix protease and matrix organization on directional persistence and translational speed in three-dimensional cell migration. *Ann. Biomed. Eng.* 35, 91–100, 2007.

30. Zaman, M.H. A multiscale probabilisitic framework to model early steps in tumor metastasis. *Mol. Cell Biomech.* 4, 133–141, 2007.

31. Ramis-Conde, I., Drasdo, D., Anderson, A.R., and Chaplain, M.A. Modeling the influence of the E-cadherin-beta-catenin pathway in cancer cell invasion: a multiscale approach. *Biophys. J.* 95, 155–165, 2008.

32. Ramis-Conde, I., Chaplain, M.A., Anderson, A.R., and Drasdo, D. Multiscale modelling of cancer cell intravasation: the role of cadherins in metastasis. *Phys. Biol.* 6, 16008, 2009.

33. Yang, T. and Zaman, M.H. Free energy landscape of receptor-mediated cell adhesion. *J. Chem. Phys.* 126, 045103, 2007.

34. Yang, T. and Zaman, M.H. Regulation of cell adhesion free energy by external sliding velocities. *Experimental Mechanics*, (2009) 49, 1, 57–63.

35. Quaranta, V., Weaver, A.M., Cummings, P.T., and Anderson, A.R. Mathematical modeling of cancer: the future of prognosis and treatment. *Clin. Chim. Acta.* 357, 173–179, 2005.

36. Anderson, A.R., Rejniak, K.A., Gerlee, P., and Quaranta, V. Microenvironment driven invasion: a multiscale multimodel investigation. *J. Math. Biol.* 58, 579–624, 2009.

37. Quaranta, V., Rejniak, K.A., Gerlee, P., and Anderson, A.R. Invasion emerges from cancer cell adaptation to competitive microenvironments: quantitative predictions from multiscale mathematical models. *Semin. Cancer Biol.* 18, 338–348, 2008.

38. Anderson, A.R. and Quaranta, V. Integrative mathematical oncology. *Nat. Rev. Cancer* 8, 227–234, 2008.

39. Anderson, A.R., Weaver, A.M., Cummings, P.T., and Quaranta, V. Tumor morphology and phenotypic evolution driven by selective pressure from the microenvironment. *Cell* 127, 905–915, 2006.

CHAPTER 9

Simulating Cancer Growth with Agent-Based Models

Zhihui Wang, Veronika Bordas,
Jonathan Sagotsky, and Thomas S. Deisboeck

CONTENTS

INTRODUCTION

Tumorigenesis, a multistage, multifactorial process originating from molecular and genetic cell abnormalities, embraces extremely diverse and complex reciprocal dialogues that tumor cells and molecules engage in en route to malignancy (Al-Hajj and Clarke 2004; Balmain et al. 2003). Despite recent advances in cancer therapies, such as molecular-targeted therapy (Sawyers 2008), the clinical outcome of highly malignant tumors (such as gliomas and non-small-cell lung cancer (NSCLC)) remains disappointing, with one in four deaths in the United States attributed to this disease (Jemal et al. 2009). New discoveries in cancer prevention, early detection, and treatment are therefore desperately needed.

Lately, interdisciplinary approaches have garnered much attention, with data-driven mathematical and computational modeling gaining recognition for its potential to integrate the immense volume of data currently available, and to simulate and analyze the behavior of complex biological systems (Deisboeck et al. 2009; Kitano 2002). Of the two main types of models currently employed in the cancer modeling community, those utilizing *continuum* techniques have been somewhat more popular than those employing *discrete* methods (see Bearer et al. 2009; Byrne and Chaplain 1995, 1996; Cristini et al. 2003; Frieboes et al. 2009; Gatenby and Gawlinski 1996; Gerisch and Chaplain 2008; Macklin and Lowengrub 2007; Silva et al. 2009; Swanson et al. 2003; Szeto et al. 2009; Wise et al. 2008 for representative examples). Some of the main reasons for this preference are that (1) continuum descriptions of tumor growth benefit from the knowledge gained in fundamental physical principles (Tracqui 2009), and (2) continuum models are capable of capturing larger-scale volumetric tumor growth dynamics (which are also accessible to conventional clinical imaging modalities) at a comparatively lesser computational cost (Araujo and McElwain 2004). For instance, such models can characterize global properties of the nutrient molecules, the extracellular matrix, and tumor mass growth with reaction-diffusion equations rather than by using discrete subunits (Schaller and Meyer-Hermann 2006). However, continuum models are a lesser choice when exploring heterogeneity in both the tumor and its surrounding microenvironment, which is an inherent feature of cancer cells (Gatenby et al. 2009). Discrete models can address these shortcomings, since they work on the scale of individual cells (see Alarcon et al. 2003; Aubert et al. 2006; Bauer et al. 2007; Drasdo and Hohme 2003; Gatenby and Gawlinski 2003; Gevertz et al. 2008; Hatzikirou and Deutsch 2008; Hogeweg 2000; Kansal et al. 2000a, 2000b; Patel et al. 2001; Turner and Sherratt 2002 for representative examples). In addition, discrete models can easily incorporate biological rules (based on biomedical data or data-driven assumptions), such as defining cell–cell and cell–matrix interactions involved in both chemotaxis and haptotaxis. Yet, a major drawback of discrete models is their compute-intense nature due to the detail that each cell is modeled in, which limits the model to a relatively small number of cells. As a result, a typical discrete model is usually designed with a submillimeter or lower domain size (Wang and Deisboeck 2008). Nowadays, in discrete models, extracellular factors are often modeled as continuous quantities, thereby rendering the models *hybrid* in nature (Anderson and Quaranta 2008; Wang and Deisboeck 2008).

An agent-based model (ABM), also referred to as an individual-based model (IBM), simulates the interactions of autonomous entities (i.e., the agents) with each other and their local environment to predict higher-level emergent phenomena (Bonabeau 2002). In biomedical research, while an agent can represent a part of a cell or a cluster of cells (Casal et al. 2005; Robertson et al. 2007), the ideal candidate for a software agent is now more commonly recognized to be an individual cell (Walker and Southgate 2009), since models can benefit from such a direct one-to-one mapping between real and virtual cells in terms of parameter acquisition from experiments and model validation. As a simulation progresses, agents (representing individual cells) interact or communicate with other agents and their common microenvironment according to a set of predefined, biomedical data-driven "rules." Because an ABM's simulation results are highly dependent on these rules, it is necessary to tightly couple these algorithms at all stages of model development with iterative *in silico* as well as in vitro and/or in vivo biological experiments in order to validate and calibrate the rules according to relevant data (Thorne et al. 2007). To date, agent-based cancer models have produced preliminary results in identifying and quantifying the relationship between individual molecular properties, their microenvironmental conditions, and the overall tumor morphology (Zhang et al. 2009b).

In this chapter, we briefly introduce this particular modeling technique and focus on the design and development of the ABMs contributed by our laboratory, most of which span molecular and microscopic scales. Other groups are also employing ABMs to simulate different aspects of cancer, such as somatic evolution in tumorigenesis (Abbott et al. 2006; Spencer et al. 2006), the growth dynamics of multicellular tumor spheroids (Pepper et al. 2007; Schaller and Meyer-Hermann 2005), and cancer cell invasion (Pearce et al. 2007; Ramis-Conde et al. 2006). Yet unlike these models, our ABMs address the role of diversity in cell populations and also within each individual cell, and have the capacity to generate related, experimentally testable hypotheses and identify biomarkers. We show how these models were used to investigate cancer growth and invasion dynamics specifically in the cases of brain and lung cancer. Finally, we discuss some of the current challenges but also the future potential of agent-based cancer models.

MODEL

In simulating cancer, a challenge that each modeler first faces is in selecting the appropriate biological scales (both spatially and temporally—from genes and proteins to individual biological cells, and tissues, up to the entire organism) to be able to capture in sufficient detail the functional aspects that the model aims to examine. Focusing on only one scale, as does the vast majority of current cancer models (Tracqui 2009), simply neglects the correlative dependence and interplay between different scales. On the other hand, examining too many scales is likely to introduce more of the uncertainties already inherent in the biological mechanisms of each scale, making the final model difficult to validate, and may also present complications with computational intensity. Thus, it is reasonable to reduce the number of (1) explicitly involved biological scales, and (2) model components or parameters on each scale, before proceeding to a more refined and complex model version.

The multiscale model architecture presented here encompasses both molecular (signaling pathway) and microscopic (multicellular) scales. We choose the molecular scale because the aberration of signaling pathways responsible for coordinating the regulation of a variety of cellular processes (including proliferation, migration, differentiation, and apoptosis (Hlavacek et al. 2006)) contributes to the initiation and progression of many solid tumors (Hanahan and Weinberg 2000). We choose to incorporate the microscopic scale because even extrinsic environmental conditions alone can induce the carcinogenic transformation of cells (Postovit et al. 2006). That is, tumor cells bidirectionally communicate with their microenvironment, not only responding to various external cues but also impacting their surroundings for instance by producing (auto- and paracrine) signals and degrading the neighboring tissue through proteases (Hendrix et al. 2007). In the following, we briefly show how to implement a signaling pathway model, construct a microenvironment for multicellular tumor cell activities, and create an explicit link between these two scales by establishing an algorithm for determining cell phenotypic transitions upon molecular changes.

Molecular Signaling Pathway

The epidermal growth factor receptor (EGFR) is mutated and overexpressed in many cancers and, thus, studying it can provide insights into the processes leading up to tumor formation and growth (Oda et al. 2005). On the molecular level, epidermal growth factor (EGF) binds EGFR and promotes dimerization and subsequent autophosphorylation, resulting in the downstream activation of a number of key cell decision-making proteins such as phospholipase Cγ (PLCγ), extracellular signal-regulated kinase (ERK), and many others (Friedl and Wolf 2003). A number of EGFR-related pathway kinetic models have been developed (Hatakeyama et al. 2003; Kholodenko et al. 1999; Schoeberl et al. 2002), but regardless of differences in their complexity, all of these models use mathematical kinetic equations to describe molecular interactions. The change in the concentration of a certain protein pathway component (X_i) over time is then determined according to the following ordinary differential equation (ODE):

$$\frac{dX_i}{dt} = \alpha \cdot X_i - \beta \cdot X_i \tag{9.1}$$

where X_i represents one of the pathway components, and α and β are the reaction rates of producing and consuming X_i, respectively. Some pathway parameters, including the initial concentrations of pathway components and reaction rate constants, are not yet available in the literature, in which case they either have to be investigated experimentally or are fitted to published time-dependent quantitative or even qualitative observations.

Each cell or agent of a particular clonal cancer population should have a self-maintained signaling network since, as a simulation progresses, cells in distinct locations are likely to experience different external microenvironmental conditions. Hence, even though their internal states (including cell phenotype and pathway component concentrations) are set to be identical initially, they will exhibit different phenotypes after a certain lapse of time due to their respective molecular changes.

Tumor Growth Environment

Tumor growth dynamics can be investigated in a two-dimensional (2D) environment made up of a discrete lattice or in a three-dimensional (3D) environment composed of a discrete cube, where each grid point is either occupied by a single cell or is empty. Although a 3D model can generate more clinically relevant simulation results, for most cases, modelers like to develop a 2D model first to examine the feasibility of the entire modeling method in a more computationally tractable setting.

Over the course of a simulation, seed cells and their progeny respond to cellular and environmental biochemical cues that determine their phenotype at each time step. External diffusive chemical cues, such as glucose, oxygen tension, and growth factors are distributed throughout the environment in accordance with the experimental setting, or by means of probability distribution methods, such as by means of a normal distribution. Moreover, taking a 3D model as an example, throughout the simulation, the chemical cues are continuously updated at a fixed rate, using the following partial differential equation (PDE) form:

$$\frac{\partial C_{ijk}}{\partial t} = D_C \cdot \nabla^2 C_{ijk}, t = 1,2,3,\ldots, \tag{9.2}$$

where C represents the concentration of the external cue, D_C corresponds to the diffusion coefficient of C, t represents the time step, and ijk is the 3D integer coordinate of a given grid location.

Cell Phenotype Decision Algorithm

Changes in cell number, location, or environment will influence the boundary conditions or sources of molecular components in the signaling pathway model, whereas the spatial distribution or intracellular concentration of key molecular or protein species will influence the fate of the cell. How does one implement this process? An algorithm to establish the link between molecular changes and cell phenotypes is essential. We exemplify an intracellular signal-driven method as follows.

PLCγ is known to be involved in directional cell movement in response to EGF (Mouneimne et al. 2004) and is activated transiently in cancer cells, to a greater extent during migration and more gradually in the proliferation mode (Dittmar et al. 2002). With this finding, it is straightforward to model PLCγ using its rates of change ($ROC_{PLC\gamma}$) to determine a cell's migratory phenotype by comparing the current $ROC_{PLC\gamma}$ to a set threshold; that is, if $ROC_{PLC\gamma}$ exceeds the threshold, the cell then has the potential to migrate. However, a cell additionally has to meet other microenvironmental requirements, such as sufficient local nutrient conditions and available adjacent space, in order to process any phenotype transitions. If any of these conditions are not sufficient, then the cell will have to remain in

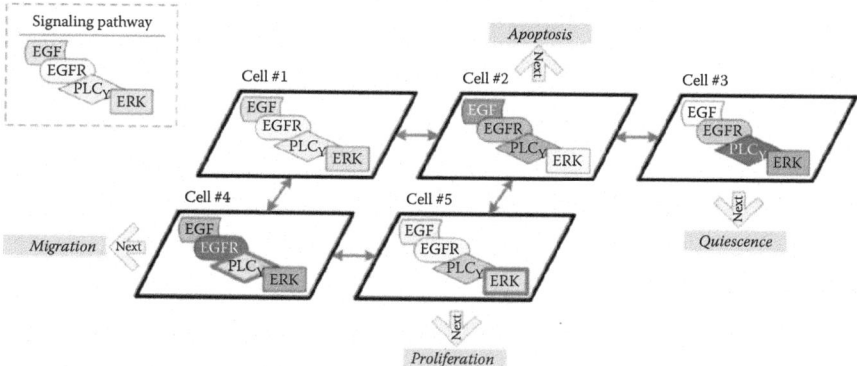

FIGURE 9.1   Schematic of the intracellularly driven cell phenotype decision algorithm. Intracellular signaling profiles for cells Nos. 1–5 are different (indicated by different grayscale levels), and thus lead to different cell phenotype output. In the next step, as illustrated, cell No. 4 will migrate because the rate of change of PLCγ has satisfied the requirements for migration. Similarly, cell No. 5 will proliferate because the rate of change of ERK has satisfied the requirements for proliferation.

its current location, waiting for the next iteration in the simulation when conditions will be reevaluated. Figure 9.1 schematically illustrates the cell phenotype decision algorithm that has been used in our cancer models. In addition to involving PLCγ to determine the cell migration fate, some of our recent models, based on experimental evidence (Santos et al. 2007), employ ERK to determine whether a cell will proliferate.

RESULTS

We have been working extensively with this modeling method to develop ABMs simulating tumor properties across multiple scales in time and space, within both brain tumors and non-small-cell lung cancer (NSCLC). A key feature of our ABMs is that tumor growth and invasion patterns due to cell proliferation and migration are neither predefined nor intuitive, but rather, they emerge as a result of intracellular signaling of individual cells and the dynamic cellular interactions within the framework of the biochemical microenvironment. In the following, we list some of the main findings of these models.

Brain Tumor Model

In a set of pilot 2D brain cancer models focusing on the simulation of glioma cell proliferation, migration, quiescence, and apoptosis on a microscopic scale (Mansury and Deisboeck 2003; Mansury et al. 2002), tumor cells were programmed to follow the path of "least resistance, lowest toxicity, and highest attraction" toward replenished and nonreplenished nutrient sources, avoiding areas of high mechanical confinement and detrimental metabolites that are the by-products of overpopulation. These models, which examined the effectiveness of treating tumors as a *self-organizing adaptive biosystem*, verified the hypothesis in favor of such a representation (Deisboeck et al. 2001). Additionally, to capture tumor growth dynamics on the pathway component level, we extended the previous model by adding a molecular scale in the form of a simplified representation of the EGFR signaling pathway (Athale et al. 2005; Athale and Deisboeck 2006). Such modeling enabled us to study and describe how the molecular profile of each individual glioma cell impacts the cell's phenotypic switch, and how such context-specific single-cell activities potentially affect the dynamics of the entire tumor system. In particular, we found that increasing the EGFR density per cell results in an acceleration of the entire tumor system's spatiotemporal expansion dynamics (Athale and Deisboeck 2006), a finding that is well supported by experimental observations (Lund-Johansen et al. 1990).

We extended our model not only to increase the biological scales of interest, but also to implement a more realistic tumor growth microenvironment. In a subsequent study, we simulated brain tumor growth in a 3D environment (Zhang et al. 2007) and implemented a simplified cell-cycle description at the subcellular scale from Alarcon et al. (2004). The simulation results not only confirmed the impact that regulation of EGFR signaling can have on tumor cell behavior, both on the single cell and multicellular level, but also indicated that over time, proliferative and migratory cell populations oscillate and have a direct effect on the entire spatiotemporal tumor expansion pattern. In a more recent study based on this 3D model (Zhang et al. 2009a), we investigated the emergence of heterogeneous tumor cell clones by introducing an element of genetic instability to analyze how heterogeneity impacts brain tumor progression patterns. Figure 9.2 demonstrates the 3D tumor growth over a series of time steps, with five cell clones that differ in their EGFR density. We found that cell clones with higher EGFR density comprise a larger migratory fraction and smaller proliferative and quiescent fractions, which corresponds well with reported experimental data (Steinbach et al. 2004). Overall, our group has been working extensively on modeling brain tumors across molecular and microscopic scales as reviewed in (Deisboeck et al. 2009; Wang and

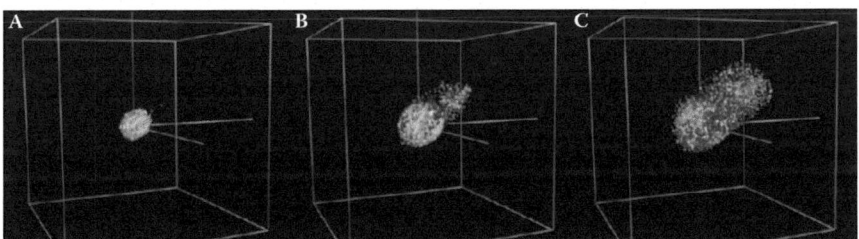

FIGURE 9.2 **(See color insert following page 40)** 3D tumor growth over three different time points. There are five different cell clones in this simulation, and each cell clone is initiated with a distinctive mutation rate and proliferation rate. Different colors represent different cell clones. (Adapted from Zhang et al. 2009a. *Math Comput. Model* 49, 307–19. With permission.)

Deisboeck 2008; Zhang et al. 2009b), effectively grounding and fostering future models incorporating multiple scales from the molecular up to the cellular level and beyond.

Non-Small-Cell Lung Cancer Model

Because the brain tumor models provided a successful computational paradigm, we began to apply these methods in parallel to the case of another tumor where EGFR plays an important role, NSCLC, with necessary modifications and extensions, and ultimately developed the first computational model of NSCLC across molecular and microscopic scales in the cancer modeling field. We first presented a 2D model with a revised EGF-induced EGFR-mediated signaling pathway that is specific to NSCLC to quantitatively understand the relationship between extrinsic chemotactic stimuli, the underlying properties of signaling networks, and the cellular biological responses they trigger in NSCLC from a systemic view (Wang et al. 2007). In addition to confirming the experimentally known fact that increasing the amount of available growth factors leads to a spatially more aggressive cancer system (Price et al. 1996; Xue et al. 2006), we found that in the cancer cell closest to the nutrient source, a minimal increase in EGF concentration can temporarily abolish its proliferative phenotype. A follow-up 2D simulation study (Wang et al. 2008) introduced a method for performing *cross-scale sensitivity analysis* to identify key model parameters that are critical in determining the output behavior of the model. While the method operated reliably over relatively large variations of most of the parameters, some parameters (three pathway components, including, e.g., ERK and eleven reaction steps) had greater impact on the system's multicellular performance (i.e., the tumor expansion rate) than others. Moreover, a small variation in the reference value (obtained from the literature) of any critical parameter appeared to result in a relatively large change in the output of the model.

More recently, we presented a 3D model in which both EGF and transforming growth factor β (TGFβ) and their interplay were taken into account (Wang et al. 2009). This physiologically and clinically motivated extension of the NSCLC modeling platform enabled us to investigate how the effects of individual and combinatorial change in EGF and TGFβ concentrations at the molecular level alter tumor growth dynamics, specifically tumor volume and expansion rate, on the multicellular level. We discovered a particular region of tumor system stability, generated by unique pairs of EGF and TGFβ concentration variations. When the variation pair of EGF and TGFβ concentrations occurred within this region, we observed that changes caused by the two growth factors did not effectively transmit to the downstream activation cascade, potentially explaining the resulting robustness of the tumor system at the multicellular level. However, the tumor system becomes sensitive to external variations in EGF and/or TGFβ when they occur outside this region, processing a phenotypic switch once the microenvironment becomes more permissive.

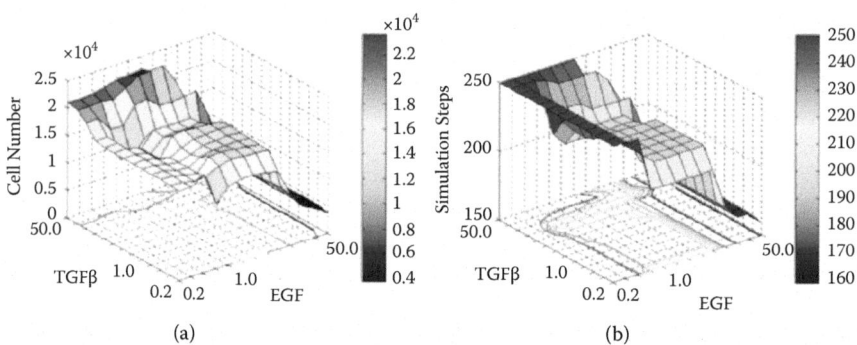

FIGURE 9.3　The effects of asynchronous combinatorial change in EGF and TGFβ concentrations on (A) tumor volume (cell number) and (B) tumor expansion rate ([inverse] simulation steps). In (A), the largest tumor volume is reached under conditions of high TGFβ and low or standard (with a variation of 1.0-fold) EGF concentrations. However, in (B), the most aggressive tumor expansion rate (fewest simulation steps) occurs under conditions of high EGF, regardless of TGFβ concentrations. (Adapted from Wang et al. 2009. *Bioinformatics* 25, 2389–96. With permission.)

Figure 9.3 shows the simulation results from changing EGF and TGFβ concentrations both simultaneously and asynchronously. As can be seen, the common stable phenotypic region is generated by 2- to 7-fold variation of EGF and 0.3- to 3-fold variation of TGFβ. The expansion rate for the standard simulation (with all kinetic parameters set to their reference values) is 2.07 μm/h, which is in very good agreement with both the modeling (Galle et al. 2005) and experimental studies (Bru et al. 2003).

Taken together, because of their cross-applicability to a variety of tumor types—beyond brain tumors and NSCLC—and their ability to integrate multiple pathways, these works have demonstrated the flexibility as well as the extensibility of our multiscale modeling architecture.

DISCUSSION AND FUTURE DIRECTIONS

The ultimate goal of *in silico* cancer research is to study the complexity of tumor progression in a reproducible setting in an effort to utilize the insights to accelerate diagnosis, to improve prediction, and to assist in treatment planning (Sanga et al. 2007). Cancer growth spans multiple spatial scales (from nanometers to centimeters) as well as temporal scales (from milliseconds to years) and, as such, developing cancer models across different biological scales is as critical (Hunter and Borg 2003) as it is daunting. We have reviewed the methods and achievements of our multiscale agent-based models—which encompass molecular as well as microscopic scales—in investigating brain tumor and NSCLC. Although still at an early stage, the models have demonstrated their ability to quantify the relationship between extracellular stimuli, intracellular signaling dynamics, and multicellular tumor growth and expansion. In these hybrid models, environmental factors such as growth factors, nutrients, and oxygen are represented by PDEs, while growth factor binding and intracellular signaling pathways are represented by ODEs. Simulation results can then be partly tested with in vitro or in vivo experiments, or verified with other theoretical studies. The method for creating the linkage of molecular and microscopic scales, first introduced by Athale et al. (2005), is novel and has stimulated a series of subsequent studies for both brain tumors and NSCLC. On the basis of these works, we argue that ABMs are highly suited to modeling complex emergent behaviors of cancerous systems, which are generated as an outcome of direct and indirect interactions between large numbers of individual cells.

The ABM framework that we have developed not only enables the monitoring of multicellular dynamics in response to molecular changes, but also facilitates the tracking of the fate of molecular components per cell and cell cluster as the entire tumor system evolves. It is now possible to ascertain the cause of a specific tumor growth pattern at the microscopic level by exploring the time-series history of intracellular signaling profiles within individual cells. For example, following tumor progression, epithelial cancer cells can transition from pursuing a collective invasion pattern toward adopting a detached and disseminated cell migration mechanism (Friedl and Wolf 2003). This process, referred to as epithelial–mesenchymal transition (EMT), is one of the fundamental mechanisms contributing to tumor infiltration and metastasis. EMT involves and requires diverse signal transduction pathways working together to initiate genetic and epigenetic changes that promote cell motility, invasiveness, and metastasis (Christiansen and Rajasekaran 2006). Using solely a continuum approach, it is difficult to (1) identify which particular cells start to move more independently of the collective, and to (2) capture the timing of EMT of those cells. However, with our ABM framework (which can track changes in the intracellular dynamics and extracellular environment conditions for each individual cell), it is straightforward to reveal which cells are undergoing EMT and to determine the cells' internal and external states at any particular time point. These simulation results can then readily be compared with, for example, experimental time-lapse video microscopy studies (Chambers et al. 2000).

It is gratifying to see that this promising approach continues to gain ground, with other cancer modeling groups recently integrating molecular features into their ABMs and providing their own cell phenotype decision algorithms. For instance, Gerlee and Anderson (Gerlee and Anderson 2007; 2009a; 2009b) use an artificial feed-forward neural network in a 2D space to investigate cancer cell motility in an evolving tumor population. They represent the microenvironmental variables such as local oxygen concentration, glucose concentration, and extracellular matrix (ECM) gradient using the input layer, regulatory genes with the hidden layer, and the response for cell phenotypes using output nodes. The weight matrix between the input and the hidden layer represents the signaling strength of cell surface receptors and, thus, changing a connection between the two layers corresponds to altering a certain receptor's level of expression. Yet this model, similar to those previously mentioned, does not adequately incorporate the effects of molecular-level variations on consequent cellular phenotypes and resulting tumor patterns, because it lacks the explicit representation of governing signaling pathways. More recently,

Ramis-Conde and co-workers integrated molecular pathways involving cell adhesion molecules, such as E-cadherin and β-catenin, into a multiscale ABM platform to study the effects of different pathways on cancer cell invasion patterns (Ramis-Conde et al. 2008; 2009). Such models have also implemented a biophysical representation of a single cell and highlighted the importance of both biological and biophysical properties in cell–cell contact formation and cell deformation.

It is noteworthy that multiscale ABMs incorporating detailed molecular dynamics have additionally been developed to aid in the understanding of biological systems other than cancer. For instance, the role of biochemical signaling in growing epithelial cell populations has been studied with multiscale lattice-free ABMs (Walker et al. 2006; 2008). Such models found that the intracellular signaling profile varied when measured across the entire cell population, and also confirmed that the local microenvironment influenced the response of individual cells. These tendencies, as well as the resulting population heterogeneity, are consistent with our findings in simulating cancer. In the case of translational systems biology of inflammation, a series of ABMs were developed to better understand acute inflammation by conceptualizing it as the interaction between endothelium and inflammatory cells (An 2008; 2009; An and Faeder 2009). Such models further highlight the generic promise of ABMs in translating the extensive mechanistic knowledge at the basic scientific level into an executable, integrated framework.

That said, a number of technical challenges exist in transitioning these ABMs to biomedical practice such as in clinics and for the pharmaceutical industry. These include the more common issues such as obtaining access to relevant data to validate simulation results, setting/streamlining standards for model definitions and, ultimately, sharing the models in environments that enable Web-based workflows (see Hunter and Borg 2003; Sagotsky et al. 2008; Thorne et al. 2007; Walker and Southgate 2009 for further discussions). Table 9.1 summarizes advantages and disadvantages of ABMs from a technical level. Most important, however, is the compute intensity associated with these discrete-based hybrid models. In modeling cancer, it is generally accepted that the higher a model's spatial and temporal resolution, the higher its compute power demand and, thus, the longer the run time (Deisboeck et al. 2009). Discrete models are more seriously affected by these problems because they are generally too detailed to simulate over a long period of time, particularly in large, 3D domains. Parallelizing the code and then running the model on

TABLE 9.1   Advantages and Disadvantages of Agent-Based Models in Modeling Cancer

Advantages	Disadvantages
• Well suited to model *emergent phenomena* resulting from the interactions of individual agents with each other and the local microenvironment • Easy incorporation and flexible change of *biological rules* for each agent in correspondence with biomedical data • Capable of investigating tumor growth dynamics at the *multicellular, microscopic,* and *molecular level* • Can examine *heterogeneity* per cell, tumor, and tumor environment • Allows for *cross-scale biomarker analysis*	• Necessity of making *simplifying assumptions* on model components and parameters • Currently limited to a relatively small scale due to *high computational demands* • Hence, necessity of coupling with *optimization approaches* to match large-scale 3D tumor growth • *Sensitive* to the predefined biological rules that govern cell properties

a cluster of supercomputers is a possible but not always practical solution that still may not resolve all the difficulties in handling the enormous amount of experimental and clinical data, which is why we and others have begun to turn to *hybrid, multiscale,* and *multiresolution* modeling (Anderson and Quaranta 2008; Sanga et al. 2007; Wang and Deisboeck 2008). In such models, multiresolution means that cells at distinct topographic regions are treated differently in terms of the modeling approach applied, achieving discretely high resolution wherever and whenever necessary to improve the model's predictive power while at the same time reducing compute intensity as much as possible to support scalability of the approach to clinically relevant levels. Alternatively, discrete modeling can be extended to incorporate different scales, from cell-scale to tissue-scale. For instance, the equation-free approach developed by Kevrekidis and co-workers (Erban et al. 2007; Gear and Kevrekidis 2003; Kolpas et al. 2007) leverages the spatiotemporal scale separation to allow for significant gains in computational efficiency by alternating short bursts of appropriately initialized microscopic simulations with accelerated result processing at the macroscopic, continuum scale. Furthermore, methods from other modeling communities, such as the Heterogeneous Multiscale Method (HMM, e.g., (E et al. 2003; Ren and E 2005)), can provide useful insight into efficient numerical methods that may be incorporated into the development of multiscale cancer models as well. By drawing on the strengths of these methods, such as scalability and multiresolution, and integrating them into the next-generation ABM models with a hierarchy of processes at varying timescales and space scales, we can produce more comprehensive, computationally efficient, and effective models to simulate tumor progression and predict treatment impact.

ACKNOWLEDGMENTS

This work has been supported in part by NIH grant CA 113004 (https://www.cvit.org) and by the Harvard–MIT (HST) Athinoula A. Martinos Center for Biomedical Imaging and the Department of Radiology at Massachusetts General Hospital.

REFERENCES

Abbott, R.G., S. Forrest, and K.J. Pienta. 2006. Simulating the hallmarks of cancer. *Artif. Life* 12: 617–34.

Al-Hajj, M. and M.F. Clarke. 2004. Self-renewal and solid tumor stem cells. *Oncogene.* 23: 7274–82.

Alarcon, T., H.M. Byrne, and P.K. Maini. 2003. A cellular automaton model for tumour growth in inhomogeneous environment. *J. Theor. Biol.* 225: 257–74.

Alarcon, T., H.M. Byrne, and P.K. Maini. 2004. A mathematical model of the effects of hypoxia on the cell-cycle of normal and cancer cells. *J. Theor. Biol.* 229: 395–411.

An, G. 2008. Introduction of an agent-based multi-scale modular architecture for dynamic knowledge representation of acute inflammation. *Theor. Biol. Med. Model* 5: 11.

An, G. 2009. A model of tlr4 signaling and tolerance using a qualitative, particle-event-based method: Introduction of spatially configured stochastic reaction chambers (scsrc). *Math. Biosci.* 217: 43–52.

An, G.C. and J.R. Faeder. 2009. Detailed qualitative dynamic knowledge representation using a bionetgen model of tlr-4 signaling and preconditioning. *Math. Biosci.* 217: 53–63.

Anderson, A.R. and V. Quaranta. 2008. Integrative mathematical oncology. *Nat. Rev. Cancer.* 8: 227–34.

Araujo, R.P. and D.L. McElwain. 2004. A history of the study of solid tumour growth: The contribution of mathematical modelling. *Bull. Math. Biol.* 66: 1039–91.

Athale, C., Y. Mansury, and T.S. Deisboeck. 2005. Simulating the impact of a molecular "decision-process" on cellular phenotype and multicellular patterns in brain tumors. *J. Theor. Biol.* 233: 469–81.

Athale, C.A. and T.S. Deisboeck. 2006. The effects of egf-receptor density on multiscale tumor growth patterns. *J. Theor. Biol.* 238: 771–9.

Aubert, M., M. Badoual, S. Fereol, C. Christov, and B. Grammaticos. 2006. A cellular automaton model for the migration of glioma cells. *Phys. Biol.* 3: 93–100.

Balmain, A., J. Gray, and B. Ponder. 2003. The genetics and genomics of cancer. *Nat. Genet.* 33(Suppl.): 238–44.

Bauer, A.L., T.L. Jackson, and Y. Jiang. 2007. A cell-based model exhibiting branching and anastomosis during tumor-induced angiogenesis. *Biophys. J.* 92: 3105–21.

Bearer, E.L., J.S. Lowengrub, H.B. Frieboes, Y.L. Chuang, F. Jin, S.M. Wise, M. Ferrari, D.B. Agus, and V. Cristini. 2009. Multiparameter computational modeling of tumor invasion. *Cancer Res.* 69: 4493–501.

Bonabeau, E. 2002. Agent-based modeling: Methods and techniques for simulating human systems. *Proc. Natl. Acad. Sci. USA* 99(Suppl. 3): 7280–7.

Bru, A., S. Albertos, J. Luis Subiza, J.L. Garcia-Asenjo, and I. Bru. 2003. The universal dynamics of tumor growth. *Biophys. J.* 85: 2948–61.

Byrne, H.M. and M.A. Chaplain. 1995. Growth of nonnecrotic tumors in the presence and absence of inhibitors. *Math. Biosci.* 130: 151–81.

Byrne, H.M. and M.A. Chaplin. 1996. Growth of necrotic tumors in the presence and absence of inhibitors. *Math. Biosci.* 135: 187–216.

Casal, A., C. Sumen, T.E. Reddy, M.S. Alber, and P.P. Lee. 2005. Agent-based modeling of the context dependency in t cell recognition. *J. Theor. Biol.* 236: 376–91.

Chambers, A.F., G.N. Naumov, S.A. Vantyghem, and A.B. Tuck. 2000. Molecular biology of breast cancer metastasis: Clinical implications of experimental studies on metastatic inefficiency. *Breast Cancer Res.* 2: 400–7.

Christiansen, J.J. and A.K. Rajasekaran. 2006. Reassessing epithelial to mesenchymal transition as a prerequisite for carcinoma invasion and metastasis. *Cancer Res.* 66: 8319–26.

Cristini, V., J. Lowengrub, and Q. Nie. 2003. Nonlinear simulation of tumor growth. *J. Math. Biol.* 46: 191–224.

Deisboeck, T.S., M.E. Berens, A.R. Kansal, S. Torquato, A.O. Stemmer-Rachamimov, and E.A. Chiocca. 2001. Pattern of self-organization in tumour systems: Complex growth dynamics in a novel brain tumour spheroid model. *Cell Prolif.* 34: 115–34.

Deisboeck, T.S., L. Zhang, J. Yoon, and J. Costa. 2009. In silico cancer modeling: Is it ready for prime time? *Nat. Clin. Pract. Oncol.* 6: 34–42.

Dittmar, T., A. Husemann, Y. Schewe, J.R. Nofer, B. Niggemann, K.S. Zanker, and B.H. Brandt. 2002. Induction of cancer cell migration by epidermal growth factor is initiated by specific phosphorylation of tyrosine 1248 of c-erbb-2 receptor via egfr. *FASEB J.* 16: 1823–5.

Drasdo, D. and S. Hohme. 2003. Individual-based approaches to birth and death in avascular tumors. *Math. Comput. Model.* 37: 1163–75.

E, W.N., B. Engquist, and Z.Y. Huang. 2003. Heterogeneous multiscale method: A general methodology for multiscale modeling. *Physical Rev. B* 67: 4.

Erban, R., T.A. Frewen, X. Wang, T.C. Elston, R. Coifman, B. Nadler, and I.G. Kevrekidis. 2007. Variable-free exploration of stochastic models: A gene regulatory network example. *J. Chem. Phys.* 126: 155103.

Frieboes, H.B., M.E. Edgerton, J.P. Fruehauf, F.R. Rose, L.K. Worrall, R.A. Gatenby, M. Ferrari, and V. Cristini. 2009. Prediction of drug response in breast cancer using integrative experimental/computational modeling. *Cancer Res.* 69: 4484–92.

Friedl, P. and K. Wolf. 2003. Tumour-cell invasion and migration: Diversity and escape mechanisms. *Nat. Rev. Cancer* 3: 362–74.

Galle, J., M. Loeffler, and D. Drasdo. 2005. Modeling the effect of deregulated proliferation and apoptosis on the growth dynamics of epithelial cell populations in vitro. *Biophys J.* 88: 62–75.

Gatenby, R.A. and E.T. Gawlinski. 1996. A reaction-diffusion model of cancer invasion. *Cancer Res.* 56: 5745–53.

Gatenby, R.A. and E.T. Gawlinski. 2003. The glycolytic phenotype in carcinogenesis and tumor invasion: Insights through mathematical models. *Cancer Res.* 63: 3847–54.

Gatenby, R.A., A.S. Silva, R.J. Gillies, and B.R. Frieden. 2009. Adaptive therapy. *Cancer Res.* 69: 4894–903.

Gear, C.W. and I.G. Kevrekidis. 2003. Telescopic projective methods for parabolic differential equations. *J. Comput. Phys.* 187: 95–109.

Gerisch, A. and M.A. Chaplain. 2008. Mathematical modelling of cancer cell invasion of tissue: Local and non-local models and the effect of adhesion. *J. Theor. Biol.* 250: 684–704.

Gerlee, P. and A.R. Anderson. 2007. An evolutionary hybrid cellular automaton model of solid tumour growth. *J. Theor. Biol.* 246: 583–603.

Gerlee, P. and A.R. Anderson. 2009a. Evolution of cell motility in an individual-based model of tumour growth. *J. Theor. Biol.* 259: 67–83.

Gerlee, P. and A.R. Anderson. 2009b. Modelling evolutionary cell behaviour using neural networks: Application to tumour growth. *Biosystems* 95: 166–74.

Gevertz, J.L., G.T. Gillies, and S. Torquato. 2008. Simulating tumor growth in confined heterogeneous environments. *Phys. Biol.* 5: 36010.

Hanahan, D. and R.A. Weinberg. 2000. The hallmarks of cancer. *Cell* 100: 57–70.

Hatakeyama, M., S. Kimura, T. Naka, T. Kawasaki, N. Yumoto, M. Ichikawa, J.H. Kim, K. Saito, M. Saeki, M. Shirouzu, S. Yokoyama, and A. Konagaya. 2003. A computational model on the modulation of mitogen-activated protein kinase (mapk) and akt pathways in heregulin-induced erbb signalling. *Biochem. J.* 373: 451–63.

Hatzikirou, H. and A. Deutsch. 2008. Cellular automata as microscopic models of cell migration in heterogeneous environments. *Curr. Top Dev. Biol.* 81: 401–34.

Hendrix, M.J., E.A. Seftor, R.E. Seftor, J. Kasemeier-Kulesa, P.M. Kulesa, and L.M. Postovit. 2007. Reprogramming metastatic tumour cells with embryonic microenvironments. *Nat. Rev. Cancer* 7: 246–55.

Hlavacek, W.S., J.R. Faeder, M.L. Blinov, R.G. Posner, M. Hucka, and W. Fontana. 2006. Rules for modeling signal-transduction systems. *Sci.* STKE 2006: re6.

Hogeweg, P. 2000. Evolving mechanisms of morphogenesis: On the interplay between differential adhesion and cell differentiation. *J. Theor. Biol.* 203: 317–33.

Hunter, P.J. and T.K. Borg. 2003. Integration from proteins to organs: The physiome project. *Nat. Rev. Mol. Cell. Biol.* 4: 237–43.

Jemal, A., R. Siegel, E. Ward, Y. Hao, J. Xu, and M.J. Thun. 2009. Cancer statistics, 2009. CA *Cancer J. Clin.*

Kansal, A.R., S. Torquato, E.A. Chiocca, and T.S. Deisboeck. 2000a. Emergence of a subpopulation in a computational model of tumor growth. *J. Theor. Biol.* 207: 431–41.

Kansal, A.R., S. Torquato, I.G. Harsh, E.A. Chiocca, and T.S. Deisboeck. 2000b. Cellular automaton of idealized brain tumor growth dynamics. *Biosystems* 55: 119–27.

Kholodenko, B.N., O.V. Demin, G. Moehren, and J.B. Hoek. 1999. Quantification of short term signaling by the epidermal growth factor receptor. *J. Biol. Chem.* 274: 30169–81.

Kitano, H. 2002. Computational systems biology. *Nature* 420: 206–10.

Kolpas, A., J. Moehlis, and I.G. Kevrekidis. 2007. Coarse-grained analysis of stochasticity-induced switching between collective motion states. *Proc. Natl. Acad. Sci. USA* 104: 5931–35.

Lund-Johansen, M., R. Bjerkvig, P.A. Humphrey, S.H. Bigner, D.D. Bigner, and O.D. Laerum. 1990. Effect of epidermal growth factor on glioma cell growth, migration, and invasion in vitro. *Cancer Res.* 50: 6039–44.

Macklin, P. and J. Lowengrub. 2007. Nonlinear simulation of the effect of microenvironment on tumor growth. *J. Theor. Biol.* 245: 677–704.

Mansury, Y. and T.S. Deisboeck. 2003. The impact of "search precision" in an agent-based tumor model. *J. Theor. Biol.* 224: 325–37.

Mansury, Y., M. Kimura, J. Lobo, and T.S. Deisboeck. 2002. Emerging patterns in tumor systems: Simulating the dynamics of multicellular clusters with an agent-based spatial agglomeration model. *J. Theor. Biol.* 219: 343–70.

Mouneimne, G., L. Soon, V. Desmarais, M. Sidani, X. Song, S.C. Yip, M. Ghosh, R. Eddy, J.M. Backer, and J. Condeelis. 2004. Phospholipase c and cofilin are required for carcinoma cell directionality in response to egf stimulation. *J. Cell Biol.* 166: 697–708.

Oda, K., Y. Matsuoka, A. Funahashi, and H. Kitano. 2005. A comprehensive pathway map of epidermal growth factor receptor signaling. *Mol. Syst. Biol.* 1: 2005 0010.

Patel, A.A., E.T. Gawlinski, S.K. Lemieux, and R.A. Gatenby. 2001. A cellular automaton model of early tumor growth and invasion. *J. Theor. Biol.* 213: 315–31.

Pearce, I.G., M.A. Chaplain, P.G. Schofield, A.R. Anderson, and S.F. Hubbard. 2007. Chemotaxis-induced spatio-temporal heterogeneity in multi-species host-parasitoid systems. *J. Math. Biol.* 55: 365–88.

Pepper, J.W., K. Sprouffske, and C.C. Maley. 2007. Animal cell differentiation patterns suppress somatic evolution. *PLoS Comput. Biol.* 3: e250.

Postovit, L.M., E.A. Seftor, R.E. Seftor, and M.J. Hendrix. 2006. Influence of the microenvironment on melanoma cell fate determination and phenotype. *Cancer Res.* 66: 7833–6.

Price, J.T., H.M. Wilson, and N.E. Haites. 1996. Epidermal growth factor (egf) increases the in vitro invasion, motility and adhesion interactions of the primary renal carcinoma cell line, a704. *Eur. J. Cancer* 32A: 1977–82.

Ramis-Conde, I., M.A. Chaplain, and A.R. Anderson. 2006. Mathematical modelling of cancer cell invasion of tissue. *Math. Comp. Modelling* 47: 533–45.

Ramis-Conde, I., M.A. Chaplain, A.R. Anderson, and D. Drasdo. 2009. Multi-scale modelling of cancer cell intravasation: The role of cadherins in metastasis. *Phys. Biol.* 6: 16008.

Ramis-Conde, I., D. Drasdo, A.R. Anderson, and M.A. Chaplain. 2008. Modeling the influence of the e-cadherin-beta-catenin pathway in cancer cell invasion: A multiscale approach. *Biophys. J.* 95: 155–65.

Ren, W.Q. and E W.N. 2005. Heterogeneous multiscale method for the modeling of complex fluids and micro-fluidics. *J. Comput. Phys.* 204: 1–26.

Robertson, S.H., C.K. Smith, A.L. Langhans, S.E. Mclinden, M.A. Oberhardt, K.R. Jakab, B. Dzamba, D.W. Desimone, J.A. Papin, and S.M. Peirce. 2007. Multiscale computational analysis of xenopus laevis morphogenesis reveals key insights of systems-level behavior. *BMC Syst. Biol.* 1: 46.

Sagotsky, J.A., L. Zhang, Z. Wang, S. Martin, and T.S. Deisboeck. 2008. Life sciences and the web: A new era for collaboration. *Mol. Syst. Biol.* 4: 201.

Sanga, S., H.B. Frieboes, X. Zheng, R. Gatenby, E.L. Bearer, and V. Cristini. 2007. Predictive oncology: A review of multidisciplinary, multiscale in silico modeling linking phenotype, morphology and growth. *Neuroimage 37 Suppl.* 1: S120–S34.

Santos, S.D., P.J. Verveer, and P.I. Bastiaens. 2007. Growth factor-induced mapk network topology shapes erk response determining pc-12 cell fate. *Nat. Cell Biol.* 9: 324–30.

Sawyers, C.L. 2008. The cancer biomarker problem. *Nature* 452: 548–52.

Schaller, G. and M. Meyer-Hermann. 2005. Multicellular tumor spheroid in an off-lattice voronoi-delaunay cell model. *Phys. Rev. E Stat. Nonlin Soft Matter Phys.* 71: 051910.

Schaller, G. and M. Meyer-Hermann. 2006. Continuum versus discrete model: A comparison for multicellular tumour spheroids. *Philos. Transact A Math. Phys. Eng. Sci.* 364: 1443–64.

Schoeberl, B., C. Eichler-Jonsson, E.D. Gilles, and G. Muller. 2002. Computational modeling of the dynamics of the map kinase cascade activated by surface and internalized egf receptors. *Nat. Biotechnol.* 20: 370–5.

Silva, A.S., J.A. Yunes, R.J. Gillies, and R.A. Gatenby. 2009. The potential role of systemic buffers in reducing intratumoral extracellular ph and acid-mediated invasion. *Cancer Res.* 69: 2677–84.

Spencer, S.L., R.A. Gerety, K.J. Pienta, and S. Forrest. 2006. Modeling somatic evolution in tumorigenesis. *PLoS Comput. Biol.* 2: e108.

Steinbach, J.P., A. Klumpp, H. Wolburg, and M. Weller. 2004. Inhibition of epidermal growth factor receptor signaling protects human malignant glioma cells from hypoxia-induced cell death. *Cancer Res.* 64: 1575–8.

Swanson, K.R., C. Bridge, J.D. Murray, and E.C. Alvord, Jr. 2003. Virtual and real brain tumors: Using mathematical modeling to quantify glioma growth and invasion. *J. Neurol. Sci.* 216: 1–10.

Szeto, M.D., G. Chakraborty, J. Hadley, R. Rockne, M. Muzi, E.C. Alvord, Jr., K.A. Krohn, A.M. Spence, and K.R. Swanson. 2009. Quantitative metrics of net proliferation and invasion link biological aggressiveness assessed by mri with hypoxia assessed by fmiso-pet in newly diagnosed glioblastomas. *Cancer Res.* 69: 4502–9.

Thorne, B.C., A.M. Bailey, and S.M. Peirce. 2007. Combining experiments with multi-cell agent-based modeling to study biological tissue patterning. *Brief Bioinform.* 8: 245–57.

Tracqui, P. 2009. Biophysical models of tumour growth. *Reports on Progress in Physics* 72.

Turner, S. and J.A. Sherratt. 2002. Intercellular adhesion and cancer invasion: A discrete simulation using the extended potts model. *J. Theor. Biol.* 216: 85–100.

Walker, D., S. Wood, J. Southgate, M. Holcombe, and R. Smallwood. 2006. An integrated agent-mathematical model of the effect of intercellular signalling via the epidermal growth factor receptor on cell proliferation. *J. Theor. Biol.* 242: 774–89.

Walker, D.C., N.T. Georgopoulos, and J. Southgate. 2008. From pathway to population—A multiscale model of juxtacrine egfr-mapk signalling. *BMC Syst. Biol.* 2: 102.

Walker, D.C. and J. Southgate. 2009. The virtual cell—A candidate co-ordinator for "middle-out" modelling of biological systems. *Brief Bioinform.* 10: 450–61.

Wang, Z., C.M. Birch, and T.S. Deisboeck. 2008. Cross-scale sensitivity analysis of a non-small cell lung cancer model: Linking molecular signaling properties to cellular behavior. *Biosystems* 92: 249–58.

Wang, Z., C.M. Birch, J. Sagotsky, and T.S. Deisboeck. 2009. Cross-scale, cross-pathway evaluation using an agent-based non-small cell lung cancer model. *Bioinformatics* 25, 2389–96.

Wang, Z. and T.S. Deisboeck. 2008. Computational modeling of brain tumors: Discrete, continuum or hybrid? *Scientific Model. Simulation* 15: 381–93.

Wang, Z., L. Zhang, J. Sagotsky, and T.S. Deisboeck. 2007. Simulating non-small cell lung cancer with a multiscale agent-based model. *Theor. Biol. Med. Model* 4: 50.

Wise, S.M., J.S. Lowengrub, H.B. Frieboes, and V. Cristini. 2008. Three-dimensional multispecies nonlinear tumor growth—I model and numerical method. *J. Theor. Biol.* 253: 524–43.

Xue, C., J. Wyckoff, F. Liang, M. Sidani, S. Violini, K.L. Tsai, Z.Y. Zhang, E. Sahai, J. Condeelis, and J.E. Segall. 2006. Epidermal growth factor receptor overexpression results in increased tumor cell motility in vivo coordinately with enhanced intravasation and metastasis. *Cancer Res.* 66: 192–7.

Zhang, L., C.A. Athale, and T.S. Deisboeck. 2007. Development of a three-dimensional multiscale agent-based tumor model: Simulating gene-protein interaction profiles, cell phenotypes and multicellular patterns in brain cancer. *J. Theor. Biol.* 244: 96–107.

Zhang, L., C.G. Strouthos, Z. Wang, and T.S. Deisboeck. 2009a. Simulating brain tumor heterogeneity with a multiscale agent-based model: Linking molecular signatures, phenotypes and expansion rate. *Math Comput. Model* 49: 307–19.

Zhang, L., Z. Wang, J.A. Sagotsky, and T.S. Deisboeck. 2009b. Multiscale agent-based cancer modeling. *J. Math. Biol.* 58: 545–59.

CHAPTER 10

Diffusional Instability as a Mechanism of Tumor Invasion

Hermann B. Frieboes, John Lowengrub, and Vittorio Cristini

CONTENTS

INTRODUCTION

Heterogeneous cell proliferation, migration, and death can be caused by genetic damage and in response to the local microenvironment. Diffusion gradients of oxygen, glucose, metabolites, and drugs, established in the microenvironment during solid tumor growth and response to treatment, can create varying local conditions for subpopulations of tumor cells. Individual cells possess a broad spectrum of survival and migration mechanisms that can be invoked in response to hostile conditions [1], such as hypoxia and hypoglycemia. The consequences of this heterogeneity and variability can be highly multiscalar. Differential cell proliferation, migration, and death along the diffusion gradients affect cell survival and motility properties as a function of these gradients. At the tumor scale, this can lead to invasive fingering and branching and even fragmentation and migration of cell clusters into the surrounding tissue. Thus, these diffusion gradients have an important role in the stability of the tumor morphology.

Numerous mathematical models have been developed to study the progression of cancer (e.g., see the reviews [2–26]). Most models fall into two broad categories, based on how the tumor tissue is represented: continuum models and discrete cell-based models. A third alternative to elucidate the complexity of cancer and the interactions among the cell- and tissue-level scales is to use a hybrid approach, coupling biological phenomena from the molecular and cellular scales to the tumor scale; for example, see the work by Kim et al. (2007) [27].

We have presented multiscale models using a continuum approach [21,26,28–36] to determine precise functional relationships among quantifiable parameters from analyses of specific phenotypic or genetic alterations in a tumor, and from in vitro experiments [30] and clinical observations [21,32,36] of tumor morphology such as cell arrangement patterns at the tumor boundary. Building upon a formulation of classical models [37–40], a breakthrough simulation of a continuum tumor model was provided by Cristini et al. (2003) [26] to study complex morphologies of solid tumor growth in the nonlinear regime using boundary-integral simulations in 2D. This work predicted that the shape of highly vascularized tumors would remain compact and without invasive fingering, even while growing unbounded. This suggested that invasive growth of vascularized tumors is associated with vascular and elastic anisotropies such as heterogeneity in oxygen and cell nutrients, thus identifying for the first time the concept of tumor "diffusional instability" as a potentially universal physical mechanism underpinning cancer morphologies. The results further suggested the possibility of tumor shape control by controlling the tumor microenvironment.

Expanding on this idea, Cristini et al. (2005) [29] formulated the hypothesis that through heterogeneous cell proliferation and migration, microenvironmental cell substrate (e.g., oxygen, nutrient, growth factor) gradients may drive tumor invasion through morphological instability with separation of cell clusters from the tumor edge and infiltration into surrounding normal tissue. Tumor morphology would be determined by the competition between heterogeneous cell proliferation caused by spatial diffusion gradients, driving shape instability and invasive tumor morphologies, and stabilizing mechanical forces, for example, cell–cell and cell–matrix adhesion.

To test this hypothesis, Frieboes et al. (2006) [30] obtained parameter-based statistics for input to the mathematical model from in vitro human and rat glioblastoma cultures. Employing a linear stability analysis of the model from Cristini et al. (2003) [26], these results predicted that glioma spheroid morphology would be marginally stable. In agreement with this prediction, for a range of parameter values, unbounded growth of the tumor mass and invasion of the environment were observed in vitro. The mechanism of tumor invasion was characterized as recursive subspheroid component development (i.e., formation of "buds") at the tumor viable rim and separation from the parent spheroid. Cristini et al. (2005) [29] further provided evidence that morphological instability could be suppressed in vivo by spatially homogeneous oxygen and nutrient supply because normoxic conditions both decrease gradients and increase cell adhesion and, therefore, the mechanical forces that maintain a well-defined tumor boundary. Taking into account the effect of the microenvironment, Macklin and Lowengrub (2007) [31] also found that tumor morphological stability could be enhanced by improving the oxygen/nutrient supply.

The results of these multiscale models predict that morphologic instability of a tumor mass, that is, morphology resulting in "roughness" or harmonic content [26,33] of the tumor margin, may provide a powerful tissue invasion mechanism since it allows tumor cells to escape growth limitations imposed by diffusion (even in vitro [30,41]) and invade the host independently of the extent of angiogenesis [29,30]. Diffusional instability may thus be a universal consideration that applies to invasion observed across tumors of different tissues (e.g., [1,42]). Experiments with various glioma models in vivo [43–46) also support these findings. For example, recently published images of rat glioblastoma in vivo [47] showed that while the bulk tumor is perfused by blood, infiltrative cell clusters are much less perfused or not at all.

MODEL

We describe a continuum model of solid tumor growth that refines and extends previous developments (e.g., [32,34,35,48–64]). The tissue is modeled as a mixture of various components (tumor and host cell species, water, and extracellular matrix), each of which moves under its own velocity field and is governed by separate mass, momentum, and energy equations. The transport of key species, such as oxygen and matrix-degrading enzymes (MDEs), are also modeled. Each of the model constituents is governed by a reaction-diffusion equation of the general form

$$\partial v / \partial t = -\nabla \cdot \mathbf{J} + \Gamma_+ - \Gamma_-, \tag{10.1}$$

where v is the evolving constituent (e.g., the density ρ_i of cell species i, the cell substrate concentration σ, the MDE concentration m, or the nondiffusable matrix macromolecule concentration c), J is the flux, and Γ_+ and Γ_- are the sources and sinks of the constituent. Specific forms for J, Γ_+, Γ_- are given in Table 10.1, where the Ds are diffusion constants; $\mathbf{u}_i$ is the velocity and $\lambda_{prolif,i}$, $\lambda_{apop,i}$, and $\lambda_{nec,i}$ are the proliferation, apoptosis, and necrosis rates of cell species i; $\mathbf{u}_w$ is the water velocity; $\lambda_{\sigma,supply}$ and $\lambda_{\sigma,uptake}$ are the cell substrate transfer and uptake rates; λ_m and $\lambda_{m,degrade}$ are the MDE secretion and degradation rates; and λ_c and $\lambda_{c,degrade}$ are the production and degradation rates for the macromolecule c. An additional equation (not shown) is posed for the mass fraction of water [34]. For simplicity, the cell substrate represents the combination of oxygen and glucose, but it is straightforward to extend the model to treat these separately or include growth promoters and inhibitors by introducing additional chemical species. The $S_{i,j}$ are mass exchange terms, and the fluxes $\mathbf{J}_{mechanics,i}$ account for the mechanical interactions between the different cell species and the extracellular matrix [34].

Mechanical forces, including cell–cell and cell–matrix adhesion, stress, and strain, can be implemented through individual contributions to a potential function $E_{mechanics,i}$ [34]. Once the key processes and variables are identified, they can be naturally incorporated by modifying the energy accordingly, and the associated parameters can be informed by available experimental data. For example, a thermodynamically consistent (e.g., [51,52]) constitutive law for the flux $\mathbf{J}_{mechanics,i}$ is obtained by taking the gradient of the variational derivative of the total energy $E_{mechanics,i}$, that is,

TABLE 10.1   Specific Forms for the Main Model Constituents

Constituent	Flux J	Sources Γ_+	Sinks Γ_-
ρ_i	$\rho_i \mathbf{u}_i + \mathbf{J}_{mechanics,i}$	$\rho_i \lambda_{prolif,i} + \Sigma_j S^+_{i,j}$	$\rho_i (\lambda_{apop,i} + \lambda_{nec,i}) + \Sigma_j S^-_{i,j}$
σ	$\sigma \mathbf{u}_w - D_\sigma \nabla \sigma$	$\lambda_{\sigma,supply}$	$\lambda_{\sigma,uptake}$
c	0	λ_c	$\lambda_{c,degrade}$
m	$m \mathbf{u}_w - D_m \nabla m$	λ_m	$\lambda_{m,degrade}$

$$J_{\text{mechanics},i} \propto \nabla(\delta E_{\text{mechanics},i}/\delta\rho_i), \qquad (10.2)$$

where $E_{\text{mechanics},i}$ is obtained by adding the contributions from each mechanism modeled, that is, adhesion, elasticity, etc.

The velocities $\mathbf{u}_i$ and $\mathbf{u}_w$ are determined from momentum equations. Following previous approaches [26,29–31,33,64,66] that reformulated and generalized the models in [2,37,39,67–69] and neglecting viscoelastic effects, Darcy's law is taken as a coarse-scale reformulation of the inertialess momentum equation, that is, instantaneous equilibrium among the following forces [32,34,36]: pressure p (isentropic stress), resistance to motion, elastic forces, forces exchanged with the extracellular matrix that lead to chemo- and haptotaxis, and other mechanical effects within $E_{\text{mechanics},i}$. This leads to

$$\mathbf{u}_i = M_i(-\nabla p + \gamma_i(\delta E_{\text{mechanics},i}/\delta\rho_i)\nabla\rho_i) + \chi_{c,i}\nabla c + \chi_{\sigma,i}\nabla\sigma, \qquad (10.3)$$

where M_i, γ_i, $\chi_{c,i}$, and $\chi_{\sigma,i}$ are the spatially inhomogeneous mobility, adhesion, hapto- and chemotaxis tensors that also take into account cell–matrix adhesion. The interaction forces γ_i $(\delta E/\delta\rho_i)\nabla\rho_i$ describe the effects of the mechanical interactions (e.g., adhesion) on the movement of the cells. The parameter M depends on the extent of cell–cell and cell–matrix adhesion in bulk regions. Interactions with rigid physical barriers (e.g., bone) have also been modeled [70,71].

The model is closed with functional relationships based on experimental data, which serve to calibrate the input parameter values, for example, rates of cell proliferation and apoptosis [32,34,36].

The tumor model is coupled nonlinearly [32,36] to a hybrid continuum-discrete, lattice-free model of tumor-induced angiogenesis that is a refinement of previous work [72,73]. This random-walk model generates vascular topology based on tumor angiogenic regulators, for example, vascular endothelial growth factor (VEGF), represented by a single continuum variable that reflects the excess of proangiogenic regulators compared to inhibitory factors. Perinecrotic tumor cells and host tissue cells close to the tumor boundary are assumed to be a source of angiogenic regulators. Endothelial cells near the sprout tips proliferate, and their migration is described by chemotaxis and haptotaxis (e.g., motion up gradients of angiogenic regulators and matrix proteins such as fibronectin). For simplicity, only leading endothelial cells are modeled and trailing cells passively follow. The tumor-induced vasculature does not initially conduct blood, as the vessels need to form loops first (anastomosis) [74]. Anastomosed vessels may provide a source of oxygen and nutrient in the tissue and may undergo spontaneous shutdown and regression during tumor growth [75], thus enhancing the diffusion gradients seen clinically and predicted *in silico*. The vasculature architecture, that is, interconnectedness and anastomoses, is captured via a set of rules, for example, a leading endothelial cell has a fixed probability of branching at each time step, while anastomosis occurs if a leading endothelial cell crosses a vessel trailing path. Parameters governing the extent of neovascularization and oxygen/nutrient supply due to blood flow are estimated in part from Dynamic Contrast Enhanced Magnetic Resonance Imaging (DCE-MRI) observations in patients [76]. We note that there are other models of tumor neovascularization (e.g. [77–82]), which have also been coupled to tumor growth (e.g., [70]). In particular, earlier versions of the model presented here, using a sharp-interface simplification [28], were coupled with an angiogenesis component by Anderson and Chaplain (1998) [83].

Recently developed adaptive numerical techniques [28,31,34,66,70,84–91] are employed to obtain numerical results. The range of length and timescales governing the tumor evolution is captured by performing local refinement in regions of rapid spatiotemporal variation, for example, near the tumor and perinecrotic boundaries and near co-opted vessels.

RESULTS

We calibrated a sharp-interface simplification (constant cell–cell adhesion) of the mixture model, based on comparing and matching growth curves and detailed morphological features predicted numerically to those measured experimentally in vitro [30]. The parameter estimates were considered sufficient once the simulation agreed with the experimental growth and morphology data, which led to measurements of proliferation rates, necrosis and cell adhesion parameters, diffusion constants, and cell substrate uptake rates.

By comparing the slope of the early (log-linear) growth curve for in vitro tumors to the analytical spherical solution of the reaction-diffusion equation [26,30], the tumor proliferation rate $\lambda_{prolif} \approx 1$ day^{-1}. Similarly, by comparing the steady-state radius to the analytical spherical solution [26], $0.26 \leq A \leq 0.38$, where $A = \lambda_{death} / \lambda_{prolif}$. The "death" parameter is the ratio of the rate of cell mass destruction (apoptosis and necrosis) to tissue creation by proliferation and its effect on tumor growth (λ_{death} combines the effects of λ_{apop} and λ_{nec} in Table 10.1). A range of values to calibrate cell–cell adhesion was determined based on linear stability analysis of the nondimensional parameter $G = \lambda_{prolif}/\lambda_{adhesion}$, which measures the relative strengths of proliferation and cell–cell adhesion in a tumor [26]. As illustrated in Figure 10.1, in the presence of cell substrate gradients, morphology can be "unstable" when cell adhesion is weak (large G), whereas for small G, tumor morphology is "stabilized" by cell adhesion [38]. The larger a tumor grows, the weaker the stabilizing effect of cell adhesion. Each G-curve describes a tumor with specific cell phenotype and divides the parameter space into stable (on the left) and unstable regions (on the right). The lower the cell adhesion, the more shifted to the left the G-curve is, thus reducing the range of sizes of tumors that will be morphologically stable. As a tumor grows, this corresponds to moving from left to right and thus may lead to eventually crossing the G-curve corresponding to that tumor's phenotype. Based on this analysis and comparing with the marginally stable morphology of the in vitro tumors, and using our earlier estimates for A, it was determined that $0.6 \leq G \leq 0.9$.

The calibrated model was consistent with other measurements [26,92]. For instance, the model gives an oxygen penetration length scale $L_{oxy} = (D_{oxy}/\lambda_{oxy,uptake})^{-1/2}$. By measuring the distance between the necrotic core and the

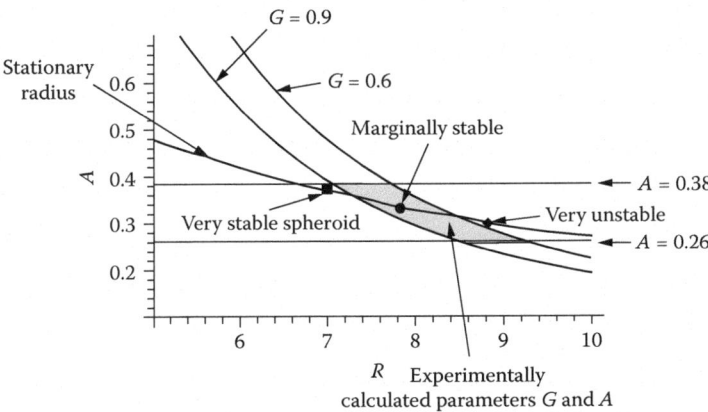

FIGURE 10.1 Morphologic stability diagram. "Stationary radius" R (unit $= 100\ \mu m$) decreases as a function of the death parameter A for different values of G. Shaded region: calibration of G and A under "stable" in vitro conditions. Tumors from the experiments (filled symbols) demonstrate predictivity of tumor morphological stability by the model. (Reprinted from Frieboes et al. 2006, *Cancer Res* 66:1597–1604. With permission from the American Association for Cancer Research.)

basement membrane, $L_{oxy} \approx 100\text{--}140$ µm; using previously published values $\lambda_{oxy,uptake} = 9.41 \times 10^{-2}\,s^{-1}$ [93] and $D_{oxy} = 1.45 \times 10^{-5}\,cm^2/s$ [94] gives L_{oxy} = 124 µm. Similar estimates were also in good agreement with the hypoxic and acidic gradients and regions of cell viability measured in vitro from immunohistochemistry [92] and with clinical histopathology data [35]. Analogous calculations were consistent for calculating the glucose penetration length and uptake rate [92,95,96], confirming that hypoxia is the limiting factor for tumor cell viability.

Adhesion was correlated with local substrate levels using this calibrated model. Adhesion decreases as glucose increases because high glucose reduces oxygen levels in tumors with diameters greater than 1 mm [97], leading to hypoxia, which can increase tumor cell motility and reduce cell–cell adhesion [1,98–100]. Using a simplified functional relationship between glucose and oxygen levels and cell adhesion (modeled here through the denominator of G), G was made an increasing function of glucose (corresponding to lower cell–cell adhesion) and a decreasing function of oxygen (corresponding to high cell–cell adhesion in well-oxygenated tumors). We compared in vitro experiments (filled symbols in Figure 10.1) with high glucose (low cell adhesion; $G > 0.9$), high serum (high proliferation, $G > 0.9$), low serum and glucose (low proliferation and high cell–cell adhesion, $G < 0.6$) with simulations and linear stability theory (shaded region in Figure 10.1). As predicted, the in vitro and *in silico* tumors corresponding to $G < 0.6$ demonstrated stable, compact morphologies, while the tumors corresponding to $G > 0.9$ were highly unstable [30]. This showed that the hypothesized functional relationships between phenotypic variables (adhesion and proliferation) and substrate levels (oxygen, glucose) were capable of correctly predicting morphology and growth for these in vitro tumors.

This conceptual framework is a description of tumor morphologic stability being regulated by diffusion gradients that promote or inhibit cell proliferation and migration. Linear stability analyses [26,33] and computer simulations [26,28–30,33] reveal that when promigratory and proliferative factors dominate, collective tumor cell migration and proliferation occur. Complex patterns (morphological "instabilities") are triggered by environmental perturbations, leading to the local invasion of the host by clusters or "fingers" of tumor cells. The presence of large substrate gradients (e.g., caused by heterogeneity in the neovasculature [101] combined with diffusion) can "destabilize" the morphology through spatially heterogeneous and time-dependent cell proliferation and migration. These

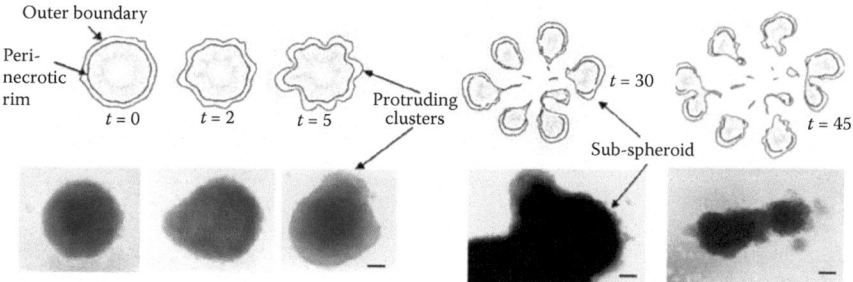

FIGURE 10.2   Cell protrusions growing into detached cell clusters and forming separate tumors as a result of intratumoral diffusion gradients of oxygen and cell nutrients. Top row: model simulation snapshots (time = days). Bottom row: in vitro observations. Bar: 130 μm. (Reprinted from Frieboes et al. 2006. *Cancer Res* 66:1597–1604. With permission from the American Association for Cancer Research.)

trends are illustrated in Figure 10.2 in a simulation calibrated to in vitro glioblastoma in standard culture conditions [30]. Cells exposed to higher substrate concentrations (outer boundary of the tumor) proliferate more quickly than those closer to the perinecrotic boundary. Under weak cell–cell adhesion, local "waves" arise in the outer boundary (also observed in [42,102] and develop into protruding clusters of cells that eventually separate from the tumor and form invasive "microsatellites" (subspheroids in the cell culture). The model results and the linear stability analyses [26] predict that stronger cell–cell adhesion would inhibit the development of protrusions, as confirmed by the in vitro experiments [30].

This mathematical modeling approach, based on biophysical instability mechanisms [26] that are predictive of tumor growth and controlled by critical model parameters, can reproduce various morphological patterns of collective cell migration and invasion that are observed in vitro and in vivo [1], including slender "finger-like" tumor cell strands and roundish, detached clusters (microsatellites). Figure 10.3 shows tumor morphologies predicted by the model using spherical initial conditions, low cell adhesion, and varied cell proliferation [36]. When the tumor evolved by taxis alone (no proliferation) **(A)**, strands of tumor cells were produced, as has been observed in vitro after inducing hypoxia in spheroids of MLP-29 rat embryo liver cells [41] **(B)**. When proliferation was significant, protrusions were formed and sometimes shed as clusters **(C)**, suggesting that the tumor relied on vessels in the nearby host tissue [103,104] and proliferated towards them, as confirmed in the brain

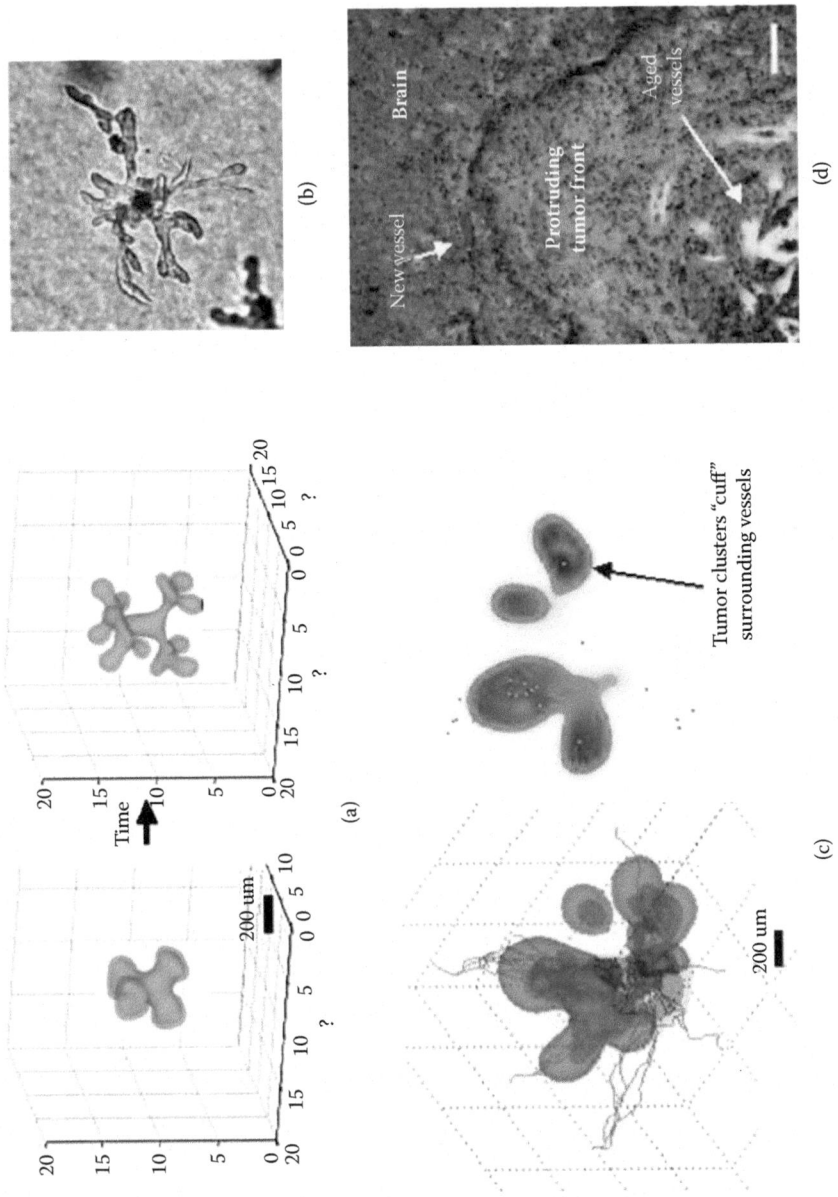

FIGURE 10.3 **(See color insert following page 40)** Variability and persistence of morphologic patterns predicted by the mathematical model simulating heterogeneity in vitro (**A**) (Li et al., 2007. *Disc. Contin. Dyn. Syst. B* 7:581–604) and in vivo (**C**) (Bearer et al., 2009. *Cancer Res.* 69:4493–4501). The tumor species is allowed to acquire a hypoxia-induced migratory phenotype clone. (**A**), proliferation is downregulated and the clone migrates up oxygen gradients toward the far-field boundary (computational box not shown; arrow, time direction). (**C**), migratory phenotype of the tumor clone (light red); and the original tumor species (dark red). Proliferation of both cell types is regulated by oxygen levels. The gray region of the 3-D graph (**C, left**) denotes necrosis. Horizontal 2-D slice (**C, right**) shows the distribution of the migratory clone; small circles indicate the cross sections of blood-conducting vessels. Morphologic instability occurs in both simulations because this clone's cell adhesion is low, resulting in cell strands in (**A**), and fingers and detached clusters in (**C**). Simulations are supported by experimental observations revealing morphologic instability after inducing hypoxia in spheroids in vitro (**B**) [reprinted from Pennacchietti et al. 2003. *Cancer Cell* 3:347–361. With permission from Elsevier], and in human glioblastoma histopathology viewed by fluorescence microscopy (H&E stain) (**D**) [reprinted from Frieboes et al. 2007. *NeuroImage* 37:S59–S70. With permission from Elsevier.] The tumor (bottom) is invading normal tissue (top) toward new conducting vessels (red). Note the demarcated margin between tumor and brain parenchyma (middle top) and the fluorescent green outlines of larger, aged vessels deeper in the tumor. Bar, 100 µm.

histopathology **(D)**. The overall tumor shape depends strongly on the vascular patterning, a result supported by animal [43–46] and clinical MRI studies (e.g., [36]). In both cases modeled, the complex tumor morphologies developed due to the interactions between cell proliferation, migration, and low cell adhesion, as modulated by the diffusion gradients of oxygen and cell nutrients.

In the work of Cristini et al. (2005) [29], the model was used to predict changes in the tumor morphology in response to perturbations in two model parameters that govern cell–cell adhesion and the density of the microvasculature in the host tissue, leading to the creation of a "morphology diagram." Figure 10.4 shows that if cell–cell adhesion is sufficiently strong (case A), then the tumor tends to maintain a compact morphology, even following angiogenesis. In contrast, when cell–cell adhesion is low (case B), the tumor tends to break into fragments [1] that invade the surrounding tissue due to substrate gradients [26,28]. When adhesion was kept low but the host vascular density is increased (case C), the substrate

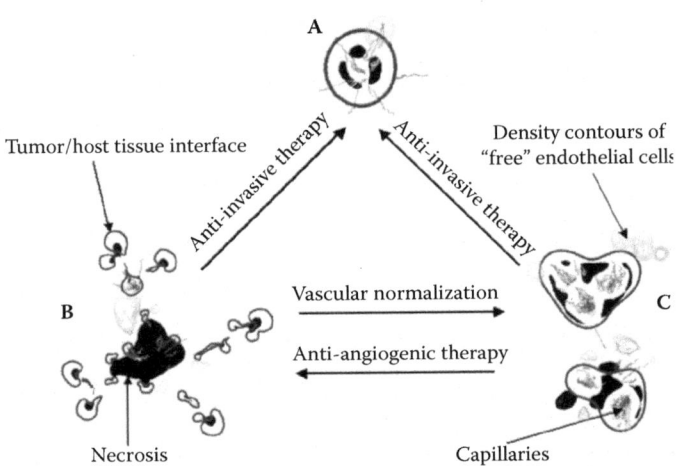

FIGURE 10.4 Tumor "morphology diagram." Solid line: calculated tumor boundary, black areas: necrosis. While anti-angiogenic therapy may induce tumor morphological instability through heterogeneity in the cellular microenvironment, including exacerbation of diffusion gradients of oxygen and cell nutrients, both anti-invasive therapy and vascular normalization lead to reduced heterogeneity and thus more compact morphologies. (Reprinted from Cristini et al. 2005. *Clin. Cancer Res.* 11:6772–6779. With permission from the American Association for Cancer Research.)

levels became more homogenous (were "normalized"), leading to a more compact morphology and reduced invasion.

This study [29] provided a preliminary quantification of the effects of three tumor therapy strategies: anti-invasive, where drugs are introduced to increase cell–cell adhesion; anti-angiogenic, where drugs target and destroy the neovasculature; and vascular-normalizing, where inefficient blood vessels are "pruned" to reduce or eliminate hypoxic gradients. The model predicts that anti-invasive therapy will lead to a transition from either case B or C to case A, while anti-angiogenic therapy causes a transition from case C to case B, and vascular-normalizing therapy reverses the transition.

DISCUSSION

The results presented in this chapter provide evidence that tumor morphogenesis may be a quantifiable function of marginally stable environmental conditions caused by diffusion gradients in cell nutrients, oxygen, and growth factors. This concept of tumor "diffusional instability" may be relevant during chemotherapy, radiotherapy, and anti-angiogenic therapy, all of which could introduce spatial and temporal variations in oxygen and cell nutrients [64]. In contrast, tumors may approach a compact, less invasive morphology when cell adhesion or other stabilizing mechanical forces (e.g., tumor encapsulation) are maximized. Compact tumor morphologies may be achievable by maintaining uniform oxygen/nutrient levels at the cellular scale and homogeneous microenvironmental conditions, thus suppressing instability [29]. A properly working tumor microvasculature could help maintain compact noninfiltrating tumor morphologies by minimizing oxygen and nutrient gradients. Anti-invasive therapy could increase cell adhesion and lead to more compact morphologies. Vascular normalization [48] would have a similar effect by making oxygen and nutrient supply more uniform. An additional benefit would be that more benign clones would be maintained, helping to keep malignant clones under control by competition for oxygen and cell nutrients.

By quantifying the link between the tumor boundary morphology and the invasive phenotype, this modeling work provides a quantitative tool for the study of tumor progression and diagnostic/prognostic applications. The results hold promise that applying biologically founded, mathematical modeling to quantify the connections between the microenvironment, tumor morphology, genotype, and phenotype may direct prognosis beyond the limitations of current methodologies. This type of modeling could be used to study system perturbations by therapeutic intervention and may aid in the design of novel clinical endpoints in therapeutic trials. By integrating the model with patient data for key tumor phenotypic and microenvironmental parameters, it is our hope that this work could help enhance clinical outcomes in the not too distant future.

ACKNOWLEDGMENTS

Grant support by the National Science Foundation (NSF) Division of Mathematical Sciences, National Institutes of Health, Department of Defense, and The Cullen Trust for Health Care.

REFERENCES

1. Friedl, P., and K. Wolf. 2003. Tumor cell invasion and migration: diversity and escape mechanisms. *Nature Rev. Cancer* 3:362–374.
2. Adam, J.A. 1996. General aspects of modeling tumor growth and the immune response. In *A Survey of Models on Tumor Immune Systems Dynamics*, ed. J.A. Adam and N. Bellomo, 15–87. Boston: Birkhäuser.
3. Chaplain, M.A.J. 2000. Pattern formation in cancer. In *On Growth and Form: Spatio-Temporal Pattern Formation in Biology*, ed. M.A.J. Chaplain, G.D. Singh, and J.C. MacLachlan, 47–70. New York: Wiley.
4. Bellomo, N., and L. Preziosi. 2000. Modelling and mathematical problems related to tumor evolution and its interaction with the immune system. *Math. Comput. Modelling* 32:413–542.
5. Moreira, J., and A., Deutsch. 2002. Cellular automaton models of tumor development: a critical review. *Adv. Complex Sys.* 5:247–267.
6. Bellomo, N., E. de Angelis, and L. Preziosi. 2003. Multiscale modeling and mathematical problems related to tumor evolution and medical therapy. *J. Theor. Medicine* 5:111–136.
7. Swanson, K.R., C. Bridge, J.D. Murray, and E.C. Alvord Jr. 2003. Virtual and real brain tumors: using mathematical modeling to quantify glioma growth and invasion. *J. Neuro. Sci.* 216:1–10.
8. Araujo, R.P., and D.L.S. McElwain. 2004. A history of the study of solid tumour growth: the contribution of mathematical modelling. *Bull. Math. Biol.* 66:1039–1091.
9. Mantzaris, N.V., S. Webb, and H.G. Othmer. 2004. Mathematical modeling of tumor-induced angiogenesis. *J. Math. Biol.* 49:111–187.
10. Friedman, A. 2004. A hierarchy of cancer models and their mathematical challenges. *Discrete Cont. Dyn. Systems Ser.* B 4:147–159.
11. Ribba, B., T. Alarcón, K. Marron, P.K. Maini, and Z. Agur. 2004. The use of hybrid cellular automaton models for improving cancer therapy. In *ACRI, LNCS*, ed. B. Chopard, P.M.A. Sloot, and A.G. Hoekstra, 444–453. Berlin: Springer.

12. Quaranta, V., A.M. Weaver, P.T. Cummings, and A.R.A. Anderson. 2005. Mathematical modeling of cancer: the future of prognosis and treatment. *Clin. Chim. Acta* 357:173–179.

13. Hatzikirou, H., A. Deutsch, C. Schaller, M. Simon, and K. Swanson. 2005. Mathematical modeling of glioblastoma tumour development: a review. *Math. Models Meth. Appl. Sci.* 15:1779–1794.

14. Nagy, J.D. 2005. The ecology and evolutionary biology of cancer: a review of mathematical models of necrosis and tumor cell diversity. *Math. Biosci. Eng.* 2:381–418.

15. Byrne, H.M., T. Alarcón, M.R. Owen, S.W. Webb, and P.K. Maini. 2006. Modeling aspects of cancer dynamics: a review. *Philos. Trans. R. Soc.* A 364:1563–1578.

16. Fasano, A. Bertuzzi, and A. Gandolfi. 2006. Mathematical modelling of tumour growth and treatment. In *Complex Systems in Biomedicine*, ed. A. Quarteroni, L. Formaggia, and A. Veneziani, 71–108. Milan: Springer.

17. van Leeuwen, I.M.M., C.M. Edwards, M. Ilyas, and H.M. Byrne. 2007. Towards a multiscale model of colorectal cancer. *World Gastroenterol.* 13:1399–1407.

18. Roose, T., S.J. Chapman, and P.K. Maini. 2007. Mathematical models of avascular tumor growth. *SIAM Rev.* 49:179–208.

19. Graziano, L., and L. Preziosi. 2007. *Mechanics in Tumor Growth*. In, ed. F. Mollica, L. Preziosi, and K.R. Rajagopal, 267–328. New York: Birkhäuser.

20. Friedman, N. Bellomo, and P.K. Maini. 2007. Mathematical analysis and challenges arising from models of tumor growth. *Math. Models Meth. Appl. Sci.* 17:1751–1772.

21. Sanga, S., H.B. Frieboes, X. Zheng, R. Gatenby, E.L. Bearer, and V. Cristini. 2007. Predictive oncology: a review of multidisciplinary, multiscale in silico modeling linking phenotype, morphology and growth. *NeuroImage* 37:S120–S134.

22. Deisboeck, T.S., L. Zhang, J. Yoon, and J. Costa. 2009. In silico cancer modeling: is it ready for prime time? *Nature Clin. Practice Oncol.* 6:34–42.

23 Anderson, A.R.A., and V. Quaranta. 2008. Integrative mathematical oncology. *Nat. Rev. Cancer* 8:227–244.

24. Bellomo, N., N.K. Li, and P.K. Maini. 2008. On the foundations of cancer modelling: selected topics, speculations, and perspectives. *Math. Models Meth. Appl. Sci.* 4:593–646.

25. Cristini, V., H.B. Frieboes, X. Li, J.S. Lowengrub, P. Macklin, S. Sanga, S.M. Wise, and X. Zheng. 2008. Nonlinear modeling and simulation of tumor growth. In *Modelling and Simulation in Science, Engineering and Technology*, ed. N. Bellomo, M. Chaplain, and E. de Angelis, xx–xx. Boston: Birkhäuser.

26. Cristini V., J. Lowengrub, and Q. Nie. 2003. Nonlinear simulation of tumor growth. *J. Math. Biol.* 46:191–224.

27. Kim, Y., M.A. Stolarska, and H.G. Othmer. 2007. A hybrid model for tumor spheroid growth in vitro I: theoretical development and early results. *Math. Meth. App. Sci.* 17:1773–1798.

28. Zheng, X., S.M. Wise, and V. Cristini. 2005a. Nonlinear simulation of tumor necrosis, neo-vascularization and tissue invasion via an adaptive finite-element/level-set method. *Bull. Math. Biol.* 67:211–259.

29. Cristini, V., H.B. Frieboes, R. Gatenby, S. Caserta, M. Ferrari, and J. Sinek. 2005. Morphologic instability and cancer invasion. *Clin. Cancer Res.* 11:6772–6779.

30. Frieboes, H.B., X. Zheng, C.-H. Sun, B. Tromberg, R. Gatenby, and V. Cristini. 2006. An integrated computational/experimental model of tumor invasion. *Cancer Res* 66:1597–1604.

31. Macklin, P., and J.S. Lowengrub. 2007. Nonlinear simulation of the effect of microenvironment on tumor growth. *J. Theor. Biol.* 245:677–704.

32. Frieboes, H.B., J.S. Lowengrub, S. Wise, X. Zheng, P. Macklin, E.L. Bearer, and V. Cristini. 2007. Computer simulation of glioma growth and morphology. *NeuroImage* 37:S59–S70.

33. Li, X., V. Cristini, Q. Nie, and J.S. Lowengrub. 2007. Nonlinear three-dimensional simulation of solid tumor growth. *Disc. Dyn. Contin. Dyn. Syst.* B 7:581–604.

34. Wise, S.M., J.S. Lowengrub, H.B. Frieboes, and V. Cristini. 2008. Three-dimensional multispecies nonlinear tumor growth—I. Model and numerical method. *J. Theor. Biol.* 253:524–543.

35. Cristini, V., X. Li, J.S. Lowengrub, and S.M. Wise. 2009. Nonlinear simulations of solid tumor growth using a mixture model: invasion and branching. *J. Math. Biol.* 58:723–763.

36 Bearer, E.L., J.S. Lowengrub, H.B. Frieboes, Y.-L. Chuang, F. Jin F, S.M. Wise, M. Ferrari, D.B. Agus, and V. Cristini. 2009. Multiparameter computational modeling of tumor invasion. *Cancer Res.* 69:4493–4501.

37. Greenspan, H.P. 1976. On the growth and stability of cell cultures and solid tumors. *J. Theor. Biol.* 56:229–242.

38. McElwain, D.L.S., and L.E. Morris. 1978. Apoptosis as a volume loss mechanism in mathematical models of solid tumor growth. *Math. Biosci.* 39:147–157.

39. Byrne, H.M., and M.A.J. Chaplain. 1996a. Modelling the role of cell-cell adhesion in the growth and development of carcinomas. *Math. Comput. Model.* 24:1–17.

40. Byrne, H.M., and M.A.J. Chaplain. 1997. Free boundary value problems associated with the growth and development of multicellular spheroids. *Eur. J. Appl. Math.* 8:639–658.

41. Pennacchietti, S., P. Michieli, M. Galluzzo, S. Giordano, and P. Comoglio. 2003. Hypoxia promotes invasive growth by transcriptional activation of the met protooncogene. *Cancer Cell* 3:347–361.

42. Debnath, J., and J. Brugge. 2005. Modelling glandular epithelial cancers in three-dimensional cultures. *Nature Rev. Cancer* 5:675–688.

43. Rubenstein, J.L., J. Kim, T. Ozawa, M. Zhang, M. Westphal, D.F. Deen, and M.A. Shuman. 2000. Anti-VEGF antibody treatment of glioblastoma prolongs survival but results in increased vascular cooption. *Neoplasia* 2:306–314.

44. Kunkel, P., U. Ulbricht, P. Bohlen, M.A. Brockmann, R. Fillbrandt, D. Stavrou, M. Westphal, and K. Lamszus. 2001. Inhibition of glioma angiogenesis and growth in vivo by systemic treatment with a monoclonal antibody against vascular endothelial growth factor receptor-2. *Cancer Res.* 61:6624–6628.

45 Bello, L., V. Lucini, F. Costa, M. Pluderi, C. Giussani, F. Acerbi, G. Carrabba, M. Pannacci, D. Caronzolo, S. Grosso, S. Shinkaruk, F. Colleoni, X. Canron, G. Tomei, G. Deleris, and A. Bikfalvi. 2004. Combinatorial administration of molecules that simultaneously inhibit angiogenesis and invasion leads to increased therapeutic efficacy in mouse models of malignant glioma. *Clin. Cancer Res.* 10:4527–4537.

46. Lamszus, K., P. Kunkel, and M. Westphal. 2003. Invasion as limitation to anti-angiogenic glioma therapy. *Acta Neurochir* Suppl. 88:169–177.

47. Madsen, S.J., E. Angell-Petersen, S. Spetalen, S.W. Carper, S.A. Ziegler, and H. Hirschberg. 2006. Photodynamic therapy of newly implanted glioma cells in the rat brain. *Lasers Surg. Med.* 38:540–548.

48 Ambrosi, D., and F. Mollica. 1992. On the mechanics of a growing tumor, *Int. J. Eng. Sci.* 40:1297–1316.

49. Preziosi, L., and A. Farina. 2002. On Darcy's law for growing porous media. *Int. J. Nonlin. Mech.* 37:485–491.

50. Ambrosi, D., and L. Preziosi. 2002. On the closure of mass balance models for tumor growth. *Math. Model. Methods Appl. Sci.* 12:737–754.

51. Byrne, H.M., and L. Preziosi. 2003. Modelling solid tumour growth using the theory of mixtures. *Math. Med. Biol.* 20:341–366.

52 Araujo, R.P., and D.S.L. McElwain. 2005a. A mixture theory for the genesis of residual stresses in growing tissues: I. A general formulation, *SIAM J. Appl. Math* 65:1261–1284.

53 Araujo, R.P., and D.L.S. McElwain. 2005b. A mixture theory for the genesis of residual stresses in growing tissues II: solutions to the biphasic equations for a multicell spheroid, *SIAM J. Appl. Math* 66:447–467.

54. Chaplain, M.A.J., L. Graziano, and L. Preziosi. 2006. Mathematical modeling of the loss of tissue compression responsiveness and its role in solid tumour development. *Math. Med. Biol.* 23:192–229.

55. Fusi, L., A.Farina, and D. Ambrosi, 2006. Mathematical modeling of a solid-liquid mixture with mass exchange between constituents. *Math. Mech. Solids* 11:575–595.

56. Ricken, T., A. Schwarz, and J. Bluhm. 2006. A triphasic theory for growth in biological tissue—basics and applications. *Materialwissenschaft und Werkstofftechnik* 37:446–456.

57. Lemon, G., J.R. King, H.M. Byrne, O.E. Jensen, and K.M. Shakesheff. 2006. Mathematical modeling of engineered tissue growth using a multiphase porous flow mixture theory. *J. Math. Biol.* 52:571–594.

58 Breward, C.J.W., H.M. Byrne, and C.E. Lewis. 2002. The role of cell-cell interactions in a two-phase model for avascular tumour growth. *J. Math. Biol.* 45:125–152.

59. Breward, C.J.W., H.M. Byrne, and C.E. Lewis. 2003. A multiphase model describing vascular tumor growth. *Bull. Math. Biol.* 65:609–640.

60. Byrne, H.M., J.R. King, D.L.S. McElwain, and L. Preziosi. 2003. A two-phase model of solid tumour growth. *Appl. Math. Lett.* 16:567–573.

61. Franks, S.J., H.M. Byrne, J.R. King, J.C.E. Underwood, and C.E. Lewis. 2003a. Modeling the early growth of ductal carcinoma in situ of the breast. *J. Math. Biol.* 47:424–452.

62. Franks, S.J., H.M. Byrne, H.S. Mudhar, J.C.E. Underwood, and C.E. Lewis. 2003b. Mathematical modeling of comedo ductal carcinoma in situ of the breast. *Math. Med. Biol.* 20:277–308.

63. Roose, T., Netti P.A., Munn LL, Boucher Y., and Jain R. 2003. Solid stress generated by spheroid growth estimated using a linear poroelastic model. *Microvascular Res.* 66:204–212.

64. Tosin, A. 2008. Multiphase modeling and qualitative analysis of the growth of tumor cords. Networks Heterogen. *Media.* 3:43–84.

65. Sinek, J.P., H.B. Frieboes, X. Zheng, and V. Cristini. 2004. Two-dimensional chemotherapy simulations demonstrate fundamental transport and tumor response limitations involving nanoparticles. *Biomed. Microdev.* 6:297–309.

66. Macklin, P., and J.S. Lowengrub. 2005. Evolving interfaces via gradients of geometry-dependent interior Poisson problems: application to tumor growth. *J. Comp. Phys.* 203:191–220.

67. Chaplain, M.A.J. 1996. Avascular growth, angiogenesis and vascular growth in solid tumours: the mathematical modelling of the stages of tumour development. *Math. Comp. Model.* 23:47–87.

68. Byrne, H.M., and M.A.J. Chaplain. 1996b. Growth of necrotic tumours in the presence and absence of inhibitors. *Math. Biosci.* 135:187–216.

69. Chaplain, M.A.J., and Anderson A.R.A. 2003. Mathematical modeling of tissue invasion. In *Cancer Modeling and Simulation*, ed. L. Preziosi, 269–297. Boca Raton, Florida: CRC Press.

70. Macklin, P., J.S. Lowengrub. 2008. A new ghost cell/level set method for moving boundary problems: application to tumor growth. *J. Sci. Comput.* 35:266–299.

71. Macklin, P., S. McDougall, A.R.A. Anderson, M.A.J. Chaplain, V. Cristini, and J.S. Lowengrub. 2009. Multiscale modeling and simulation of vascular tumour growth. *J. Math. Biol.* 58:765–798.

72. Plank, M.J., and B.D. Sleeman. 2003. A reinforced random walk model of tumour angiogenesis and anti-angiogenic strategies. *Math. Med. Biol.* 20:135–181.

73. Plank, M.J., and B.D. Sleeman. 2004. Lattice and non-lattice models of tumour angiogenesis. *Bull. Math. Biol.* 66:1785–1819.

74. Augustin, H.G. 2001. Tubes, branches, and pillars: the many ways of forming a new vasculature. *Circ. Research* 89:645–647.

75. Padera, T.P., B.R. Stoll, J.B. Tooredman, D. Capen, E. di Tomaso, and R. Jain. 2004. Cancer cells compress intratumour vessels. *Nature* 427:695.

76. O'Connor, J.P.B., A. Jackson, G.J.M. Parker, and G.C. Jayson. 2007. DCE-MRI biomarkers in the clinical evaluation of antiangiogenic and vascular disrupting agents. *Br. J. Cancer* 96:189–195.

77. Sun, S., M.F. Wheeler, M. Obeyesekere, and C.W. Patrick Jr. 2005. A deterministic model of growth factor induced angiogenesis. *Bull. Math. Biol.* 67:313–337.
78. Stephanou, A., S.R. McDougall, A.R.A. Anderson, and M.A.J. Chaplain. 2005. Mathematical modeling of flow in 2D and 3D vascular networks: applications to anti-angiogenic and chemotherapeutic drug strategies. *Math. Comp. Modeling* 41:1137–1156.
79. McDougall, S.R., A.R.A. Anderson, and M.A.J. Chaplain. 2006. Mathematical modeling of dynamic adaptive tumour-induced angiogenesis: clinical applications and therapeutic targeting strategies. *J. Theor. Biol.* 241:564–589.
80. Lee, D.S., H. Rieger, and K. Bartha. 2006. Flow correlated percolation during vascular remodeling in growing tumors. *Phys. Rev. Lett.* 96: 058104.
81. Welter, M., K. Bartha, and H. Rieger. 2008. Emergent vascular network inhomogeneities and resulting blood flow patterns in a growing tumor. *J. Theor. Biol.* 250:257–280.
82. Owen, M.R., T. Alarcon, P.K. Maini, and H.M. Byrne. 2009. Angiogenesis and vascular remodeling in normal and cancerous tissues. *J. Math. Biol.* 58:689–721.
83 Anderson, A.R., and M.A.J. Chaplain. 1998. Continuous and discrete mathematical models of tumor-induced angiogenesis. *Bull. Math. Biol.* 60:857–899.
84. Cristini, V., J. Blawzdziewicz, and M. Loewenberg. 2001. An adaptive mesh algorithm for evolving surfaces: simulations of drop breakup and coalescence. *J. Comp. Phys.* 168:445–463.
85. Anderson, A., X. Zheng, and V. Cristini. 2005. Adaptive unstructured volume remeshing—I: the method. *J. Comp. Phys.* 208:616–625.
86. Zheng, X., J.S. Lowengrub, A. Anderson, and V. Cristini. 2005b. Adaptive unstructured volume remeshing—II: application to two- and three-dimensional level-set simulations of multiphase flow. *J. Comp. Phys.* 208:626–650.
87. Wise, S.M., J.S. Kim, and J.S. Lowengrub. 2007. Solving the regularized, strongly anisotropic Cahn–Hilliard equation by an adaptive nonlinear multigrid method. *J. Comp. Phys.* 226:414–446.
88. Kim, J.S., K. Kang, and J.S. Lowengrub. 2004a. Conservative multigrid methods for Cahn–Hilliard fluids. *J. Comp. Phys.* 193:511–543.
89. Kim, J.S., K. Kang, and J.S. Lowengrub. 2004b. Conservative multigrid methods for ternary Cahn–Hilliard systems. *Comm. Math. Sci.* 12:53–77.
90. Yang, X., A.J. James, J.S. Lowengrub, X. Zheng, and V. Cristini 2006. An adaptive coupled level-set/volume-of-fluid interface tracking method for unstructured triangular grids. *J. Comp. Phys.* 217:364–394.
91. Macklin, P., and J.S. Lowengrub. 2006. An improved geometry-aware curvature discretization for level-set methods. *J. Comp. Phys.* 215:392–401.
92. Gatenby, R.A., K. Smallbone, P.K. Maini, F. Rose, J. Averill, R.B. Nagle, and R.J. Gillies. 2007. Cellular adaptations to hypoxia and acidosis during somatic evolution of breast cancer. *Br. J. Cancer* 97:646–653.
93. Casciari, J.J., and S.V. Sotirchos, R.M. Sutherland. 1992. Variations in tumor cell growth rates and metabolism with oxygen concentration, glucose concentration, and extracellular pH. *J. Cell Physiol.* 151:386–394.

94. Nichols, M.G., and T.H. Foster. 1994. Oxygen diffusion and reaction kinetics in the photodynamic therapy of multicell tumour spheroids. *Phys. Med. Biol.* 39:2161–2181.

95. Kallinowski, F., P. Vaupel, S. Runkel, G. Berg, H.P. Fortmeyer, K.H. Baessler, K. Wagner, W. Mueller-Klieser, and S. Walenta. 1988. Glucose uptake, lactate release, ketone body turnover, metabolic milieu and pH distributions in human cancer xenografts in nude rats. *Cancer Res.* 48:7264–7272.

96. Groebe, K., S. Erz, and W. Mueller-Kleiser. 1994. Glucose diffusion coefficients determined from concentration profiles in EMT6 tumor spheroids incubated in radioactively labeled l-glucose. *Adv. Exp. Med. Biol.* 361:619–625.

97. Sutherland, R.M., J. Carlsson, R.E. Durand, and J. Yuhas. 1981. Spheroids in cancer research. *Cancer Res.* 41:2980–2994.

98. Friedl, P., E.B. Brocker, and K.S. Zanker. 1998. Integrins, cell matrix interactions and cell migration strategies: fundamental differences in leukocytes and tumor cells. *Cell Comm. Adhesion* 6:225–236.

99. Jain, R.K. 2001. Normalizing tumor vasculature with anti-angiogenic therapy: a new paradigm for combination therapy. *Nature Med.* 7:987–989.

100. Jain, R.K. 2005. Normalization of tumor vasculature: an emerging concept in antiangiogenic therapy. *Science* 307:58–62.

101. Jain, R.K. 1988. Determinants of tumor blood flow: A review. *Cancer Res.* 48:2641–2658.

102. Debnath, J., K. Mills, N. Collins, M. Reginato, S. Muthuswamy, and J. Brugge. 2002. The role of apoptosis in creating and maintaining luminal space within normal and oncogene-expressing mammary acini. *Cell* 111:29–40.

103. Bartels, U., C. Hawkins, M. Jing, M. Ho, P. Dirks, J. Rutka, D. Stephens, and E. Bouffet. 2006. Vascularity and angiogenesis as predictors of growth in optic pathway/hypothalamic gliomas. *J. Neurosurg.* 104:314–320.

104. Preusser, M., H. Heinzl, E. Gelpi, K. Schonegger, C. Haberler, P. Birner, C. Marosi, M. Hegi, T. Gorlia, and J.A. Hainfellner. 2006. Histopathologic assessment of hot-spot microvessel density and vascular patterns in glioblastoma: Poor observer agreement limits clinical utility as prognostic factors: a translational research project of the European Organization for Research and Treatment of Cancer Brain Tumor Group. *Cancer* 107:162–170.

105. Wurzel, M., C. Schaller, M. Simon, and A. Deutsch. 2005. Cancer cell invasion of brain tissue: guided by a prepattern? *J. Theor. Med.* 6:21–31.

106. Benjamin, R., J. Capparella, and A. Brown. 2003. Classification of glioblastoma multiforme in adults by molecular genetics. *Cancer J.* 9:82–90.

107. Ambrosi, D., A. Duperray, V. Peschetola, and C. Verdier. 2009. Traction patterns of tumor cells. *J. Math. Biol.* 58:163–181.

108. Maher, E., F. Furnari, R. Bachoo, D. Rowitch, D. Louis, W. Cavenee, and R. De-Pinho. 2001. Malignant glioma: genetics and biology of a grave matter. *Genes Dev.* 15:1311–1333.

CHAPTER 11

Continuum Models of Mesenchymal Cell Migration and Sprouting Angiogenesis

Michael Bergdorf, Florian Milde, and Petros Koumoutsakos

CONTENTS

INTRODUCTION

Cell migration is fundamental to a number of physiological processes such as embryogenesis, organ growth, inflammation, wound healing, and tumor-induced angiogenesis. In embryonic development, the blastocyst cells migrate to form layers (gastrulation) and later migrate to target destinations in the developing embryo to specialize and become components of organs. This process of developmental migration continues in the adult, as some cells in our bodies are born, migrate, and die on a daily basis. Wound healing is another example of collective cell migration, involving the proliferation of existing cells and the migration of cells close to the wound towards each other to close the cleft. In the pathological context, enhanced cell migration is key to invasive growth of tumor cells, which is the basis of metastasis. This invasion involves the detachment of individual cells from their originating tissue and is followed by their propulsion through existing tissue.

A particular aspect of cell migration that has attracted increased attention recently is related to the process of angiogenesis as induced by tumor cells emitting vascular endothelial growth factors (VEGFs). Tumor-induced formation of new blood vessels in the context of sprouting angiogenesis is based on the orchestrated migration of clusters or cords of endothelial cells. The endothelial cells migrate through the extracellular matrix (ECM) following chemotactic and haptotactic cues. For these endothelial cells, the most prominent chemoattractant is VEGF [1]. VEGF gradients can be established in a variety of ways; soluble VEGF isoforms can be secreted by tumor cells, macrophages, or astrocytes [2]. This secretion can lead to long-range chemotactic signals. At the same time, some chemokines can also be sequestered in the ECM, where they represent highly localized cues. Another subset can be cleaved by matrix proteinases [3] and made soluble, or released as the matrix itself is degraded by heparinases or plasmin [4,5].

All these processes are characterized by a collective, directed motion of the cells in play. In many cases, the migration direction is aligned with a chemical gradient, for example, of nutrient or a growth factor: in this case, the migration is referred to as *chemotaxis*. There has been much debate on how cells are able to persistently sense a chemical gradient of a *chemoattractant*. Mathematical models have been used largely as qualitative indicators of cell motion. Their predictive value can be traced to their capability of differentiating between various mechanisms by which a cell might be able to sense a gradient and migrate in response to it, in a sustained fashion. A number of mathematical models have been proposed for gradient-driven cell motions. These models are predominantly based on mechanisms of "Local Excitation-Global Inhibition" (LEGI) [6] and its extension to "balanced inactivation" [7] (see Reference [8] for a review of chemotaxis models). The effect of noise on the directional sensing ability of these models has been studied by Ueda and Shibata [9], and Rappel and Levine [10], and may lead to testable predictions.

Morphogenesis *In Silico*

The simulations of sprouting angiogenesis can be broadly categorized as simulations of morphogenetic patterns that in turn correspond to a variety of computational models. These models often focus on a particular aspect of the patterning process and, in turn, their computational parameters are tuned to the process under consideration. For example, a vast body of literature exists on the mathematical description of bacterial chemotaxis (see the comprehensive review by Tindall et al. [11]). The patterning observed in developmental vasculogenesis has been successfully reproduced *in silico* by simulations that independently considered either purely chemical signaling [12] or purely mechanical effects [13].

While much of the theoretical work published on chemotactic systems focuses on modeling the gradient sensing response, the representation and interaction of the cells, or in reproducing final, static patterns, little attention has been paid to the environment in which the migration takes place. Models of vasculogenesis [13–15], either consider the ECM as a homogeneous viscoelastic tissue, or do not consider it at all (see review [16] for a survey of different vasculogenesis models). Although reported mathematical models of sprouting angiogenesis continue to achieve a great degree of sophistication [17], in these models the ECM plays a mostly secondary role and is assumed to be homogeneous. Heterogeneity is introduced through random components in the behavioral rules of the cells; for example, cells branch with a predefined probability [18,19]. Exceptions include the deterministic model of Sun et al. [20], where the ECM affects the migration by deflecting the migration velocity through a random anisotropy. On a microscopic scale, the ECM was structurally considered in [21] a Cellular

Potts approach. Due to the detailed cell representation, this model is only suitable for studying the dynamics of a handful of cells.

A grand majority of the models that describe to some extent the dynamics of collective cellular motion are based on cell-based representations. These formulations are appealing because they allow researchers to impose behavioral rules in an intuitive, cell-centered fashion. Also, the interpretation of the results is straightforward, as there is no abstraction layer; for example, instead of integrating a mean cell density, one can simply count the cells in the plot. A smaller subclass does employ continuum formulations; however, at a coarse macroscopic scale, where the output of the simulation has to be interpreted as a probability density function for the cell density, and therefore, morphological information is all but completely buried in the description. One reason why continuum models that can recover detailed morphological features are scarce is because some cellular interactions are difficult to model at the macroscopic scale. One basic problem is the representation of cell–cell adhesion. In cell-based models, cell–cell adhesion can conveniently be expressed in the shape of an intercell attraction. In mesoscale continuum models, the notion of a single cell does not exist, and we have to model the macroscopic effect of cell–cell adhesion. A handful of approaches have been proposed; in the context of tumor growth, cell–cell adhesion is often represented by a surface tension term that acts on the interface between tissues of different cell types [22,23]. Other continuum models have been proposed [24] based on the concept of deducing collective continuum behavior from an underlying discrete model [25], or on medium-range attractive forces [26]. One common drawback of these models is that they are computationally expensive. Surface tension calculations in continuum models require solving a global problem. The approach of Armstrong et al. [26] is local; it, however, requires performing a convolution with a kernel involving $\mathcal{O}\, 20^d$ neighbors.

In a recent work, we introduced the first hybrid model of sprouting angiogenesis, which explicitly considers the structure of the ECM [27]. In this model, matrix fibers are explicitly represented as a collection of fiber bundles that direct cell migration, and can sequester chemical cues leading to branching of blood vessels without requiring a priori defined branching probabilities. Painter [28] employed a similar approach to study cell organization and the incipience of fingering patterns. In the present article, we focus on the continuum formulation of a computationally efficient cell–cell adhesion model, and the interaction of *in silico* cells with the underlying artificial ECM.

Sprouting Angiogenesis

Sprouting angiogenesis, the process of new capillaries forming from existing vessels, can be observed in the human body under various conditions. Apart from angiogenesis in a physiological context, which mainly takes place during embryogenesis and fetal development, angiogenesis can be observed under pathological conditions, such as wound healing, thrombosis, and tumor growth [29]. In the case of wound healing and thrombosis, newly formed capillaries grow in a controlled manner and stop growing once the pathological condition has been alleviated; this is, however, not the case for tumor-induced angiogenesis [29].

Tumor-induced angiogenesis can persist for years, involving a disorganized, inefficient, and leaky vasculature [29]. Nonetheless, this vasculature supplies the tumor with nutrients and growth factors, which enable increased tumor cell proliferation and thus enhanced tumor growth. Furthermore, angiogenesis enables hematogenous spread of cancer: single cancer cells or cell clusters that detach from the primary tumor may enter the leaky vessels and use the vasculature to metastasize to remote organs. Regulating, or even inhibiting tumor-induced angiogenesis, can affect tumor growth. Inhibition of angiogenesis restrains nutrient supply, reducing the growth rate of the tumor and hindering migrating cell clusters from entering the vasculature, thus reducing the risk of metastasis [2]. However, a complete inhibition-promoting hypoxia (state of oxygen shortage) within the tumor could increase the occurrence of aggressive migrating tumor cell phenotypes [30,31]. Regulating capillary growth may help in establishing a more efficient pathway for drug delivery, as the leaky vessels and the high interstitial pressure in the proximity of the tumor prevent effective supply of drugs through the blood vessels into the core of the tumor [32]. Although anti-angiogenic therapy is comparatively young, it has already been established as a novel form of chemotherapy in cancer treatment [2] (next to surgery, radiation, and conventional chemotherapy).

Tumor-Induced Sprouting Angiogenesis

The maximal size a tumor can assume without relying on a vasculature for nutrient supply is restricted to 1 mm^3 [2]. In this case—which is referred to as avascular growth—nutrients reach the tumor by the sole means of diffusion through the surrounding tissue. If the tumor grows beyond this size of about 1 mm^3, cells in the core of the tumor cannot obtain enough oxygen to survive. These cells become hypoxic and eventually starve, forming a necrotic region at the core of the tumor. Tumors can reside in this avascular state for a long time [2]. However, one of the many responses of tumor cells to hypoxia is that they start to secrete angiogenic growth factors that are responsible for initiating sprouting angiogenesis. Several growth factors are involved in the process of angiogenesis. VEGFs have been identified to be one of the main driving forces [1]. VEGFs, upon release by hypoxic tumor cells, diffuse through the ECM occupying the space between tumor and existing vasculature in the proximity of the tumor and establish a chemical gradient between the tumor and nearby vessels. Once VEGF has reached a vessel, it binds to receptors located on endothelial cells (EC), which line the blood vessel walls. This binding sets off a cascade of events.

In the early phase of angiogenesis, ECs stimulated by VEGF start releasing proteases that degrade the basal lamina, a fibril structure building the outermost layer of the vessel wall. This enables ECs to leave the vessel wall and enter the ECM. Further signaling pathways downstream of VEGF lead to an increase in EC proliferation and coordinate the selection of migrating ECs located at the tip of outgrowing sprouts. Migrating sprout tip cells probe their environment by extending filopodia and migrate along the VEGF gradient toward regions of higher concentration, a directed motion referred to as *chemotaxis*. ECs located initially behind the migrating tip cells proliferate, thus extending the sprouting blood vessel. Fibronectin, which is distributed in the ECM and at the same time released by the migrating tip cells, establishes an adhesive gradient that serves as another cue for the ECs following behind. Fibronectin released by the ECs binds to integrins located on collagens and other fibers, which occupy roughly 30% of the ECM. Matrix-bound fibronectin in turn can bind to transmembrane receptors located on the EC membrane. This autocrine signaling pathway, promoting cell–cell and cell–matrix adhesion, accounts for a movement referred to as haptotaxis. In addition to chemotactic and haptotactic cues, the fibrous structures itself present in the ECM also influence cell migration by facilitating movement in fiber directions.

After the initial sprouts have extended into the ECM for some distance, repeated branching of the tips can be observed. Sprout tips approaching others may fuse and form loops, a process called *anastomosis*. Anastomosis can be observed to occur in two ways, either sprout tips fusing with other sprout tips, stopping their migration completely, or sprout tips fusing with already established sprouts at some distance behind the tip. Along with anastomosis, the formation of lumen within the strands of endothelial cells establishes a network that allows the circulation of blood. Maturation, the final stage of angiogenesis, incorporates the rebuilding of a basal lamina and the recruitment of pericytes and smooth muscle cells to stabilize the vessel walls.

In tumor-induced angiogenesis, the newly built vasculature is often disorganized and leaky, leading to high interstitial pressure and inefficient blood supply. Together with a growing tumor, which exerts pressure on the fragile capillaries and thus suppresses temporal and local blood delivery, ever-new regions of acute hypoxia arise, releasing VEGF, which sets off the process of angiogenesis anew. The process therefore never comes to a stop, and full maturation is impaired.

Vascular Endothelial Growth Factors (VEGFs)
In addition to the soluble isoform of VEGF, the presence of other VEGF isoforms has been identified to significantly influence morphology of capillary network formation [3,33]. Some VEGF isoforms express a binding site for heparan sulfate proteoglycans that is found on cell surfaces, in the ECM, and in body fluids; thus, such isoforms can be bound by the ECM. These "matrix-bound" VEGF isoforms can be cleaved from the ECM by matrix metalloproteinases (MMPs) [3], establishing very localized chemotactic cues. MMPs are expressed both by tumors and migrating ECs. Inflammatory cells stimulated by the tumor can also release VEGF and contribute to the chemotactic cues ECs react to.

Extracellular Matrix (ECM)
The ECM describes any material that occupies the space between cells in metazoans, including the space between the initial vasculature and the tumor. It plays an important role in cell migration. Fibers such as collagen, laminin, and fibrillin are distributed throughout the ECM, occupying roughly 30% of it. These fibers form bundles that serve as guiding structures for migrating cells [34,35]. The structures have been shown to be subject to remodeling by endothelial tip cells [36], facilitating migration through the matrix and playing a crucial role in lumen formation. The ECM further presents binding sites for fibronectin and matrix-bound VEGF isoforms that can be cleaved by MMPs.

COMPUTATIONAL MODELING OF ANGIOGENESIS

Computational models of tumor-induced angiogenesis address a limited number of the involved biological processes. The choice of the modeled

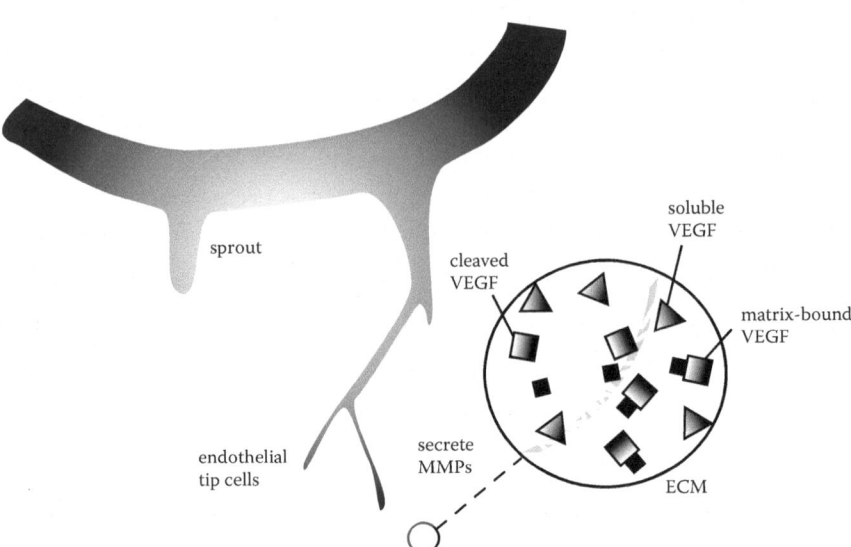

FIGURE 11.1 Conceptual sketch of the different VEGF isoforms present in the ECM. Soluble and cleaved VEGF isoforms freely diffuse through the ECM. Matrix-bound VEGF isoforms stick to the fibrous structures comprising the ECM and can be cleaved by MMPs secreted by the sprouting tips.

processes is dictated by the availability of biological data and by the understanding of the key processes for the phenomena under investigation. In the present model, we consider the motion of the ECs as affected by chemical gradients induced by VEGF, cell–cell adhesion, and the structure of the ECM.

In the present model, VEGF appears in soluble and matrix-bound isoforms. The soluble VEGF is released from an implicit tumor source, and diffuses freely through the ECM. The matrix-bound VEGF isoform can be cleaved by MMPs released at the sprout tips (section titled, "Vascular Endothelial Growth Factors"), and contributes to the migration cues of the ECs (see Figure 11.1). As an extension to the explicit modeling of matrix-bound growth factors, we present a subgrid-scale approach to account for the influence of matrix-bound VEGF on cell migration. Existing models of sprouting angiogenesis account for chemotaxis induced by the soluble isoform of VEGF [18,20,21,37], and a matrix-bound isoform of VEGF has been implicitly accounted for in a recent work by Bauer et al. [21], in which the ECM consists of fiber bundles, structural cells, and interstitial fluid, influencing EC migration through adhesive forces. In [20], the matrix is represented by a random anisotropic conductivity field that affects the migration velocity of the sprout tips. Models of angiogenesis can be classified in three broad categories:

1. Discrete, cell-based models that aim to capture the behavior of individual biological cells [21]

2. Continuum models that describe the large-scale, averaged behavior of cell populations [37,38]

3. Discrete models that model explicitly vascular networks determined by the migration of tip cells [18,20].

Bauer et al. [21] developed a two-dimensional, Cellular-Potts-based model to simulate migration, division, and adhesion of endothelial cells establishing an interconnected network. They distinguished two types of ECs: migrating tip cells that degrade the matrix fibers, and proliferating stalk cells located behind the tip cells. This allows for branching and anastomosis of blood vessels without explicitly defined rules. Anderson and Chaplain [37] presented two-dimensional, continuum models of angiogenesis, with a probabilistic modeling of capillaries and a discrete extension to this model, with capillaries as masked points on a grid. Capillary branching is modeled through a branching probability, depending on the sprout age, the tumor angiogenic growth factor (TAF) level, and the endothelial cell density. The model was later extended to three dimensions [18]. Sun et al. [20] presented a deterministic, discrete two-dimensional model of sprouting angiogenesis. They used a capillary indicator function to describe the network structure and formulated branching as a function of the sprout age and the anisotropy of the ECM. In summary, cell-based models aim to describe angiogenesis at cell-level resolution, but they are difficult to extend to macroscopic systems due to their computational cost. Continuum-based models bypass these limitations by modeling the evolution of cell densities at the expense of detailed cell–cell interactions.

In the present paper, we report our investigations of angiogenesis-like growth using a continuum model, which does not rely on any heuristic rules (e.g., branching rules) to obtain blood vessel morphologies. This experiment might provide some hints on the relative significance of the forces at the core of angiogenesis and mesenchymal motion in general. The following presentation considers 2D systems. These systems are relevant as many in vitro experiments of angiogenesis are essentially 2D. We note that all the techniques described herein have been extended to 3D.

A Continuum Representation of Cell Migration

There are two basic choices to represent the endothelial cells: similar to the representation of multiphase flow, we can chose to represent the cells by a density function (diffuse interface approach), or by a level set that represents the boundary to the extracellular domain (sharp interface approach). In the case of cell migration, we need to be able to represent agglomerates of cells that are highly elongated. Therefore, the level set approach is less favorable as we always require a narrow band of several grid spacings around the level set, and the requirements for the resolution are much more demanding than for a corresponding diffuse interface approach.

A single cell population is represented by a density ρ. This density evolves in time according to

$$\frac{\partial \rho}{\partial t} + \nabla \cdot (\mathbf{a}\rho) = d\Delta\rho, \tag{11.1}$$

where $\mathbf{a}$ denotes the cumulative effect of cell–cell adhesion ($\mathbf{a}^{c/c}$), pressure ($\mathbf{a}^{p}$), and migration cues ($\mathbf{a}^{ecm,\phi}$) as defined in the following sections. The term on the right-hand side accounts for random fluctuations on the cell population modeled at a macroscopic scale by a diffusion term with diffusion coefficient d.

If more then one cell type is present, one density is used per type, that is, $\{\rho_i\}_{i=1}^{\#CellTypes}$.

Cell–Cell Adhesion

Cell–cell adhesion is a fundamental biophysical mechanism. It is responsible for tissue formation, stability, and breakdown. It is involved in tissue invasion and metastasis of tumor cells. It is a crucial mechanism in embryogenesis, as it is the driving force of cell-sorting processes.

Cell adhesion to another cell or the ECM is established by specific adhesion receptors on the cell membrane, such as integrins, which may bind to collagens and fibronectin (in the ECM), and intercellular adhesion molecules such as cadherins. This reaction is very local, as it happens upon contact. In order to model cell adhesion, we state a set of requirements that reflect the main characteristics of the process: (1) cell adhesion is a short-range force; (2) cell adhesion will give rise to a movement of the cells toward the entity they adhere to; and (3) this cell movement will decrease as the cell density approaches the close-packing density, and at close-packing density we expect to find no residual movement caused by adhesion.

Given these characteristics, we can model cell adhesion as an auto-crine- (in the case of cell–cell adhesion), or paracrine-like signal f (in the case of cell adhesion to the ECM) acting as an adhesive force $\mathbf{a}^{c/c}$ on the cell population ρ. Consider the case of cell–cell adhesion: in the absence of other influences, we can model cell–cell adhesion forces as

$$\mathbf{a}^{c/c} = \kappa_f L(f, df) \nabla f,$$

$$\frac{\partial f}{\partial t} = -\mu f + \alpha \left(1 - \frac{f}{f_{\max}}\right)\rho + D_f \Delta f,$$
(11.2)

Parameters μ, α, and D_f are the decay, release, and diffusion parameters of the adhesion signal that define the range of the adhesive forces. The parameter $f_{\max}$ denotes the threshold value for the release of f. $L(f, df)$ is a cutoff function that keeps the magnitude of the gradient bounded by df in order to limit the migration velocity of the cells; here, we chose

$$L(f, df) = df \left(\max(df, |\nabla f|)\right)^{-1}.$$
(11.3)

The factor κ_f determines the influence of the adhesive force on the cell population. In the case where we have populations of different cells ρ_i, the model (11.2) is easily extended to

$$\mathbf{a}_i^{c/c} = \sum_j^{\# CellTypes} \kappa_{ij} L(f_j, df_j) \nabla f_j,$$

$$\frac{\partial f_i}{\partial t} = -\mu_i f_i + \alpha_i \left(1 - \frac{f_i}{f_{i,\max}}\right)\rho_i + D_i \Delta f_i,$$
(11.4)

where κ_{ij} describes the heterotypic ($i \neq j$) and homotypic ($i = j$) adhesion strength.

Close-Packing Density

The models (11.2) and (11.4) do not incorporate any repulsive effects that might limit the local cell density. Such effects are easily introduced by adding the following pressure-like term to the velocity:

$$\mathbf{a}^p = -\kappa_p H(\rho - \bar{\rho}) \nabla \rho |\nabla \rho|^{-1,}$$
(11.5)

where $\rho \equiv \Sigma_i \rho_i, \kappa_p$ is a constant that determines the cell population response to pressure, $\bar{\rho}$ is the cell close-packing density, and H is the Heaviside function.

The Extracellular Matrix

Continuum simulations of cellular motion rarely explicitly consider effects of the ECM on the migration. These effects are, however, crucial, as the ECM serves as a scaffolding with adhesive sites that the cells can use to exert forces and propel themselves. We propose to model the ECM as a collection of bundles of adhesive fibers that facilitate but also bias migration. The matrix is constructed as follows: We create N_f fibers as lines of thickness b_p from $(x^{\text{start}}, y^{\text{start}})$ to $(x^{\text{end}}, y^{\text{end}})$. The start point is drawn from a uniform distribution in the computational domain Ω. The end point is given as

$$\begin{pmatrix} x_p^{\text{end}} \\ y_p^{\text{end}} \end{pmatrix} = \begin{pmatrix} x_p^{\text{start}} + l_p \sin(2\pi\alpha_p) \\ y_p^{\text{start}} + l_p \cos(2\pi\alpha_p) \end{pmatrix}, \tag{11.6}$$

where, α_p $u.a.r \in [0,1)$ and l is the length of the fiber, which is obtained as

$$l_p = l2^{mz}, \text{ with } z \in \mathcal{N}(0,1), \tag{11.7}$$

The base fiber length l and the fiber length variation m are parameters of the simulation. The fiber thickness is given as b_p $u.a.r. \in [b_{\min}, b_{\max})$. These fibers are then discretized onto the ECM grid e using Bresenham's line rasterization algorithm (See Figure 11.2). In order to get a differentiable field we filter e N_{filter}- times with a second-order B-spline kernel.

Chemotaxis and Adhesion inside the ECM

In our model of chemotaxis, the migrating cells follow the concentration gradient of a chemoattractant ϕ. The model extension presented here to account for cell–ECM adhesion is a formulation of the following assumptions: in order to maximize its migration velocity, a cell will crawl along fibers, if these fibers are not transverse to the chemotactic cue ($\nabla\phi$). If there are no fibers in its environment, that is, e = 0, then a cell will not be able to migrate efficiently ($e_o \ll 1$); if the fiber density is too high ($e \approx e_\infty$),

FIGURE 11.2 An example matrix e in the domain $\Omega = [0,8] \times [0,1]$: base fiber length $l = 0.201$, $N_{filter} = 3$, $b_{min} = 4 \times 10^{-3}$, and $b_{max} = 2.7 \times 10^{-2}$.

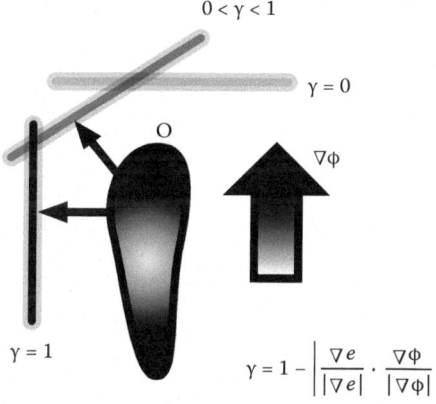

FIGURE 11.3 A cell will move "onto" a fiber if the fiber direction is not transverse to the chemotactic gradient; that is, the gradient of adhesion is not aligned with the chemotactic direction.

then cells have to degrade the matrix before they are able to migrate. These assertions are represented by

$$\mathbf{a}^{ecm,\phi} = \left[\left(1 - \left|\frac{\nabla e}{|\nabla e|} \cdot \frac{\nabla \phi}{|\nabla \phi|}\right|\right)\nabla e + \nabla \phi\right](e+e_o)(e_\infty - e),$$

(11.8)

and illustrated in Figure 11.3. We would like to point out here that the modeling of the chemotactic response $\mathbf{a}_{ecm,\phi}$ is but the most simple one and ignores many effects, such as the saturation of receptor sites on the cell membranes. The parameter e_o defines the minimal migrative response in the total absence of an ECM, and e_∞ defines the matrix density threshold that completely blocks migration.

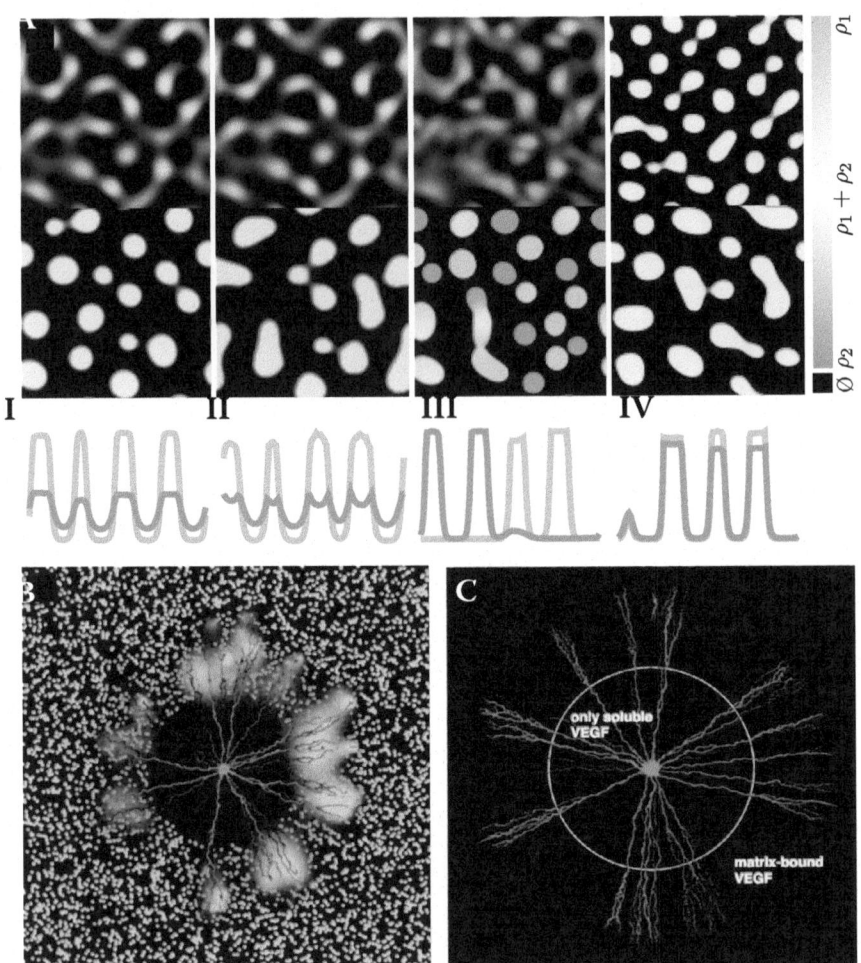

FIGURE 11.4 **(See color insert following page 40)** (A) Cell sorting simulations of two distinct cell populations ρ_1 and ρ_2. Top row depicts solution at $t = 40$, middle row at $t = 160$ bottom row depicts a cut through the domain at $t = 160$. Columns: I engulfment, $\kappa_{11} = 0.25$, $\kappa_{22} = 0.225$, and $\kappa_{12} = \kappa_{21} = 0.05$; II engulfment with pressure, $\kappa_{11} = 0.25$, $\kappa_{22} = 0.025$, and $\kappa_{12} = \kappa_{21} = 0.05$; III sorting, $\kappa_{11} = 0.25$, $\kappa_{22} = 0.025$, and $\kappa_{12} = \kappa_{21} = 0.00$; IV mixing, $\kappa_{11} = 0.25$, $\kappa_{22} = 0.09$, and $\kappa_{12} = \kappa_{21} = 0.2$. (B) Simulation with matrix-bound growth factors using pockets of matrix-bound VEGF distributed in the matrix. The endothelial cells release MMPs that cleave the bound growth factors and make them soluble (diffuse blue cues). (C) Simulation with matrix-bound growth factors by the "subscale" model. Within the white circle, there are only soluble growth factors present, and outside of the circle a constant concentration of growth factors is bound to the matrix. As is apparent from the network structure, the matrix-bound growth factors lead to distinctive increased branching.

MODEL APPLICATIONS

We present simulation results for different scenarios of cell sorting and cell migration processes. We consider both dynamics of a single cell type as well as interactions among different cell types with varying adhesive properties.

Cell Sorting

In order to test our model of cell–cell adhesion, we performed simulations of cell-sorting processes. According to the "Differential Adhesion Hypothesis" (see References [39,40] and references therein), cellular sorting is induced by differences in intercellular adhesiveness. We consider here two different types of cells that differ in their adhesion parameters k_{ij}. Depending on the choice of adhesion within and across cell types, we obtain different sorting behaviors (see Figure 11.4A) such as engulfment of one population by the other, complete sorting, and mixing. We also considered the pressure/repulsion effects as introduced earlier in one of the cases (see case AII in Figure 11.4). As we model contact-adhesion through auto/paracrine signaling, the proposed model successfully recovers the different sorting behaviors, although indirectly.

Angiogenesis-Like Migration

We now assemble a system modeling the chemotactically driven migration of cells through an ECM. This model consists of (1) cell–cell adhesion, (2) cell–matrix adhesion, and (3) chemotaxis. Accordingly, we substitute **a** in Equation 11.1 by $\mathbf{a} = \mathbf{a}^{c/c} + \mathbf{a}^{\phi 2}$ and define

$$\mathbf{a}^{\phi 2} = \kappa_e \left(1 - \left| \frac{\nabla e}{|\nabla e|} \cdot \frac{\nabla \phi}{|\nabla \phi|} \right| \right) \nabla e + \kappa_\phi \nabla \phi. \tag{11.9}$$

Note that for greater lucidity the system (11.9) ignores the effects of fiber density that we have introduced in Equation 11.8. Parameters κ_e and κ_e define the influence of cell–matrix adhesion and the response to chemotaxis.

Figures 11.5–11.7 illustrate the effects of modifying the model parameters on the resulting vessel morphology. The simulation outcome can be quantified by branch point and vessel section length dynamics as shown in Figures 11.8 and 11.9 in the case of varying matrix density and cell–cell adhesion.

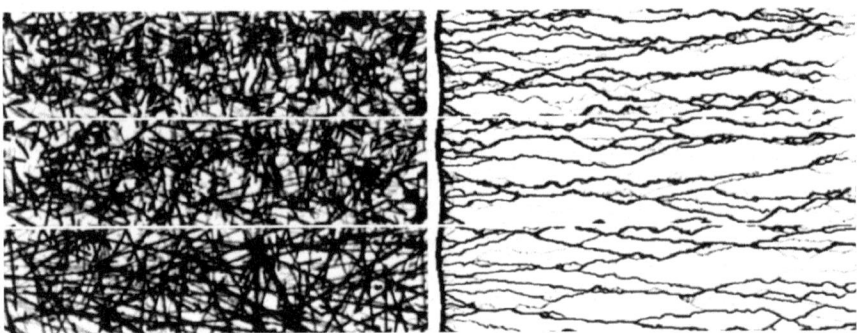

FIGURE 11.5 Effect of increasing the length of the fibers (in Equation 11.7), $m = 0.25$, 1.0, and 4.0.

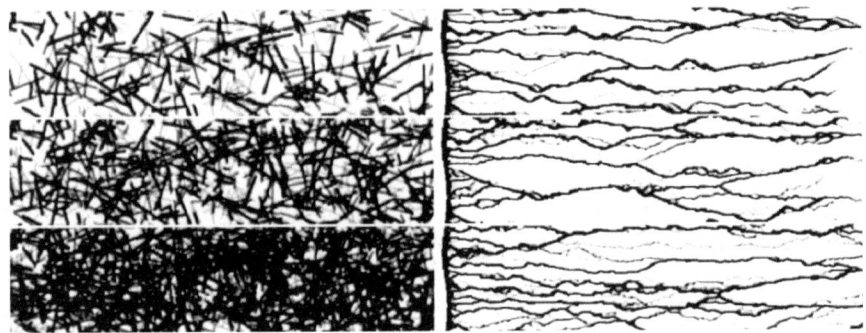

FIGURE 11.6 Effect of increasing the matrix density: 31%, 51%, and 75%.

In the presented simulation, we assume a linear increase in the chemoattractant concentration ϕ from the left side of the domain to the right.

A Multiscale Modeling Approach for Matrix-Bound Growth Factors

So far we have assumed that the growth factors are soluble and freely diffuse through the ECM. However, not all growth factors exist in an a priori soluble form; for example, it is well established that there are several different isoforms of VEGF, some that are soluble and some that can bind to heparin sites, which can be found in the matrix. These isoforms can bind to the matrix and do not diffuse freely. Endothelial cells secrete matrix metalloproteinases (MMPs) during angiogenesis. These MMPs have been shown [3] to cleave such matrix-bound VEGF isoforms, thus freeing them. We modeled this process by distributing small pockets of matrix-bound VEGF and by having the ECs secrete a compound that cleaves that matrix-bound VEGF. Once cleaved, this VEGF adds to the gradient established

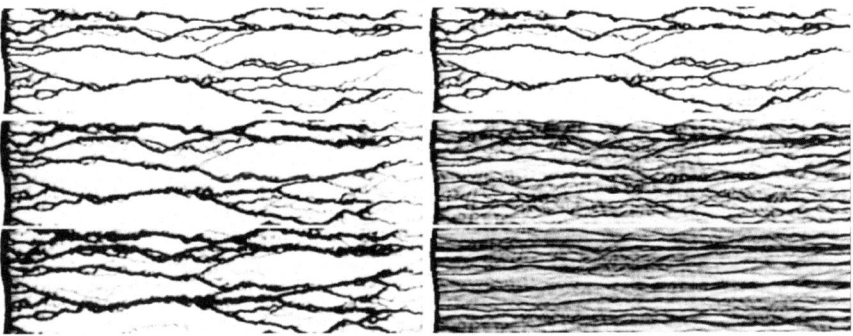

FIGURE 11.7 *Left:* Creating thicker vessels by decreasing the close-packing density $\bar{\rho}$. *Right:* Effect of reducing cell adhesion (cell–cell and cell–matrix).

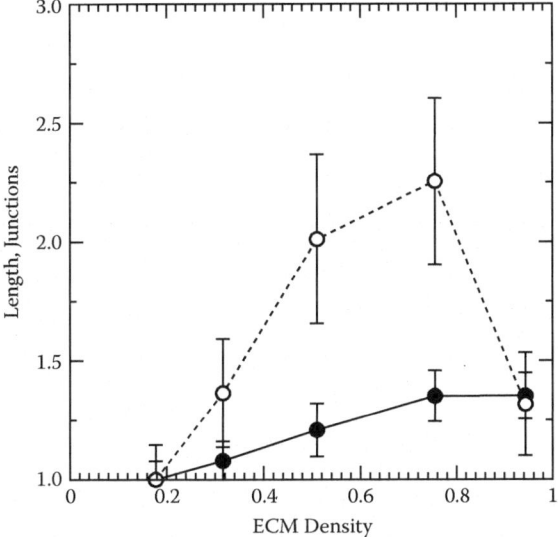

FIGURE 11.8 Sprouting angiogenesis: Effect of the ECM density on the total vessel network length (filled circles) and the number of junction points (empty circles).

by the soluble VEGF present in the domain. Figure 11.4B shows such a situation.

A setting such as this is able to provide localized chemotactic cues. However, we do not observe an increase in branching, which is what is observed in both *in vitro* and *in vivo* models of angiogenesis and vasculogenesis [3,33].

If we look at the distribution of matrix-bound VEGF, we must realize that it is very unlikely to find pockets of VEGF of that size in a real ECM.

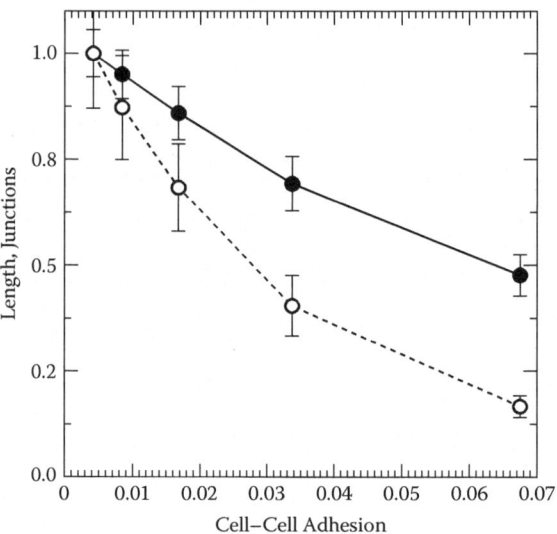

FIGURE 11.9   Sprouting angiogenesis: Effect of cell–cell adhesion on the total vessel network length (filled circles) and the number of junction points (empty circles).

Focus points of matrix-bound VEGF must be quite a bit smaller than the cell scale. In our continuum description, it is, however, not possible to explicitly incorporate localized structures that are truly microscopic. We must, therefore, resort to subgrid-scale modeling. From a mesoscopic point of view, what will be the effect of localized chemotactic queues that are smaller than the description scale? We will clearly not be able to distinguish any residual localized movement due to these cues. What we can expect to see is the cumulative effect on one cell, which from a mesoscopic viewpoint will be increased (apparently) random motion. It is well known that microscopic random motion manifests itself as diffusion from a macroscopic viewpoint. We therefore model the presence of matrix-bound VEGF by an increase in the spatially varying diffusion coefficient d in Equation 11.1 for the EC density ρ. That is, if only soluble factors are present, the diffusion is zero; in the local presence of matrix-bound factors, the diffusion term is increased depending on the concentration of matrix-bound isoforms. So both the release of MMPs and the cleaving of matrix-bound VEGF are modeled implicitly by increasing the diffusion of the ECs. The result of this modification of the model is depicted in Figure 11.4C: this type of modeling of matrix-bound VEGF does lead to increased branching (see Figure 11.10). We employ the automated imaging

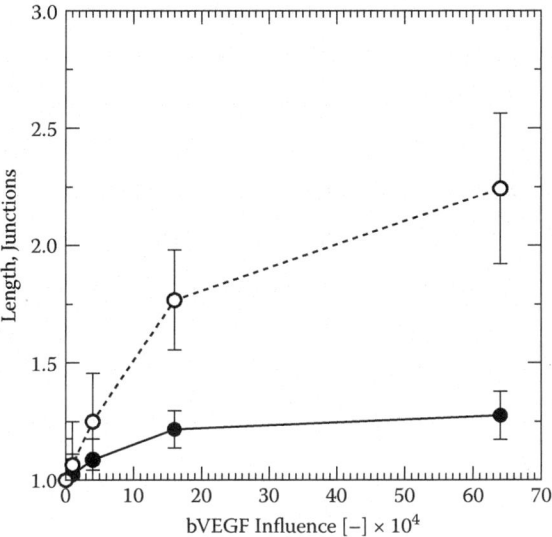

FIGURE 11.10   Sprouting angiogenesis: Influence of bound VEGF distributed in the second half of the domain on the total vessel network length (filled circles) and the number of junction points (empty circles).

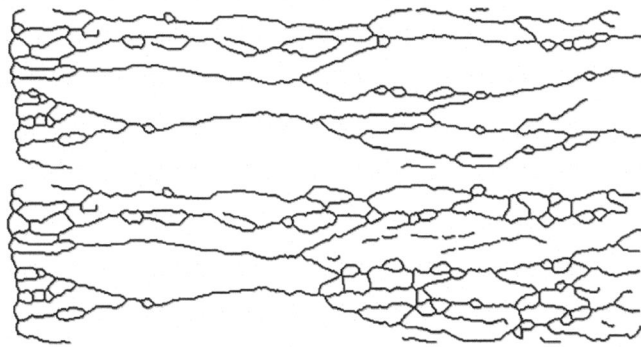

FIGURE 11.11   Extracted networks (using AngioQuant) from simulations with a bVEGF inluence factor of 0 (top) and 1.6×10^{-4} in the right half of the computational domain.

software AngioQuant [41] to extract the network and branching statistics from the simulation results (see Figure 11.11).

DISCUSSION

We have presented a pure continuum model of sprouting angiogenesis. This model considers several core aspects of mesenchymal motion: (1) cell–cell

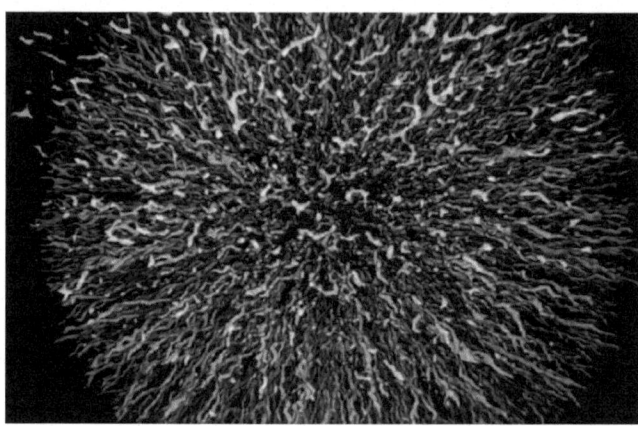

FIGURE 11.12   3D simulations of cells shed from the surface of a spherical tumor, and their invasion of the surrounding ECM.

adhesion, (2) the structure of the ECM, (3) cell–matrix adhesion, (4) chemotaxis, and (5) the effect of matrix-bound growth factors.

In this model, we simulated cell–cell adhesion by modeling it as an autocrine signal. Although indirect, this formulation is very simple and manages to recover aspects of the sorting behavior of cells. Compared to existing continuum models of cell–cell adhesion [26], the present model is less intuitive; however, it is more efficient and easier to implement. In the present work, the ECM is represented explicitly as a collection of fibers that exert adhesive forces. This method of representing the mechanical aspects of the cell microenvironment enables us to recover branching behavior as an output of the model, and hence we do not need to call on heuristic branching events. In addition to soluble chemotactic cues, we also consider matrix-bound cues, which are modeled using a subgrid-scale approach. The incorporation of these localized cues yields an increase in vessel branching, and thus results in the same morphological chances as observed in experiments. In its general form, the methods presented earlier can be used to simulate mesenchymal motion in general. The invasion of cancer cells into healthy tissue is another tumor-related phenomenon that is dominated by mesenchymal motion, and to which the present model may be applied (see Figure 11.12).

ACKNOWLEDGMENTS

We wish to acknowledge financial support by the Swiss National Science Foundation and by the Systems-X program.

REFERENCES

1. Ferrara, N., Gerber, H.-P., LeCouter, J. The biology of VEGF and its receptors. *Nature Medicine* 9, 669–676 (2003).
2. Folkman, J. Angiogenesis. *Annual Review of Medicine* 57(1), 1–18 (2006).
3. Lee, S., Jilani, S. M., Nikolova, G. V., Carpizo, D., Iruela-Arispe, M. L. Processing of VEGF-A by matrix metalloproteinases regulates bioavailability and vascular patterning in tumors. *Journal of Cell Biology* 169(4), 681–691 (2005) 10.1083/jcb.200409115.
4. Houck, K., Leung, D., Rowland, A., Winer, J., Ferrara, N. Dual regulation of vascular endothelial growth factor bioavailability by genetic and proteolytic mechanisms. *Journal of Biological Chemistry* 267(36), 26031–26037 (1992).
5. Plouet, J., Moro, F., Bertagnolli, S., Coldeboeuf, N., Mazarguil, H., Clamens, S., Bayard, F. Extracellular cleavage of the vascular endothelial growth factor 189-amino acid form by urokinase is required for its mitogenic effect. *Journal of Biological Chemistry* 272(20), 13390–13396 (1997).
6. Parent, C. A., Devreotes, P. N. A cell's sense of direction. *Science* 284(5415), 765–770 (1999).
7. Levine, H., Kessler, D. A., Rappel, W.-J. Directional sensing in eukaryotic chemotaxis: A balanced inactivation model. *Proceedings of the National Academy of Sciences* 103(26), 9761–9766 (2006).
8. Iglesias, P. A., Devreotes, P. N. Navigating through models of chemotaxis. *Current Opinion in Cell Biology* 20(1), 35–40 (2008).
9. Ueda, M., Shibata, T. Stochastic signal processing and transduction in chemotactic response of eukaryotic cells. *Biophysical Journal* 93(1), 11–20 (2007).
10. Rappel, W.-J., Levine, H. Receptor noise limitations on chemotactic sensing. *Proceedings of the National Academy of Sciences* 105(49), 19270–19275 (2008).
11. Tindall, M., Maini, P., Porter, S., Armitage, J. Overview of mathematical approaches used to model bacterial chemotaxis II: Bacterial populations. *Bulletin of Mathematical Biology* 70(6), 1570–1607 (2008).
12. Gamba, A., Ambrosi, D., Coniglio, A., de Candia, A., Di Talia, S., Giraudo, E., Serini, G., Preziosi, L., Bussolino, F. Percolation, morphogenesis, and burgers dynamics in blood vessels formation. *Physical Review Letters* 90(11), 118101–118104 (2003).
13. Manoussaki, D. A mechanochemical model of angiogenesis and vasculogenesis. *ESAIM: Mathematical Modelling and Numerical Analysis* 37(4), 581–599 (2003).
14. Tosin, A., Ambrosi, D., Preziosi, L. Mechanics and chemotaxis in the morphogenesis of vascular networks. *Bulletin of Mathematical Biology* 68(7), 1819–1836 (2006).
15. Filbet, F., Shu, C.-W. Approximation of hyperbolic models for chemosensitive movement. *SIAM Journal on Scientific Computing* 27(3), 850–872 (2005) 10.1137/040604054.
16. Ambrosi, D., Bussolino, F., Preziosi, L. A review of vasculogenesis models. *Computational and Mathematical Methods in Medicine* 6(1), 1–19 (2005).

17. McDougall, S. R., Anderson, A. R. A., Chaplain, M. A. J. Mathematical modelling of dynamic adaptive tumour-induced angiogenesis: Clinical implications and therapeutic targeting strategies. *Journal of Theoretical Biology* 241(3), 564–589 (2006).

18. Chaplain, M. A. Mathematical modelling of angiogenesis. *Journal of Neurooncology* 50(1–2), 37–51 (2000).

19. Plank, M., Sleeman, B. Lattice and non-lattice models of tumour angiogenesis. *Bulletin of Mathematical Biology* 66(6), 1785–1819 (2004).

20. Sun, S., Wheeler, M. F., Obeyesekere, M., Patrick, J. A deterministic model of growth factor-induced angiogenesis. *Bulletin of Mathematical Biology* 67, 313–337 (2005).

21. Bauer, A. L., Jackson, T. L., Jiang, Y. A cell-based model exhibiting branching and anastomosis during tumor-induced angiogenesis. *Biophysical Journal* 92(9), 3105–3121 (2007).

22. Byrne, H. M., Chaplain, M. A. J. Modelling the role of cell-cell adhesion in the growth and development of carcinomas. *Mathematical and Computer Modelling* 24(12), 1–17 (1996).

23. Macklin, P., Lowengrub, J. Evolving interfaces via gradients of geometry-dependent interior Poisson problems: Application to tumor growth. *Journal of Computational Physics* 203, 191–220 (2005).

24. Turner, S., Sherratt, J. A., Painter, K. J., Savill, N. J. From a discrete to a continuous model of biological cell movement. *Physical Review* E 69(2), 021910 (2004) 10.1103/PhysRevE.69.021910.

25. Erban, R., Othmer, H. G. From individual to collective behavior in bacterial chemotaxis. *SIAM Journal on Applied Mathematics* 65(2), 361–391 (2004) 10.1137/S0036139903433232.

26. Armstrong, N. J., Painter, K. J., Sherratt, J. A. A continuum approach to modelling cell-cell adhesion. *Journal of Theoretical Biology,* accepted (2006).

27. Milde, F., Bergdorf, M., Koumoutsakos, P. A hybrid model for 3D simulations of sprouting angiogenesis. *Biophysical Journal,* submitted (2007).

28. Painter, K. Modelling cell migration strategies in the extracellular matrix. *Journal of Mathematical Biology* 58(4), 511–543 (2009).

29. Folkman, J. Angiogenesis: an organizing principle for drug discovery? *Nature Reviews Drug Discovery* 6(4), 273–286 (2007).

30. Axelson, H., Fredlund, E., Ovenberger, M., Landberg, G., Pahlman, S. Hypoxia-induced dedifferentiation of tumor cells—A mechanism behind heterogeneity and aggressiveness of solid tumors; Biology of hypoxia and myogenesis and muscle disease. *Seminars in Cell and Developmental Biology* 16(4–5), 554–563 (2005).

31. Pennacchietti, S., Michieli, P., Galluzzo, M., Mazzone, M., Giordano, S., Comoglio, P. M. Hypoxia promotes invasive growth by transcriptional activation of the met protooncogene. *Cancer Cell* 3, 347–361 (2003).

32. Saharinen, P., Alitalo, K. Double target for tumor mass destruction. *Journal of Clinical Investigation* 111(9), 1277–1280 (2003).

33. Ruhrberg, C., Gerhardt, H., Golding, M., Watson, R., Ioannidou, S., Fujisawa, H., Betsholtz, C., Shima, D. T. Spatially restricted patterning cues provided by heparin-binding VEGF-A control blood vessel branching morphogenesis. *Genes and Development* 16(20), 2684–2698 (2002) 10.1101/gad.242002.

34. Davis, G. E., Senger, D. R. Endothelial extracellular matrix: Biosynthesis, remodeling, and functions during vascular morphogenesis and neovessel stabilization. *Circulation Research* 97(11), 1093–1107 (2005).

35. Friedl, P., Bröcker, -B. The biology of cell locomotion within three-dimensional extracellular matrix. *Cellular and Molecular Life Sciences (CMLS)* 57(1), 41–64 (2000).

36. Kirkpatrick, N. D., Andreou, S., Hoying, J. B., Utzinger, U. Live imaging of collagen remodeling during angiogenesis. *AJP—Heart and Circulatory Physiology*, 01234.2006 (2007).

37. Anderson, A. R. A., Chaplain, M. A. J. Continuous and discrete mathematical models of tumor-induced angiogenesis. *Bulletin of Mathematical Biology* 60, 857–900 (1998).

38. Levine, H. A., Sleeman, B. D., Nilsen-Hamilton, M. Mathematical modeling of the onset of capillary formation initiating angiogenesis. *Journal of Mathematical Biology* V42(3), 195–238 (2001).

39. Steinberg, M. S., Takeichi, M. Experimental specification of cell sorting, tissue spreading, and specific spatial patterning by quantitative differences in cadherin expression. *Proceedings of the National Academy of Sciences* 91(1), 206–209 (1994).

40. Steinberg, M. Differential adhesion in morphogenesis: A modern view. *Current Opinion in Genetics and Development* 17(4), 281–286 (2007).

41. Niemisto, A., Dunmire, V., Yli-Harja, O., Zhang, W., Shmulevich, I. Robust quantification of in vitro angiogenesis through image analysis. *IEEE Transactions on Medical Imaging* 24(4), 549–553 (April 2005) 10.1109/TMI.2004.837339.

CHAPTER 12

Do Tumor Invasion Strategies Follow Basic Physical Laws?

Caterina Guiot, Pier Poalo Delsanto, and Antonio Salvador Gliozzi

CONTENTS

INTRODUCTION

Physical and mathematical models can be very useful in many subfields of biomedicine and, in particular, in oncology. Physical models, based on a comparison with well-known phenomena, which present formal analogies with some aspects of tumoral development, may be extremely helpful, since they may suggest new mechanisms to be tested and analyzed. Mathematical and computational models allow researchers to vary the details of the proposed model or their parameters, for example, by adding new ingredients and/or eliminating ineffective ones. Thus, it becomes possible to perform virtual experiments of selected therapies and to predict or optimize the outcome of suggested therapeutic protocols. The current relevance of physical, mathematical, and computational modeling is due to a combination of related factors (Deisboeck et al. 2007, 2009; Liu et al. 2006; Kitano, 2002; Hornberg et al. 2006; Coffey 1998).

As an example of physical models, consider a study of tumor invasiveness based on the analogy with two well-known physical mechanisms, that is, the mechanical insertion of a solid inclusion in an elastic material specimen or the impinging of a water drop on a solid surface (Guiot et al. 2007). As far as the computational models are concerned, there exists a large number of simulation techniques, such as cellular automata, finite difference methods, LISA (Local Interaction Simulation Approach), etc. (Scalerandi et al. 1999, 2001; Delsanto et al. 2000). They generally consist of "mesoscopic" formulations that help us to connect the macroscopic and microscopic points of view, that is, what is mainly of clinical interest from what can be learned from the bio-chemo-physics of the cells, for example, by means of ab initio calculations (Chignola et al. 2007). Such an understanding is necessary not only to predict the emergence of macroscopic phenomena from microscopic laws, but also to correlate microscopic and macroscopic parameters (Delsanto et al. 2005, 2008).

Closer to real clinical tumors, many biological models have been proposed and investigated. Since the complexity of tumor growth dynamics makes this task very difficult in in vivo or even observations ex vivo, a convenient experimental tool that captures some of the most relevant features of tumor growth kinetics, while allowing for a manageable description, are the multicellular tumor spheroids (MTSs) (Mueller Klieser 1993; Delsanto et al. 2004). MTSs are spherical aggregations of tumor cells that may be grown under strictly controlled conditions. Their simple geometries and the ability to produce them in large quantities have led to interesting new insights into cancer research. At a higher level of complexity, tumor models on laboratory animals, mostly mice (normal, or genetically modified to have various genes "knocked out") are currently being developed (Talmadge 2007).

MODEL: CANCER GROWTH

In order to understand the "basics" of tumor evolution, we recall that it is generally assumed that tumors originate from a "seed" and grow by cell duplication. The notion that a tumor develops from a single cell is called *monoclonality*, and was proposed already in 1862 by Virchow (1892). His hypothesis is today supported by data from the majority of human tumors, even multicentric or detected in paired organs (Friberg and Mattson 1997). Heteroclonality is likely to occur later during the lifespan of a tumor.

We describe the volumetric tumor growth with the equation:

$$\frac{dV}{dt} = cV \qquad (12.1)$$

where c can depend on both V (nonlinearity) and t. However, if we assume $c = c(V, t)$ in its complete generality, we cannot go too far. For the sake of a quantitative analysis, it is necessary to impose some constraint, which, although arbitrary, at least should be independent of the particular field of application.

Following the Phenomenological Universalities (PUN) approach, recently proposed by P.P. Delsanto and collaborators (Delsanto 2007; Castorina et al. 2006), we assume that c and its time derivative are related through the simple power expansion:

$$dc/dt = \beta c + \gamma c^2 + \delta c^3 + \ldots \qquad (12.2)$$

All the datasets $V(t)$, which are well described by Equations 12.1 and 12.2, with the latter truncated at the Nth term, are said to belong to the Phenomenological Universality (PUN) class UN. For example, if we retain only the first term (βc), we have U1. Retaining the second term also, γc^2, yields U2, etc.

Equations 12.1 and 12.2 describe in general all tumor developmental phases, but the various terms (with coefficients β, γ, δ) are expected to increase their relevance during tumor growth, for example, when suitable thresholds in nutrients and oxygen availability are reached.

$$\text{PUN CLASS U0:} \quad \beta, \gamma, \delta = \ldots = 0 \qquad (12.3)$$

Here the growth rate c is constant:

$$dV/dt = cV, \text{ that is,} \quad V = V_0 \exp (ct), \qquad (12.4)$$

and therefore, it follows an exponential growth law. As long as no mechanical or nutritional restrictions apply, tumoral cells keep on replicating with a constant duplication time. Contrary to expectations, such a model is not totally unrealistic. In fact, U0 represents the first phase of any in vitro experiment, when there is complete availability of nutrients and oxygen for all tumoral cells, and also the initial phase of all tumors.

An interesting biological model, in which the duration of such an unrestricted growth is extended to days, months, or even years, is the multipassage tumor (MPT), where tumors grow following the subcutaneous implantation on the back of a lab animal (usually mice) of ~ 10^6 tumor cells (from cell cultures or surgical resection). Tumor cells are then passaged from one mouse to another by harvesting them from a growing tumor and implanting a given number of them into another healthy animal. Once the tumor has grown above a certain volume, it is harvested again. This passage of tumor cells is repeated for multiple rounds (Steel 1977; McCredie and Sutherland 1971). For example, McCredie et al. reported the case of a spontaneous mammary tumor in a C3H mouse, from which the first syngenic transplant was done in 1946 and which has been serially transplanted into the C3H/HeJ strain, reaching the 900th generation in 1971.

As shown in Figure 12.1, tumors grow with a larger rate at each successive transplant; that is, the growth curves become progressively steeper.

Before looking for complex explanations, for example, those based on multiclonality, we should recall that we always transplant "small" tumors. This implies that the new seed is reimplanted each time after a short time T (e.g., 10 days) into a new healthy, nutrient-rich environment;

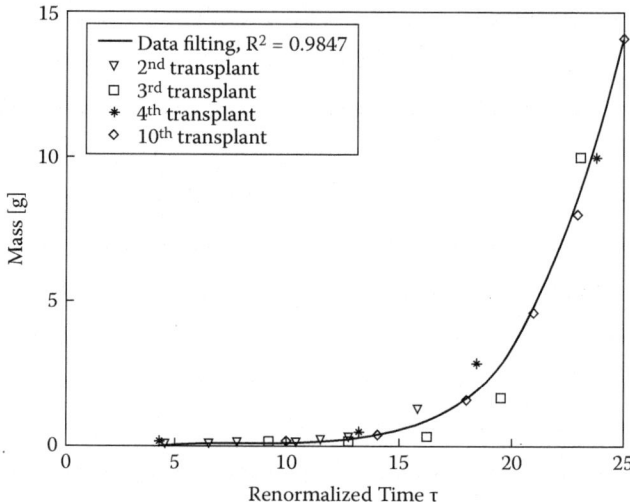

FIGURE 12.1 Experimental data relative to the growth of successively transplanted tumor cells in rodents. (From Steel, G.G. 1977. *Growth Kinetics of Tumors*, Clarendon Press, Oxford.)

thus, we can assume at each "passage" an experimental growth law with approximately the same rate c. The tumor mass, after n passages in which the same amount m_0 of tumor is reimplanted, and at a time from the beginning of the whole procedure equal to $t = nT + \Delta t$, is given by Gliozzi et al. (2009):

$$m(t) = m_0 \, (1+exp \, (ncT) \, (exp(c\Delta t)\text{-}1)) \tag{12.5}$$

Equation 12.5 shows that the exponential trend is corrected by a term that accounts for the real age of the tumor and increases at each transplant, thus accelerating the growth. Correspondingly, the rate of growth will be

$$dm/dt = m_0 \, c \, exp(ncT), \tag{12.6}$$

that is, larger for larger n. Or, equivalently, it remains the same, provided that the time is properly renormalized to the "scaled" time τ

$$\tau = t/r \tag{12.7}$$

where the "acceleration parameter" r increases with the number of transplants n (see Figures 12.2 and 12.3).

If we follow the tumor growth curve for a longer time (even in vitro), a necrotic core will develop, due to the "screening" of the nutrient flow

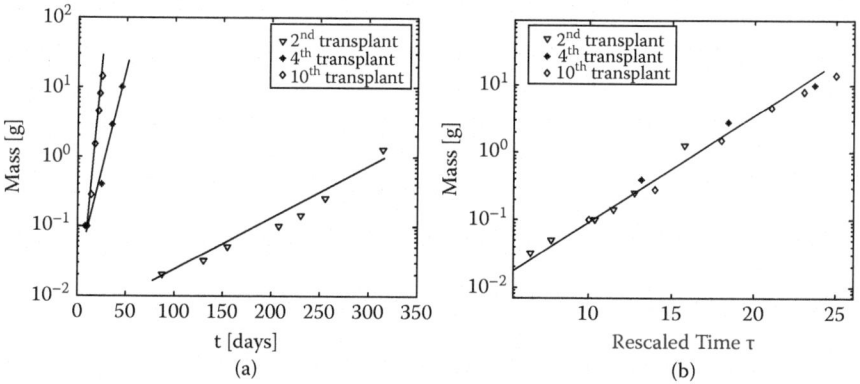

FIGURE 12.2 Renormalization by means of Equation 12.7 of the data of Figure 12.1 and PUN fitting. (Data from Steel, G.G. 1977. *Growth Kinetics of Tumors*, Clarendon Press, Oxford.)

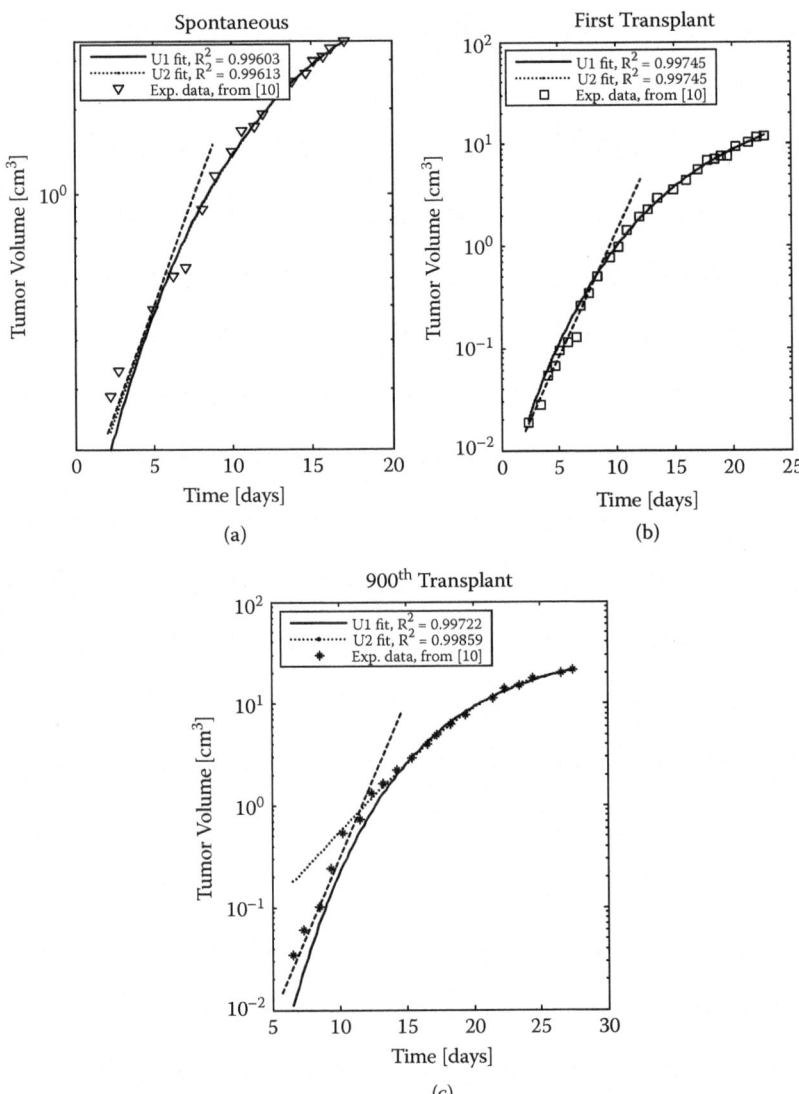

FIGURE 12.3 Renormalization by means of Equation 12.7 of the data of Figure 12.1 and PUN fitting. (Data from McCredie, J.A. and Sutherland, R.M. 1971. *Cancer* 27:635–642.)

from the cells of the outer layers, so that a saturation level is reached. This corresponds to the

$$\text{PUN CLASS U1: } \beta \neq 0 \ \gamma, \ \delta = 0 \tag{12.8}$$

The class is defined by

$$dc/dt = \beta c \tag{12.9}$$

with $\beta < 0$. By integration, we easily find

$$c(t) = c_0 \exp(\beta t) \tag{12.10}$$

and

$$V(t) = \frac{c_0}{\beta}\left(\exp(\beta t) - 1\right) \tag{12.11}$$

and with a saturation value

$$V_\infty = V_0 \exp(c_0/\beta) \tag{12.12}$$

Such saturation is normally seen in MTSs (see Figure 12.4).

For benign neoplasies, this is the end of the story, but most tumors nevertheless keep on growing, becoming malignant.

MODEL: MALIGNANT NEOPLASIAS

Most aggressive tumors overcome nutrient deprivation by means of angiogenesis, and the neovascular network partly supports growth, as discussed by C. Guiot et al. (2006), following the model of G.B. West and collaborators (West et al. 1999, 2001, 2004). Often, this late phase is complemented, in in vivo or ex vivo tumors by the processes of tumor invasion and metastasis. Starting from West's law, a generalized expression for growth is then

$$dV/dt = a \, S - b \, V \tag{12.13}$$

where S is the embedding layer acting as boundary with the host (i.e., the effective surface from which nutrients come and waste products are

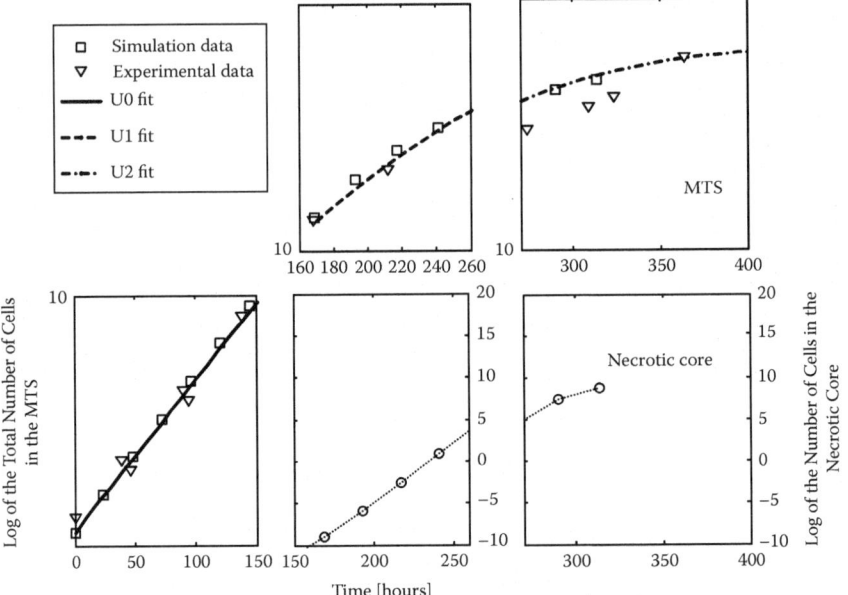

FIGURE 12.4 The three phases of growth of EMT6/Ro mouse mammary carcinoma cells MTSs. (Data from Scalerandi et al. 1999. *Phys. Rev. E,* 63:11901 and Delsanto et al. 2004. *Appl. Phys. Lett.* 85:4225–4227.)

removed), and V the volume, to which the energy consumption for metabolism is assumed to be proportional. (Note: S is not, in general, a 2D surface.)

It follows that growth implies a favorable ratio S/V, or at least $S/V > b/a$. In fact, tumor invasion, neoangiogenesis, and metastases can be interpreted as different surface-maximizing strategies for the tumor, to improve oxygen and nutrient exchanges (Deisboeck et al. 2006). They also are fractal-like self-similar processes.

This stage of tumor growth corresponds to the

$$\text{PUN CLASS U2: } dc/dt = \beta c + \gamma c^2 \tag{12.14}$$

By direct integration, we find:

$$c(t) = c_0 \left[\left(1 + \frac{c_0 \gamma}{\beta} \right) e^{-\beta t} - \frac{c_0 \gamma}{\beta} \right]^{-1} \tag{12.15}$$

$$V(t) = -\frac{1}{\gamma} \ln\left[1 + \frac{c_0\gamma}{\beta}\left(1 - e^{\beta t}\right)\right] \tag{12.16}$$

with a saturation volume that is given by

$$V_\infty = V_0\left(1 + c_0\frac{\gamma}{\beta}\right)^{-1/\gamma} \tag{12.17}$$

The class U2 includes, besides Gompertz (1825) as a special case, all the growth models proposed to date in all fields of research, that is, besides the already mentioned model of West et al. (1999, 2001), also the exponential, logistic, thetalogistic, potential, von Bertalanffy, etc. (for a review, see de Vladar et al. 2006).

The explicit expression of U2 ($V(t) = \exp[z(t)]$) satisfies the ordinary differential equation (ODE):

$$dV/dt = a\,V^p - b\,V \tag{12.18}$$

where V is the tumor volume and a, b, and p are suitable parameters expressed by real numbers.

By comparison with the relation $dV/dt = a\,S - b\,V$, we see that p depends on the fractal characteristics of the effective surface of the tumor. The value of p may vary, but in the case of a spheroid, $S = V^{2/3}$; hence, $p = 2/3$. In the case investigated by West et al. for living organisms and Guiot et al. (2003) for tumors, the optimalization of the fractal-like distributive system yields $p = \frac{3}{4}$. Proliferation is possible up to $V < V_{max} = (a/b)^3$.

The choice of the specific strategy for maximizing the exchange surface depends on many factors, according to the characteristics of the host and the molecular weapons available. In physical terms, the strategy followed by an infiltrative tumor has been investigated using both the propagating fracture and the water-drop analogies. First, all such invasion processes based on fingering can be described as fractal processes.

The analogy with the water drop (see Figure 12.5), and in particular the identification of the relevance of the surface tension entering our physical analogy, is actually very suggestive. Let us introduce an equivalent tumor-host tension σ, which measures the balance between

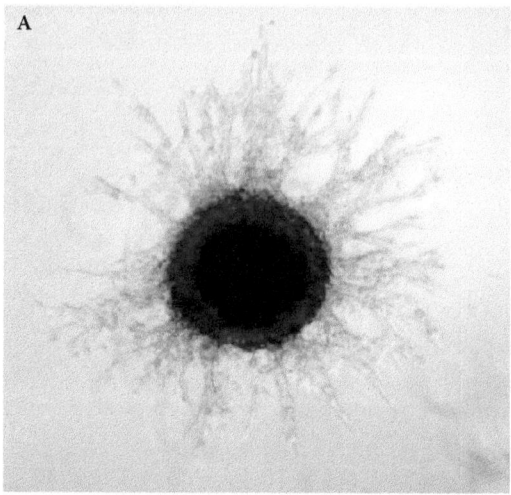

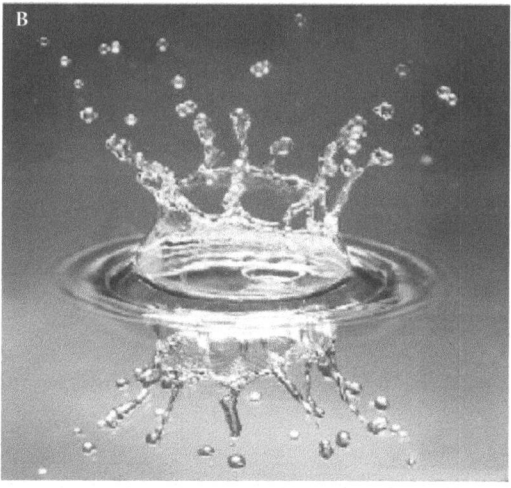

FIGURE 12.5 **(See color insert following page 40)** (A) MTS tumor (image reprinted from Habib et al. *Physica* A 327 501, 2003. With permission). (B) Water drop (image courtesy of Professor A Davidazy, Imaging and Photographic Technology, Rochester Institute of Technology, Rochester, New York).

adhesive and cohesive forces exerted by the tumor and its host. On the basis of both analogies, we can define a dimensionless Invasion Parameter IP (Guiot et al. 2007):

$$IP = P\,R/\sigma \qquad (12.19)$$

where P is the excess pressure (between outside and inside), R the radius, and σ the tumor-host tension. Invasive behavior is expected whenever

IP > 1, and can be prevented, for example, by increasing the tumor-host tension, (e.g., using Dexamethasone). Such an equation can be rewritten as a generalization of Laplace's law:

$$P = IP\,\sigma/R \tag{12.20}$$

which states that the tumor-host tension can counteract the excess pressure (e.g., given by rapid tumor proliferation) only when *IP ≤ 1*. When IP locally increases, then invasive branching can develop (see Figure 12.6a).

The preceding relation can be easily extended to ellipsoids (of semiaxes R_1 and R_2, respectively):

$$P = IP\,\sigma\,(1/R_1 + 1/R_2) \tag{12.21}$$

and to cylinders (if one allows one of the two semiaxes to go to infinity; see Figure 12.6b).

Such a model could be suitable to describe the evolution from breast ductal carcinoma in situ (DCIS) to invasive ductal carcinoma (IDC). In fact, modifications in cohesion and adhesion properties have been recognized in several studies, for example, Min Hu et al. (2008) and Haj et al. (2007). Also, such an approach explains the complex clinical management of gliomas (Deisboeck and Guiot 2008). A specular situation can be found when malignant tumors strongly increase their growth rate by expressing molecular weapons acting on the endothelial cells of the preexisting

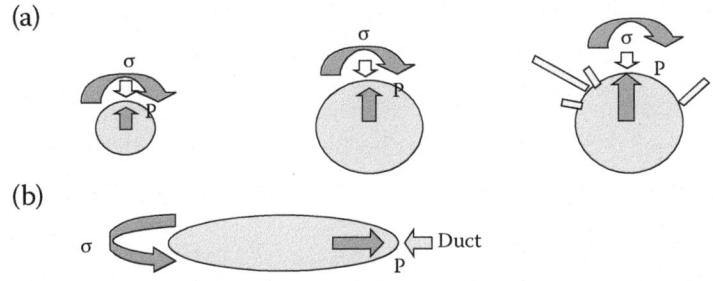

FIGURE 12.6   (a) Schematic drawing of a growing tumor spheroid where surface tension counteracts the increase of internal pressure up to a given fingering threshold, (b) schematic drawing of a tumor ellipsoid spheroid growing in a duct where surface tension counteracts the increase of internal pressure.

vasculature, generating a new vascular network inside the tumor. Such a process is called *neoangiogenesis*, and is common to almost all malignant tumors.

A vascular network is the basis for growth in all living beings (West et al. 1999, 2001) and is the main limiting factor for cellular growth. For most living organisms, it develops according to some optimality principles [34], provided eight assumptions are made:

1. Distribution network scaling

2. Distribution network hierarchical

3. Equivalence of vessels at the same hierarchical level

4. Constant branching ratio

5. Space-filling network

6. Minimized energy losses in the network

7. Same capillary characteristics

8. Capillaries are the only exchange surfaces across which oxygen is supplied

Under these assumptions, it can be formally proved that the scaling parameter is $p = 3/4$ (as already empirically proposed by Kleiber in 1932). Deviations are possible if some of the foregoing assumptions are relaxed, and this is certainly the case of tumor neoangiogenic vessels, which are known to be very irregular, randomly distributed, and inefficient. In particular, a recent paper (Mazzone et al. 2009) showed that the morphology of the neovasculature may vary from disordered and inefficient to almost normally layered from endothelial cells when tumors develop in normal animals or in mice deprived of one of the two copies of the PHD2 gene. In other terms, at least one protein (PHD2) is involved as oxygen sensor, not only in the short-term response but in vascular plasticity itself. This encompasses the possibility that the tumor evolves by changing the degree of fractality of its nutrient input system and may imply that the fractal parameter p varies dynamically through tumor growth (Guiot et al. 2006).

In particular, by fitting the experimental data for a series of short time intervals, it was found that p varies with time (see Figure 12.7).

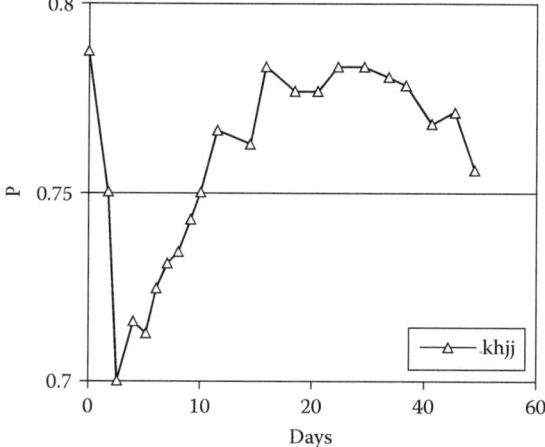

FIGURE 12.7   Fitting of the data (obtained from Steel 1977 and related to the experimental tumor cell line khjj transplanted in mice) with a dynamic exponent model. (From Guiot et al. 2006b. With permission.)

DISCUSSION: WHAT NEXT?

$$\text{PUN CLASS U3}: \beta, \gamma, \delta \neq 0 \text{ is presently under study.} \quad (12.22)$$

A proposed mechanism to further enhance tumor metabolism is the expression of hormonal receptors on the tumor surface. It is well known that breast carcinoma is initially deprived of estrogen receptors (ER-), but in most cases such receptors develop later (ER+), being the biological base for effective hormonotherapy (Colleoni et al. 2008). Also, other recent papers focus on the role of hormones in enhancing tumor metabolism: for instance, it has been shown that protein such as Myc can regulate the number of cellular mitochondria and the ability of the tumor of metabolize glucose and glutamine (Gao et al. 2009); or similarly, high levels of the protein ATP synthase, involved in the production of the energy-rich molecules of ATP, have been found on the surface of cancer cells (Tsui-Chin Huang et al. 2008). Such gaps in the tumor cell metabolism, responsible for tumor growth enhancement, maybe perhaps be still hidden in the yet unexplored class U3 (or other PUN classes).

ACKNOWLEDGMENTS

CG kindly acknowledges the contribution from Regione Piemonte (Progetti Sanitari Finalizzati 2008) and Università di Torino.

REFERENCES

Castorina, P., Delsanto, P.P., and Guiot, C. 2006. Classification scheme for phenomenological universalities in growth problems in physics and other sciences, *Phys. Rev. Lett.* 96:188701.

Chignola, R., Del Fabbro, A., Dalla Pellegrina, C., and Milotti, E. 2007. Ab initio phenomenological simulation of the growth of large tumor cell populations, *Phys. Biol.* 4:114–133.

Coffey, D.S. 1998. Self-organization, complexity and chaos: the new biology for medicine. *Nat. Med.* 4:882–885.

Colleoni, M. et al. 2008. Expression of ER, PgR, HER1, HER2, and response: a study of preoperative chemotherapy. *Ann. Oncol.* 19:465–472.

de Vladar, H.P. 2006. Density-dependence as a size-independent regulatory mechanism. *J. Theor. Biol.* 238:2:245–256.

Deisboeck, T.S., Guiot, C., Delsanto, P.P., and Pugno, N. 2006. Does cancer growth depend on surface extension? *Medical Hypotheses* 67:1338–1341.

Deisboeck, T.S. et al. 2007. Advancing cancer systems biology: introducing the Center for the Development of a Virtual Tumor, CViT. *Cancer Informatics*:1–8.

Deisboeck, T.S. and Guiot, C. 2008. Surgical impact on brain tumor invasion: a physical perspective. *Annals of Surgical Innovation and Research*, 2:1.http://www.asir-journal.com/content/2/1/1.

Deisboeck, T.S. et al. 2009. In silico cancer modeling: is it ready for prime time. *Nat. Clinical Practice* 6:34.

Delsanto, P., Romano A., Scalerandi M., and Pescarmona, G.P. 2000. Analysis of a phase transition between tumor growth and latency. *Phys. Rev. E.* 62:2547–2554.

Delsanto, P.P., Guiot, C., Degiorgis, P.G., Condat, A.C., Mansury. Y., Desboeck, T.S. 2004. Growth model for multicellular tumor spheroids. *Appl. Phys. Lett.* 85:4225–4227.

Delsanto, P., Griffa, M., Condat, C.A., Delsantom S., and Morra, L. 2005. Bridging the gap between mesoscopic and macroscopic models: the case of multicellular tumor spheroids. *Phys. Rev. Lett.* 94:148105.

Delsanto, P.P. (Ed.) 2007. *Universality of Nonclassical Nonlinearity with Applications to NDE and Ultrasonics.* Springer, New York.

Delsanto, P.P., Condat, C.A., Pugno, N., Gliozzi, A.S., and Griffa, M. 2008. A multilevel approach to cancer growth modeling. *J. Theor. Biol.* 250:16–24.

Friberg, S. and Mattson, S. 1997. On the growth rates of human malignant tumors: implications for medical decision making. *J. Surg. Oncol.* 65:284–297.

Gao, P., Tchernyshyov, I., Chang, T.C. et al. 2009. c-Myc suppression of miR-23a/b enhances mitochondrial glutaminase expression and glutamine metabolism. *Nature* 458:762–U100.

Gliozzi, A.S., Guiot, C., and Delsanto, P.P. 2009. A new computational tool for the phenomenological analysis of multipassage tumor growth curves. *PLoS ONE* 4(4):e5358.doi:10.1371/journal.pone.0005358.

Gompertz, B. 1825. On the nature of the function expressive of the law of human mortality and a new mode of determining life contingencies. *Philos. Trans. R. Soc.* 115:513–585.

Guiot, C., Degiorgis, P.G., Delsanto, P.P., Gabriele, P., and Deisboeck, T.S. 2003. Does tumor growth follow a "universal law"? *J. Theor. Biol.* 225:147–151.

Guiot, C., Delsanto, P.P., and Pugno, N. 2006a. Elastomechanical model of tumor invasion. *Appl. Phys. Lett.* 8:1.

Guiot, C., Delsanto, P.P., Carpinteri, A., Pugno, N., Mansury, Y., and Deisboeck, T.S. 2006b. The dynamic evolution of the power exponent in a universal growth model of tumors. *J. Theor. Biol.* 240:459–463.

Guiot, C., Pugno, N., Delsanto, P.P., and Deisboeck, T.S. 2007. Physical aspect of cancer invasion. *Phys. Biol.* 4:1–6. doi:10.1088/1478-3975/4/4/P01.

Haj, M. et al. 2007. Loss of intercellular adhesion molecules in breast cancer: Does it predict a poor prognosis? *Breast Care* 2:378–383.

Hornberg, J.J. et al. 2006. Cancer: a systems biology disease. *Biosystems* 83:81–90.

Kitano, H. 2002. Computational systems biology. *Nature* 420:206–210.

Kleiber, M. 1932. Body size and metabolism. *Hilgardia* 6:315–353.

Liu, E.T. et al. 2006. In the pursuit of complexity: systems medicine in cancer biology. *Cancer Cell* 9:245–247.

Mazzone, M., Dettori, D., de Oliveira, R.L. et al. 2009. Heterozygous deficiency of phd2 restores tumor oxygenation and inhibits metastasis via endothelial normalization. *Cell* 136:839–851.

McCredie, J.A. and Sutherland, R.M. 1971. Differences in growth and morphology between the spontaneous C3H mammary carcinoma in the mouse and its syngeneic transplants. *Cancer* 27:635–642.

Min Hu et al 2008. Progression to invasion was promoted by fibroblasts and inhibited by normal myoepithelial cells. *Cancer Cell* 13:394–406.

Mueller Klieser, W. 1993. Three-dimensional cell cultures: from molecular mechanisms to clinical applications. *Am. J. Physiol. Cell Physiol.* 273:C1109–C1123.

Savage, V.M., Deeds, E.J., and Fontana, W. 2009. Sizing up allometric scaling theory. *PLOS Comput. Biol.*, 4:e1000171.

Scalerandi, M., Romano, A., Pescarmona, G.P., Delsanto, P., and Condat, C.A. 1999. Nutrient competition as a determinant for cancer growth. *Phys. Rev. E* 59:2206–2217.

Scalerandi, M., Pescarmon, G.P., Delsanto, P., and Capogrosso Sansone, B. 2001. A LISA model of the response of the vascular system to metabolic changes of the cells behavior. *Phys. Rev. E*, 63:11901.

Steel, G.G. 1977. *Growth Kinetics of Tumors.* Clarendon Press, Oxford.

Talmadge, J.E., Singh, R.K., Fidler, I.J. et al.2007. Murine models to evaluate novel and conventional therapeutic strategies for cancer. *Am. J. Path.* 170:793–804.

Tsui-Chin Huang, et al. 2008. Targeting therapy for breast carcinoma by ATP syntase inhibitor Aurovertin B. *J. Proteome. Res.* 7:1433–1444.

Virchow, R. *Vorlesungen uber Pathologie.* 1892. Berlin A Hirchwald Verlag.

West, G.B., Brown, J.H., and Enquist, B.J. 1999. The fourth dimension of life: fractal geometry and allometric scaling of organisms. *Science* 284:1677–1679.

West, G.B., Brown, J.H., and Enquist, B.J. 2001. A general model for ontogenetic growth. *Nature* 413:628–631.

West, G.B. and Brown, J.H. 2004. Life's universal scaling laws. *Physics Today* 57:36–43.

CHAPTER 13

Multiscale Mathematical Modeling of Vascular Tumor Growth

An Exercise in Transatlantic Cooperation

Mark A.J. Chaplain, Paul Macklin, Stephen McDougall, Alexander R.A. Anderson, Vittorio Cristini, and John Lowengrub

CONTENTS

INTRODUCTION

Cancer growth, and as a particular example in this paper, solid tumor growth, is a complicated phenomenon involving many interrelated processes across a wide range of spatial and temporal scales, and as such presents the mathematical modeler with a correspondingly complex set of problems to solve. The aim of this paper is to formulate a multiscale mathematical model of solid tumor growth, incorporating three key features: the avascular growth phase, the recruitment of new blood vessels by the tumor (angiogenesis), and the vascular growth and host tissue invasion phase.

Solid tumors are known to progress through two distinct phases of growth: the avascular phase and the vascular phase. The initial avascular growth phase can be studied in the laboratory by culturing cancer cells in the form of three-dimensional *multicell spheroids*. It is well known that these spheroids, whether grown from established tumor cell lines or actual in vivo tumor specimens, possess growth kinetics that are very similar to in vivo solid tumors. Typically, these avascular nodules grow to a few millimeters in diameter. Cells toward the center, being deprived of vital nutrients, die and give rise to a necrotic core. Proliferating cells can be found in the outer cell layers. Lying between these two regions is a layer of quiescent (or hypoxic) cells, a proportion of which can be recruited into the outer layer of proliferating cells. Much experimental data have been gathered on the internal architecture of spheroids, and studies regarding the distribution of vital nutrients (e.g., oxygen) and metabolites within the spheroids have been carried out. See, for example, the recent reviews by Walles et al. (2007), Kim (2005), Kunz-Schugart et al. (2004), and Chomyak and Sidorenko (2001), and the references therein.

The transition from the relatively harmless and confined dormant avascular state to the vascular state, in which the tumor possesses the ability to invade surrounding tissue and metastasize to distant parts of the body, depends on the ability of the tumor to induce new blood vessels from the surrounding tissue to sprout toward and then gradually surround and penetrate the tumor, thus providing it with an adequate blood supply and microcirculation. Tumor-induced angiogenesis, the process by which new blood vessels develop from an existing vasculature, through endothelial cell sprouting, proliferation, and fusion, is therefore a crucial part of solid tumor growth. Sustained angiogenesis is a hallmark of cancer (Hanahan and Weinberg 2000). Mature endothelial cells are normally quiescent and, apart from certain developmental processes (e.g., embryogenesis and wound healing), angiogenesis is generally a pathological process implicated in arthritis, some eye diseases, and solid tumor development, invasion, and metastasis. Tumor-induced angiogenesis is believed to start when a small avascular tumor exceeds a critical diameter (~2 mm), above which normal tissue vasculature is no longer able to support its growth. At this stage, the tumor cells lacking nutrients and oxygen become hypoxic. In response, the tumor cells secrete a number of diffusible chemical substances—tumor angiogenic factors (TAFs)—into the surrounding tissues and extracellular matrix (ECM). The TAF diffuses into the surrounding tissue and eventually reaches the endothelial cells (EC) that line nearby blood vessels. ECs subsequently respond to the TAF concentration gradient by degrading the basement membrane surrounding the parent vessel, forming sprouts, proliferating, and migrating towards the tumor. It takes approximately 10 to 21 days for the growing network to link the tumor to the parent vessel, and this vascular connection subsequently provides all the nutrients and oxygen required for continued tumor growth. An excellent summary of all the key cell-biological processes involved in angiogenesis can be found in the comprehensive review article of Paweletz and Knierim (1989). See also the recent review by Carmeliet (2005). Once vascularized, the solid tumors grow rapidly as exophytic masses. In certain types of cancer, for example, carcinoma arising within an organ, this process typically consists of columns of cells projecting from the central mass of cells and extending into the surrounding tissue area. The local spread of these carcinoma often assume an irregular jagged shape. By the time a tumor has grown to a size whereby it can be detected by clinical means, there is a strong likelihood that it has already reached the vascular growth phase.

Cancers also possess the ability to actively invade the local tissue and then spread throughout the body. Invasion and metastasis are the most insidious and life-threatening aspects of cancer (Liotta and Stetler-Stevenson, 1991; Liotta and Clair, 2000). Indeed, the prognosis of a cancer is primarily dependent on its ability to invade and metastasize. Many steps that occur during tumor invasion and metastasis require the regulated turnover of ECM macromolecules. The breakdown of these barriers is catalyzed by proteolytic enzymes released from the invading tumor. Most of these proteases belong to one of two general classes: many are metalloproteases (Parsons et al., 1997), while others are serine proteases (Andreasen et al., 1997, 2000). Proteases give cancers their defining characteristic—the ability of malignant cells to break out of tissue compartments. Motility, coupled with regulated, intermittent adhesion to the ECM and degradation of matrix molecules, allows an invading cell to move through the ECM (Liotta and Stetler-Stevenson, 1991; Lauffenburger and Horwitz, 1996; Friedl and Wolf, 2003). However, proteolytic degradation of the ECM is essential for the key processes involved in tissue remodeling as well. These processes take place in a number of distinct physiological events in the healthy organism, such as trophoblast invasion, mammary gland involution, angiogenesis, and wound healing.

The most significant turning point in cancer, however, is the establishment of metastasis. The metastatic spread of tumor cells is the predominant cause of cancer deaths, and with few exceptions, all cancers can metastasize. Metastasis is defined as the formation of secondary tumor foci at a site discontinuous from the primary tumor (Liotta and Stetler-Stevenson, 1991; Liotta and Clair, 2000). Metastasis unequivocally signifies that a tumor is malignant, and this is, in fact, what makes cancer so lethal. In principle, metastases can form following invasion and penetration into adjacent tissues followed by dissemination of cells in the blood vascular system (hematogeneous metastasis) and lymphatics (lymphatic metastases).

Metastases can appear shortly after surgery, but can also remain undetected for more than a decade before manifesting themselves clinically (King, 2000; Chambers et al., 2002; Fidler, 2002). This indicates that disseminated cancer cells can persist in a dormant state (either individually or as an avascular tumor spheroid), unable to form a progressively increasing tumor mass (Chambers et al., 2002). Such heterogeneity of outcome indicates that the fate of tumor cells that disseminate to distant organs before surgery must be regulated by either inherent cancer cell properties or the milieu of the target organs, or both. Identifying the mechanisms that keep metastases in their dormant, occult state is one of the most challenging and important avenues of cancer research (Chambers et al., 2002; Fidler, 2002).

Since the seminal work of Greenspan (1976), the mathematical modeling of avascular solid tumor growth, similar to its subject, has been rapidly expanding. Most models in this area consist of systems of nonlinear partial differential equations, and may be described as macroscopic. The review paper of Araujo and McElwain (2004) provides an excellent overview. See also the recent reviews by Quaranta et al. (2005), Byrne et al. (2006), Sanga et al. (2006), Graziano and Preziosi (2007), and Roose et al. (2007). Likewise, modeling tumor-induced angiogenesis has a well-established history, beginning with the work of Balding and McElwain (1983). The review paper of Mantzaris et al. (2004) provides an excellent overview of the work in this area. However, unlike avascular growth and angiogenesis, vascular tumor growth has received considerably less attention in the mathematical modeling literature.

Recently, Zheng et al. (2005) developed and coupled a level-set method for solid tumor growth with a hybrid continuous-discrete model of angiogenesis originally developed by Anderson and Chaplain (1998). This work served as a building block for studies of chemotherapy (Sinek et al. (2004)) and morphological instability and tumor invasion (Cristini et al. (2005); Frieboes et al. (2006)). Hogea et al. (2006) have also begun to investigate tumor-induced angiogenesis and vascular growth using a level-set method coupled with a continuous model of angiogenesis. Following the strategy pioneered by Zheng et al., Frieboes et al. (2007) coupled a mixture model with a lattice-free continuous-discrete model of angiogenesis (originally developed by Planck and Sleeman (2004)) and studied vascular tumor growth in three dimensions. In these works, however, the effects of blood flow through, and subsequent remodeling of, the vascular network were not included. Recently, the effects of blood flow through a vascular network on tumor growth were considered by Alarcon et al. (2005), Lee et al (2006), Bartha and Rieger (2007), and Welter et al. (2007) using cellular automaton (CA) tumor growth models coupled with network models for the vasculature. These authors investigated vascular network inhomogeneities, the stress-induced collapse of blood vessels, and the implications for therapy. Because of the computational cost of simulating cell growth using CA, these studies are limited to small scales.

In this paper, we couple an improved continuum model of solid tumor invasion (following Macklin and Lowengrub 2007) that is capable of spanning the 10^2_m-cm scale and accounts for cell–cell, cell–ECM adhesion, ECM degradation, and tumor cell migration, proliferation, and necrosis with a model of tumor-induced angiogenesis (following McDougall et al. (2006)) that accounts for blood flow through the vascular network, non-Newtonian effects and vascular network remodeling, due to wall shear stress and mechanical stresses generated by the growing tumor, to produce a new multiscale model of vascular solid tumor growth. As in Zheng et al. (2005), the invasion and angiogenesis models are coupled through the tumor angiogenic factors (TAFs) that are released by the tumor cells and through the nutrient extravasated from the neovascular network. As the blood flows through the neovascular network, nutrients (e.g., oxygen) are extravasated and diffuse through the ECM, triggering further growth of the tumor, which in turn influences the TAF expression. In addition, the extravasation is mediated by the hydrostatic stress generated by the growing tumor and, as mentioned earlier, the hydrostatic stress also affects vascular remodeling by restricting the radii of the vessels. The vascular network and tumor progression are also coupled via the ECM as both the tumor cells and ECs upregulate matrix-degrading proteolytic enzymes, which cause localized degradation of the ECM, which in turn affects haptotactic migration.

We perform simulations of the multiscale model that demonstrate the importance, on tumor invasion of the host tissue, of the nonlinear coupling between the growth and remodeling of the vascular network, the blood flow through the network, and the tumor progression. Consistent with clinical observations, the hydrostatic stress generated by tumor cell proliferation shuts down large portions of the vascular network, dramatically affecting the flow, the subsequent network remodeling, the delivery of nutrients to the tumor, and the subsequent tumor progression. In addition, ECM degradation by tumor cells is seen to have a dramatic affect on both the development of the vascular network and the growth response of the tumor. In particular, when the ECM degradation is significant, the newly formed vessels tend to encapsulate, rather than penetrate, the tumor and are thus less effective in delivering nutrients.

The outline of the chapter is as follows. In the following section, the mathematical models are presented and the numerical methods are briefly described. In the section titled "Results," numerical results are presented, and conclusions and future work are discussed in the last section. Further details of the mathematical modelling and numerical methods can be found in Macklin et al. (2009).

MODEL

In this exposition, we present the nondimensional model, starting first with the model of tumor invasion in the section titled "The Tumor Invasion Model" and followed by the model of tumor-induced angiogenesis in the section titled "Angiogenesis Model." Here, time is nondimensionalized by the characteristic tumor cell proliferation time (i.e., $1/\lambda_m$, where $\lambda_m \approx 2/3\,\mathrm{day}^{-1}$ is the mitosis rate), and space is nondimensionalized by the characteristic diffusion penetration length (i.e., $\left(D_\sigma^* / \lambda_\sigma^*\right)^{1/2} \approx 200\mu m$, where D_σ^* and λ_σ^* are characteristic values of the oxygen diffusion coefficient and uptake rate in the proliferating tumor region, respectively). In the following, quantities defined with an overbar correspond to nondimensional constants. The nondimensionalization of the parameters and the corresponding values used in the numerical simulations are presented in Tables 13.4–13.6. For the nondimensionalization and further biological discussion of the models, the reader is referred to Macklin and Lowengrub (2006, 2007) and McDougall et al. (2006).

The Tumor Invasion Model

To accurately model tumor growth in heterogeneous tissues, we develop a mathematical model that accounts for spatially dependent cell necrosis, cell apoptosis, cell–cell and cell–matrix adhesion, matrix degradation, cell proliferation and cell migration. The model is based on continuum reaction-diffusion equations that describe these processes and is a generalization and improvement of earlier models (see the reviews listed previously and recent work by Macklin and Lowengrub [2007b, 2007c]). We present the model in two dimensions, but it is equally valid for the three-dimensional case as well.

Let Ω denote a tumor mass, and let Σ denote its boundary. The tumor can be divided into three regions: a proliferating rim ΩP where the tumor cells have sufficient nutrient levels for proliferation; a hypoxic/quiescent region ΩH where the nutrient levels are too low for normal metabolic

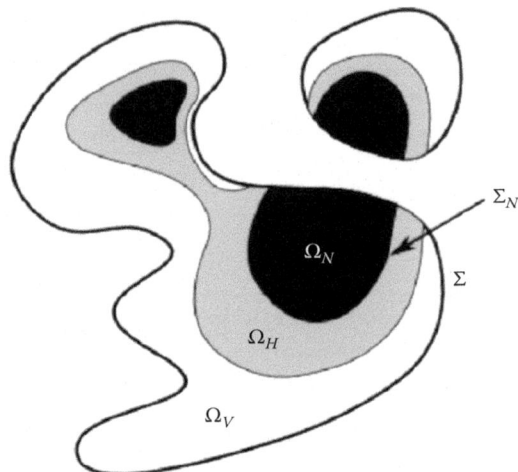

FIGURE 13.1 Schematic of the tumor regions. Ω_P, Ω_H, and Ω_N are the proliferating, quiescent/hypoxic, and necrotic regions, respectively.

activity but not so low that the cells begin to die; and a necrotic region ΩN where the nutrient level has dropped so low that the tumor cells have died and are degraded. Because necrosis is irreversible, we track the necrotic core and its interface ΣN separately of the tumor interface (see Figure 13.1).

Nutrient Transport

We model the net effect of nutrients (e.g., oxygen and glucose) and growth-promoting and growth–inhibiting factors with a single nutrient σ. Here, we focus our attention on the role of oxygen, which is supplied by the vascular network via the red blood cells. This can be modeled using the haematocrit, which represents the volume fraction of red blood cells contained in the blood. Oxygen, and other nutrients, are supplied by the preexisting bulk vasculature and the neovasculature at rates $\lambda\sigma_{bulk}$ and $\lambda\sigma_{neo}$, diffuses throughout the cancerous and noncancerous tissue, is uptaken in the non-necrotic portions of the tumor, and decays elsewhere (see the following text). Wherever the oxygen level inside the tumor drops below a threshold value σH, the tumor cells become hypoxic (quiescent), cease proliferating, and uptake nutrient at a lower rate. If the oxygen level falls further below a threshold value σN, then the tumor cells become necrotic. Inside the necrotic core, oxygen reacts with cellular debris to form reactive oxygen species (Kloner and Jennings 2001; Galaris et al. 2006), which we model by a decay term. Since oxygen diffusion occurs more rapidly than cell mitosis (the time scale on which the equations are nondimensionalized), these processes are described by the quasi-steady reaction-diffusion equation

$$0 = \nabla \cdot (D\nabla\sigma) - \lambda^\sigma(\sigma)\sigma + \lambda_{pre}^\sigma(\mathbf{x}, t, B_{pre}, P, \sigma) + \lambda_{neo}^\sigma(\mathbf{x}, t, B_{neo}, P, \sigma, h), \quad (13.1)$$

where D is the diffusion coefficient, and the parameter λ^σ combines the effects of oxygen uptake and decay and takes the form

$$\lambda^\sigma = \begin{cases} \overline{\lambda}_{tissue} p_\sigma(E_0) & \text{outside } \Omega \\ \overline{\lambda}_\sigma & \text{in } \Omega_P \\ q_\sigma(\sigma) & \text{in } \Omega_H \\ \overline{\lambda}_N & \text{in } \Omega_N \end{cases}, \quad (13.2)$$

where p_σ and q_σ are smooth interpolating functions and E_0 the density of the original ECM, which is used to assess changes in uptake/decay in the host microenvironment (see the section titled "Tumor–Microenvironment Interaction"). The interpolating function $q_\sigma(\sigma_H + \sigma_N/2) = \overline{\lambda}_H$, where σ_H and σ_N are the oxygen concentration thresholds for quiescence and necrosis, respectively, and $\overline{\lambda}_H$ is the rate of oxygen uptake by quiescent cells in the hypoxic tumor. Further, $\overline{\lambda}_{tissue}$ and $\overline{\lambda}_\sigma$ are the rates of oxygen uptake in the host microenvironment and in the proliferating tumor regions, respectively, and $\overline{\lambda}_N$ is the rate of oxygen decay in the necrotic portion of the tumor. We note that because the location of the viable, hypoxic, and necrotic tumor regions depends on the past and present values of the oxygen level σ, the uptake/decay term λ^σ introduces nonlinearity.

The two remaining sources $\lambda_{pre}^\sigma(\mathbf{x}, t, B_{pre}, P, \sigma)$ and $\lambda_{neo}^\sigma(\mathbf{x}, t, B_{neo}, P, \sigma, h)$ in Equation 13.1 reflect the oxygen-tissue transfer from the preexisting and neovascular blood vessels, respectively, and are given by:

$$\lambda_{pre}^\sigma = \overline{\lambda}_{pre}^\sigma B_{pre}(\mathbf{x}, t)(1 - \sigma)(1 - 1_\Omega) \quad (13.3)$$

and

$$\lambda_{neo}^\sigma = \overline{\lambda}_{neo}^\sigma B_{neo}(\mathbf{x}, t) \left(\frac{h}{H_D} - \overline{h}_{min} \right)_+ (1 - c(P_{vessel}, P))(1 - \sigma), \quad (13.4)$$

where $\overline{\lambda}^\sigma_{pre}$ and $\overline{\lambda}^\sigma_{neo}$ are constant transfer rates from the preexisting and neovessels. Here, B_{pre} is the (nondimensional) blood vessel density of the preexisting vessels, whose locations are assumed to be unchanging in time. In fact, we take a uniform distribution of preexisting vessels in the host tissue, and B_{pre} satisfies Equation 13.19, where MDE is assumed to degrade the preexisting vasculature. The function $B_{neo}(\mathbf{x},t) = \mathbf{1}_{neo}$ is the characteristic or indicator function of the neovessels (i.e., equal to 1 at the locations of the neovessels), and $\mathbf{1}_\Omega$ is the characteristic function of the tumor region Ω (i.e., equal to 1 inside the tumor and is 0 in the tumor exterior). Further, P is the oncotic (solid/mechanical) pressure, P_{vessel} and h are the dimensional pressure and the haematocrit in the neovascular network, respectively. The constants $\overline{H}_D$ and $\overline{h}_{min}$ reflect the normal value of haematocrit in the blood (generally about 0.45) and the minimum haematocrit needed to extravasate oxygen, respectively. The haematocrit is modeled via the blood flow in the vascular network and is determined from the angiogenesis model. This provides one aspect of the *coupling* between the tumor growth and angiogenesis models. A second mode of coupling between the two models occurs through the cutoff function $c(P_{vessel}, P)$, which is given by

$$c(P_{vessel}, P) = \begin{cases} 0 & \Delta P < 0 \\ p_{cutoff}(\Delta P) & 0 \le \Delta P \le 1, \\ 1 & \Delta P > 1 \end{cases}$$

(13.5)

where p_{cutoff} is a cubic, interpolating polynomial. That is, large oncotic pressures may prevent extravasation and transfer of oxygen from the vessels into the tissue. Later, we will discuss how the oncotic pressure may also constrict the neovessels. Further, in Equation 13.5,

$$\Delta P = \left(P - P_{vessel} / \overline{P}_{vessel} \right) / \overline{P}_{scale},$$

(13.6)

where $\overline{P}_{vessel}$ is a characteristic pressure scale and $\overline{P}_{scale}$ is a scale factor. Note that we could have analogously taken the oxygen transfer rate from the preexisting vessels to also be coupled to the haematocrit and blood vessel pressure. This will be explored in a future work.

The oxygen source terms in Equations 13.3 and 13.4 are designed such that for sufficiently large transfer rates $\overline{\lambda}^\sigma_{pre}$ and $\overline{\lambda}^\sigma_{neo}$, the oxygen concentration $\sigma \approx 1$ at the spatial locations of the preexisting vessels and neovessels. In practice, we will take $\overline{\lambda}^\sigma_{neo}$ large but $\overline{\lambda}^\sigma_{pre}$ small, which models the supply of only a small amount of oxygen in the host tissue from preexisting vessels. We will assume a parent vessel, located at the boundary of the computational microenvironment domain as discussed later, supplies the bulk of the oxygen in the host tissue. Note that oxygen flux conditions across the preexisting vessels and neovessels could be imposed (e.g., Alarcon et al. 2005).

The boundary conditions for Equation 13.1 are taken to be a combination of Dirichlet, Neumann conditions. In particular, in the simulations we present later, we assume that a parent vessel coincides with the upper boundary of the computational domain. Therefore, a Dirichlet condition, $\sigma = 1$, is posed along the upper boundary. Zero Neumann conditions, $\partial\sigma / \partial n = 0$, are imposed along the other boundaries of the computational domain.

Tumor Mechanics and the Cell Velocity

The tumor cells, the ECM, and host noncancerous cells are influenced by a combination of forces that contribute to the cellular velocity field. The proliferating cells generate an oncotic mechanical pressure (hydrostatic stress) that also exerts force on the ECM and host cells. The cells respond to pressure variations by overcoming cell–cell and cell–ECM adhesion and migrating through the microenvironment. The ECM may also deform, degrade, and remodel in response to pressure and to the chemical factors released by the cells. The cells may respond haptotactically to adhesion gradients in the ECM.

Following previous work, we assume that all solid phases move with a single cellular velocity field, and we model the cellular motion within the ECM as incompressible fluid flow in a porous medium. In the future, we plan to use mixture models (e.g., Ambrosi and Preziosi 2002; Byrne and Preziosi 2003; Araujo and McElwain 2005a, 2005b) to relax these assumptions. In this simplified description of tumor mechanics used here, Darcy's law is taken as the constitutive assumption and, thus, the velocity is proportional to the forces in the problem. See Ambrosi and Preziosi (2002) and Byrne and Preziosi (2003) for a motivation of this approach from a mixture modeling perspective. Accordingly, the nondimensional velocity is given by

$$\mathbf{u} = -\mu\nabla P + \chi_E \nabla E, \tag{13.7}$$

where μ is the cell mobility, which models the net effects of cell–cell and cell–matrix adhesion, E is the ECM density (e.g., a nondiffusible matrix macromolecule such as fibronectin, collagen, or laminin), and χ_E is the haptotaxis coefficient. Models for μ and χ_E are given below in the next section. Further, assuming that the density of tumor cells is constant in the viable region, the growth of the tumor is then associated with the rate of volume change:

$$\nabla \bullet \mathbf{u} = \lambda_p, \tag{13.8}$$

where λ_p is the nondimensional net proliferation rate. This implies that the nondimensional pressure satisfies

$$-\nabla \cdot (\mu \nabla P) = \lambda_p - \nabla \bullet (\chi_E \nabla E) \tag{13.9}$$

We assume that in the proliferating region, cell mitosis is proportional to the amount of nutrient present and that apoptosis may occur. Volume loss may occur in the necrotic core, and there is no proliferation in either the host microenvironment or the hypoxic/quiescent regions. We therefore take

$$\lambda_p = \begin{cases} 0 & \text{if } \mathbf{x} \notin \Omega \\ \sigma - A\Omega & \text{if } \mathbf{x} \in \Omega_P \\ 0 & \text{if } \mathbf{x} \in \Omega_H \\ -G_N\Omega & \text{if } \mathbf{x} \in \Omega_N \end{cases}, \tag{13.10}$$

where A is the nondimensional apoptosis rate ("preprogrammed" cell death); and G_N is the nondimensional rate of volume loss in the necrotic core as water is removed and cellular debris is degraded. Assuming a uniform cell–cell adhesion throughout the tumor, cell–cell adhesion can be incorporated as a surface-tension-like jump boundary condition at the tumor–host interface Σ:

$$[P] = (P_{\text{inner}} - P_{\text{outer}}) = \frac{1}{G}\kappa, \tag{13.11}$$

where G is a nondimensional parameter that measures the aggressiveness of the tumor (the strength of cell proliferation relative to cell–cell adhesion), and κ is the mean curvature of the interface. At the necrotic boundary Σ_N, we assume P is continuous. We assume that no voids form and, therefore, we take

$$[\mathbf{u}\cdot\mathbf{n}]=0 \text{ which implies that } [\mu\nabla P\cdot\mathbf{n}]=[\chi_E\nabla E\bullet\mathbf{n}], \qquad (13.12)$$

where $\mathbf{n}$ is the unit outward normal to Σ. For simplicity, we will also assume that $[\nabla E\bullet\mathbf{n}]=0$. At the necrotic boundary, we assume analogous conditions. The velocity of the tumor–host interface Σ is then given by

$$V=-\mu\nabla P\bullet\mathbf{n}+\chi_E\nabla E\bullet\mathbf{n}, \qquad (13.13)$$

and the velocity of the necrotic boundary Σ_N is

$$V_N=-\mu\nabla P\bullet\mathbf{n}_N+\chi_E\nabla E\bullet\mathbf{n}_N, \qquad (13.14)$$

where $\mathbf{n}_N$ is the outward unit normal vector along Σ_N. In the far field at the boundaries of the computational domain, the pressure is assumed to satisfy zero Neumann boundary conditions $\partial P/\partial n=0$.

Tumor–Microenvironment Interaction

We model the tumor microenvironment by introducing an ECM density E that represents the density of nondiffusible matrix macromolecules such as fibronectin, collagen, elastin, and laminin, etc. In addition, as mentioned earlier, we keep track of the density E_0 of the original ECM and the preexisting blood vessel density B_{pre} to assess the level of oxygen uptake and supply, respectively, in the microenvironment.

The tumor interacts with the microenvironment by responding to the nutrients supplied by the preexisting and the neovasculature (e.g., see Equation 13.1), remodeling the ECM locally by secreting both MDE and ECM macromolecules and by a heterogeneous response to pressure and ECM adhesion gradients through nonconstant cell mobility and haptotaxis coefficients. In order for tumors cells to migrate into the porous matrix, they must overcome cell–matrix adhesion. However, in experiments, a maximum migration speed is obtained that depends on the level of integrin expression (e.g., Palecek et al. (1997), DiMilla et al. (1991)) and, correspondingly, a nonmonotonic dependence of cell migration velocity on integrin expression and adhesion gradients in the ECM has been predicted (DiMilla et al. (1991); Dickinson and Tranquillo (1993)). This has been explained by the fact that while some integrins are required for focal adhesion-based migration, too much focal contact strength can retard the detachment of the cell's trailing edge from the ECM. While we do not model integrin expression directly here, we take this effect into account by making the haptotaxis coefficient a nonmonotone function of E:

$$\chi_E = \begin{cases} \overline{\chi}_{E,\min} & E < \overline{E}^{\chi}_{\min\text{cutoff}} \\ p_{\chi}(E) & \overline{E}^{\chi}_{\min\text{cutoff}} \leq E \leq \overline{E}^{\chi}_{\max\text{cutoff}} \\ \overline{\chi}_{E,\min} & E > \overline{E}^{\chi}_{\max\text{cutoff}} \end{cases}, \qquad (13.15)$$

where $\overline{\chi}_{E,\min}$ are the nondimensional haptotaxis in low-/high-density ECM, and p_{χ} is a nonmonotone interpolating function with a maximum $\overline{\chi}_{E,\min}$ located at $E = (\overline{E}^{\chi}_{\min\text{cutoff}} + \overline{E}^{\chi}_{\max\text{cutoff}})/2$. Although the mobility μ may also be nonmonotone, for simplicity, we take a monotone decreasing function of E here:

$$\mu = \begin{cases} \overline{\mu}_{\max} & E < \overline{E}^{\mu}_{\min\text{cutoff}} \\ p_{\mu}(E) & \overline{E}^{\mu}_{\min\text{cutoff}} \leq E \leq \overline{E}^{\mu}_{\max\text{cutoff}} \\ \overline{\mu}_{\min} & E > \overline{E}^{\mu}_{\max\text{cutoff}} \end{cases}, \qquad (13.16)$$

where p_{μ} is a smooth interpolating function. In a future work, we will investigate nonmonotonic cell mobilities μ. In addition, the mobility and chemotaxis parameters may also be functions of oxygen concentration σ as hypoxic conditions may result in upregulation of HIF-1 alpha target genes that may result in decreased cell–cell and cell–matrix adhesion, among other effects, and therefore enable cells to more easily migrate through and invade the tumor microenvironment (e.g., see Kaur et al. (2005); Erler et al. (2006); Pouyssegur et al. (2006)). These effects will also be explored in a forthcoming work.

In order to migrate through the ECM and invade the host tissue, tumor cells secrete matrix-degrading proteolytic enzymes (MDEs), for example, matrix metalloproteases and urokinase plasminogen activators, which cause the degradation of the ECM, provide space for the cells, and enhance the attachment of the cells to ECM macromolecules, enabling the cells to exert traction forces to propel themselves through the ECM. In addition, the tumor cells remodel the ECM by secreting insoluble matrix macromolecules and possibly reorienting them. We note that during the angiogenetic response of the host vasculature, an analogous molecular cascade occurs as tumor angiogenesis factors (TAFs) and ECM macromolecules (e.g. fibronectin, collagen, and laminin) bind to specific membrane receptors on ECs and activate the cell's migratory machinery. This leads to a remodeling of ECM similar to that described earlier for tumor cells. Here, we will not consider the effect of orientational reordering. We model the remaining processes as follows. For the MDE, we take

$$\frac{\partial M}{\partial t} = \nabla \cdot (D_M \nabla M) + \bar{\lambda}_{\text{production}}^M (1 - M) \mathbf{1}_{\Omega_V} - \bar{\lambda}_{\text{decay}}^M M + \bar{\lambda}_{\text{sprout production}}^M \mathbf{1}_{\text{sprout tips}}$$

$$(13.17)$$

where M is the nondimensional MDE concentration, $D_M = \bar{D}_M$ is the diffusion coefficient (assumed to be constant), $\bar{\lambda}_{\text{production}}^M$ and $\bar{\lambda}_{\text{sprout production}}^M$ are the nondimensional rates of production of MDE by the viable tumor cells ($\Omega_V = \Omega_P \cup \Omega_H$) and the sprout tip ECs, respectively. Further, $\bar{\lambda}_{\text{decay}}^M$ is the rate of decay (it is assumed that MDE is not used up as a result of the interaction with the ECM (Quaranta, private communication)). Finally, $\mathbf{1}_{\text{sprout tips}}$ is the characteristic function of the sprout tips. Because the diffusion coefficient of MDE, D_M, is much smaller than that for oxygen diffusion, the full time-dependent diffusion equation is used (Stephanou et al., 2006). In the far field (boundary of the computational domain), we take the zero Neumann boundary conditions $\partial M / \partial \mathbf{n} = 0$.

The ECM density satisfies

$$\frac{\partial E}{\partial t} = -\bar{\lambda}_{\text{degradation}}^E EM + \bar{\lambda}_{\text{production}}^E (1 - E) \mathbf{1}_{\Omega_V} + \bar{\lambda}_{\text{sprout production}}^E \mathbf{1}_{\text{sprout tips}}, \quad (13.18)$$

where $\bar{\lambda}_{\text{production}}^E$ and $\bar{\lambda}_{\text{sprout production}}^E$ are the nondimensional rates of production of ECM by the viable tumor cells and sprout tip ECs and the $\bar{\lambda}_{\text{degradation}}^E$ is the nondimensional rate of matrix degradation by the MDE.

Finally, the original ECM and the preexisting blood vessel density are assumed to be degraded by the MDE:

$$\frac{\partial B_{\text{pre}}}{\partial t} = -\bar{\lambda}_{\text{degradation}}^B M B_{\text{pre}} \quad \text{and} \quad \frac{\partial E_0}{\partial t} = -\bar{\lambda}_{\text{degradation}}^E M E_0, \quad (13.19)$$

where $\bar{\lambda}_{\text{degradation}}^B$ and $\bar{\lambda}_{\text{degradation}}^E$ are nondimensional degradation rates.

Tumor Angiogenic Factors

When tumor cells become hypoxic/quiescent, they are assumed to secrete TAFs, which diffuse into the surrounding tissue and attract ECs. ECs respond to the TAF by binding with it, proliferating and chemotaxing up the TAF gradient. The diffusion coefficient of TAF is similar to that of oxygen, and so we model the production, diffusion, decay, and binding of TAF by

$$0 = \nabla \cdot (D_T \nabla T) + \bar{\lambda}^T_{\text{production}} (1 - T) \mathbf{1}_{\Omega_H} - \bar{\lambda}^T_{\text{decay}} T - \bar{\lambda}^T_{\text{binding}} T \mathbf{1}_{\text{sprout tips}} \quad (13.20)$$

where T is the nondimensional TAF concentration, $D_T = \bar{D}_T$ is the diffusion coefficient (assumed to be constant), and $\bar{\lambda}^T_{\text{production}}$, $\bar{\lambda}^T_{\text{decay}}$, and $\bar{\lambda}^T_{\text{binding}}$ denote the nondimensional production, natural decay, and binding rates of TAF. In the far field at the boundary of the computational domain, we also take zero Neumann boundary conditions $\partial T / \partial n = 0$.

Angiogenesis Model

We begin with a description, in the next section, of an initial mathematical model for the growth of a hollow capillary network in the absence of any blood flow; this follows Anderson and Chaplain (1998). Then, in the two succeeding sections, following McDougall et al. (2006), we will add the effects of blood flow and vascular network remodeling, respectively.

Basic Network Model

As described earlier, TAF and ECM macromolecules bind to specific membrane receptors on ECs and activate the cell's migratory machinery. The model of EC migration given in the following text describes how capillary sprouts emerging from a parent vessel migrate toward a tumor, leading to the formation of a vascular network that supplies nutrients for continued development (see Figure 13.2).

At this level, since there is no flow or vessel remodeling, this model may perhaps be considered more appropriate at describing in vitro endothelial cell migration and capillary sprout formation. The model, inspired by the tumor angiogenesis model developed by Anderson and Chaplain (1998), assumes that endothelial cells migrate through (1) random motility, (2) chemotaxis in response to TAF released by the tumor, and (3) haptotaxis in response to ECM gradients. If we denote by n the nondimensional endothelial cell density per unit area, then the nondimensional equation describing EC conservation is given by

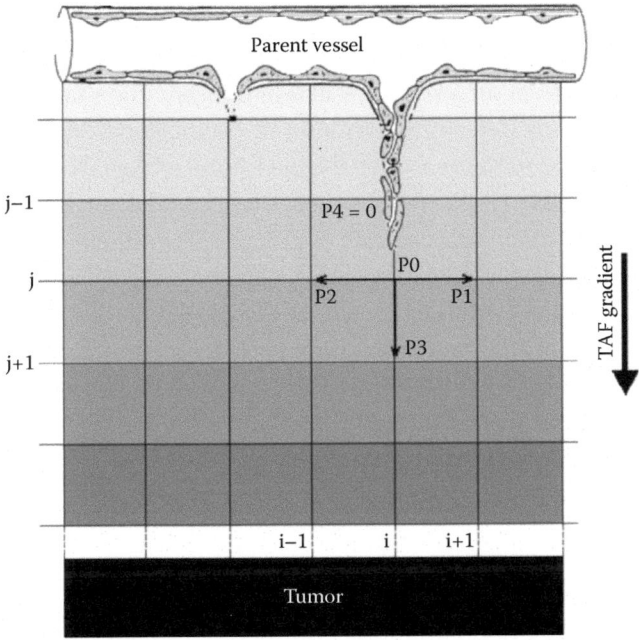

FIGURE 13.2 A schematic diagram of the basic network model of tumor-induced angiogenesis. P1, P2, P3, and P4 denote the probabilities of the sprout EC moving in the coordinate directions. P0 denotes the probability of its remaining stationary. (Reprinted with permission from McDougall et al. 2006. *Theor. Biol.* 241, 564–589.)

$$\frac{\partial n}{\partial t} = \nabla \cdot (D \nabla n) - \nabla \cdot \left(\chi^T_{\text{sprout}}(T) n \nabla T \right) - \nabla \cdot \left(\chi^E_{\text{sprout}} n \nabla E \right). \tag{13.21}$$

See McDougall et al. (2006) for the nondimensionalization. The diffusion (random migration) coefficient is $D = \bar{D}$ (assumed to be constant), and the chemotactic and haptotactic migration are characterized by the functions $\chi^T_{\text{sprout}} = \bar{\chi}^T_{\text{sprout}} / (1 + \delta \cdot T)$, which reflects the decrease in chemotactic sensitivity with increased TAF concentration and $\chi^E_{\text{sprout}} = \bar{\chi}^E_{\text{sprout}}$, where for simplicity we have taken the haptotactic migration parameter to be constant. In a future work, we will investigate the heterogeneous response of the ECs to the ECM and oxygen gradients as discussed earlier in the section titled "Tumor–Microenvironment Interaction." The coefficients $\bar{D}$, $\bar{\chi}^T_{\text{sprout}}$, and $\bar{\chi}^E_{\text{sprout}}$ characterize the nondimensional random, chemotactic, and haptotactic cell migration, respectively.

The displacement of each individual EC, located at the tips of each growing sprout, is given by the discretized form of the EC mass conservation equation (13.21) on a regular Cartesian mesh. The migration of each cell is consequently determined by a set of coefficients (P0–P4) emerging from this equation, which relate to the likelihood of the cell remaining stationary, moving left, right, up, or down. These coefficients incorporate the effects of random, chemotactic, and haptotactic movement and depend on the local chemical environment (ECM density and TAF concentration). Proliferation of the endothelial cells at the capillary tips and branching at capillary tips are implemented in the model at the discrete level. Tip branching depends on the TAF concentration at a given spatial location (see Table 13.1 and Anderson and Chaplain (1998) for details.). Using the foregoing model, it is possible to generate "hollow" capillary networks that are structurally similar to those observed experimentally.

Modeling Blood Flow in the Developing Capillary Network
Blood is a complex multiphase medium composed of many different constituents, including red blood cells (erythrocytes), white blood cells (leukocytes), and platelets involved in clotting cascades. These solid elements represent approximately 45% of the total blood composition—red cells are predominant—and are carried in the plasma, which constitutes the fluid phase. A measure of the solid phase is given by the blood *haematocrit*, which represents the volume fraction of red blood cells contained in the blood. The average human haematocrit has a value of around 0.45. Because of its multiphasic nature, blood does not behave as a continuum, and the viscosity measured while flowing at different rates in microvessels is not constant. The direct measurement of blood viscosity in living microvessels is very difficult to achieve with any degree of accuracy. However, by

TABLE 13.1   Vessel branching probabilities according to the local TAF (T) concentration and to the magnitude of the local wall shear stress (τw). TAF$_{max}$ (T_{max}) is the maximum TAF concentration and τ_{max}= 5 Pa (50 dynes/cm^2), the maximum shear stress derived from preliminary flow simulations

		WSS/τ_{max}				
		[0.0,0.2)	[0.2,0.4)	[0.4,0.6)	[0.6,0.8)	[0.8,1.0)
	[0,0,0.3)	0.00	0.00	0.00	0.00	0.00
[TAF]/TAF$_{max}$	[0.3,0.5)	0.00	0.02	0.04	0.06	0.08
	[0.5,0.7)	0.00	0.03	0.06	0.09	0.12
	[0.7,0.8)	0.00	0.04	0.08	0.12	0.16
	[0.8,1.0)	0.00	0.10	0.20	0.30	0.40

comparing the flow distribution in a numerical network (generated by a mathematical model) with a similar experimental system, Pries et al. (1996) determined a relationship between the apparent viscosity of blood, the blood haematocrit, and the radius of the vessel, through which the blood is flowing that provides a good fit with microvascular experimental data. The relationship is given as follows:

$$\mu_{apparent} = \mu_{plasma} \cdot \mu_{rel},$$

$$\mu_{rel}(R,h) = \left(1 + \left(\mu_{0.45} - 1\right) f(h, R/R^*) \left(\frac{2R/R^*}{2R/R^* - 1.1}\right)^2\right) \left(\frac{2R/R^*}{2R/R^* - 1.1}\right)$$

$$(13.22)$$

where μ_{plasma} is the plasma viscosity, μ_{rel} is the relative viscosity, $\mu_{0.45}$ is the (nondimensional) viscosity corresponding to the normal value of the discharge haematocrit (i.e., $h = \bar{H}_D = 0.45$), R is the dimensional vessel radius (in μm), R^* is a radius scale factor (taken to be equal to 1 μm), and $f(h,R)$ is a modulating function of the haematocrit and vessel radius. These effects are modeled by:

$$\mu_{0.45} = 6e^{-0.17R/R^*} - 2.44e^{-0.06(2R/R^*)^{0.645}} + 3.2,$$

$$f(h, R/R^*) = \frac{(1-h)^{C(R/R^*)} - 1}{(1 - \bar{H}_D)^{C(R/R^*)} - 1},$$

$$C(R/R^*) = \left(e^{-1.5 \times 10^{-7} R/R^*} + 0.8\right) \left(\frac{1}{1 + 10^{-23}(2R/R^*)^{12}} - 1\right)$$

$$+ \frac{1}{1 + 10^{-23}(2R/R^*)^{12}}.$$

$$(13.23)$$

The apparent blood viscosity (i.e., Equation 13.22) generally increases with decreasing capillary radius, although the precise relationship is nonlinear since it is actually haematocrit dependent.

In order to calculate flow within the entire interconnected network of capillaries, it is first necessary to decide upon a local relationship between the pressure gradient ΔP_{vessel} and flow rate $\dot{Q}$ at the scale of a single capillary element of length L and radius R. Such a relationship in the case of a non-Newtonian fluid can be approximated by the following Poiseuille-like expression:

$$\dot{Q} = \frac{\pi R^4 \Delta P_{vessel}}{8\mu_{app}(R,h)L}.$$

(13.24)

where $\mu_{app} = \mu_{apparent}$ from Equation 13.22. In order to determine the pressure (and flow rate) and in the vascular network of interconnected capillary elements having distributed radii, one simply conserves mass (or flow if the fluid is incompressible) at each junction where capillary elements meet (see Figure 13.3).

Hence, for each node the following expression can be written:

$$\sum_{k=1}^{N} \dot{Q}_{(i,j),k} = 0$$

(13.25)

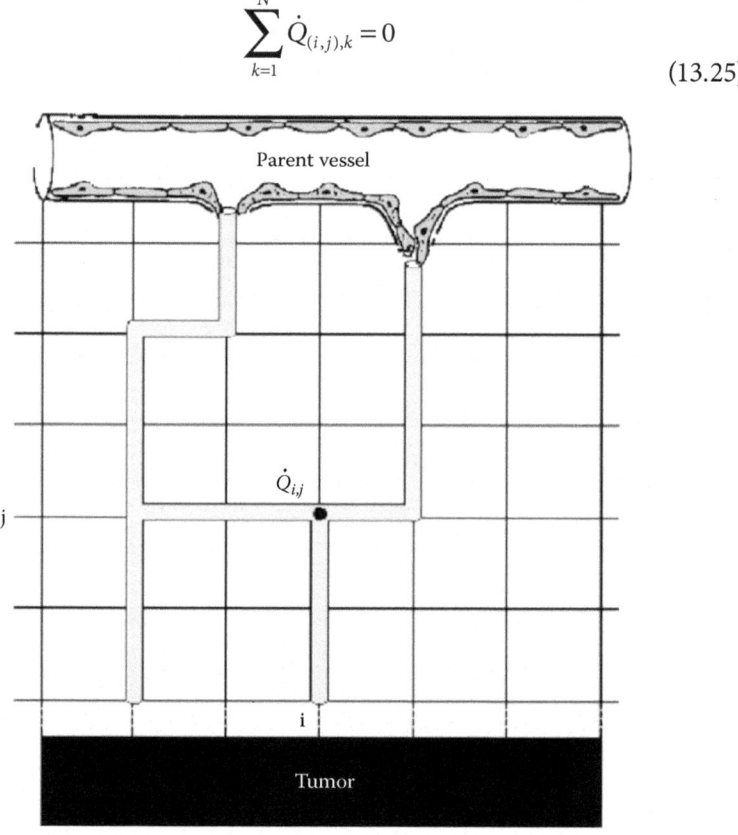

FIGURE 13.3 A schematic representation of the neovessels superimposed on the computational grid used for the flow calculation. (After McDougall et al. 2006. *Theor. Biol.* 241, 564–589.)

where the index k refers to adjacent nodes and $N = 4$ in a fully connected regular 2D grid as considered in this paper (or $N = 6$ in 3D).
This procedure leads to a set of linear equations for the nodal pressures
(P_{vessel}, i) that can be solved numerically using any of a number of different algorithms including successive over-relaxation (SOR). Once the
nodal pressures are known, Equation 13.24 can be used to calculate the
flow in each capillary element in turn. A more complete discussion of
the procedure can be found in McDougall et al. (2002). The evolution of
haematocrit h in the vessels is also calculated using mass conservation
once the flow is determined.

Capillary Vessel Adaptation and Remodeling

Blood rheological properties and microvascular network remodeling are
interrelated issues, as blood flow creates stresses on the vascular wall
(shear stress, pressure, tensile stress) that lead to adaptation of the vascular diameters via either vasodilatation or constriction. In turn, blood
rheology (viscosity, haematocrit, etc.) is affected by the new network
architecture—consequently, we should expect adaptive angiogenesis to be
a highly dynamic process. We follow the work of Pries et al. (1995, 1996,
1998, 2001a) in incorporating vessel adaptation into our model. In particular, we consider a number of stimuli that affect the vessel diameters.
Specifically, we account for the influence of the wall shear stress ($Swss$),
the intravascular pressure (Sp), a metabolic mechanism depending on
the blood haematocrit (Sm), as well as the natural tendency for vessels to
shrink (Ss). These stimuli form a basic set of requirements for obtaining
stable network structures with realistic distributions of vessels diameters
and flow velocities. The theoretical model for vessel adaptation assumes
that the change in a flowing vessel radius _R over a time step __, where
time is scaled by the rate of the response of the vessel to wall shear stress
(k_w), is proportional to both the global stimulus acting on the vessel and
to the initial vessel radius R. We refer the reader to McDougall et al. (2006)
for the definitions of the stimuli and a brief discussion.

After the radius of the vessel is updated, the effect of the mechanical
pressure P, generated by the proliferating and invading tumor on the
vessel radius is then taken into account. The tendency of the pressure to
shrink the vessel is modeled by the simple cutoff:

$$R \to R_{min} + (R - R_{min}) \cdot (1 - c(P, P_{vessel})), \tag{13.26}$$

where $c(P,P_{vessel})$ is the cutoff function introduced earlier in Equation 13.5 and R_{min} is a threshold minimum radius. This provides another means of coupling tumor invasion (and mechanics) with the angiogenic response and the developing neovascular network. In particular, the solid/mechanical pressure may constrict and cutoff vessels in the neovasculature. To prevent singularities in practice, the radius of the vessel is constrained to lie between 2.0 and 14 μm, which is the size of the parent capillary.

Inclusion of the foregoing mechanisms into our modeling framework now allows us to simulate dynamic remodeling of a flowing vasculature. This significant improvement in angiogenesis modeling, introduced by McDougall et al. (2006), allows us to describe vascular growth in a far more realistic manner, with areas of the capillary network dilating and constricting in response to variations in perfusion-related stresses, stimuli and pressure mechanical forces exerted on the host microenvironment by the invading tumor. The final step in the development of the complete *dynamic adaptive tumor-induced angiogenesis* (DATIA) model is to couple the network flow modeling approach outlined in this section to the "hollow capillary" model derived from the endothelial cell migration equations described earlier. This is achieved through the role of wall shear stress.

Wall shear stress is known to play a leading role in the growth and branching of capillary vessel networks (Pries et al. (2001a, 2001b)). In order to "bring the morphological and the physiological concepts together" (Thompson, 1917), the cell migration and flow models are coupled by incorporating the mechanism of *shear-dependent vessel branching* in addition to sprout-tip branching via local TAF concentrations. This enables the capillary network structures to adapt dynamically through adjuvant vessel branching in areas of the network experiencing increased shear stresses following anastomosis elsewhere in the system. We note that because the shear stress is due to the blood flowing through the capillaries, vessel branching can only occur after some degree of anastomosis has taken place. Therefore, the early stages of angiogenesis are primarily characterized by branching at the capillary tips, which depends only on the TAF concentration.

The combined effects of the local wall shear stress and TAF concentration on vessel branching probability have been implemented in the model as described in Table 13.1. In the absence of quantitative experimental data, the probabilities chosen for the vessel branching process have been defined on a qualitative basis and reflect the combined influence of the wall shear stress (WSS) and local TAF concentration. High values of WSS

TABLE 13.2   Sprout tip branching probabilities as
a function of the local TAF concentration

TAF concentration	Sprout tip branching probability
≤0.3	0.0
[0.3–0.5]	0.2
[0.5–0.7]	0.3
[0.7–0.8]	0.4
>0.8	1.0

in tandem with high local TAF concentrations lead to a higher branching probability, while lower values of one or both of WSS and TAF concentration lead to lower branching probability. For each range of WSS (linearly distributed in the interval [0,1]), the corresponding TAF probability profile has been obtained via a linear scaling of the values reported in McDougall et al. (2002) and Stéphanou et al. (2005a, 2005b).

As mentioned earlier, in the absence of WSS, TAF-dependent sprout tip branching is the only means by which a migrating vessel can bifurcate. Sprout tip branching is performed using the algorithm developed by Anderson and Chaplain (1998) and the corresponding tip branching probabilities are shown in Table 13.2.

Numerical Schemes

Tumor Invasion Model

The tumor invasion model described in the section titled "The Tumor Invasion Model" consists of a coupled system of nonlinear, elliptic, and parabolic (reaction-diffusion) differential equations that must be solved on a complex, moving domain where the motion of the tumor/host boundary depends on gradients of the solutions to these equations. Further, one of these solutions—the pressure—is discontinuous across the tumor–host interface where the discontinuity depends on the geometry (i.e., the curvature) of the interface, which is an additional source of nonlinearity. Therefore, standard finite difference methods cannot be used to accurately solve the system. Instead, specialized methods that can accurately take into account discontinuities in solutions and complex domains must be used. Here, we use a ghost-cell/level-set method and adapt and extend the numerical techniques we recently developed (Macklin and Lowengrub 2005, 2006, 2007) to solve this system. In this approach, the equations are discretized on a regular Cartesian mesh and the difference stencils

TABLE 13.3 The variables in the tumor invasion model and their nondimensionalization

Biological quantity	Nondimensional variable	Scaled by
Oxygen concentration	σ	σ^*, the oxygen level in well-oxygenated tissue (assumed to be the same as the oxygen level in the blood vessels)
Proliferation-induced biomechanical pressure	P	$\lambda_M l^2 / \mu^*$, where μ^* is a characteristic mobility value, and λ_m is the mitosis rate
Tumor-secreted angiogenic growth factor (TAF)	T	T^*, the concentration of TAF secreted by tumor cells
Matrix-degrading enzyme (MDE)	M	M^*, the concentration of MDE secreted by the tumor cells
Extracellular matrix (ECM)	E	E^*, the concentration of ECM secreted by tumor cells
Original extracellular matrix	E_0	E^*, the concentration of ECM secreted by tumor cells
Endothelial cell (EC) density	n	n^*, a characteristic density of ECs
Preexisting blood vessel density	B_{pre}	B_{pre}^*, a characteristic density of preexisting vessels (e.g., value at the initial time)

near discontinuities are modified. We note that other alternatives exist (see the discussion in Macklin and Lowengrub 2007), but an advantage of our approach is that it can be implemented in a dimension-by-dimension manner, making the extension to 3D straightforward, and our algorithm is simpler to implement than the alternative approaches.

In this approach, the interface is captured as the zero set of an auxiliary function (the level-set function) φ satisfying $\varphi < 0$ inside Ω, $\varphi > 0$ outside Ω, and $\varphi = 0$ on the tumor–host interface Σ. Typically, φ is taken to be an approximation to the signed distance function; that is, $|\nabla \varphi| \approx 1$. The interface normal and curvature can easily be calculated from φ. The interface ΣN separating viable tumor cells from the necrotic cells is also captured using additional level-set function boundary φ_N that satisfies the same properties as φ, only with Ω_N and Σ_N in place of Ω and Σ.

Away from Σ, the elliptic/parabolic equations can be discretized using centered finite differences. However, near the interface, the difference stencils need to be modified to account for possible jumps in solutions and in their normal derivatives. To do this, ghost cells on either side of the interface are introduced, and the variables are extrapolated across the interface to ensure that the difference stencil effectively does not include nodes on the other side of the interface. The resulting nonlinear system is solved using an iterative algorithm.

The Dynamic Adaptive Tumor-Induced Angiogenesis Model

For a fixed tumor geometry and TAF distribution, the tumor vasculature is first grown using the basic network model given in the section titled "Basic Network Model"; capillary tips may branch or anastomose during this stage. Further, the Cartesian mesh for the tumor growth system coincides with that used for the neovascular network. After a certain period of time, referred to as the capillary growth duration time, the fluid flow is solved in the fixed neovascular network and then the network is dynamically remodeled, following the algorithm described in sections titled "Modeling Blood Flow in the Developing Capillary Network" and "Capillary Vessel Adaptation and Remodeling," respectively. During the simulation of the flow, a CFL condition is imposed on the time step: $\Delta\tau \cong \min(V_{cap} / \dot{Q}_{cap})$, where V_{cap} and $\dot{Q}_{cap}$ are the velocity and flow rate in a capillary element. The minimum is taken over the neovascular network. This ensures haematocrit remains conserved during the simulation (e.g., McDougall et al. 2002). Then, the process of blood flow, followed by remodeling, is repeated for an amount of time referred to as the *flow duration time*.

Overall Computational Solution Technique

Initially, the avascular tumor, the preexisting vascular network, the oxygen, and ECM and MDE concentrations are given. We will consider a single parent vessel placed at the top of the computational domain (see Figures 13.1 and 13.2). The algorithm then consists of iterating the following steps:

1. Solve Equation 13.1 for the oxygen concentration, where the oxygen source in Equation 13.4 is obtained from the haematocrit and the pressure in the existing vascular network and the tumor mechanical pressure from the previous time step. We then use the solution σ to update the position of the necrotic core:

$$\Omega_N^{updated} = \Omega_N^{previous} \cup \left(\{ \mathbf{x} : \sigma(\mathbf{x},t) < \sigma_N \} \cap \Omega \right)$$
,

and to identify the hypoxic region ΩH. As described earlier, the necrotic core is expanded to include previously necrotic tissue as well as any tumor tissue where the oxygen level has dipped below the necrotic threshold σN. We then rebuild φN as a level-set function that represents the updated region ΩN. (See the Appendix 2, Macklin and Lowengrub (2007) and the level-set references earlier for information on initializing a new level-set function.)

2. Solve Equation 13.20 for the tumor angiogenic growth factor (TAF), and update the MDE and ECM according to Equations 13.17 and 13.18, respectively.

3. Determine the cellular mobility, and solve for the tumor biomechanical pressure from Equation 13.9.

4. Update the position of the tumor–host interface Σ and the necrotic/viable Σ_N by advecting the level-set functions φ and φ_N with the appropriate velocities. If necessary, the level-set functions are reinitialized to be local distance functions to Σ and Σ_N.

5. From the updated tumor position, TAF, MDE, and ECM fields, the neovascular network is grown using the basic network model.

6. Steps (1)–(5) are repeated until the growth duration time interval is reached. At this point, the fluid flow in the neovascular network is determined, and the network is adapted. The hydrostatic pressure P and the TAF are held fixed during this process. The flow and network adaption are repeated (for fixed tumor and capillary tip positions) until the flow duration time is reached.

7. Go to Step 1, and repeat the algorithm.

RESULTS

In this work, we shall focus on tumor growth coupled to angiogenesis in a square 4 × 4 mm region. Although we solve the nondimensional equations, we present dimensional results using the length scale $l \approx 200\,\mu m$ and the timescale $1/\lambda_m \approx 1.5\,\text{day}$. A parent capillary vessel is located at the top of the computational domain. A preexisting vasculature is assumed to exist and provides a small level of nutrient uniformly throughout the host tissue domain. Initially, a small cluster of proliferating cells is placed approximately 3 mm from the parent vessel. The initial

ECM is taken to be nearly constant (=1) but with small random perturbations uniformly distributed throughout the computational domain
(see the time $t = 0$ plot in Figure 13.7a. Accordingly, whenever we calculate gradients of E, we actually calculate the gradient of smoothed E,
where a Gaussian smoothing with standard deviation 3.0 is used (see
Macklin and Lowengrub 2005). We begin by demonstrating that in the
absence of tumor-induced angiogenesis, the small tumor cluster grows
to an avascular tumor (2D) spheroid. Actually, since there is a preexisting vasculature, this is a misuse of notation; however, we still refer
to this case as avascular since there is no neovascular network. Then,
tumor-induced angiogenesis is initiated and we present several simulations of angiogenesis and vascular growth. Finally, we examine the effect
of increased ECM degradation by MDE and its affect on avascular and
vascular growth. The parameters, and nondimensionalization, used in
the simulations are given in Tables 13.3–13.6.

Avascular Growth to a Multicellular (2D) Spheroid

In Figure 13.4a, we present the growth of an avascular tumor. The spatial grid is 200×200 and the time step $\Delta t = 0.05$ and is adapted to satisfy the Courant–Friedrichs–Lewy (CFL) condition (see Macklin and
Lowengrub 2006, 2007). The red, blue and brown colors denote Ω_P, Ω_H,
and Ω_N the proliferating, hypoxic/quiescent, and necrotic regions, respectively. The nondimensional oxygen and ECM concentrations and the solid
(oncotic) pressure are also shown. The oxygen diffuses only a short distance
(about 0.2 mm from the diffusion length) from the parent vessel as can
be observed from the figure. However, the preexisting vasculature (which
yields a background oxygen concentration of approximately 0.4) provides
enough oxygen for the tumor to grow. As the tumor grows, the pressure
in the proliferating region increases, the oxygen is depleted in the tumor,
and the ECM is degraded. A hypoxic/quiescent core forms at about 9 days,
when the tumor radius is approximately 0.34 mm. While the tumor continues to grow and degrade the ECM, the pressure decreases and the tumor
growth starts to slow, as can be seen in Figure 13.4b. A necrotic core forms
around day 15, when the radius of the tumor is approximately 0.5 mm.
The pressure drops significantly to reflect the volume loss in the necrotic
core associated with the breakdown of the necrotic cells, and the growth
of the tumor slows even further as the tumor approaches a steady state.
As the growth of the tumor slows, the ECM degradation becomes more
pronounced. This actually causes a competition between two effects: the

TABLE 13.4   Values of the parameters for the tumor invasion model—Equations 13.1–13.20

Biological quantity	Nondimensional parameter	Scaled by	Value used in simulations
Uptake rate of oxygen in proliferating tumor region	$\bar{\lambda}_\sigma$	λ_σ^*, the characteristic rate of oxygen uptake in the proliferating tumor region	1
Uptake rate of oxygen in host microenvironment	$\bar{\lambda}_{tissue}$	λ_σ^*	0.25
Uptake rate oxygen in quiescent (hypoxic) tumor region	$\bar{\lambda}_H$	3.5×10^{-4}	0.5
Decay rate of oxygen in necrotic tumor region	$\bar{\chi}_{sprout}^T$	δ	0.25
Rate of blood-tissue oxygen transfer (extravasation) from preexisting vessels	$\bar{\chi}_{sprout}^E$	k_p	0.25
Rate of blood-tissue oxygen transfer (extravasation) from neovessels	k_w	$1 \times 10^{-6} m$	20
Lower cutoff of original ECM, used in uptake in host microenvironment	τ_w^*	E^*, the concentration of ECM secreted by the tumor cells	1.0
Upper cutoff of original ECM, used in uptake in host microenvironment	τ_{ref}	E^*	0.625
Characteristic value of discharge haematocrit	τ_e^*	Already nondimensional	0.45
Minimum value of haematocrit needed for extravasation	τ_s^*	Already nondimensional	0.05
Characteristic dimensional value of pressure in the neovessels (Equation 13.6)	P_{vessel}^*	—	5000 Pa
Apoptosis rate	A	λ_m, the mitosis rate	0

TABLE 13.4   Values of the parameters for the tumor invasion model—Equations 13.1–13.20 (Continued)

Biological quantity	Nondimensional parameter	Scaled by	Value used in simulations
Scaling factor for hydrostatic (mechanical) pressure cutoff (Equation 13.6)	$\dot{Q}_{ref}$		0.33
Necrosis rate	*GN*	μ_m	*0.3*
Tumor aggressiveness (adhesion) rate	*G*	μ_m	*40*
Minimum haptotaxis rate	$1.909 \times 10^{-11} \, m^3/s$	$6 \times 10^{-6} m$, where l is the length scale	0
Maximum haptotaxis rate	$\bar{\chi}_{E,max}$	$\chi_E^* E^* / (l^2 \lambda_m)$	0.25
Maximum value of ECM used in haptotaxis coefficient (Equation 13.15)	$\bar{E}^{\chi}_{maxcutoff}$	E^*	0.1
Minimum value of ECM used in haptotaxis coefficient (Equation 13.15)	$\bar{E}^{\chi}_{mincutoff}$	E^*	0.9
Minimum tumor mobility (response to hydrostatic/ mechanical pressure)	$\bar{\mu}_{min}$	μ^*, a characteristic value of mobility	1.0
Maximum tumor mobility (response to hydrostatic/ mechanical pressure)	$\bar{\mu}_{max}$	μ^*	4.0
Minimum value of ECM used in mobility coefficient (Equation 13.16)	$E^{\mu}_{mincutoff}$	E^*	0
Maximum value of ECM used in mobility coefficient (Equation 13.16)	$E^{\mu}_{maxcutoff}$	E^*	1.0
Oxygen diffusion coefficient	$\bar{D}_{\sigma}$	D_{σ}^*, a characteristic dimensional oxygen diffusion coefficient	1.0

(Continued)

TABLE 13.4   Values of the parameters for the tumor invasion model—Equations 13.1–13.20 (Continued)

Biological quantity	Nondimensional parameter	Scaled by	Value used in simulations
MDE diffusion coefficient	$\bar{D}_M$	$l^2\,\lambda_m$	1.0
Production rate of MDE by tumor cells	$\bar{\lambda}^M_{production}$	λ_m	100
Natural decay of MDE	$\bar{\lambda}^M_{decay}$	λ_m	10
Production rate of MDE by EC sprout tips	$\bar{\lambda}^M_{sprout\ production}$	$\lambda_m\,M^*$	1.0
Rate of degradation of ECM by MDE	$\bar{\lambda}^E_{degradation}$	$\lambda_m\,/M^*$	(Figures 13.4–13.7) 1.0 (Figures 13.8–13.9)
Rate of production of ECM by tumor cells	$\bar{\lambda}^E_{production}$	λ_m	(Figures 13.4–13.7) 2.72 Figures 13.8–13.9)
Rate of production of ECM by EC sprout tips	$\bar{\lambda}^E_{sprout\ production}$	$\lambda_m\,/E^*$	0.1
Rate of degradation of preexisting blood vessels	$\bar{\lambda}^B_{degradation}$	$\lambda_m\,/M^*$	(Figures 13.4–13.7) 1.0 (Figures 13.8–13.9)
TAF diffusion coefficient	$\bar{D}_T$	D^*_σ	1.0
Rate of TAF production by hypoxic tumor cells	$\bar{\lambda}^T_{production}$	λ^*_σ	100
Rate of natural decay of TAF	$\bar{\lambda}^T_{decay}$	λ^*_σ	0.01
Rate of binding of TAF by EC sprout tips	$\bar{\lambda}^T_{binding}$	λ^*_σ	0.025

pressure-induced motion, which becomes more effective since the mobility increases when the ECM decreases; and haptotaxis, which tends to inhibit growth of the tumor into the less dense ECM outside the tumor (recall that haptotaxis induces motion up ECM gradients). Further, the MDE also degrades the preexisting vessels, which results in a reduction in the supply of oxygen. As a result of haptotaxis and the reduced oxygen supply, the tumor actually shrinks slightly after reaching a maximum radius of about 0.64 mm (see Figure 13.4b).

TABLE 13.5   Values of the nondimensional parameters for the angiogenesis model—Equations 13.21–13.26

Biological quantity	Nondimensional parameter	Scaled by	Value used in simulations
Random motion of EC	D	$l2\,\lambda_m$	3.5×10^{-4}
Chemotactic response of EC sprout tip	$\bar{\chi}^T_{sprout}$	$l^2\,\lambda_m/T^*$	0.38
Decrease in chemotactic sensitivity	δ	$1/T^*$	0.6
Haptotactic response of EC sprout tip	$\bar{\chi}^E_{sprout}$	$l^2\,\lambda_m/E^*$	0.16
Response rate of neovessel radius to intravascular pressure	k_p	k_w, the response rate of the neovessel radius to wall shear stress	0.1
Response rate of neovessel radius to metabolic stimulus	$1 \times 10^{-6}\,\mathrm{m}$	τ_w^*	0.07
Natural shrinking tendency of neovessel radius	τ_{ref}	Already nondimensional	0.35

Tumor-Induced Angiogenesis and Vascular Growth: No Solid Pressure-Induced Neovascular Response

We next consider tumor-induced angiogenesis where there is no effect of the solid pressure on either the radius of the neovessels or the extravasation of nutrient. In particular, we take $c(P_{vessel}, P) = 0$ in Equations 13.4 and 13.26. Angiogenesis is initiated from the avascular tumor configuration at $t = 45$ days from Figure 13.4. At this time, 10 sprout tips are released from the parent vessel. The initial vessel radii are $6\,\mu m$. The inlet pressure and outlet pressures in the parent vessel are $P_{vessel,in} = 3660\,Pa$ and $P_{vessel,out} = 2060\,Pa$, respectively. The growth duration is $t = 0.05$, which means that the intravascular flow and vessel adaption algorithms are called nearly every tumor growth time step. The flow duration is $\tau = 0.25$ with a time step approximately equal to $\Delta\tau = 0.005$ (again, $\Delta\tau$ is adaptive to satisfy an intravascular CFL condition). This means that 50

TABLE 13.6 Values of the dimensional parameters for the angiogenesis model—
Equations 13.21–13.26

Biological quantity	Dimensional parameter	Value used in simulations
Dimensional Neovascular radius scale factor (Equation 13.22)	R^{*}	1×10^{-6} m
Dimensional Wall shear stress scale factor	τ_{w}^{*}	0.1 Pa
Dimensional Wall shear stress regularization factor	τ_{ref}	0.0103 Pa
Dimensional Intravascular pressure stress scale factor	τ_{e}^{*}	0.1 Pa
Dimensional Natural shrinking tendency scale factor	τ_{s}^{*}	1 Pa
Dimensional Intravascular pressure scale factor	P_{vessel}^{*}	103 Pa
Flow rate in the parent vessel	$\dot{Q}_{ref}$	1.909×10^{-11} m³ / s
Threshold minimum neovessel radius for pressure cutoff (Equation 13.27)	R_{min}	6×10^{-6} m

iterations of the flow and vascular adaptation algorithms are performed every tumor growth time step. By flowing and adapting the vascular network so frequently, we hoped that a relatively short flow duration time could be used to get a reasonable approximation of the blood flow in the network. Indeed, preliminary simulations showed that increasing the flow duration did not change the results qualitatively or, in some cases depending on the vascular network configuration, quantitatively. In a future work, we will quantify the effect of the flow duration on the results.

The evolution of the tumor and the neovascular network is shown in Figures 13.5a and 13.5b. As can be seen from the figures, it takes some time for flow to develop after angiogenesis is initiated. Further, flow first occurs after about 7 days in a region near the parent vessel. This can be seen from the plots of haematocrit and oxygen, which are signatures of blood flow. Little additional oxygen diffuses to the tumor. Accordingly, the tumor maintains a steady size (or shrinks a little due to the reasons described earlier). This may be seen in Figure 13.5c. Some of the vessels continue to lengthen, branch, and migrate toward the tumor, heading in particular for the hypoxic region where TAF is released.

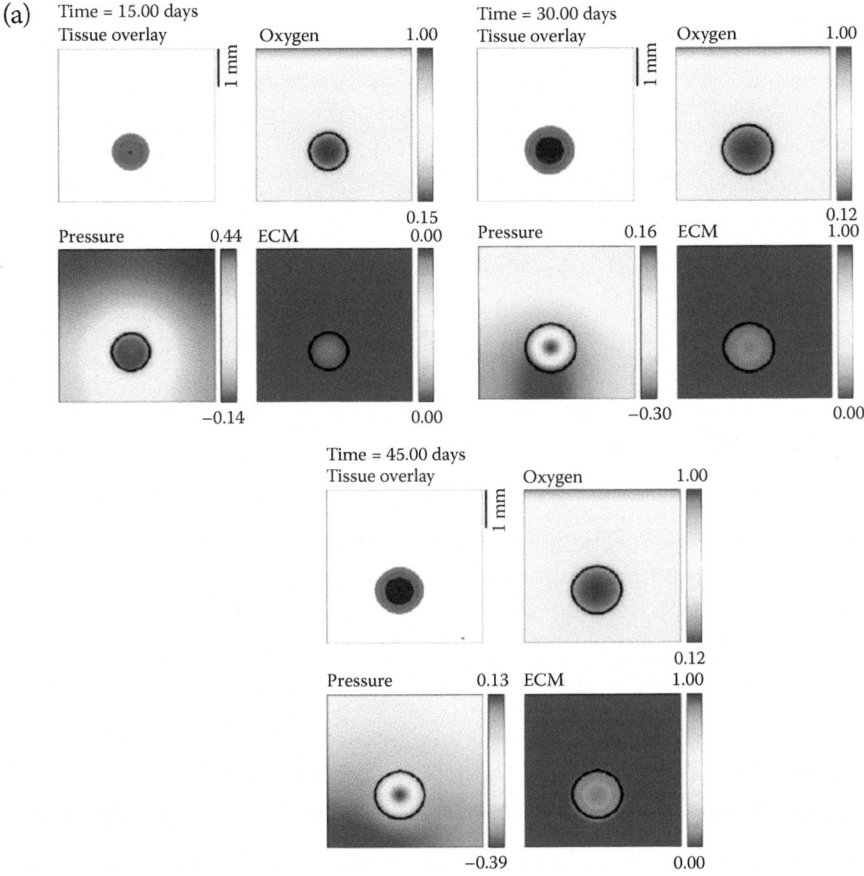

FIGURE 13.4 **(See color insert following page 40)** (a) The evolution toward a steady-state avascular multicell (2D) spheroid. The tumor regions (red—proliferating Ω_P, blue—hypoxic/quiescent Ω_H, brown—necrotic Ω_N), the oxygen, mechanical pressure, and ECM are shown at the times indicated.

After about 10 days, a large loop forms through which blood flows. The loop penetrates the tumor and provides the tumor cells with a direct source of oxygen. The tumor responds by rapidly growing along the oxygen source and co-opts the neovessels, and the hypoxic region shrinks and changes shape. As the tumor grows, the hypoxic and necrotic regions start to grow again as well and the neovessels near the tumor–host interface branch in response to wall shear stresses and increased TAF levels. This

(b)

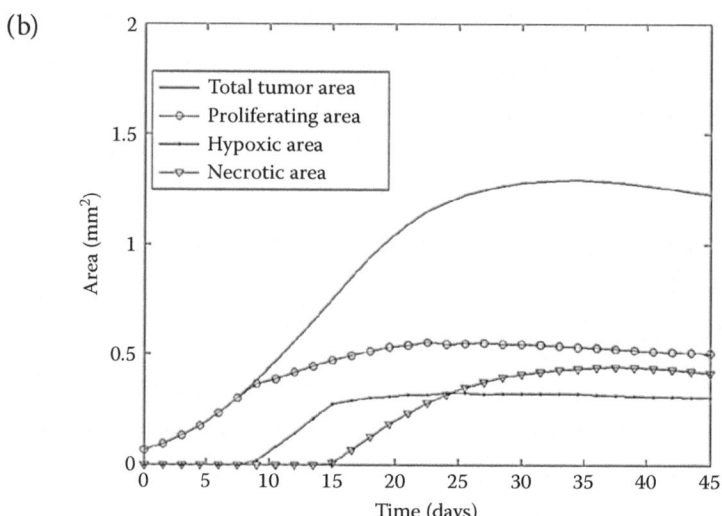

FIGURE 13.4   (Continued) (b) The areas (mm²) of the total tumor (solid line), proliferating region ("o"), hypoxic region ("●"), and the necrotic region ("inverted triangle") as a function of time for the simulation in Figure 13.4a.

results in increased anastomosis and blood flow. The increased oxygen supply in turn causes large pressures to form in the proliferating region and the tumor to grow even more rapidly, enhancing this effect. Because there is no response of the neovessels to these large pressures, the tumor simply continues to co-opt the vessels, creating an effective tumor microvasculature. This microvasculature provides a nearly uniform source of nutrient in the upper two-thirds of the tumor; the lower third is primarily hypoxic and quiescent. As a consequence, the tumor shape remains compact as the tumor grows.

In Figure 13.5b, the dimensional neovessel radii (in m) and intravascular pressures (in *Pa*) are shown together with the nondimensional ECM and TAF concentrations. At early times, the radii are small and TAF diffuses from the quiescent zone. The ring of lowered ECM surrounding the tumor is clearly seen. The pressure is highest in the neovessels closest to the inlet of the parent capillary, where the highest pressures are. As blood flow starts, the radii increase and the overall pressure decreases, while the pressure in some vessels increases as blood spreads throughout the network. This process continues as the tumor grows and the vasculature continues to branch, anastomose, and carry more and more flow. As the

(a)

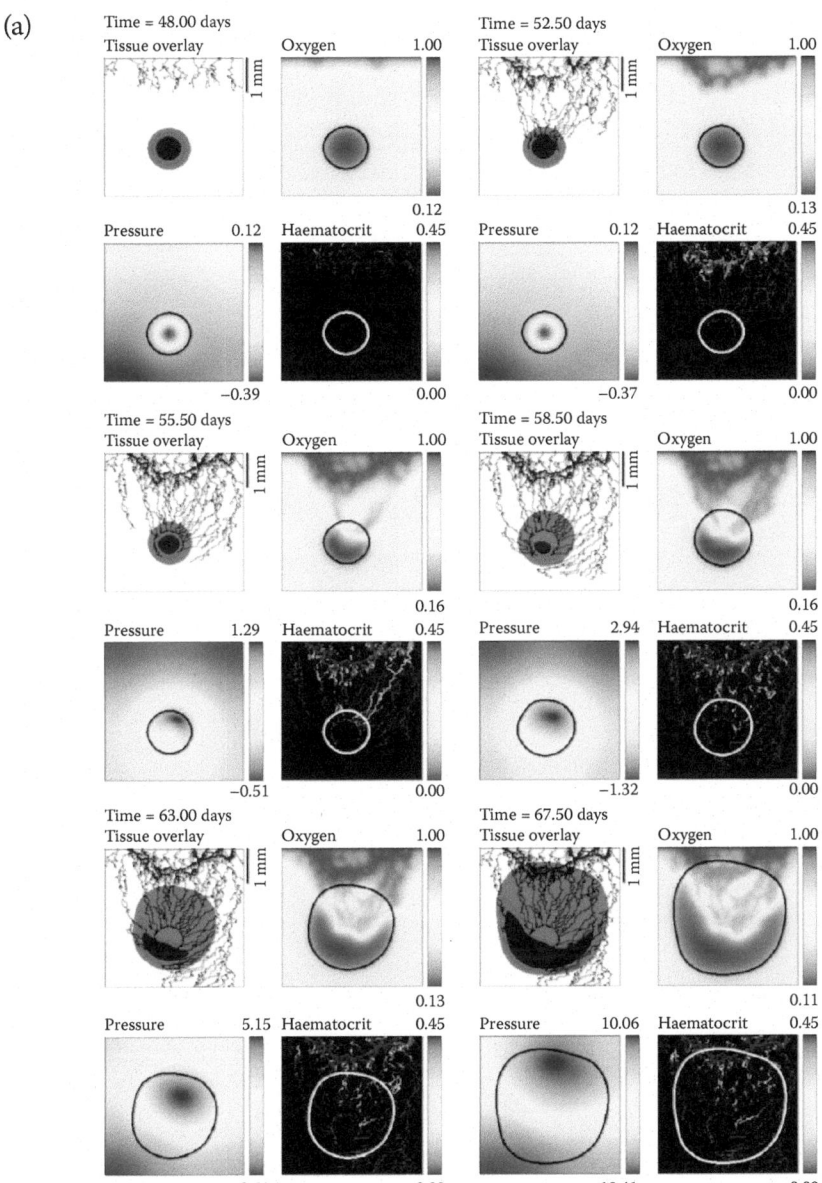

FIGURE 13.5 (a) Tumor-induced angiogenesis and vascular tumor growth. The vessels do not respond to the solid pressure generated by the growing tumor. The tumor develops a microvascular network that provides it with a direct source of oxygen and results in rapid growth with a compact (sphere-like) shape. The color scheme is the same as in Figure 13.4a.

(b)

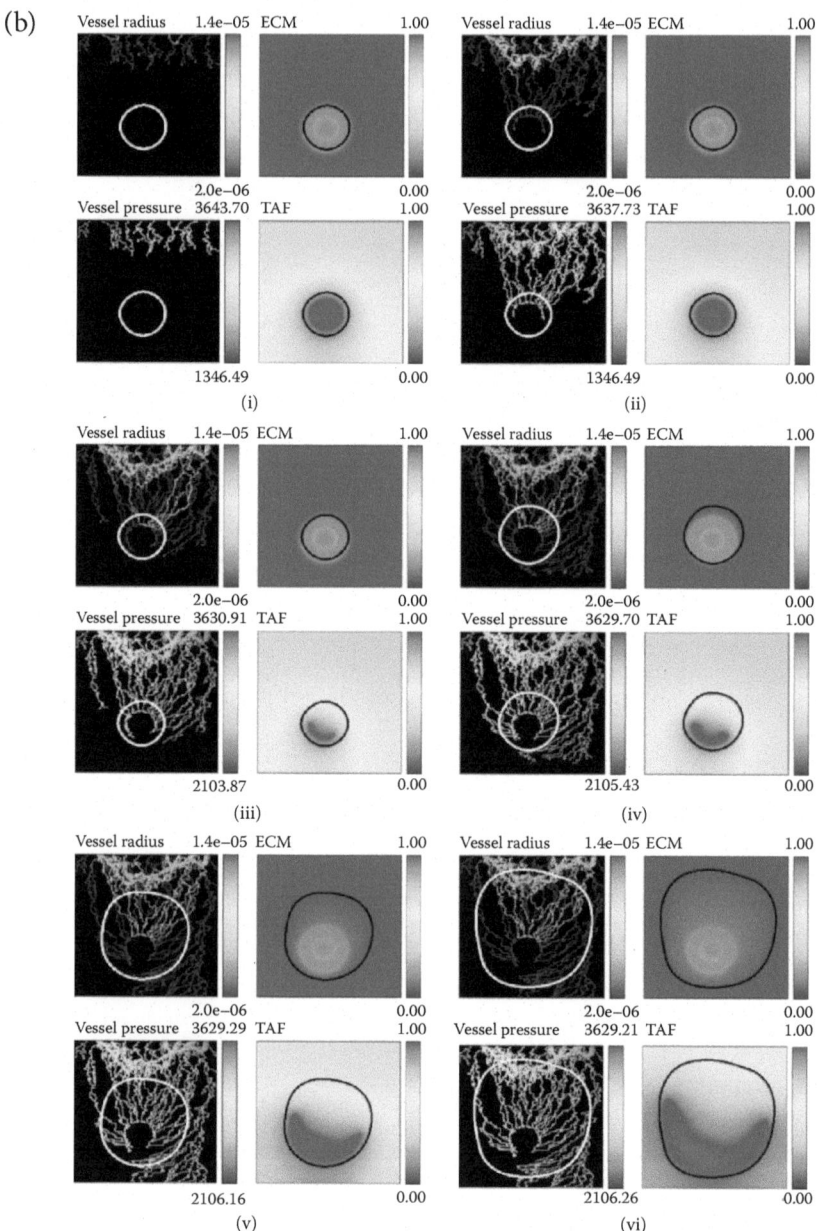

FIGURE 13.5 (Continued) (b) The dimensional intravascular radius (m) and pressure (Pa) together with the nondimensional ECM and TAF concentrations at different times from the simulation shown in Figure 13.5a. (1) Time (days) = 48, (2) 52.5, (3) 55.5, (4) 58.5, (5). 63.0, (6) 67.5.

(c)

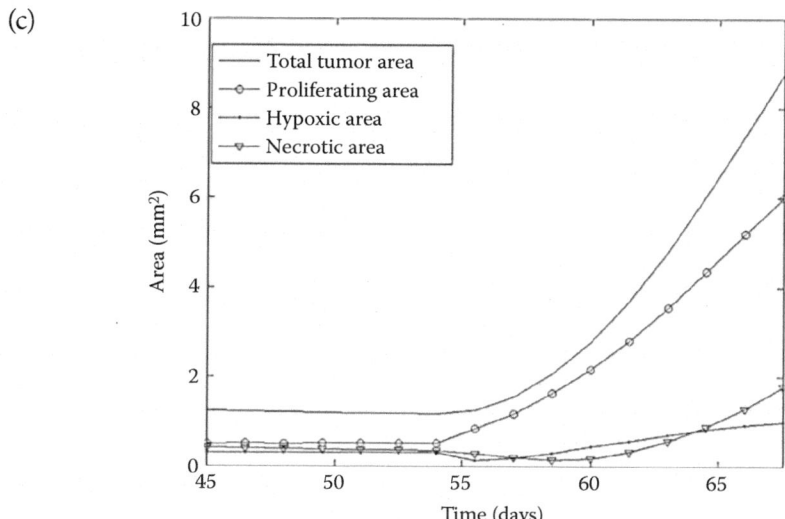

FIGURE 13.5 (Continued) (c) The areas (mm²) of the total tumor (solid line), proliferating region ("o"), hypoxic region ("•"), and the necrotic region ("inverted triangle") as a function of time for the simulation in Figure 13.5a.

hypoxic and necrotic regions shrink, the TAF distribution changes, and the vessels respond accordingly. Observe that the degraded ECM just outside the tumor does not prevent the vessels from penetrating the tumor even though the sprout tips have to migrate up ECM gradients to accomplish this.

The first vessels that penetrate the tumor do not carry blood and, thus, the tumor does not respond to their penetration. Instead, these vessels migrate toward the hypoxic region, where they tend to get stuck. This occurs because the TAF concentration is nearly uniform ($T = 1$) and so the sprout tips move randomly and tend to collide with their own trailing vessel, preventing further migration. At later times though, new vessels grow into the tumor center and anastomose. This leads to blood flow and oxygen extravasation deep in the tumor interior. Further, observe that the tumor grows so fast that it outruns the ring of degraded ECM around its boundary and is growing into only very slightly degraded ECM. The ECM in the tumor interior degrades rather slowly, and the ECM signature of the original avascular tumor spheroid can still be seen at late times.

This simulation shows that when the neovessels are not affected by the tumor solid pressure, dramatic growth occurs as the tumor co-opts the host vasculature to create its own microvasculature and receives a direct source of oxygen. In addition, the tumor growth and angiogenesis processes are nonlinearly coupled as the vasculature responds to the growth by migrating towards the ever-changing TAF distributions and by branching and anastomosing near the tumor–host interface. This leads to increased blood flow. At the same time, the increased blood flow in the vascular network affects how the tumor grows and, in particular, speeds growth up. This then affects the response of the vasculature.

Tumor-Induced Angiogenesis and Vascular Growth: The Effect of Solid Pressure-Induced Neovascular Response

Next, we consider, in Figures 13.6a–13.c, the effect of solid/mechanical pressure-induced vascular response on tumor-induced angiogenesis and vascular growth. We repeat the simulation in the section titled "Angiogenesis Model" except with $c(P_{vessel}, P)$ nonzero as given in Equation 13.5. This means that the transfer of oxygen from the neovessels to the tissue may be significantly reduced, and the vessel radii may be correspondingly constricted. With the values of the parameters used here (Tables 13.4–13.6), a solid pressure-induced vascular response begins to occur when the solid pressure $P \approx 0.8$.

At early times, the angiogenic response and the tumor growth are similar to the case presented earlier in Figures 13.5a–13.5c. The newly developing vessels migrate, proliferate, branch, and anastomose. It also takes some time for flow to begin with significant flow developing only after about 10 days. Blood flow in the neovasculature starts near the parent capillary and, eventually, the flow reaches the tumor. Because the initial ECM is slightly different from that in Figure 13.5 (due to the random component) and due to the random component of the sprout tip motion, the vascular network at early times is not identical to that obtained previously in Figure 13.5.

In contrast to the case considered in Figure 13.5, here the solid pressure prevents any delivery of oxygen internally to the tumor and, thus, the delivery of oxygen is heterogeneous, and significant oxygen gradients persist in the tumor interior. There is no functional microvasculature internal to the tumor. While the tumor responds by growing toward the oxygen-delivering neovessels, the solid pressure generated by tumor cell proliferation also constricts the neovessels in the direction of growth (where pressure is highest) and also correspondingly inhibits the transfer of oxygen from those vessels. As a consequence, the overall solid

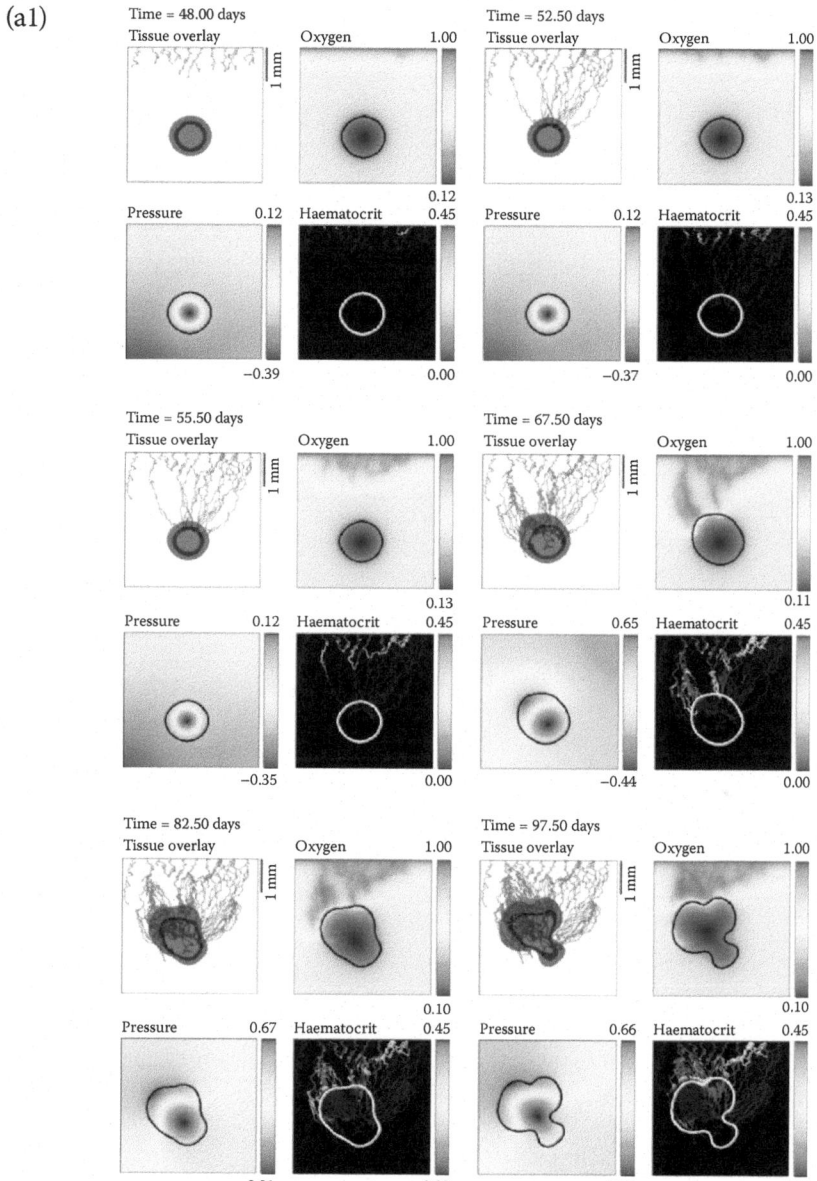

FIGURE 13.6 (a1) Tumor-induced angiogenesis and vascular tumor growth. The vessels respond to the solid pressure generated by the growing tumor. Accordingly, strong oxygen gradients are present that result in strongly heterogeneous tumor cell proliferation and shape instability. The color scheme is the same as in Figure 13.4a.

(a2)

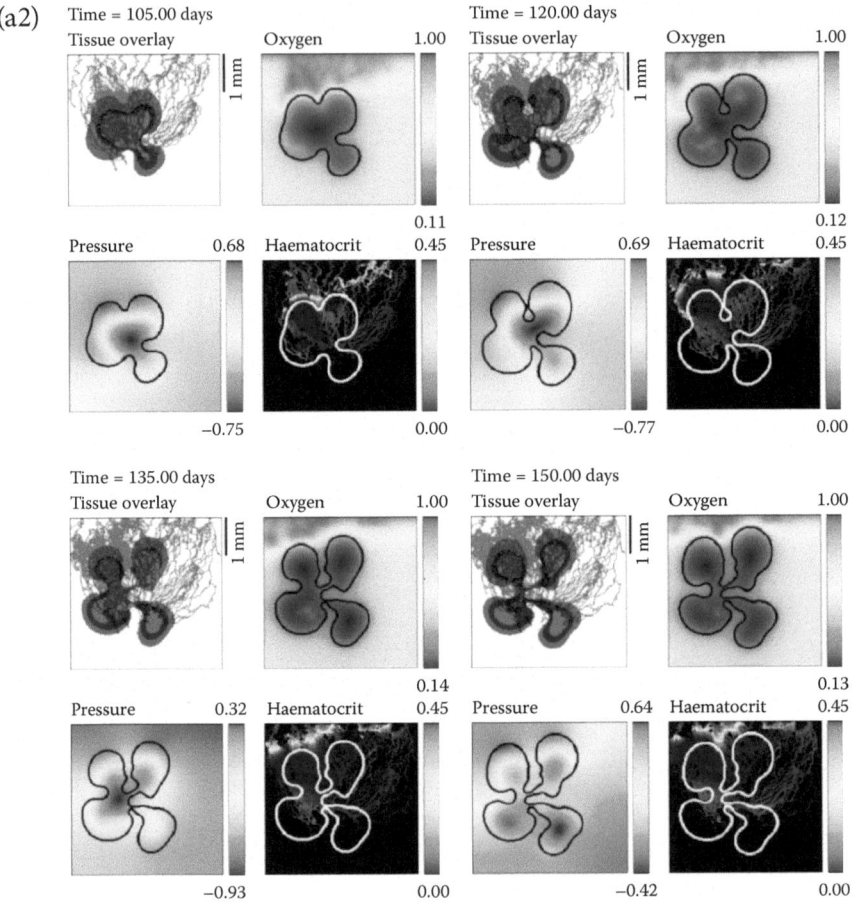

FIGURE 13.6 (Continued) (a2) Tumor-induced angiogenesis and vascular tumor growth. The vessels respond to the solid pressure generated by the growing tumor. Accordingly, strong oxygen gradients are present that result in strongly heterogeneous tumor cell proliferation and shape instability. The color scheme is the same as in Figure 13.4a.

pressure is significantly lower than that in Figure 13.5. This makes the tumor grow much more slowly than that in Figure 13.5, as can be seen in Figure 13.6c. Note that the vertical scale in Figure 13.6c is one half of that in Figure 13.5c.

The neovessels in other areas of the host microenvironment then provide a stronger source of oxygen. This triggers tumor–cell proliferation and growth in regions where proliferation had been decreased previously.

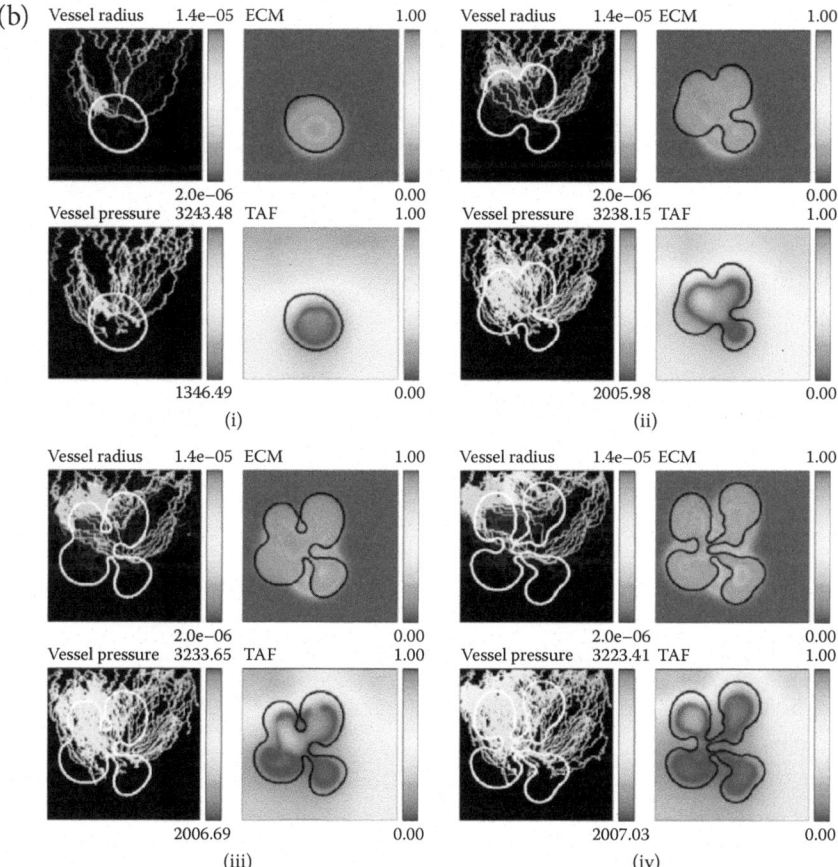

FIGURE 13.6 (Continued) (b) The dimensional intravascular radius (m) and pressure (Pa) together with the nondimensional ECM and TAF concentrations at different times from the simulation shown in Figure 13.6a. (1) Time (days) = 67.5, (2) 105, (3) 120, (4) 150.

The heterogeneity of oxygen delivery and the associated oxygen gradients cause heterogeneous tumor cell proliferation. Unlike the case in Figure 13.5, proliferation is confined to regions close to the tumor–host interface. This results in morphological instability that leads to the formation of invasive tumor clusters (e.g., buds) and a complex tumor morphology. This result is consistent with the theory and predictions made earlier (see, for example, Cristini et al. (2003); Cristini et al. (2005); Anderson (2005); Macklin and Lowengrub (2005, 2006, 2007); Anderson et. al. (2006); and Gerlee and Anderson (2007)) that substrate inhomogeneities

(c)

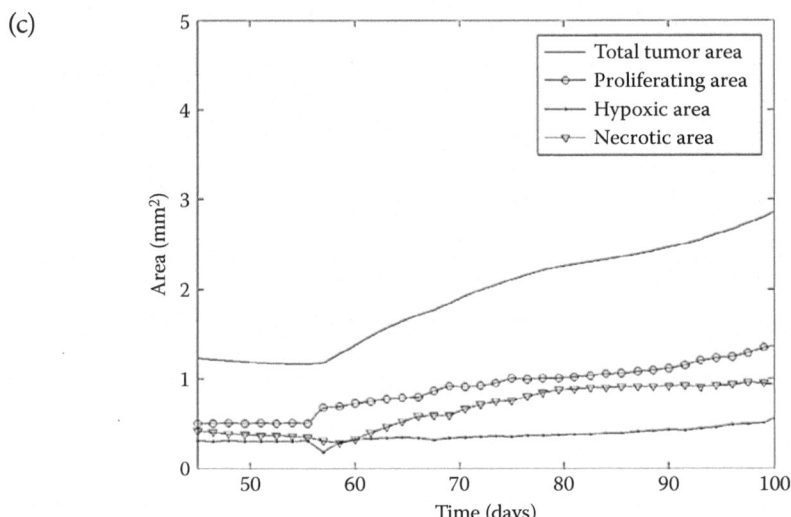

FIGURE 13.6 (Continued) (c) The areas (mm²) of the total tumor (solid line), proliferating region ("o"), hypoxic region ("•"), and the necrotic region ("inverted triangle") as a function of time for the simulation in Figure 13.6a.

in the tumor microenvironment tend to cause morphological instabilities in growing tumors.

Although nutrient-providing, functional vessels are not able to penetrate the tumor during growth, the growth of the tumor elicits a strong branching and anastomosis response from nearby neovessels in the host microenvironment. Although there is an analogous neovascular response seen in Figure 13.5, the effect here is much more pronounced as the levels of TAF are higher in these regions (because tumor hypoxia is increased) and, thus, the wall shear stresses initiate more significant branching.

In Figure 13.6b, the dimensional neovessel radii (in m) and intravascular pressures (in Pa) are shown together with the nondimensional ECM and TAF concentrations. As before, blood flow causes a dilation of the vessels and an overall decrease of pressure as branching, anastomosis, and increased blood flow occur throughout the neovascular network. The constriction of neovessels in response to the solid pressure is clearly seen.

The tumor-secreted MDE degrades the ECM in the host microenvironment near the tumor and in the tumor interior. As before (recall Figure 13.5b), the neovessels are still able to migrate through the region of lower ECM even though this acts against haptotaxis. Because the tumor grows more slowly than that in Figure 13.5b, only the tips of the invasive

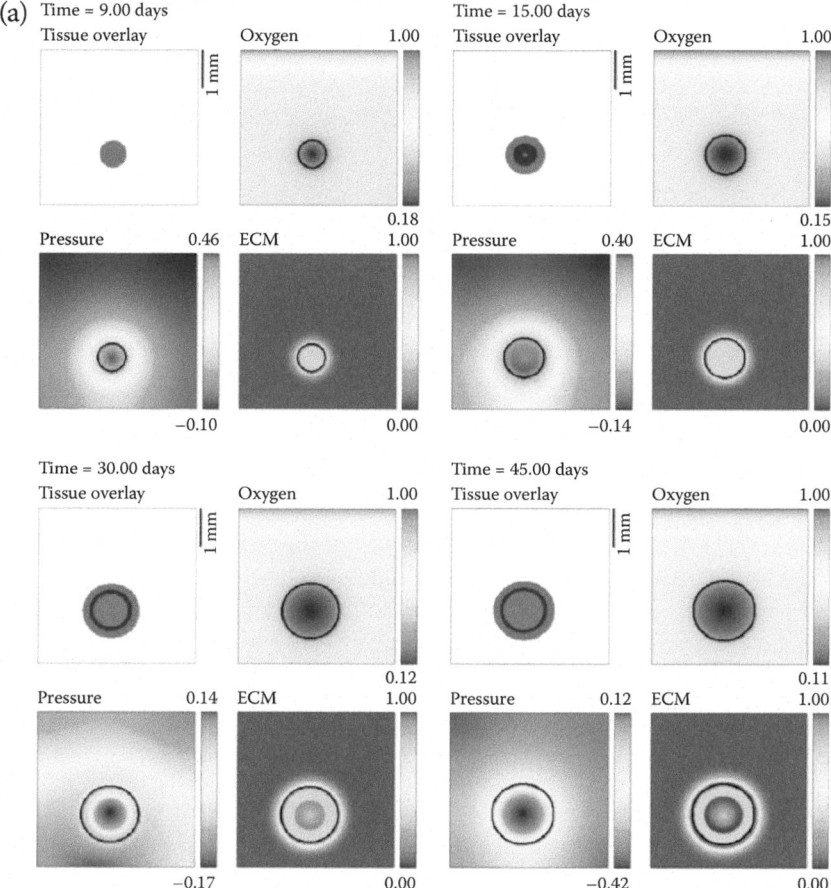

FIGURE 13.7 (a) The evolution toward a steady-state avascular multicell (2D) spheroid with enhanced ECM degradation. The MDE production and degradation parameters are larger than those used in Figure 13.4a; see Table 13.4.

clusters outrun the degraded ECM. As can be seen in Figure 13.6c, the host ECM is degraded in the region between the invading clusters. The ECM signature of the original avascular tumor spheroid can no longer be seen at later times.

This simulation shows even stronger nonlinear coupling between the tumor-induced angiogenesis and the progression of the tumor compared to the prior case shown in Figure 13.5. The pressure-induced vascular response of constricting neovessel radii and inhibiting blood-tissue oxygen transfer not only affects the tumor growth dramatically, but also significantly affects the growth of the neovascular network, and vice versa.

(b)

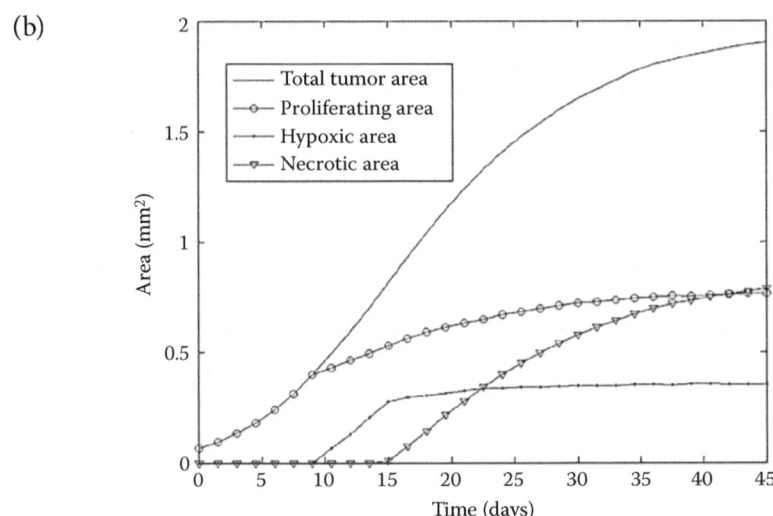

FIGURE 13.7   (Continued) (b) The areas (mm²) of the total tumor (solid line), proliferating region ("o"), hypoxic region ("•"), and the necrotic region ("inverted triangle") as a function of time for the simulation in Figure 13.7a.

Avascular Growth to a Multicellular (2D) Spheroid with Enhanced ECM Degradation and Production

We next examine the effect of ECM degradation on the results. In Figure 13.7a, we repeat the simulation in the section titled "The Tumor Invasion Model" except that both the MDE degradation and production parameters are increased (see Table 13.4). The tumor grows by uptaking oxygen delivered by the preexisting (uniform) vasculature and growth is more rapid than that for the avascular tumor shown in the aforementioned section (Figure 13.4a). This occurs because the mobility is greater here due to the enhanced degradation of ECM. This effect overcomes the tendency of haptotaxis to keep the tumor away from the degraded ECM.

The tumor reaches a nearly steady size, containing both a hypoxic and a necrotic core, that is significantly larger than that shown in Figures 13.4a and 13.4b; the radius at 45 days is approximately 0.78 mm (see Figure 13.7b). At the final time shown (45 days), the ECM is significantly degraded in the host microenvironment and in the tumor necrotic core to the point that there is even a thin annular "hole" in the ECM immediately surrounding the spheroid, and a circular hole in the necrotic region, where the density of ECM $E \approx 0$.

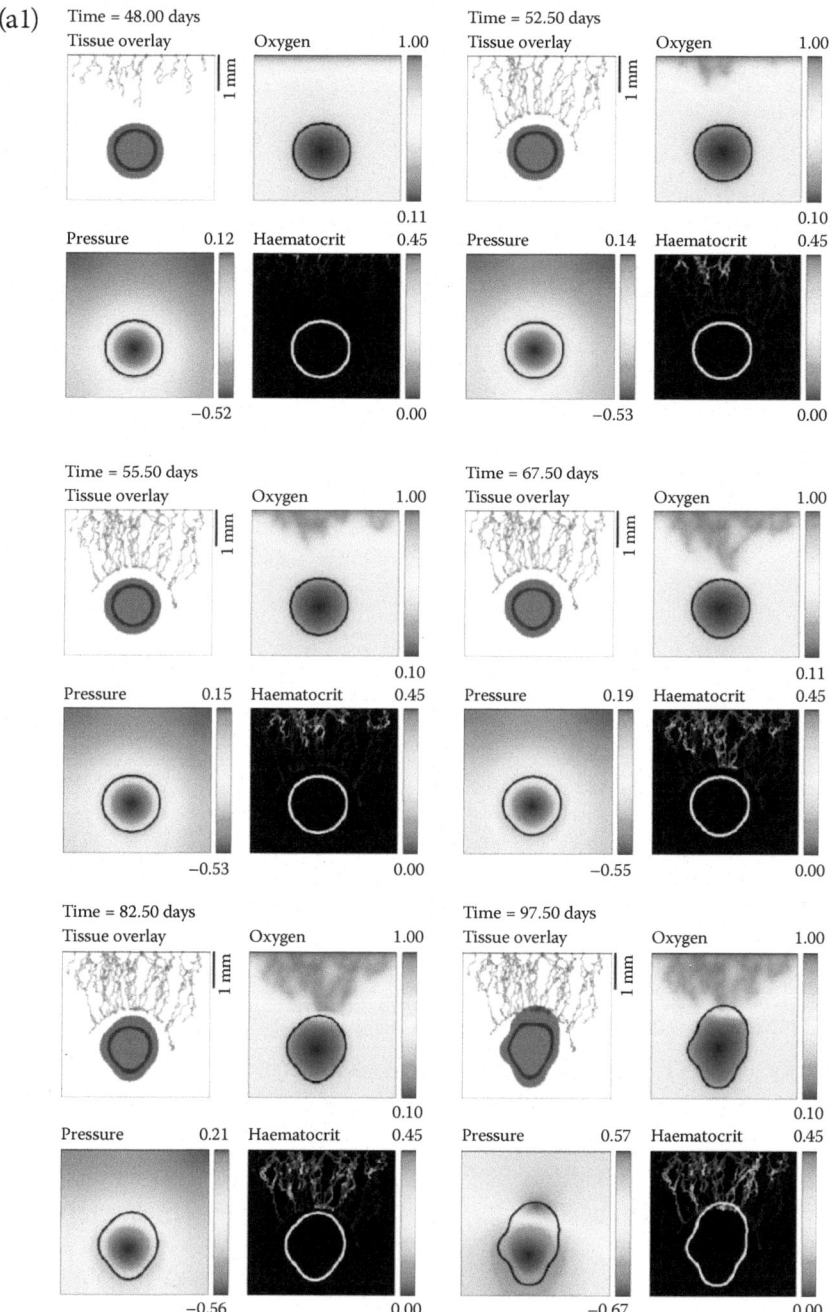

FIGURE 13.8 (a1) Tumor-induced angiogenesis and vascular tumor growth with enhanced ECM degradation.

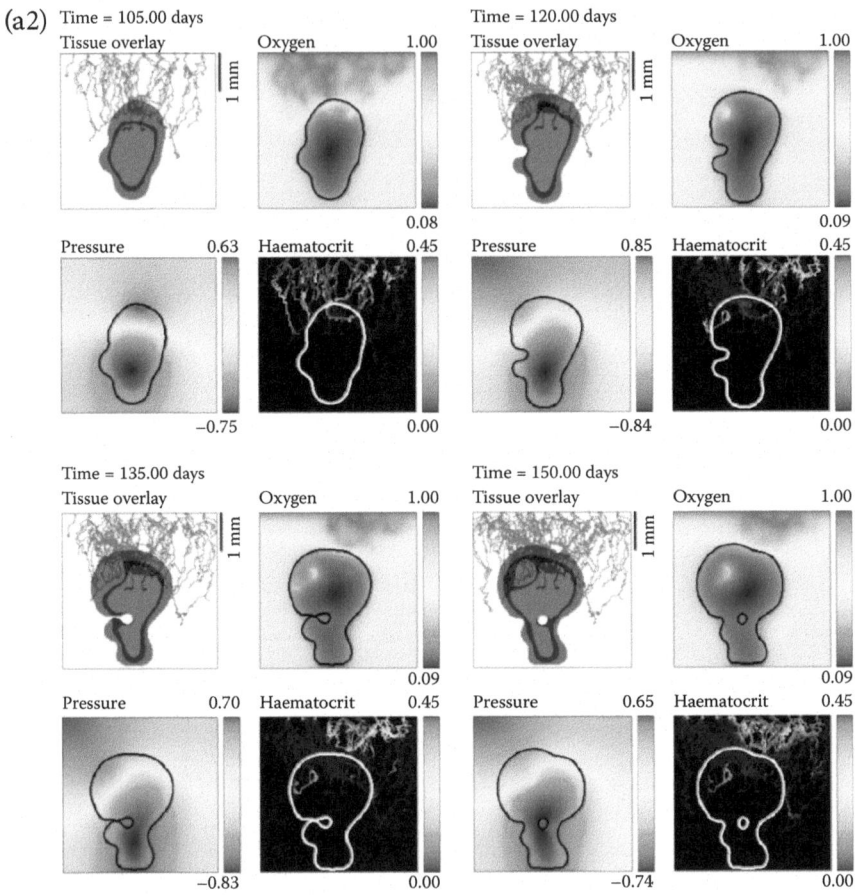

FIGURE 13.8 (Continued) (a2) Tumor-induced angiogenesis and vascular tumor growth with enhanced ECM degradation.

Tumor-Induced Angiogenesis and Vascular Growth: The Effect of Solid Pressure-Induced Neovascular Response and Enhanced ECM Degradation and Production

Next, we consider, in Figures 13.8a–13.8c, the effect of enhanced ECM degradation on tumor-induced angiogenesis and vascular growth. We repeat the simulation shown in the section titled "Numerical Schemas" except that the initial condition is the $t = 45$ day simulation from Figure 13.7a and the MDE parameters are the same as in the previous section (see Tables 13.4–13.6).

As in the simulation shown in the section titled "The Tumor Invasion Model," the neovessels grow and form loops near the parent capillary.

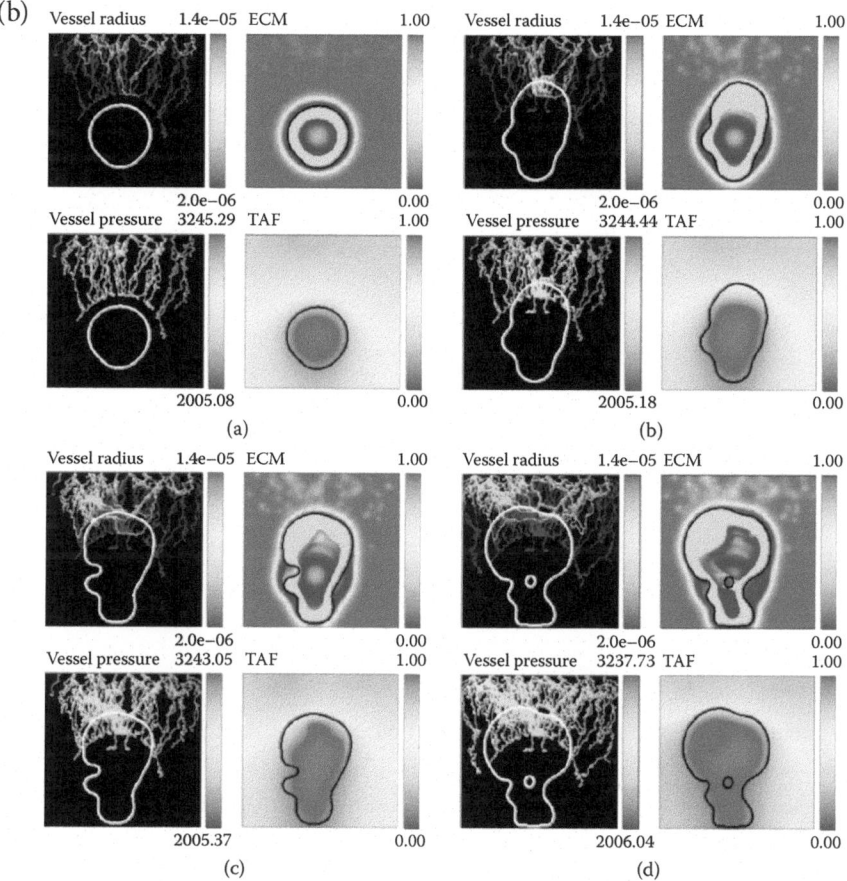

FIGURE 13.8 (Continued) (B) The dimensional intravascular radius (m) and pressure (Pa) together with the nondimensional ECM and TAF concentrations at different times from the simulation shown in Figure 13.8a. (1) Time (days) = 67.5, (2) 105, (3) 120, (4) 150.

However, now because of the growing ECM annular hole surrounding the tumor, the neovessels are not able to reach the tumor and are instead trapped by the ECM hole due to haptotaxis. The vessels then encapsulate roughly the upper half of the tumor.

As blood flows through the neovascular network and approaches the tumor, the tumor responds by growing toward the flowing neovessels that provide the oxygen source, as described in the section titled "Numerical Schemas." The tumor elongates, constricts the neovessels in its path, and

(b)

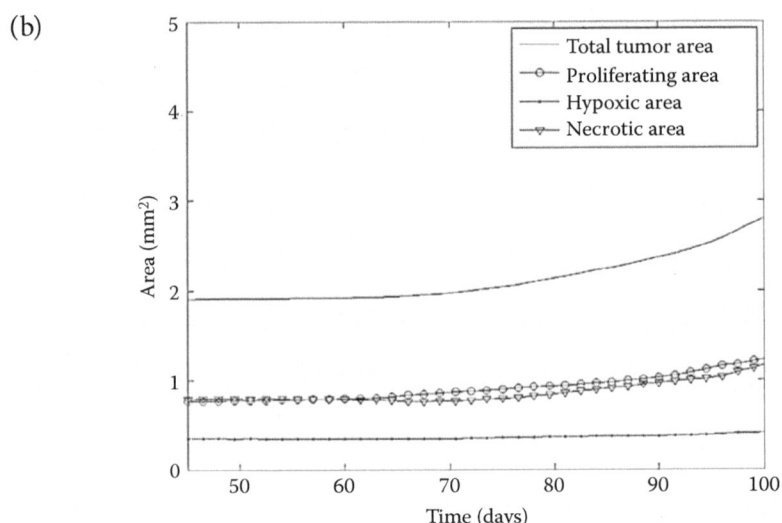

FIGURE 13.8 (Continued) (c) The areas (mm²) of the total tumor (solid line), proliferating region ("o"), hypoxic region ("•"), and the necrotic region ("inverted triangle") as a function of time for the simulation in Figure 13.8a.

prevents the transfer of oxygen from the neovessels to the host. This limits tumor cell proliferation and results in a roughly steady maximum solid pressure. Correspondingly, there are heterogeneous oxygen supply, heterogeneous tumor cell proliferation, and strong oxygen gradients. As in the aforementioned section, this results in a morphological instability of the growing solid tumor.

As the tumor continues to grow, the neovessels respond by increasing branching and anastomosing near the tumor–host interface. This results in a broader supply of oxygen in the part of the tumor closest to the parent capillary. Proliferation is increased, and the top of the tumor flattens. The increased proliferation leads to large solid pressures, which then constrict the nearby neovessels and inhibit oxygen supply. The tumor then begins to grow toward other vessels near the parent capillary, and the top of the tumor becomes unstable. Further, there is instability along the side of the tumor that leads to the encapsulation of host domain inside the tumor. Also, observe that a small amount of oxygen is able to be delivered to the tumor at very late times as haematocrit is trapped in a constricted vessel at a location where the pressure is sufficiently low to allow extravasation.

Figure 13.8b shows the dimensional neovessel radii (in m) and the intravascular pressures (in Pa) together with the nondimensional ECM and TAF concentrations. The results are similar to those obtained before except that the tumor does not outrun the ECM hole, although at the top of the tumor, the hole is quite shallow.

Interestingly, even though the initial tumor in Figure 13.8a is larger than that in Figure 13.6a, the final tumor size at $t = 150$ days is roughly the same for both cases (see Figures 13.8c and 13.6c). The ECM hole present in the simulation in Figure 13.8 prevents the neovessels from getting close to the tumor during the early stages of growth; this allows the tumor in Figure 13.6 to catch up and even grow slightly larger than that in Figure 13.8.

Finally, in Figure 13.9, we compare the average radii in the neovascular networks for the simulations in Figures 13.5, 13.6, and 13.8. At early times, the radii for the simulation in Figure 13.5, where the neovasculature does not respond to solid pressure, grows the fastest as blood flows uninhibited through the network. Later, however, the simulation with lower ECM degradation shows the most rapid radii increase. This occurs because the EC sprout tips are able to move more freely through the host domain and do not get caught by degraded ECM. This provides the vascular network with a more widely varying flow response.

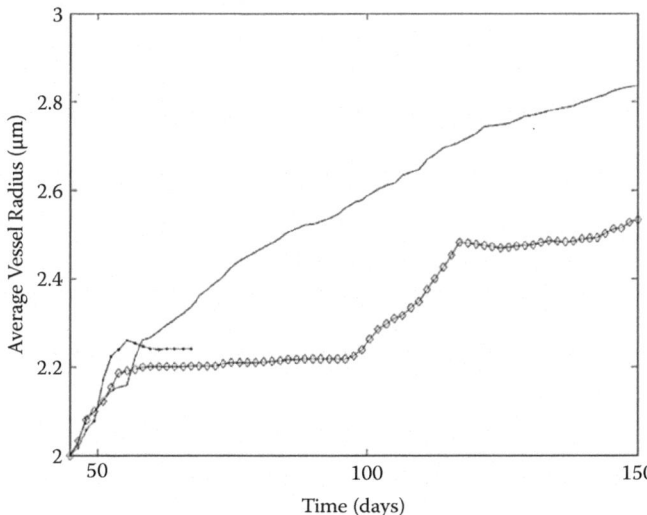

FIGURE 13.9 Average vessel radii. Simulation from Figure 13.6 (solid), simulation from Figure 13.5. ("•"), simulation from Figure 13.8 ("diamond").

DISCUSSION

In this paper, we have coupled an improved continuum model of solid tumor invasion (following Macklin and Lowengrub (2007)) with a model of tumor-induced angiogenesis (following McDougall et al. (2006)) to produce a new multiscale model of vascular solid tumor growth. The invasion and angiogenesis models were coupled through the tumor angiogenic factors (TAFs) released by the tumor cells and through the nutrient extravasated from the neovascular network. As the blood flows through the neovascular network, nutrients (e.g., oxygen) are extravasated and diffuse through the ECM, triggering further growth of the tumor, which in turn influences the TAF expression. In addition, the extravasation is mediated by the hydrostatic stress (solid pressure) generated by the growing tumor. The solid pressure also affects vascular remodeling by restricting the radii of the vessels and, thus, the flow pattern and wall shear stresses. The vascular network and tumor progression were also coupled via the ECM as both the tumor cells and ECs upregulate matrix-degrading proteolytic enzymes, which cause localized degradation of the ECM that in turn affects haptotactic migration.

We performed simulations of the multiscale model that demonstrated the importance of the nonlinear coupling between the growth and remodeling of the vascular network, the blood flow through the network and the tumor progression. The solid pressure generated by tumor cell proliferation effectively shuts down large portions of the vascular network, dramatically affecting the flow, the subsequent network remodeling, the delivery of nutrients to the tumor and the subsequent tumor progression. In addition, ECM degradation by tumor cells was seen to have a dramatic affect on both the development of the vascular network and the growth response of the tumor. In particular, when the ECM degradation is significant, the newly formed vessels tended to encapsulate, rather than penetrate, the tumor and were thus less effective in delivering nutrients.

There are many directions in which this work will be taken in the future both in terms of modeling additional biophysical effects as well as algorithmic improvements. Regarding the algorithm, we plan to upgrade the solid pressure/nutrient solver by solving for P and _ as a coupled system. This will prevent oscillations that may occur by lagging P in the source term for nutrient. We also plan to accelerate the solver for the intravascular pressure to improve performance of the coupled algorithm.

Regarding the model, we plan to develop a more detailed model of the effect of solid pressure on the constriction and collapse of vessels in the microvasculature and on the corresponding response of the microvascular network. We also plan to include, for the first time, the effects of the venous system. Other features, such as the recruitment of pericytes by the vascular ECs, will also be investigated. In addition, we will incorporate more realistic models for soft tissue mechanics.

The work presented here demonstrates that nonlinear simulations are a powerful tool for understanding phenomena fundamental to solid tumor growth. A biophysically justified computer model could provide an enormous benefit to the clinician, the patient, and society by efficiently searching parameter space to identify optimal, or nearly optimal, individualized treatment strategies involving, for example, chemotherapy and adjuvant treatments such as antiangiogenic or anti-invasive therapies. This is a direction we plan to explore in the future.

ACKNOWLEDGMENTS

MC was supported by a Leverhulme Trust Personal Research Fellowship and NIH Grant CA 113004. PM acknowledges partial support from a U.S. Department of Education GAANN (Graduate Assistance in Areas of National Need) fellowship. JSL acknowledges partial support from the National Science Foundation (NSF) through grants DMS-0352143 and DMS-0612878 and from the National Institutes of Health (NIH) grant P50GM76516 for a National Center of Excellence in Systems Biology at UCI. VC acknowledges partial support from the NSF and the (NIH) through grants NSF-DMS-0314463 and NIH-5R01CA093650-03. Further, PM, VC, and JL are grateful to Herman Frieboes for valuable discussions.

REFERENCES

Adalsteinsson, D., Sethian, J.A. 1999. The fast construction of extension velocities in level-set methods. *J. Comput. Phys.* 148:2–22.

Alarcon, T., Byrne, H.M., Maini, P.K. 2003. A cellular automaton model for tumour growth in a heterogeneous environment. *J. Theor. Biol.* 225:257–274.

Alarcon, T., Byrne, H.M., Maini, P.K., 2005. A multiple scale model for tumor growth, *Multiscale Mod. Simul.* 3:440–475.

Ambrosi, D., Preziosi, L. 2002. On the closure of mass balance models for tumor growth. *Math. Mod. Meth. Appl. Sci.* 12:737–754.

Anderson, A.R.A., Chaplain, M.A.J. 1998. Continuous and discrete mathematical models of tumor-induced angiogenesis. *Bull. Math. Biol.* 60: 857–899.

Anderson, A.R.A., Chaplain, M.A.J., Newman, E.L., Steele, R.J.C., and Thompson, A.M., 2000. Mathematical modelling of tumour invasion and metastasis. *J. Theor. Med.* 2:129–154.

Anderson, A.R.A., 2005. A hybrid mathematical model of solid tumour invasion: the importance of cell adhesion. *Math. Med. Biol.* 22: 163–186.

Araujo, R.P., McElwain, D.L.S. 2004. A history of the study of solid tumor growth: the contribution of mathematical modelling. *Bull. Math. Biol.* 66:1039–1091.

Araujo, R.P., McElwain, D.L.S. 2005a. A mixture theory for the genesis of residual stresses in growing tissues I: a general formulation. *SIAM J. Appl. Math.* 65:1261–1284.

Araujo, R.P., McElwain, D.L.S. 2005b. A mixture theory for the genesis of residual stresses in growing tissues II: solutions to the biphasic equations for a multicell spheroid. *SIAM J. Appl. Math.* 66:447–467.

Ausprunk, D.H., Folkman, J., 1977. Migration and proliferation of endothelial cells in preformed and newly formed blood vessels during tumour angiogenesis. *Microvasc. Res.* 14:53–65.

Balding, D., McElwain, D.L.S. 1985. A mathematical model of tumour-induced capillary growth. *J. Theor. Biol.* 114:53–73.

Byrne, H.M., Chaplain, M.A.J. 1995. Mathematical models for tumour angiogenesis: numerical simulations and nonlinear wave solutions. *Bull. Math. Biol.* 57:461–486.

Byrne, H.M., Chaplain, M.A.J., Pettet, G.J., and McElwain, D.L.S. 1999. A mathematical model of trophoblast invasion. *J. Theor. Med.*1:275–286.

Byrne, H.M., Preziosi, L. 2003. Modeling of solid tumour growth using the theory of mixtures. *Math. Med. Biol.* 20:341–366.

Byrne, H.M., Alarcon, T., Owen, M.R., Webb, S.D., and Maini, P.K. 2006. Modeling aspects of cancer dynamics: a review. *Phil. Trans. R. Soc. A* 364:1563–1578.

Carmeliet, P. 2005. Angiogenesis in life, disease and medicine. *Nature* 438:932–936.

Chaplain, M.A.J., Stuart A.M. 1993. A model mechanism for the chemotactic response of endothelial cells to tumour angiogenesis factor. *IMA J. Math. Appl. Med. Biol.* 10:149–168.

Chaplain, M.A.J., 2000. Mathematical modelling of angiogenesis. *J. Neuro-Oncol.* 50:37–51.

Chaplain, M.A.J., Lolas, G. 2005. Mathematical modelling of cancer cell invasion of tissue: the role of the urokinase plasminogen activation system. *Math. Modell. Methods. Appl. Sci.* 15:1685–1734.

Chaplain, M.A.J., McDougall, S.R., Anderson, A.R.A. 2006. Mathematical modelling of tumour-induced angiogenesis. *Annu. Rev. Biomed. Eng.* 8:233–257.

Chaplain, M.A.J., Lolas, G. 2006. Mathematical modelling of cancer invasion of tissue: dynamic heterogeneity. *Net. Hetero. Med.* 1:399–439.

Chomyak, O.G., Sidorenko, M.V. 2001. Multicellular spheroids model in oncology. *Exp. Oncol.* 23:236–241.

Cristini, V., Lowengrub J., Nie, Q. 2003. Nonlinear simulation of tumour growth. *J. Math. Biol.* 46:191–224.

Deakin, A.S. 1976. Model for initial vascular patterns in melanoma transplants. *Growth* 40: 91–201.

Dickinson, R.B., Tranquillo, R.T. 1993. A stochastic model for adhesion-mediated cell random motility and haptotaxis, *J. Math. Biol.* 31:563–600.

DiMilla, P.A., Barbee, K., Lauffenburger, D.A. 1991. Mathematical model for the effects of adhesion and mechanics on cell migration speed, *Biophys. J.* 60:15–37.

Erler, J.T., Bennewith, K.L., Nicolau, M., Dornhoefer, N., Kong, C.L., Chi, Q-T., Jeffrey, S.S., and Siaccia, A.J. 2006. Lysyl oxidase is essential for hypoxia-induced metastasis. *Nature* 440:1222–1226.

Folkman, J., 1971. Tumor angiogenesis: therapeutic implications. *N. Eng. J. Med.* 285:1182–1186.

Folkman, J., 1995. Angiogenesis in cancer, vascular, rheumatoid and other disease. *Nature Med.* 1:21–31.

Folkman, J., Klagsbrun, M., 1987. Angiogenic factors. *Science* 235:442–447.

Friedl, P., Wolf, K. 2003. Tumor-cell invasion and migration: diversity and escape mechanisms. *Nature Rev.* 3:362–371.

Galaris, D., Barbouti, A., Korantzopoulos, P. 2006. Oxidative stress in hepatic ischemia—reperfusion injury: the role of antioxidants and iron chelating compounds. *Current Pharma. Design* 12:2875–2890.

Gerlee, P., Anderson, A.R.A. 2007. Stability analysis of a hybrid cellular automaton model of cell colony growth. *Phys. Rev. E* 75:0151911.

Gimbrone, M.A., Cotran, R.S., Leapman, S.B., and Folkman, J., 1974. Tumor growth and neovascularization: an experimental model using the rabbit cornea. *J. Natl. Cancer Inst.* 52:413–427.

Godde, R., Kurz, H., 2001. Structural and biophysical simulation of angiogenesis and vascular remodeling. *Dev. Dynam.* 220:387–401.

Graziano, L., Preziosi, L. 2007. Mechanics in tumor growth, in *Modelling of Biological Materials*, ed. F. Mollica, K.R. Rajagopal, and L. Preziosi, Birkhauser.

Greenspan, H.P. 1976. On the growth and stability of cell cultures and solid tumours. *J. Theor. Biol.* 56:229–242.

Hanahan, D., Weinberg, R.A. 2000. The hallmarks of cancer. *Cell* 100:57–70.

Jain, R.K., 1987. Transport of molecules in the tumor interstitium: a review. *Cancer Res.* 47:3039–3051.

Jain, R.K., 1988. Determinants of tumour blood flow: a review. *Cancer Res.* 48:2641–2658.

Jain, R.K., 2001. Normalizing tumor vasculature with anti-angiogenic therapy: a new paradigm for combination therapy. *Nature Med.* 7:987–989.

Jain, R.K., 2003. Molecular regulation of vessel maturation. *Nature Med.* 9:685–693.

Jain, R.K., 2004. Recognition of tumor blood vessel normalization as a new antiangiogenic concept. *Nature Med.* 10:329–330.

Kamiya, A., Bukhari, R., Togawa, T., 1984. Adaptive regulation of wall shear stress optimizing vascular tree function. *Bull. Math. Biol.* 46:127–137.

Kaur, B., Khwaja, F.W., Severson, E.A., Matheny, S.L., Brat, D.J., and VanMeir, E.G. 2005. Hypoxia and the hypoxia-inducible-factor pathway in glioma growth and angiogenesis. *Neuro-Oncol.* 7:134–153.

Kim, J.B. 2005. Three-dimensional tissue culture models in cancer biology. *Semin. Cancer Biol.* 15:365–377.

Kloner, R.A., Jennings, R.B. 2001. Consequences of brief ischemia: stunning, preconditioning and their clinical implications: Part 1. *Circulation* 104:2981–2989.

Krenz, G.S., Dawson, C.A., 2002. Vessel distensibility and flow distribution in vascular trees. *J. Math. Biol.* 44:360–374.

Kunz-Schughart, L.A., Freyer, J.P., Hofstaedter, F., and Ebner, R. 2004. The use of 3-D cultures for high-throughput screening: the multicellular spheroid model. *J. Biomol. Screening* 9:273–285.

Levine, H.A., Pamuk, S., Sleeman, B.D., Nielsen-Hamilton, M. 2001. Mathematical modeling of the capillary formation and development in tumor angiogenesis: penetration into the stroma. *Bull. Math. Biol.* 63:801–863.

Liotta, L.A., Saidel, G.M., Kleinerman, J. 1977. Diffusion model of tumor vascularization and growth. *Bull. Math. Biol.* 39:117–129.

Macklin, P., McDougall, S.R., Anderson, A.R.A., Chaplain, M.A.J., Cristini, V., and Lowengrub, J., 2009. Multiscale modelling and nonlinear simulation of vascular tumour growth. *J. Math. Biol.* 58:765–798.

Mantzaris, N.V., Webb, S., Othmer, H.G., 2004. Mathematical modeling of tumor-induced angiogenesis. *J. Math. Biol.* 49:111–187.

McDougall, S.R., Sorbie, K., 1997. The application of network modelling techniques to multiphase flow in porous media. *Petrol. Geosci.* 3:161–169.

McDougall, S.R., Anderson, A.R.A., Chaplain, M.A.J., and Sherratt, J.A., 2002. Mathematical modelling of flow through vascular networks: implications for tumour-induced angiogenesis and chemotherapy strategies. *Bull. Math. Biol.* 64:673–702.

McDougall, S.R., Anderson, A.R.A., Chaplain, M.A.J. 2006. Mathematical modelling of dynamic adaptive tumour-induced angiogenesis: clinical implications and therapeutic targeting strategies. *J. Theor. Biol.* 241:564–589.

Morikawa, S., Baluk, P., Kaidoh, T., Haskell, A., Jain, R.K., and McDonald, D.M., 2002. Abnormalities in pericytes on blood vessels and endothelial sprouts in tumors. *Am. J. Pathol.* 160:985–1000.

Muthukkaruppan, V.R., Kubai, L., Auerbach, R., 1982. Tumor-induced neovascularization in the mouse eye. *J. Natl. Cancer Inst.* 69:699–705.

Olsen, L., Sherratt, J.A., Maini, P.K., and Arnold, F. 1997. A mathematical model for the capillary endothelial cell-extracellular matrix interactions in wound-healing angiogenesis. *IMA J. Math. Appl. Med. Biol.* 14:261–281.

Orme, M.E., Chaplain, M.A.J. 1996. A mathematical model of vascular tumour growth and invasion. *Math. Comp. Modelling* 23:43–60.

Orme, M.E., Chaplain, M.A.J., 1997. Two-dimensional models of tumour angiogenesis and anti-angiogenesis strategies. *IMA J. Math. Appl. Med. Biol.* 14:189–205.

Paweletz, N., Knierim, M., 1989. Tumor-related angiogenesis. *Crit. Rev. Oncol. Hematol.* 9:197–242.

Palecek, S.P., Loftus, J.C., Ginsberg, M.H., Lauffenburger, D.A., and Horwitz, A.F. 1997. Integrin-ligand binding properties govern cell migration speed through cell-substratum adhesiveness. *Nature* 385:537–540.

Plank, M.J., Sleeman, B.D., 2004. Lattice and non-lattice models of tumour angiogenesis. *Bull. Math. Biol.* 66:1785–1819.

Pouyssegur, J., Dayan, F., Mazure, N.M. 2006. Hypoxia signalling in cancer and approaches to enforce tumour regression. *Nature* 441:437–443.

Pries, A.R., Secomb, T.W., Gaehtgens, P., 1996. Biophysical aspects of blood flow in the microvasculature. *Cardiovasc. Res.* 32:654–667.

Pries, A.R., Secomb, T.W., Gaehtgens, P., 1998. Structural adaptation and stability of microvascular netwoks: theory and simulation. *Am. J. Physiol. Heart Circ. Physiol.* 275:H349–H360.

Pries, A.R., Reglin, B., Secomb, T.W., 2001a. Structural adaptation of microvascular networks: functional roles of adaptive responses. *Am. J. Physiol. Heart Circ. Physiol.* 281:H1015–H1025.

Pries, A.R., Reglin, B., Secomb, T.W., 2001b. Structural adaptation of vascular networks: role of the pressure response. *Hypertension* 38:1476–1479.

Quaranta, V., Weaver, A.M., Cummings, P.T., and Anderson, A.R.A. 2005. Mathematical modelling of cancer: the future of prognosis and treatment, *Clin. Chim. Acta* 357:173–179.

Risau, W., 1997. Mechanisms of angiogenesis. *Nature* 386:671–674.

Roose, T., Chapman, S.J., Maini, P.K. 2007. Mathematical models of avascular cancer. *SIAM Reviews* 49:179–208.

Secomb, T.W. 1995. Mechanics of blood flow in the microcirculation. In *Biological Fluid Dynamics*. ed. C.P. Ellington and T.J. Pedley. Company of Biologists, Cambridge.

Stéphanou, A., McDougall, S.R., Anderson, A.R.A., and Chaplain, M.A.J., 2005. Mathematical modelling of flow in 2D and 3D vascular networks: applications to anti-angiogenic and chemotherapeutic drug strategies. *Math. Comp. Model.* 41:1137–1156.

Stéphanou, A., McDougall, S.R., Anderson, A.R.A., and Chaplain, M.A.J., 2006. Mathematical modelling of the influence of blood rheological properties upon adaptive tumour-induced angiogenesis. *Math. Comp. Model.* 44:96–123.

Sternlicht, M.D., Werb, Z., 2001. How matrix metalloproteinases regulate cell behavior. *Annu. Rev. Cell Dev. Biol.* 17:463–516.

Stokes, C.L., Lauffenburger, D.A., 1991. Analysis of the roles of microvessel endothelial cell random motility and chemotaxis in angiogenesis. *J. Theor. Biol.* 152:377–403.

Thompson, D.W., 1917. *On Growth and Form*. Cambridge University Press, Cambridge.

Walles, T., Weimer, M., Linke, K., Michaelis, J., and Mertsching, H. 2007. The potential of bioartificial tissues in oncology research and treatment. *Onkologie* 30:388–394.

Yan, L., Moses, M.A., Huang, S., and Ingber, D., 2000. Adhesion-dependent control of matrix metalloproteinase-2 activation in human capillary endothelial cells. *J. Cell Sci.* 113:3979–3987.

Ylä-Herttuala, S., Alitalo, K., 2003. Gene transfer as a tool to induce therapeutic vascular growth. *Nature Med.* 9:694–701.

Zhao, H.K. 2005. A fast sweeping method for Eikonal equations. *Math. Comp.* 74:603–627.

Zheng, X., Wise, S.M., Cristini, V. 2005. Nonlinear simulation of tumor necrosis, neo-vascularization and tissue invasion via an adaptive finite-element/level-set method. *Bull. Math. Biol.* 67:211–259.

CHAPTER 14

A Multiscale Simulation Framework for Modeling Solid Tumor Growth with an Explicit Vessel Network

Sven Hirsch, Bryn Lloyd, Dominik Szczerba, and Gabor Székely

CONTENTS

INTRODUCTION

According to the World Cancer Report, 12 million new cancer diagnoses are expected worldwide this year, and by 2010 it will be the leading cause of death. Better understanding of tumor formation is of utmost social, economic, and political importance, and finding more effective therapies may be regarded as one of the biggest challenges of our time.

Most tumors are caused by abnormalities in the genetic material of the transformed cells. Cancer progression seems to involve the accumulation of such abnormalities, leading to changes in the behavior of the tumor cells and making them more invasive (Hanahan and Weinberg 2000). Initially, the transformed cells proliferate and form a cluster, which gradually increases in size. In this early phase, tumors vivo tend to develop without a dedicated vascular network, relying on diffusion from the neighboring healthy vascularized tissue for the supply of oxygen and nutrients and for the removal of wastes (e.g., CO_2). Diffusion from the host tissue alone, however, does not suffice to support larger tumors, which often develop hypoxic and eventually necrotic regions. In response to acute hypoxia the neoplasm produces a cascade of specialized agents, collectively called *tumor angiogenesis growth factors* (TAF). These biochemicals promote the proliferation and migration of endothelial cells, thereby creating a specialized vascular network, which is able to supply the tumor. This formation of a vascular system is called *tumor-induced angiogenesis* (Carmeliet and Jain 2000), enabling extensive growth and leading to increasing pressure and traction on the tumor's micro-environment (Gordon et al. 2003). It is believed that pressure might have an impact on the transport of nutrients (and drugs), tissue composition and proliferation. In this paper we restrict our attention to solid (i.e., benign) tumors, although some concepts could also be applied to invasive (i.e., malignant) ones. A solid tumor is an abnormal mass of tissue that usually does not contain cysts or liquid areas.

The complexity and quantity of biological knowledge, combined with rapidly increasing computational power and the vision of more efficient therapy and drug design, has generated an increasing interest in mathematical and computational models of tumor growth, angiogenesis, tissue mechanics, blood flow, and other relevant fields in the past three decades. At the different spatial scales, various types of models have been developed, including continuous, discrete, deterministic, stochastic, analytical, and numerical methods. Exhaustive surveys can be found in Araujo and McElwain (2004), in Byrne et al. (2006), and in the recent paper by Bellomo et al. (2008). Additional literature on specific aspects relevant to our model was presented in our previous works (Szczerba and Székely 2002, 2005; Lloyd et al. 2008) and will not be repeated here.

One of the main limitations of many current modeling approaches is that they neglect the important interactions between the individual underlying processes, covering a broad spectrum of spatial and temporal scales. As documented in Bellomo et al. (2008), there is a long list of different models operating at the characteristic (natural) scale of the phenomenon they represent. If we are to build models that capture the interdependence between different processes involved, it is inevitable that we develop methods to integrate the corresponding scales into one consistent model. One of the first tumor models incorporating multiple scales in a systematic way (Alarcón et al. 2004, 2005) included three modules: oxygen and TAF diffusion, cell division dynamics, and blood flow. The cell cycle dynamics is represented by a cellular automaton, which shares the same domain as a preexisting hexagonal vascular network. They do not include the process of sprouting angiogenesis, instead replacing it by remodeling due to (flow-induced) shear stress and TAF, based on the work by Pries et al. (1998). More recently they extended their approach to include cell crowding (Betteridge et al. 2006). The model described by Frieboes et al. (2006) treats the cell dynamics and transport of oxygen and growth factors using convection-diffusion-reaction equations (Macklin and Lowengrub 2007). Cell-to-cell adhesive forces are modeled by surface tension, treating the tissue as an incompressible fluid. Their recent work (Sanga et al. 2007) includes angiogenesis using a model based on the approach by Plank and Sleeman (2004). The paper by Kim et al. (2007) presents a hybrid model of avascular tumor growth, which treats the actively proliferating rim using a discrete approach, while employing a continuum viscoelastic material response for the remaining regions. This reduces the computational burden compared to a purely discrete approach.

We have been working on a framework that allows us to consider multiple phenomena in a coupled way by interfacing components both horizontally (component-wise) and vertically (scale-wise). A first attempt for computational implementation has been presented in Lloyd et al. (2007). A more detailed description, with a special focus on multiscale aspects of nutrient transport was given in Lloyd et al. (2008). Specifically, we explicitly model the sprouting tips of the developing vasculature and the convective transport of oxygen in the circulatory system. Transport and blood-tissue exchange phenomena are an essential model component also with respect to drug delivery (Jang et al. 2003). Furthermore, we have tested alternative biomechanical models in order to investigate their impact on the morphogenesis of a specific type of solid tumor, a so-called uterine (leio-) myoma (Szczerba et al. 2009). In order to account for tissue necrosis and heterogeneities we have recently extended our model to a compartmentalized approach, which makes it possible to simulate the effect of therapy strategies (Hirsch et al. 2009). Since the focus of these last two works has been on tissue mechanical and heterogeneity aspects rather than angiogenesis, we have employed a simplified implicit representation of vascular growth, similar to the work of Anderson and Chaplain (1998).

The framework is modular and allows any finite element- based module (i.e., mesh) to be added horizontally by giving it access to the global underlying information such as the domain representation (geometry, topology, and constraints) and expecting its contribution to a global coefficient matrix, or performing a requested number of iterations in case no global assembly is required or the problem is nonlinear. The dialogue between the different components does not need to be restricted to the same resolution. For example, the oxygen source in a tissue-level reaction-diffusion model can be estimated from a cellular-scale blood flow simulation. The interaction is realized via constitutive laws (e.g., effective responses to loads, tractions, or constraints modulated by other components of the model).

A comprehensive package will allow simulating the underlying physical and biochemical processes, covering mutual relationships between mechanical and biochemical stimuli, transport flow, diffusive transport, active cell migration or possibly even gene expression, resulting in tissue formation, deformation, or removal.

TUMOR SIMULATION BASED ON AN EXPLICIT VESSEL MODEL

Our model combines several previously identified components in a consistent framework, including neoplastic tissue growth, blood and oxygen transport, and angiogenic sprouting. We use a hybrid approach combining continuum scale reaction-diffusion equations and microscale transport for addressing the oxygen delivery problem. Sprouting angiogenesis is treated by explicitly modeling sprouting vessels' tips, proliferating and migrating up the gradient of angiogenic growth factors, which are produced in hypoxic regions of the tumor. Finally, a continuum model of mechanical tissue response and growth is added to the multiphysics, multiscale simulation.

The major difference in time scales at which the diffusion of molecules takes place (ns-s) and at which the cells proliferate (days) allows us to treat the processes in a quasistatic manner, that is, at a given time step we compute a concentration map of oxygen according to the sources, and consumption by the tissue based on the actual tumor and vascular geometry. In the following section we give a more detailed description of the individual model components.

Tissue Growth

The growth model was presented in Lloyd et al. (2007) and will only briefly be described here in order to keep this chapter self-contained. Solid tumor development is assumed to be relatively slow, therefore, it is reasonable to surmise that over time the neighboring tissue will accommodate for the stretching by increased growth or cell migration. The volumetric growth due to cell proliferation is prescribed by an initial strain ε_0 (Zienkiewicz and Taylor 2000). We compute the new deformation caused by a volume expansion using the finite element method. In our initial implementation we neglected the accumulation of stress during growth, instead assuming that the residual stress caused by inhomogeneous volume increase dissipates completely, constituting an elasto-plastic growth law. From biological research it is known that cells need certain conditions in order to undergo mitosis. One mechanism, which regulates proliferation through changes in cell cycle dynamics, is the dependency on oxygen and glucose (Jiang et al. 2005). At low concentrations, oxygen consumption and cell proliferation are reduced according to Freyer and Sutherland (1985), who explain this observation by the increased number of cells in a quiescent state. Under acute hypoxia, cells will eventually die, leading to necrotic regions in the tumor. Based on these observations we have defined a functional relationship between the prescribed initial strain (i.e., tissue growth) and the availability of oxygen:

$$\varepsilon_0(t) = f(c_{O_2}(t)), \tag{14.1}$$

where f is selected as a piece-wise linear function. For high oxygen concentrations ($>h_1$), the cells reach the maximum proliferative potential, which decreases at lower levels. In necrotic regions with oxygen concentrations below a threshold h_2, we allow for removal of tissue debris by prescribing a negative strain leading to volume loss.

Vascular System and Oxygen Transport

Oxygen is transported by the blood and diffuses through the vessel walls and into the tissue. For computational reasons, and since we are mainly interested in the development of the tumor and not the healthy tissue, we choose to use a dual representation of the oxygen source. Inside the tumor, an explicitly modeled vasculature supplies the tissue with oxygen, while in the healthy tissue we postulate a homogeneous source of oxygen from an implicit preexisting regular vascular system. In the implicit representation, oxygen diffuses into the tumor from the boundary. At a macroscopic scale this process can be modeled by a reaction-diffusion equation

$$\frac{\partial c_{O_2}}{\partial t} = D_{O_2} \nabla^2 c_{O_2} + R_{O_2}^-(c_{O_2}) + R_{O_2}^+(c_{O_2}), \qquad (14.2)$$

where c_{O_2} is the oxygen concentration and D_{O_2} the diffusion constant. The oxygen consumption rate $R_{O_2}^-$ depends on the tissue type and can vary, depending on the metabolic activity. The source term $R_{O_2}^+$ is governed by the blood flow inside the vascular system and the local tissue oxygen content, which influences how much oxygen diffuses through the vessel wall. Under the assumption that the oxygen levels in the healthy tissue are not affected by the presence of the tumor, we can model this as a reaction-diffusion problem (14.2) with no flux boundary conditions at a distant surface $\delta\Omega_1$ (Figure 14.1).

In order to estimate the oxygen source term $R_{O_2}^+$, we first need to solve the blood flow problem. The vessel network generated by our angiogenesis simulation consists of straight vessel segments (pipes) connected in a network. Following our previous work on tumor-induced angiogenesis we compute the pressure distribution and flow in the network (Szczerba and Székely 2002, 2005). For laminar, steady flow in stiff and straight tubes, the flow is computed according to Poiseuille's law

$$Q_{ij} = G_{ij}(P_i - P_j) = \frac{\pi}{8} \frac{R_{ij}^4}{\mu_{ij} L_{ij}} (P_i - P_j), \qquad (14.3)$$

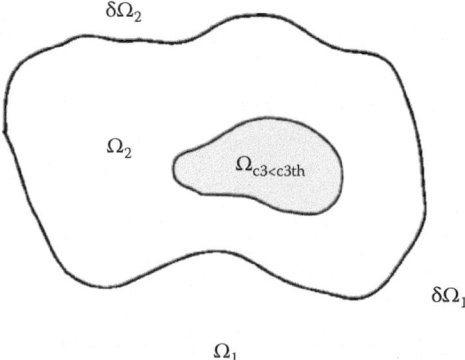

FIGURE 14.1 Compartments of the model. The tumor ($\Omega2$) consisting of necrotic and viable parts, is embedded into the host tissue ($\Omega1$). δ denotes the respective interfaces.

where Q_{ij}, R_{ij}, L_{ij}, and μ_{ij} are the flow, the radius, the length, and the viscosity between nodes i and j. The variables P_i, P_j are the pressure values at the nodes, and G_{ij} is the conductance. The flow fulfills mass balance at the junctions between pipe segments (i.e., $\sum_j Q_{ij} = 0$ for all nodes i). Mass balance and Poiseuille's law lead to a system of linear equations, with pressure as an unknown quantity. For large vessels the blood viscosity μ_{ij} is approximately constant, while for small diameters the Fahraeus–Lindqvist effect causes non-Newtonian behavior and has to be taken into account. Therefore, in order to correct for small vessels, we use an apparent viscosity, empirically estimated under in vivo conditions by Pries et al. (1996).

Oxygen Supply by the Vascular System

As blood flows through the vessels, oxygen diffuses through the vessel walls into the tissue. Using a mass balance principle for the blood oxygen content and Fick's first law, we compute the amount of oxygen, delivered to the tissue

$$Qc_{O_2}(x) - Qc_{O_2}(x+dx) = dS \cdot D_{O_2}^w \cdot \nabla^2 c_{O_2}, \tag{14.4}$$

where Q is the blood flow, c_{O_2} is the oxygen concentration, dS the surface of the vessel between x and $x+dx$, and w is the wall thickness. $D_{O_2}^w$ is the diffusion constant of oxygen in the vessel wall; the position x is upstream of $x+dx$ (Figure 14.2). The oxygen concentration c_{O_2} in the tissue is proportional to the oxygen partial pressure P_{O_2} according to Henri's Law

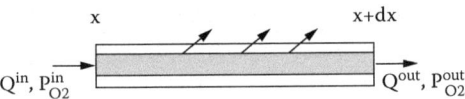

FIGURE 14.2 Schematic illustration of the convective transport of oxygen in a pipe segment.

$$P_{O_2} = k_{O_2}^{tissue} c_{O_2} = \frac{1}{\alpha_{O_2}^{tissue}} c_{O_2} .$$ (14.5)

This linear relationship is only valid for dilute solutions as in the case of oxygen dissolved in the tissue interstitial fluid. The parameters $k_{O_2}^{tissue}$, $\alpha_{O_2}^{tissue}$. are called *Henri's constant* and *solubility*, respectively. Assuming that Henri's Law is valid inside the vessel wall, the gradient term in Equation 14. can be replaced and approximated by a difference quotient

$$Qc_{O_2}(x) - Qc_{O_2}(x+dx) = dS . D_{o_2}^w . \alpha_{O_2}^{tissue} (P_{O_2}^{blood} - P_{O_2}^{tissue}) \frac{1}{w},$$ (14.6)

where $P_{O_2}^{tissue}$ is the pressure on the interface between the vessel and the tissue. The partial pressure on the surface of the vessel lumen is given by $P_{O_2}^{blood}$. Basically, oxygen is removed downstream, eventually leading to a small difference in the oxygen partial pressure between the tissue and the blood (i.e., at low blood P_{O_2} very little oxygen will be delivered to the tissue.) Additionally, we can see that the blood gradient along the vessel will be smaller if the flow is higher.

Because only a small proportion of blood oxygen content is dissolved and the major part is bound to hemoglobin, the sigmoidal shape of the P_{O_2}–c_{O_2} curve is dominated by the oxygen–hemoglobin dissociation relationship. In general, the oxygen content in the blood can be given by

$$c_{O_2} = \alpha^b P_{O_2} + \beta S_{O_2}(P_{O_2}),$$ (14.7)

where α^b is the solubility of oxygen in blood and S_{O_2} is the oxygen saturation of the hemoglobin (i.e., the percentage of hem sites that are combined with oxygen). We use Hill's equation for the saturation function S_{O_2} (Ji et al. 2006). Using the same nomenclature as for the blood flow calculation, we can rewrite Equation 14.6 as

$$Q_{ij}c_i - Q_{ij}c_j = 2R_{ij} \cdot L_{ij} \cdot D_{O_2}^w \cdot \alpha_{O_2}^w (P_{O_2}^{blood} - P_{O_2}^{tissue}) \frac{1}{w} \qquad (14.8)$$

Because, according to Equation 14.7 the blood partial pressure depends on the oxygen concentration in a nonlinear fashion, the above relation is a nonlinear function in c_i and c_j. If c_i is known, the unknown downstream value c_j can be calculated using a Newton–Raphson scheme.

Finally, if the oxygen concentration in the arterial inflow pipes is known, we can compute the c_{O_2} distribution in the entire vascular system by propagating through the network in the direction of the blood flow. By integrating over the surface of the vessels, we estimate the local amount of oxygen delivered by the explicit vessel network. This result is used in the macroscopic-level reaction-diffusion in Equation 14.2, which we solve using a finite element method (Huebner et al. 2001). Note, that the actual amount of oxygen flux across the vessel walls depends on the oxygen concentrations in the tissue via $P_{O_2}^{tissue}$. Therefore, the macroscopic and microscopic models are coupled and must be solved iteratively.

Sprouting Angiogenesis

The formation of blood vessels by sprouting angiogenesis is a process where capillary sprouts depart from preexisting parent vessels in response to externally supplied chemical stimuli. By means of endothelial cell proliferation, migration, and remodeling, the sprouts then organize themselves into a branched, connected network structure. The preliminary capillary plexus does not necessarily require blood flow to form.

It is widely agreed on that a major cause of elevated angiogenesis growth factors (TAF) is hypoxia. Hypoxia upregulates many genes, but the induction of vascular endothelial growth factors (VEGF) is perhaps the most remarkable—up to 30-fold within minutes (Carmeliet 2003). As the growth factors diffuse into the surrounding tissue, there is some uptake and binding by the cells (Ausprunk and Folkman 1977). Therefore, the process can be modeled by a reaction-diffusion equation with a natural decay term

$$\frac{\partial c_{taf}}{\partial t} = S_{taf} + D_{taf}\nabla^2 c_{taf} - \Theta_0 c_{taf}, \qquad (14.9)$$

where c_{taf} is the TAF concentration, D_{taf} the diffusion coefficient, and Θ_0 the decay rate. The source term depends on the local tissue oxygen concentration. If the oxygen tension is below a threshold, growth factors are secreted by the affected cells. We can model this by a function $S_{taf}(c_{O_2})$, e.g., $S_{taf}(c_{O_2}) = S_0 \cdot H(c_{O_2} - h_2)$, where $H(\cdot)$ is the Heaviside function, S_0 is a constant production rate, and h_2 is the oxygen concentration threshold, below which the tissue becomes hypoxic.

Once a sufficient concentration of growth agents has been established, endothelial cells respond to the stimulus by sprouting. In our previous work, we have developed a model of sprouting angiogenesis, which generates functional flow networks, consisting of interconnected straight pipes (Szczerba and Székely 2002, 2005). An example of vascular flow network sprouting towards higher TAF concentrations is depicted in Figure 14.3. Similarly to our previous method we model the dynamics of the sprouting vessel tips using a Lagrangian approach. In Lloyd et al. (2008) we couple the vascularization to the tissue growth via the time-dependent changes in growth factor concentration, which typically decreases with increasing vascularization, since the initially hypoxic region receives a new source of oxygen. Additionally, we allow for the capability of the vascular system to adapt to current hemodynamic

FIGURE 14.3 **(See color insert following page 40)** Example of angiogenic sprouting toward the higher concentration of TAF (shown using volume rendering). The color-coding corresponds to the vessel radius.

conditions by remodeling the vessels diameters in order to decrease the flow resistance.

Tissue Heterogeneity

Tumorous tissue and in fact tissue in general is a complex compound of various cell types, embedded in an extracellular environment. In reality there is a plenitude of different cell types and developmental states for both tumor and healthy tissue, such as proliferating, quiescent, necrotic, transient cells and even these may be characterized with subtypes. This diversity has to be taken into account for developing a more realistic picture of the tumor development.

As the explicit vessel representation described in the previous section titled "Vascular System and Oxygen Transport" is computationally demanding, we have developed an alternative implicit (density) description, which includes the motility of endothelial cells and the resulting oxygen transport and delivery. A potential advantage of this approach is that only a reduced set of (lumped) parameters is required (Szczerba et al. 2009; Hirsch et al. 2009). Here we regard the vessel system as a density distribution of endothelial cells (EC). A relation is established estimating the diffusion of oxygen through the vessels for a given vessel diameter. Hence, the density of EC is directly linked to an oxygen supply. The model consists of mass and force balance equations; the mass transport of all constituents is modeled with reaction-convection-diffusion equations. Concentrations of growth factor c_{taf}, endothelial cells c_{ec}, and oxygen c_{O_2} are transported through the tissue, and may enter chemical reactions anywhere in the whole domain.

Healthy tissue is in a dynamic balance between cell proliferation and death. The process of controlled cell death—apoptosis—is an integral part of the constant renewal of tissue in the natural cell cycle. The control mechanism is part of the homeostasis required by living organisms to maintain their internal states within certain limits. After apoptosis, the cell may be phagocytosed completely without exposing the tissue to potentially harmful intracellular debris. In contrast, necrosis is an uncontrolled death, for example, due to hypoxia or toxic agents, and is characterized by an uncontrolled bursting of the cell membrane with only partial resorption of the debris. As the apoptotic cells are removed, the overall density of cells inside the tumor N can be decomposed into two compartments, viable N^+ and necrotic N^- cells:

$$N(t,x,y) = N^+(t,x,y) + N^-(t,x,y). \qquad (14.10)$$

In our model, proliferation is constrained by the availability of oxygen (Graziano and Preziosi 2007; Anderson and Chaplain 1998) and the spatial conditions (Chaplain et al. 2006). We account for tumor cell proliferation, apoptosis, and necrosis. Necrotic cells originate from hypoxic tumor cells. Tissue may be assumed to be in homeostasis under normal conditions and thus the healthy tissue is assumed to have a zero net growth. g^+, g^{--}, and g^- are the individual growth rates for each mechanism:

$$growth: g^+ = h^+_{c_{O_2}}(c_{O_2})h^+_\sigma(\sigma)\frac{\ln 2}{T^+_2}, \tag{14.11}$$

$$necrosis: g^- = h^-_{c_{O_2}}(c_{O_2})\frac{\ln 2}{T^-_2}, \tag{14.12}$$

$$apoptosis: g^{--} = \frac{\ln 2}{T^{--}_2}, \tag{14.13}$$

where the variables T_2 are the doubling times of the respective processes. This formulation seamlessly introduces a dependency to mechanical stress and oxygen concentrations via the factors h^+_σ, $h^+_{c_{O_2}}$, and $h^-_{c_{O_2}}$. However, on a cellular level it is still very difficult to quantify the response to stress, hence we neglect this particular effect ($h^+_\sigma = 1$). The populations N^+ and N^- are governed by

$$\frac{\partial N^+(t,x,y)}{\partial t} = N^+(t,x,y)\cdot\left[g^+ - g^- - g^{--}\right], \tag{14.14}$$

$$\frac{\partial N^-(t,x,y)}{\partial t} = N^+(t,x,y)\cdot g^-. \tag{14.15}$$

Tumor progression and regression is modeled as initial strain condition, governed by the relation in the cell populations. For the time-discrete formulation the volumetric strain at timestep i is:

$$\varepsilon_0(t) = \frac{\partial V(t)}{\partial V} \approx \frac{(N^+_i + N^-_i) - (N^+_{i-1} + N^-_{i-1})}{(N^+_{i-1} + N^-_{i-1})}, \tag{14.16}$$

and can also be negative, meaning cell degradation and removal.

Tissue Mechanics

Tissue is a highly complex material which, depending on the time scale and applied loads, can be hyper-, viscoelastic, or plastic (Gladilin et al. 2003). It is certainly a major simplification to treat the tissue response to volumetric growth with a linear constitutive law. When, for example, studying the formation of myomas, it is apparent that a viscoelastic model is necessary to describe the mechanical behavior (Szczerba et al. 2009). Uterine leiomyomas (fibroids, a typical exponent of a benign tumor) are the most common uterine neoplasm. In general, a myoma grows slowly but continuously, an increase of volume by a factor of two usually takes several months or years. A myoma has a much stronger tendency to keep its shape than any of the tissues surrounding it, as it is composed of very dense fibrotic tissue. There is no real capsule around a that which is only surrounded by a clustered myometrium. The endometrium is a reactive tissue carrying viscoelastic characteristics covering the entire uterine cavity including protruding myomas of any degree.

We base our modeling on experimental observations of mechanical tissue responses to chemical growth factors (see, e.g., Gordon et al. [2003]). To quantify this stress one needs to model the tissue response to strain (Cristini et al. 2003). To describe the tissue behavior under some strain $\varepsilon(t)$ we take the standard viscoelastic model after Humphrey and DeLange (2004):

$$\frac{d}{dt}\sigma(t) + \frac{E\sigma(t)}{\mu} = \frac{EE_0\varepsilon}{\mu} + (E + E_0)\frac{d}{dt}\varepsilon(t). \tag{14.17}$$

Solving symbolically for $\sigma(t)$ leads to an expression for stress relaxation:

$$\sigma(t) = \varepsilon_0(E_0 + Ee^{-Et/\mu}) \tag{14.18}$$

with $\varepsilon_0(E_0 + E)$ being the linear contribution to total stress and μ the relaxation time. Under the assumption of tissue incompressibility and approximate spherical symmetry, the thickness of the tissue rim surrounding the myoma from the cavity side has to get thinner with the increase of the pathology radius in order to compensate for the volume change. Perfectly elastic wall (hoop) stress in a thin-walled sphere is given as:

$$\sigma(r) = \frac{1}{2}\frac{pr}{-r + \sqrt{r^2 + 2w_0 r_0 + w_0^2}} \tag{14.19}$$

with r_0 and w_0 being the rest-state radius of the sphere and the thickness of its wall, respectively.

Even if the strain-stress relation is linear the wall stress is not, because of the variable wall thickness. Solution for a general stress (including viscous effects) will be of a similar form, multiplied by a relaxation factor. As a consequence, we are able to simulate a spectrum of tissue behaviors: from perfectly elastic ($\mu \to 0$) to perfectly plastic ($\mu \to \infty$), regulating the relaxation contribution with μ. Of course these model parameters will be different for the tumor and the surrounding tissue.

Drug Therapy Prediction

It is one of the strengths of this fully integrated model that many agents may be incorporated easily, given that they may be expressed within the reaction-diffusion-convection scheme. We present a first approach for virtual therapies, which lies within our general direction to develop usable tumor models for clinical research.

Angiostatin (AST) is a potent inhibitor of angiogenesis, and by this mechanism it influences tumor cell proliferation. For simulating the effect of an externally introduced anti-angiogenetic drug, the influence of diffusion and a natural decay on its concentration c_{ast} has to be taken into account:

$$\frac{\partial c_{ast}}{\partial t} = D_{ast} \nabla^2 c_{ast} - R_{ast} c_{taf} \qquad (14.20)$$

with *no flux* boundary condition on $\delta\Omega_1$. As the actual mechanism of AST is not fully understood, we anticipate that this agent directly neutralizes the growth factor. Hence, Equation (14.9) is extended by a reaction term depending on AST concentration

$$\frac{\partial c_{taf}}{\partial t} = S_{taf} + D_{taf} \nabla^2 c_{taf} - \Theta_0 c_{taf} - R_{taf,ast} c_{ast}, \qquad (14.21)$$

The realism of the simulation is obviously very sensitive to the selection of the governing parameters and their determination is a challenging part of tumor simulation. Further information on the parameter choice may be found in an earlier publication (Hirsch et al. 2009).

RESULTS

In the preceding section we have presented a modeling framework, which can incorporate multiple physical, biochemical, and biological aspects of tumor growth and angiogenesis. The basic model relies on explicit modeling of sprouting angiogenesis and has been implemented in three dimensions. Variations of this model, relying on the same or similar equations employ an implicit representation of the growing vessels, that is, it cannot account for blood flow and the related convective transport of nutrients. Nevertheless, these extensions allow the computationally efficient investigation of (a) the compartmentalization of tissue in separate states and the transition between these states (proliferation, necrosis, apoptosis); (b) the influence of mechanical tissue response on tumor growth; and (c) the therapeutic effects of antiangiogenic agents in cancer therapy. In this section we first present results corresponding to the basic model. In the following, simulation results for the mentioned studies will be shown.

Basic Coupled Tumor Model

To initialize the pathology growth inside the hosting tissue, we place a small avascular ball inside a larger domain representing the healthy tissue. The initial tumor cluster does not contain any necrotic tissue and is small enough to receive enough oxygen for rapid growth. At a certain distance from the tumor, we define several preexisting vessels, which will be the origin of the new sprout formation. The growth process is iteratively solved and updated, solving quasi-static problems using the most recent solution the from the other simulation components in each time step as described in the section titled "Tumor Simulation Based on an Explicit Vessel Model."

Figures 14.4a and 14.4b presents the simulation results at two different stages of tumor development. The figure depicts a volume rendering of the angiogenesis growth factor concentration and the developing vascular system. The first stage represents the avascular phase, with diffusion from the neighboring healthy tissue as the main source of oxygen. In Figure 14.4b the vasculature has penetrated hypoxic areas and is already dense enough to support a dramatic increase in growth rate.

The formation of a vascular shell around the tumor characterized by the rather chaotic geometric arrangement of the individual vessel segments can be well explained by the diminishing gradient of TAF concentration resulting in decreasing influence of the directed motility during tip migration. The resulting geometry resembles qualitative experimental observations made on casts of tumor neovascularization (Walocha et al. 2003). The spatial distributions of some governing factors and characteristic system descriptors are depicted in Figure 14.4, corresponding to the same time step as in (b). The oxygen concentration is plotted in (c); in (d) the local growth expansion in percent and in (e) the TAF concentration is shown.

(a)

(b)

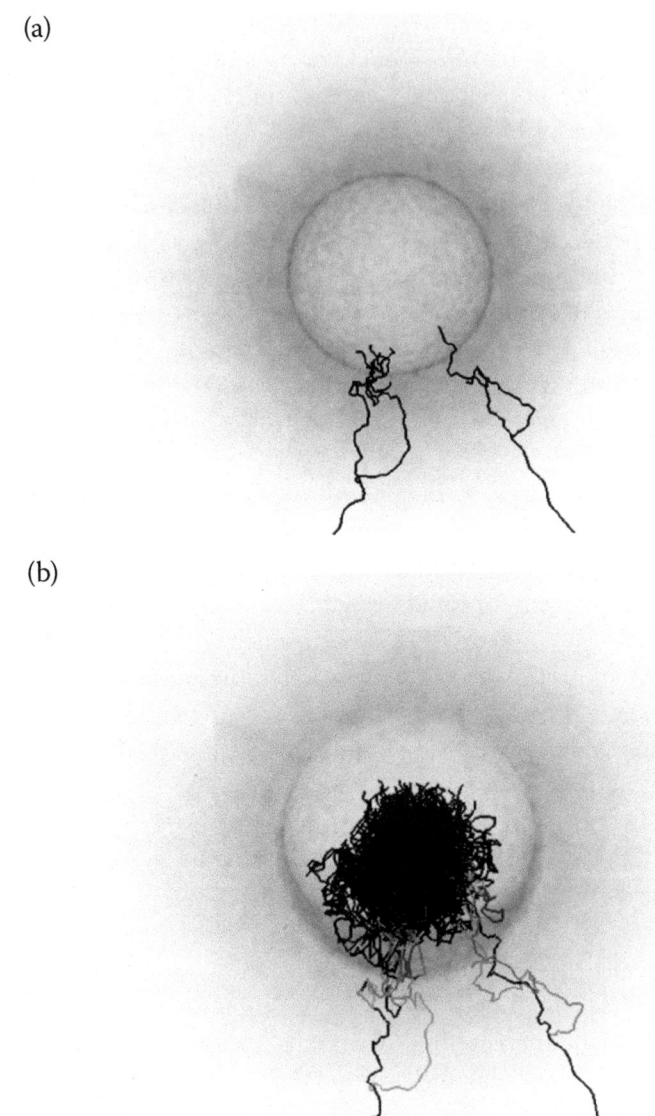

FIGURE 14.4 (a) and (b) show a volume rendering of TAF concentration distribution, together with the developing vascular system at two different time steps.

(c)

(d)

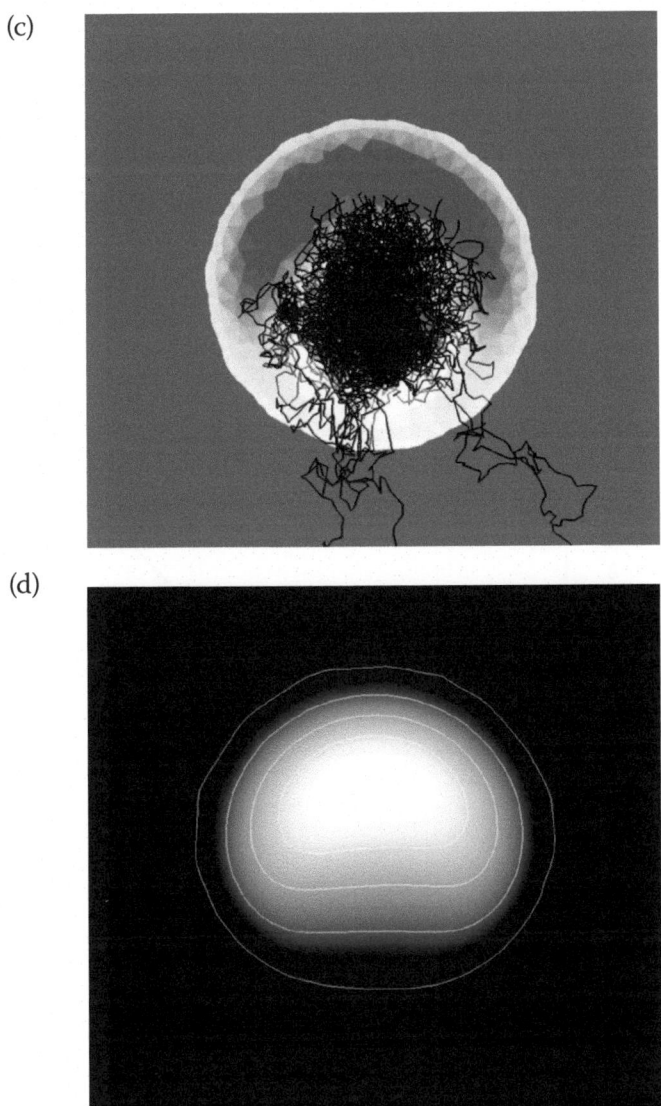

FIGURE 14.4 (Continued) In (c) the growth rate and in (d) the TAF distribution is depicted at the second development stage for a slice cutting through the tumor.

In Figure 14.5 we have computed various entities, which quantitatively describe the development process and allow the identification of characteristic events during the process. Initially, the growth seems to be more or less linear until the angiogenic switch is turned on. The last third of the process shown exhibits a nearly exponential volumetric growth. This correlates well with the onset of the availability of additional oxygen supply driven by the flow through the newly formed vessels, as shown in Figure 14.4b. In (d) we see that the volume of the hypoxic region of the

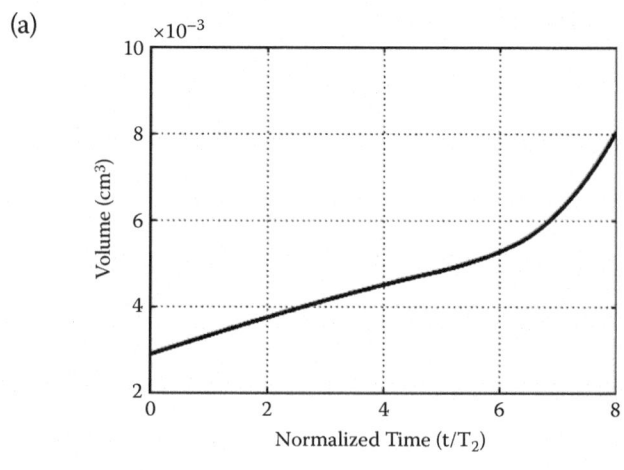

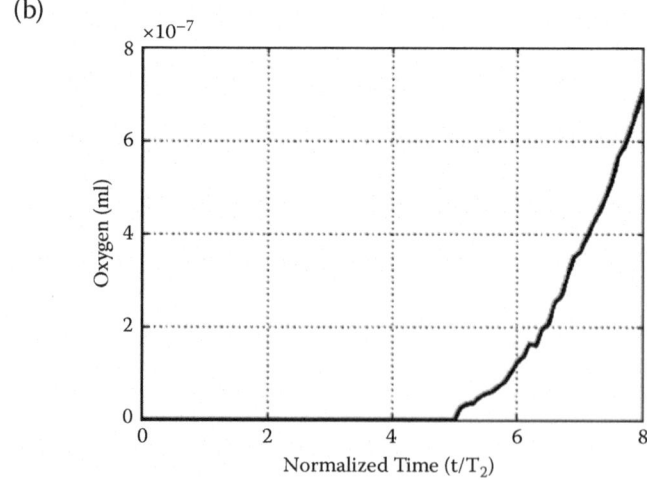

FIGURE 14.5 The temporal evolution of some quantities characterizing the progress of growth and vascularization during simulation are shown in (a)–(d).

(c)

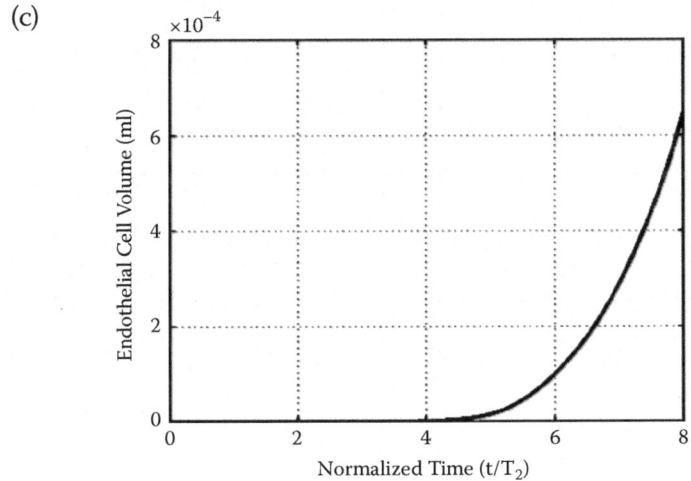

(d)

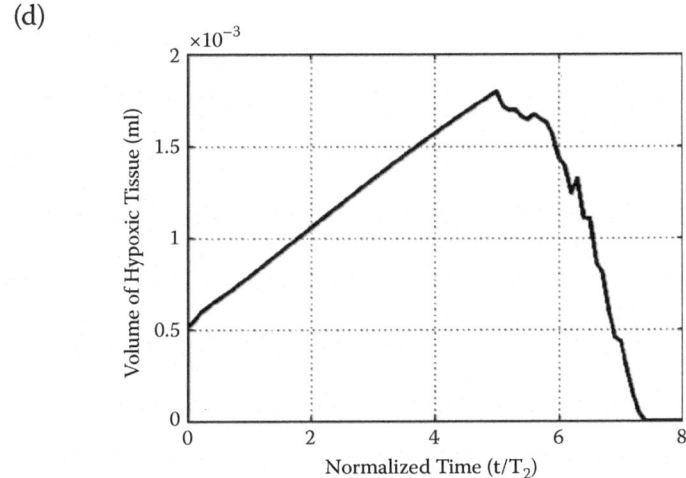

FIGURE 14.5 (Continued) The temporal evolution of some quantities character-izing the progress of growth and vascularization during simulation are shown in (a)–(d).

tumor increases at the beginning, but as the new oxygen source becomes available it gradually decreases again until it reaches zero.

Tissue Heterogeneity

Again, the simulations are initiated with a small cluster of tumor cells in the center of the host tissue surrounded by vessels on each side. The unregulated tumor growth is presented in Figure 14.6 as an oxygenation

(a)

(b)

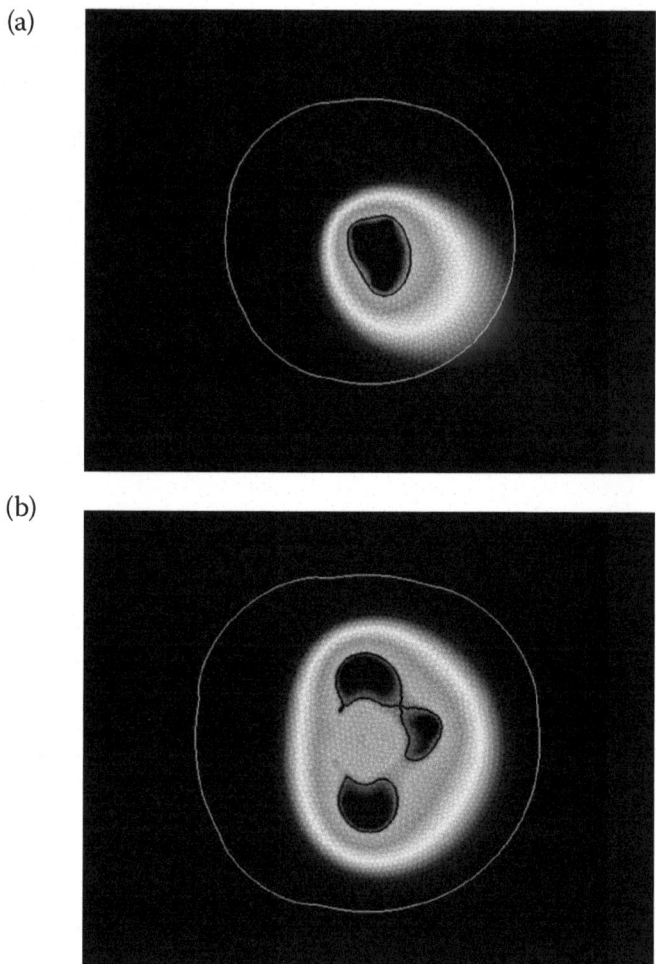

FIGURE 14.6   Two examples of compartment formation within the tumor (white outline with mesh) of necrotic and viable tissue for different necrosis thresholds. The oxygenation map is grey coded, the inner areas denote oxygen iso-contours corresponding to the threshold level.

map and in Figure 14.8a as the EC density. The corresponding high EC density is visible in Figure 14.8 as a decaying gray shadow gradient at the box walls. The tumor promotes directed vessel growth via the secretion of tumor angiogenic factors (TAF), leading to EC migration from the adjacent parent vessels. In this vascular phase the tumor expands virtually unbounded and will eventually cause physiological problems due to compression of the surrounding tissue. The elevated EC density leads to the increase of the corresponding vessels' diameter and wall thickness in time, which in turn modulates their oxygen delivery rate. For large vessels, blood flow increases and diffusion through the wall decreases. Nonsymmetrical compartments develop despite the initial boundary symmetry. EC density is realized with a typical capillary buildup in the center of the host tissue in the form of a frequently observed vascular capsule. Four branches of vessels are clearly visible, connecting the tumor to the feeding arteries on the periphery. The tumor itself is penetrated by a dense network of capillaries, corresponding to experimental findings.

Effects of Mechanical Tissue Response

With an extended mechanical model we can investigate, how the morphology of a growing myoma is influenced by the local mechanical conditions. A development of a virtual tumor is shown in Figure 14.7 coded with stress, using identical grayscale maps to facilitate magnitude comparisons. The neoplasm started growing as a small oval pathology in the myometrium and continues to grow into the surrounding tissue that is largely dissipating the generated strain energy. As shown, the tissue gradually relaxes after being exposed to the critical stress of around 1 MPa as in Figure 14.7e (initial stress value beyond the scale bar), allowing the pathology to continue expanding and therefore only weakly constraining its progression. The formation of an experimentally observed myoma protruding the uterine cavity is inevitable. Should the behavior of the surrounding tissue be dominated by elasticity (i.e., the expansion induced stress in the surrounding thin tissue slab would be mostly preserved), the stretched layer of the myometrium would very strongly hinder the growth. This is because the reaction force would be equal to—or even greater than—the accumulated stress, comparable in magnitude to the elastic modulus of the pathology, eventually constraining further growth in a stage, where no acute angle is yet formed between the myoma and the uterine wall (as in Figure 14.7e).

Prediction of Therapy Outcome

Clinical studies clearly demonstrated that angiostatin (AST) is a potent inhibitor of angiogenesis. Several pathways for the AST influence on the tumor are proposed, yet the underlying mechanisms are not fully understood (Sim et al. 2000; Folkman 2002). We have been investigating the effect of its application based on the model described in the section titled

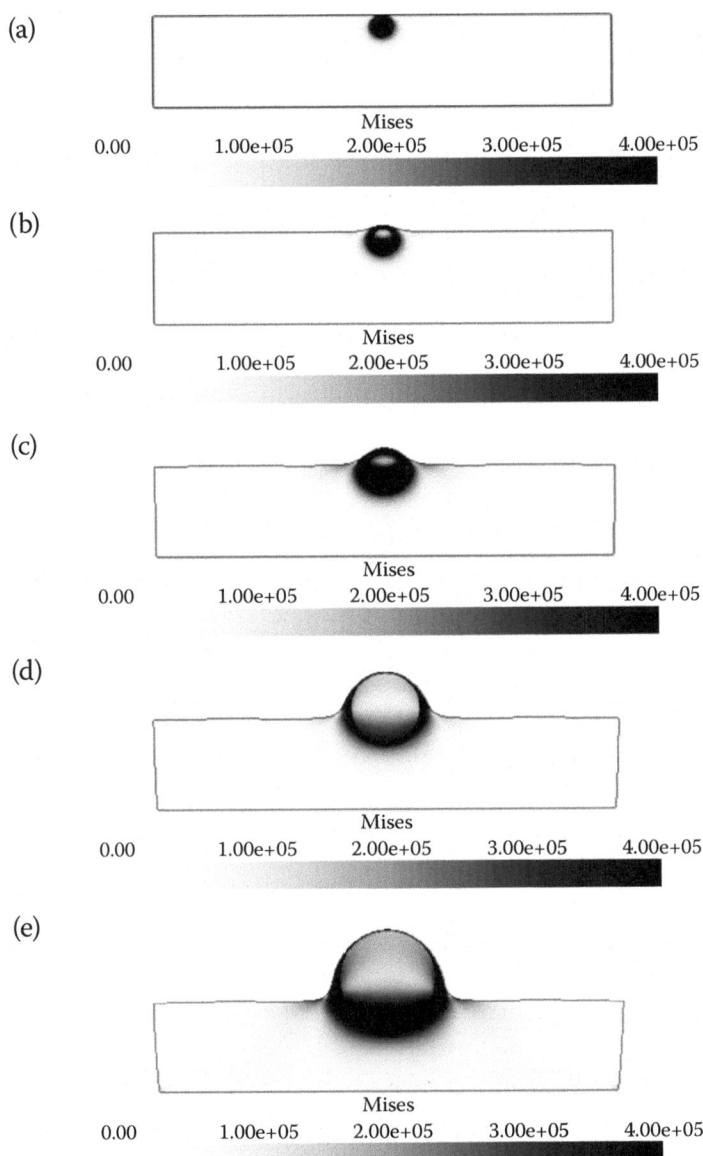

FIGURE 14.7 Simulation results for a myoma (type I) developing at the border of the myometrium. Due to stress dissipation there is no significant resistance from the tissue layer on the cavity side (upwards on the image). As a result the pathology envelope forms an acute angle with the basement layer.

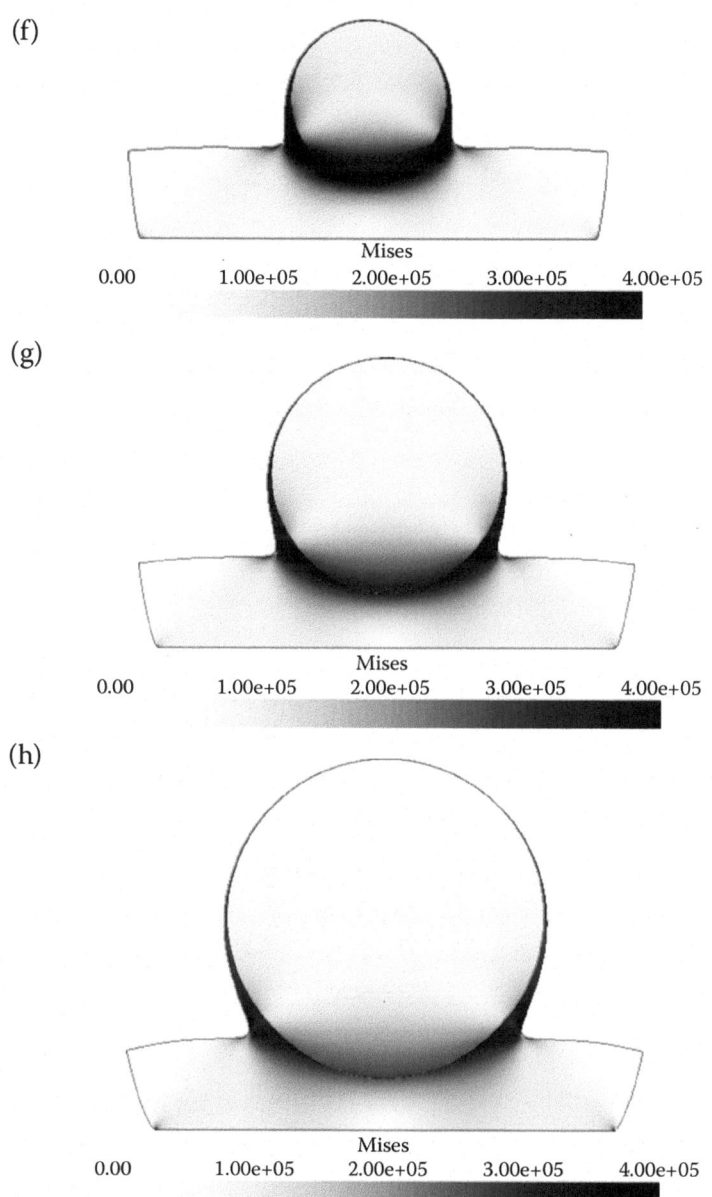

FIGURE 14.7 (Continued) Simulation results for a myoma (type I) developing at the border of the myometrium. Due to stress dissipation there is no significant resistance from the tissue layer on the cavity side (upwards on the image). As a result the pathology envelope forms an acute angle with the basement layer.

(i)

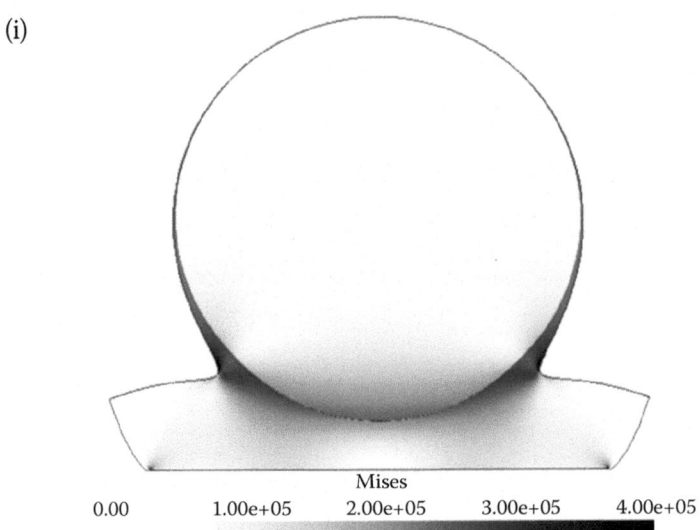

FIGURE 14.7 (Continued) Simulation results for a myoma (type I) developing at the border of the myometrium. Due to stress dissipation there is no significant resistance from the tissue layer on the cavity side (upwards on the image). As a result the pathology envelope forms an acute angle with the basement layer.

"Drug Therapy Prediction," accounting for the direct neutralization of the growth factor as described by Equation 14.21.

In our numerical study AST has been applied off-center in the lower left quadrant of a quadratic tissue domain (Figure 14.8a) and not inside the tumor, in order to prevent the vessels from reaching the tumor. In Figure 14.8b we notice a strongly asymmetric EC density distribution, where the AST supplied area is excluded from any vessel growth. The local concentration of AST does not actually prevent the vessels from connecting to the tumor. Instead, the EC density indicates clearly defined feeding vessels around the AST supplied area. The EC density accumulates outside the AST application region and is able to supply a dense capillary capsule. Similar observations have been made when studying the effects of embolization, which are reported in detail in (Hirsch et al. 2009).

DISCUSSION

We have presented an integrated framework to model solid tumor development, handling some of the major processes, that is, the mechanical deformation due to growth, the biochemical response to hypoxia, blood

(a)

(b)

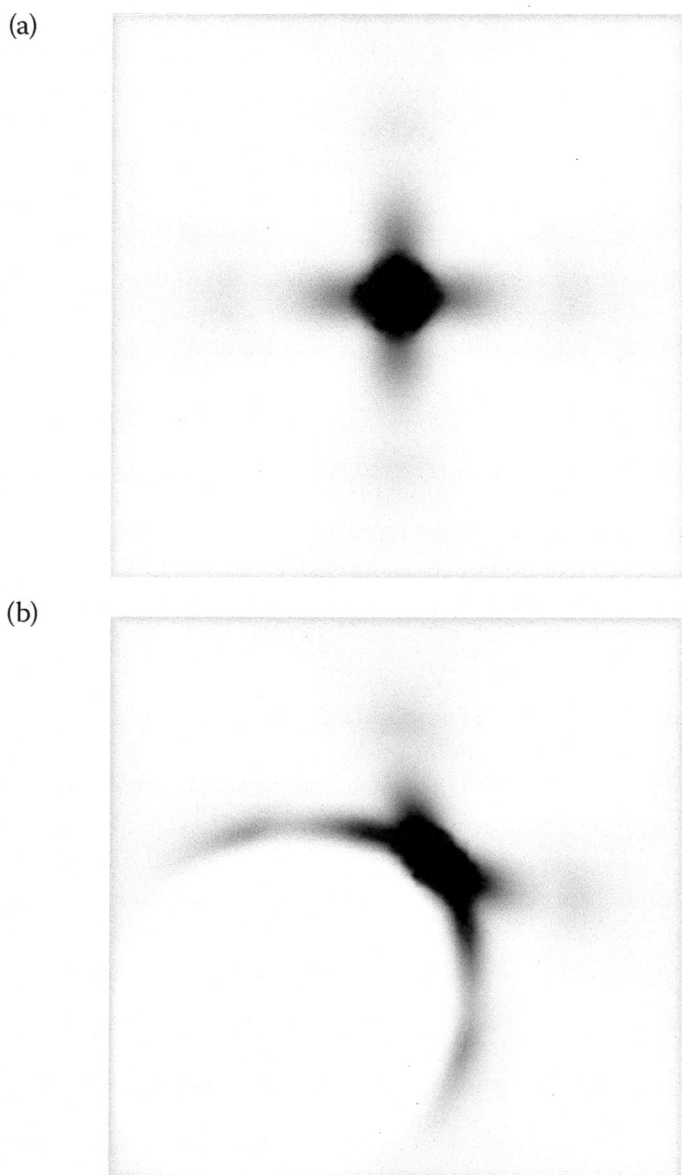

FIGURE 14.8 Endothelial cell density within the host tissue resulting from unregulated growth (a) and after the application of angiostation in the lower left quadrant of the domain (b).

flow, oxygenation, and the explicit development of a vascular system in a coupled way. The results demonstrate the feasibility of the approach. Unlike previous methods to simulate angiogenesis, we do not treat the vascular sprouting remodeling as a process, which takes place within a static domain, or as a static distribution of growth factors. Instead, our angiogenesis and remodeling approach is coupled to a dynamically growing domain, with evolving TAF concentrations. Whereas in a fixed domain vascular growth and remodeling eventually reach a static equilibrium, this is not the case for vascularized tumors. With our modeling framework we are now able to perform *in silico* studies to identify factors, which could possibly stabilize this process.

We showed possible extensions to our approach dealing with the compartmentalization or different material properties, and we illustrated how our framework can be used to evaluate therapeutic effects. These experiments were conducted using a simplified implicit vessel representation. According to our experience this approach is still feasible for investigating several problems, while being computationally much more effective than relying on explicit representation of the developing vascularity. We demonstrated the ability of this simplified model to describe capillary penetration of the tumor, tumor compartmentalization, and some therapeutic effects. Therefore, it seems to be a very convenient testing ground for possible refinements and we plan extensions towards incorporating further therapeutically relevant aspects like thermal effects, hormonal therapy, radiation or chemotherapy. Once, first experience has been collected, some of the concepts can be incorporated into the explicit vessel model.

We reproduced the growth of a myoma by using an advanced viscoelastic tissue model. We have shown that the morphology of the mature myoma depends strongly on the tissue properties. Here, a point of discussion is the viscoelasticity of the surrounding tissue. This could correspond, for example, to increased proliferation or loss of elasticity due to structural fiber damage under prolonged exposure to stress. From histological stains we know that the healthy myometrium mesh and the membranaceous endometrium may be stretched by a stiff-growing myoma to a multiple of their original dimensions, due to hypertrophy and/or hyperplasia. At a certain moment, thinning of healthy structures is not compatible with sufficient blood supply anymore, leading to necrosis in the most stressed area first, causing, for example, pain and bleeding disorders.

The tumor model is generally difficult to validate since many of the parameters are unknown or carry large measurement errors. It is the most difficult challenge to quantify these effects on the cellular level, which is the reason why the simulation outcome currently cannot be rigorously compared against physiological findings. Our ultimate goal is to develop a comprehensive framework allowing creating, testing, and validating models of tumor development, which can be used as a research tool to test hypotheses and perform *in silico* experiments for drug and therapy design. We are currently working on setting up a corresponding experimental environment based on animal models. In order to be able to collect comprehensive spatiotemporal in vivo data, mouse tumor models will be investigated using multimodal noninvasive imaging approaches addressing various aspects involved in the formation of neovasculature. These longitudinal in vivo studies will be complemented by high-resolution x-ray tomographic analysis of the tumor vascular system. The obtained data will not only provide the experimental environment of model building and simulation but will also serve as a basis for the quantitative validation of the developed methods.

ACKNOWLEDGMENT

This work is part of the Swiss National Center of Competence in Research on Computer Aided and Image Guided Medical Interventions (NCCR Co-Me), supported by the Swiss National Science Foundation.

REFERENCES

Alarcón, T., H. M. Byrne, and P. K. Maini (2004). Towards whole-organ modelling of tumour growth. *Prog. Biophys. Mol. Biol. 85*(23), 451–472.

Alarcón, T., H. M. Byrne, and P. K. Maini (2005). A multiple scale model for tumor growth. *Multiscale Model. Simulation 3*(2), 440–475.

Anderson, A. R. and M. A. Chaplain (1998). Continuous and discrete mathematical models of tumor-induced angiogenesis. *Bull. Math. Biol. 60*(5), 857–899.

Araujo, R. P. and D. L. S. McElwain (2004). A history of the study of solid tumour growth: the contribution of mathematical modelling. *Bull. Math. Biol. 66*(5), 1039–1091.

Ausprunk, D. H. and J. Folkman (1977). Migration and proliferation of endothelial cells in preformed and newly formed blood vessels during tumor angiogenesis. *Microvasc. Res. 14*(1), 53–65.

Bellomo, N., N. Li, and P. Maini (2008). On the foundation of cancer modelling: selected topics, speculations, and perspectives. *Math. Models Method. Appl. Sci. 18*(4), 593–646.

Betteridge, R., M. R. Owen, H. M. Byrne, T. Alarcón, and P. K. Maini (2006). The impact of cell crowding and active cell movement on vascular tumour growth. *Networks Heterogeneous Media 1*(4).

Byrne, H. M., T. Alarcon, M. R. Owen, S. D. Webb, and P. K. Maini (2006). Modelling aspects of cancer dynamics: a review. *Philos. Trans. A Math. Phys. Eng. Sci. 364*(1843), 1563–1578.

Carmeliet, P. (2003). Angiogenesis in health and disease. *Nat. Med. 9*(6), 653–660.

Carmeliet, P. and R. K. Jain (2000). Angiogenesis in cancer and other diseases. *Nature 407*(6801), 249–257.

Chaplain, M. A. J., L. Graziano, and L. Preziosi (2006). Mathematical modelling of the loss of tissue compression responsiveness and its role in solid tumour development. *Math. Med. Biol. 23*(3), 197–229.

Cristini, V., J. Lowengrub, and Q. Nie (2003). Nonlinear simulation of tumor growth. *J. Math. Biol. V46*(3), 191–224.

Folkman, J. (2002). Role of angiogenesis in tumor growth and metastasis. *Semin. Oncol. 29*(6 Suppl. 16), 15–18.

Freyer, J. P. and R. M. Sutherland (1985). A reduction in the in situ rates of oxygen and glucose consumption of cells in EMT6/Ro spheroids during growth. *J. Cell Physiol. 124*(3), 516–524.

Frieboes, H. B., X. Zheng, C.-H. Sun, B. Tromberg, R. Gatenby, and V. Cristini (2006). An integrated computational/experimental model of tumor invasion. *Cancer Res. 66*(3), 1597–1604.

Gladilin, E., S. Zachow, P. Deuflhard, and H. C. Hege (2003). On constitutive modeling of soft tissue for the long-term prediction of cranio-maxillofacial surgery outcome. *International Congress Series 1256*, 343–348.

Gordon, V. D., M. T. Valentine, M. L. Gardel, D. Andor-Ardo, S. Dennison, A. A. Bogdanov, D. A. Weitz, and T. S. Deisboeck (2003). Measuring the mechanical stress induced by an expanding multicellular tumor system: a case study. *Experimental Cell Res. 289*(1), 58–66.

Graziano, L. and L. Preziosi (2007). Mechanics in tumor growth. In F. Mollica, K. Rajagopal, and L. Preziosi (Eds.), *Modelling of Biological Materials*, pp. 267–328. Birkhäuser, Boston.

Hanahan, D. and R. A. Weinberg (2000). The hallmarks of cancer. *Cell 100*(1), 57–70.

Hirsch, S., D. Szczerba, B. Lloyd, M. Bajka, N. Kuster, and G. Székely (2009). A mechano-chemical model of a solid tumor for therapy outcome predictions. Lecture Notes in Computer Science, LNCS Vol. 5544, pp. 715–724, Springer-Verlag, Berlin/Heidelberg.

Huebner, K. H., D. Dewhirst, D. E. Smith, and T. G. Byrom (2001). *The Finite Element Method for Engineers*. John Wiley & Sons, New York.

Humphrey, J. D. and S. DeLange (2004). *An Introduction to Biomechanics. Solids and Fluids, Analysis and Design*. Springer, New York.

Jang, S. H., M. G. Wientjes, D. Lu, and J. L. S. Au (2003). Drug delivery and transport to solid tumors. *Pharmaceutical Res. 20*(9), 1337–1350.

Ji, J. W., N. M. Tsoukias, D. Goldman, and A. S. Popel (2006). A computational model of oxygen transport in skeletal muscle for sprouting and splitting modes of angiogenesis. *J. Theor. Biol. 241*(1), 94–108.

Jiang, Y., J. Pjesivac-Grbovic, C. Cantrell, and J. P. Freyer (2005). A multiscale model for avascular tumor growth. *Biophys. J. 89*(6), 3884–3894.

Kim, Y., M. Stolarska, and H. Othmer (2007). A hybrid model for tumor spheroid growth in vitro I: theoretical development and early results. *Math. Models Methods Appl. Sci. 17*, 1773–1798.

Lloyd, B. A., D. Szczerba, M. Rudin, and G. Székely (2008). A computational framework for modelling solid tumour growth. *Philos. Transact. A Math Phys. Eng. Sci.* 366, 3301–3318.

Lloyd, B. A., D. Szczerba, and G. Székely (2007). A coupled finite element model of tumor growth and vascularization. In N. Ayache, S. Ourselin, and A. Maeder (Eds.), *Medical Image Computing and Computer-Assisted Intervention— MICCAI 2007: 10th International Conference*, Lecture Notes in Computer Science 4792, pp. 874–881. Springer-Verlag, Berlin, Heidelberg.

Macklin, P. and J. Lowengrub (2007). Nonlinear simulation of the effect of microenvironment on tumor growth. *J. Theoretical B. 245*(4), 677–704.

Plank, M. J. and B. D. Sleeman (2004). Lattice and non-lattice models of tumour angiogenesis. *Bull. Math. Biol. 66*(6), 1785–1819.

Pries, A. R., T. W. Secomb, and P. Gaehtgens (1996). Biophysical aspects of blood flow in the microvasculature. *Cardiovasc. Res. 32*(4), 654–667.

Pries, A. R., T. W. Secomb, and P. Gaehtgens (1998). Structural adaptation and stability of microvascular networks: theory and simulations. *Am. J. Physiol. 275*(2 Pt 2), H349–H360.

Sanga, S., H. B. Frieboes, X. Zheng, R. Gatenby, E. L. Bearer, and V. Cristini (2007). Predictive oncology: a review of multidisciplinary, multiscale in silico modeling linking phenotype, morphology and growth. *Neuroimage. 37* (Suppl. 1), S120–S134.

Sim, B. K., N. J. MacDonald, and E. R. Gubish (2000). Angiostatin and endostatin: endogenous inhibitors of tumor growth. *Cancer Metastasis Rev. 19*(1–2), 181–190.

Szczerba, D., B. Lloyd, M. Bajka, and G. Székely (2009). A multiphysics model of myoma growth. *Int. J. Multiscale Computational Eng. 7*, 17–27.

Szczerba, D. and G. Székely (2002). Macroscopic modeling of vascular systems. In *Lecture Notes in Computer Science (LNCS)*, Volume 2489 of *Lecture Notes in Computer Science (LNCS)*, pp. 284–292. Springer, Heidelberg.

Szczerba, D. and G. Székely (2005). Simulating vascular systems in arbitrary anatomies. In *MICCAI, 8th International Conference*, Volume 3750 of *LNCS*, pp. 641–648. Springer, Berlin/Heidelberg.

Walocha, J. A., J. A. Litwin, and A. J. Miodoński (2003). Vascular system of intramural leiomyomata revealed by corrosion casting and scanning electron microscopy. *Hum. Reproduction 18*(5), 1088–1093.

Zienkiewicz, O. and R. Taylor (2000). *The Finite Element Method, Volume 1—The Basics*. Elsevier, Oxford.

CHAPTER 15

Building Stochastic Models for Cancer Growth and Treatment

Natalia L. Komarova

CONTENTS

INTRODUCTION

Drug resistance in cancer is one of the major causes of cancer treatment failure. Many types of resistance are thought to be associated with genetic events that modify cellular phenotype inside the tumor. In this chapter we will formulate a mathematical model that allows for a systematic study of drug resistance in cancer and its effects on treatment. The goal of this approach is to aid in optimal treatment strategy design.

The first stochastic model of drug resistance was created by Goldie and Coldman (1979), who developed a whole new approach to mathematical treatment of resistance in their subsequent work (see, e.g., Goldie and Coldman 1983; Goldie and Coldman 1985a; Goldie and Coldman 1985b; Goldie and Coldman 1986). A number of important theoretical and numerical results have been obtained by the authors for the case of one or more drugs. Since this groundbreaking work, a lot of mathematical models of drug resistance in cancer have been proposed. Several models, including stochastic branching models for stable and unstable gene amplification and its relevance to drug resistance, were explored (Axelrod et al. 1994; Harnevo and Agur 1991; Harnevo and Agur 1993; Kimmel and Axelrod 1990; Kimmel and Stivers 1994). Stem cell dynamics is explicitly incorporated in models (see Glauche et al. 2009; Roeder and Glauche 2008). Methods of optimal control theory were used to analyze drug dosing and treatment strategies (Cojocaru and Agur 1992; Kimmel et al. 1998). Models for tumor growth incorporating age-structured cell-cycle dynamics, in application to chemotherapy scheduling, have been developed by Gaffney (2004, 2005). Mechanistic mathematical models developed to improve the design of chemotherapy regimens are reviewed in Gardner and Fernandes (2003). Jackson and Byrne (2000) extended an earlier partial differential equation (PDE) model of Byrne and Chaplain (1995) to study the role of drug resistance and vasculature in tumors' response to chemotherapy; in this class of spatial models, the tumor is treated as a continuum of different types of cells that include susceptible and resistant cells. Another class of models is based on the Luria–Delbruck mutation analysis (Jaffrezou et al. 1994; Kendal and Frost 1988; Komarova et al. 2007; Michor et al. 2006).

In this paper we review some of the recent work where the stochastic dynamics of drug resistance in cancer was studied. Biological applications of the model include: (i) the number of drugs needed for successful treatment; (ii) designing the optimal therapy where the number of drugs is reduced in the course of treatment, and (iii) treatments with cross-resistant drugs. Applications are discussed in the context of chronic myeloid leukemia (CML).

MODEL

Mutation Networks and Cellular Dynamics

Following the method developed in Komarova (2006), and Komarova and Wodarz (2005), we start by building a mutation network of the resistance types of interest. This network's nodes denote cancer cell phenotypes that have different characteristics with respect to their drug susceptibility. For example, if two drugs are used to treat the tumor, then potentially there could be at least four different cell types: those that are fully susceptible, characterized by the binary index $s = 00$; those resistant to drug 1 and susceptible to drug 2 ($s = 10$); those resistant to drug 2 and susceptible to drug 1 ($s = 01$), and those resistant to both drugs (or fully resistant), with $s = 11$ (see Figure 15.1). In general, if m drugs are applied in the course of the therapy (separately or in combination), we have 2^m combinatorial resistance types. The binary index s has m positions corresponding to the m drugs; "1" in a given position denotes resistance to the corresponding drug, while "0" means susceptibility. In principle, there could be other cell types that differ from the ones listed above by the *level* of their susceptibility. Such types can be included; the methodology described in this chapter will work as long as the types can be annotated by a discrete index.

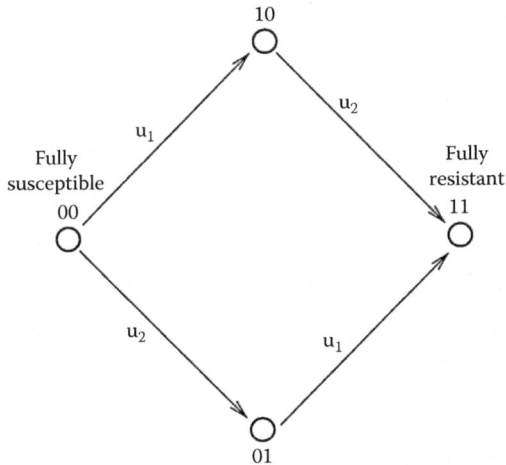

FIGURE 15.1 Mutation diagram for $m = 2$ drugs.

The nodes of the network are connected with arrows corresponding to mutation processes that transform one cell type into another. By mutations we mean a wide range of heritable cellular transformations that includes point mutations per se as well as larger-scale genetic modifications and epigenetic events. The mutation rates are marked by the arrows and denote the probability to produce one daughter cell of the modified type upon a division of the cell of the given type.

The dynamics of cells includes the following events: a faithful cell division (such that both daughter cells are of the original type), a division with a mutation (whereby one of the daughter cells acquires a different phenotype, in agreement with the mutation network), cell death, and other events such as cellular quiescence and awakening from the state of quiescence (Komarova and Wodarz 2007a, 2007b). The present framework would have to be modified considerably to include spatial cell migration.

To set up the stochastic description of cell dynamics, we denote by i_s the number of cells of resistance class s. Let $\varphi_{i_0,\ldots,i_n}(t)$ denote the probability that at time t there are i_s cells of resistance class s, for all classes s, with $0 \le s \le n = 2^m - 1$. Suppose that cancerous cells divide with rate l_s, and die with rate D_s. These kinetic rates can depend on the treatment dose; in particular, the death rate of cells is comprised of the "natural" rate of cell death in an untreated tumor and the action of the drug(s), if any, upon the cells. We assume an exponential time distribution for various events (such as cell reproduction and death), and set up the following Kolmogorov forward equation for a linear birth-death process: $\dot{\xi}_{i_0,\ldots,i_n} = \sum_{s=0}^{n} Q_s$, with

$$Q_s = \xi_{i_0,\ldots,i_s-1,\ldots,i_n}(i_s-1)l_s\left(1 - \sum_j u_{s\to j}^{out}\right)$$
$$+ i_s l_s \sum_j \xi_{i_0,\ldots,i_s,\ldots,i_j-1,\ldots,i_n} u_j^{s,out} + \xi_{i_0,\ldots,i_s+1,\ldots,i_n}(i_s+1)D_s; \qquad (15.1)$$

here, $u_{s\to j}^{out}$ denotes the mutation rates for all arrows originating at s.

Probability Generating Functions

The following probability generating function can be defined: $\Psi(\vec{\xi};t) = \sum_{s=0}^{n} \varphi_{i_0,\ldots,i_n}(t)\xi_0^{i_0}\ldots\xi_n^{i_n}$, with $\vec{\xi} = (\xi_0,\ldots,\xi_n)$. This function satisfies the following partial differential equation (obtained by standard methods from the Kolmogorov forward equation above):

$$\frac{\partial \Psi}{\partial t} = \sum_{s=0}^{n} \frac{\partial \Psi}{\partial \xi_s}\left(l_s(1-\sum_{j} u_{s\to j}^{out})\xi_s^2 + \left[\sum_{j} l_s u_{s\to j}^{out}\xi_j^{out} - (l_s + D_s)\right]\xi_s + D_s \right).$$

(15.2)

Following the standard technique for transport-type equations, we have a system of equations for characteristics:

$$\dot{\xi}_s = -F_s(\vec{\xi};t),$$ (15.3)

where

$$F_s(\vec{\xi};t) = l_s(1-\sum_{j} u_{s\to j}^{out})\xi_s^2 + \left[\sum_{j} l_s u_{s\to j}^{out}\xi_j^{out} - (l_s + D_s)\right]\xi_s + D_s.$$ (15.4)

The dependence on time in the right hand side comes from the fact that the kinetic rates such as birth and death rate can, in general, be functions of time. In order to evaluate the function $\Psi(\vec{\xi}';t')$ at some point $\vec{\xi}' = (\xi_0',\dots,\xi_n')$, we "reverse the time" by the change of variables, $t \to t'-t$ and solve the following system:

$$\dot{\xi}_s = F_s(\vec{\xi};t), \qquad \xi_x(0) = \xi_s'.$$ (15.5)

Then the desired function is given by

$$\Psi(\vec{\xi}';t') = \prod_{s=0}^{n} [\xi_s(t')]^{M_s},$$ (15.6)

where in the right-hand side the functions $\xi_s(t)$ are solutions of the above system, and the constants M_s are the initial abundances of mutants of type s. In the rest of this chapter we will assume for simplicity that, initially, there are M_0 cells of type $s = 00$, and zero cells of any other types.

In the particular case of $m = 2$ drugs and four mutant classes, we have four equations:

$$\dot{\xi}_{00} = l_{00}(1-u_1-u_2)\xi_{00}^2 + (l_{00}(u_1\xi_{10}+u_2\xi_{01})-(l_{00}+D_{00}))\xi_{00} + D_{00},$$ (15.7)

$$\dot{\xi}_{10} = l_{10}(1-u_2)\xi_{10}^2 + (l_{10}u_2\xi_{11} - (l_{10}+D_{10}))\xi_{10} + D_{10}, \qquad (15.8)$$

$$\dot{\xi}_{01} = l_{01}(1-u_1)\xi_{01}^2 + (l_{01}u_1\xi_{11} - (l_{01}+D_{01}))\xi_{01} + D_{01}, \qquad (15.9)$$

$$\dot{\xi}_{11} = l_{11}\xi_{11}^2 - (l_{11}+D_{11})\xi_{11} + D_{11}, \qquad (15.10)$$

where we denoted by u_1 (u_2) the mutation rate with which resistance to drug 1 is created (drug 2). The time-dependence of the coefficients in Equations 15.7–15.10 is implicit.

Treatment Strategies

Different treatment strategies define the values of the death rates, D_s, at different moments of time:

$$D_s = d_s + h_s(t),$$

where the coefficients d_s are natural death rates of the cancer cells, and h_s (t) are the drug-induced cell death rates. The functions $h_s(t)$ depend on the particular strategy used. As different drugs are applied, the "strength" of each drug, which depends on the concentration of the drug in the patient's blood, changes as some smooth functions of time. The exact shape of these functions and, therefore, the shape of $h_s(t)$, depends not only on the treatment strategy (that is, whether drugs are applied in combination or cyclically), but also on the way the drugs are administered, and on how quickly they are absorbed. For example, it can be assumed that $h_s(t)$ for a susceptible class reaches a maximum sometime after the drug is taken, and decays until the next administration of the drug. However, in this paper we simplify this picture by assuming that the functions $h_s(t)$ are piecewise constant. They are assumed to have a constant nontrivial value for all the susceptible classes as long as the patient is treated with a given drug, and they become zero after the drug is discontinued.

Different resistance types are characterized by different drug-induced killing rates, and those in turn depend on which drugs are applied. In Table 15.1 we present the drug-induced killing rates for different resistance types, under treatments by drug 1 only, drug 2 only, and drugs 1 and 2 in combination. We assume that when applied alone in the context of susceptible types, the two drugs induce death rates h_1 and h_2, respectively. The

TABLE 15.1   Drug-induced death rates of different resistance types under different treatment conditions

	Type 00	Type 10	Type 01	Type 11
Drug 1	h_1	0	h_1	0
Drug 2	h_2	h_2	0	0
Drugs 1 and 2	$F(h_1 + h_2)$	h_2	h_1	0

effect of the two drugs in combination upon the fully-susceptible type is described by the function $F(h_1 + h_2)$ which satisfies $max(h_1, h_2) \leq F(h_1 + h_2) \leq h_1 + h_2$. In other words, the combined action of two drugs can range from the single action of the stronger of the drugs to the sum of the two actions.

Let us suppose that the natural history of a tumor can be split into several time intervals, according to the treatment strategies used. In the case of a single treatment stage with drugs 1 and 2 in combination, which starts at time t_1 and ends at time t_2, we could have three time intervals: pretreatment: $0 \leq t < t_1$; treatment: $t_1 \leq t < t_2$; and posttreatment: $t \geq t_2$. We will further assume that the division rates, l_s, and the death rates, d_s, of cells are time independent, and also that some of the rates are the same for all types, namely, that $l_s = l$ and $d_s = d$ for all s. Then, the values of the time-dependent coefficients are given by:

$$D_{00} = D_{10} = D_{01} = D_{11} = d, \text{ for } 0 \leq t < t_1$$

$$D_{00} = d + F(h_1 + h_2), D_{10} = d + h_2, D_{01} = d + h_1, D_{11} = d, \text{ for } t_1 \leq t < t_2, \text{ and}$$

$$D_{00} = D_{10} = D_{01} = D_{11} = d, \text{ for } t \geq t_2.$$

Colony Extinction and Probability of Treatment Success

In this chapter we will be concerned with the probability of treatment success, which is the same as the probability of extinction of the colony. This quantity is given by

$$\varphi_{0,\dots,0}(t) = \Psi(\vec{0}; t). \tag{15.11}$$

To evaluate this function, we will use general formula (15.2) with $M_s = 0$ for all (partially) resistant types:

$$\Psi(\vec{0}; t) = \xi_{00}^{M_0}(t), \tag{15.12}$$

where ξ_{00} (t) is the solution of system (15.3–15.6) with the initial conditions

$$\xi_{00} = \xi_{01} = \xi_{10} = \xi_{11} = 0 \tag{15.13}$$

To exclude the scenarios where the colony goes extinct spontaneously before treatment starts, we will be studying the following slightly modified quantity:

$$P_{success}(t) = \frac{\xi_{00}^{M_0}(t) - (d/l)^{M_0}}{1 - (d/l)^{M_0}}. \tag{15.14}$$

To obtain solution ξ_{00} (t), system (15.7–15.10) with piecewise-constant coefficients can be solved numerically. Because of the time-reversal procedure used in the derivation of the ODEs, the following simulation algorithm applies. Suppose that there are k stages in the natural history of the tumor: $0 \leq t < t_1$, $t_1 \leq t < t_2$, ..., $t_{k-1} \leq t < t_k$, where the value t_k can be infinite. We will denote the constants corresponding to the coefficient values during stage i as $D_s^{(i)}$ and the length of each time interval $\Delta t_i = t_i - t_{i-1}$. To start the solution process, we will integrate system (15.7–15.10) with constant coefficients $D_s^{(k)}$ and with initial conditions (15.13) to find the solution corresponding to the time Δt_k. Then, we will plug these values as initial conditions for system (15.7–15.10) with coefficients $D_s^{(k-1)}$ and solve the equations for time duration Δt_{k-1}. We will then use the resulting values as the initial conditions and repeat the process the total of k times. This procedure corresponds to the ("reversed"') time-variable changing from the end of treatment (physical time t_k) back to time $t = 0$. The obtained function $\xi_{00}(t)$ is used in Formula 15.14.

The limiting value of the probability of treatment success,

$$\lim_{t \to \infty} P_{success}(t),$$

is of a particular interest. This approximation corresponds to long-term treatment strategies where the drugs are used long enough for all the susceptible types to be eliminated. The limiting procedure in our simulations corresponding to the length of stage k, Δt_k, increasing to infinity.

The probability of treatment success defined above depends on all the parameters of the system. Of special importance are, however, the mutation rates at which resistant types are generated, and tumor size at start of treatment. The latter quantity is not included explicitly in the calculation. In our studies, we will assume that the colony size, N, and the time when treatment begins, t_1, are related by means of a deterministic equation for the average population size. In particular, if the growth and death rates of all the resistant types are the same in the absence of treatment, we have $t_1 = \log(N/M_0)/(l-d)$.

RESULTS

In this section we will review several applications of the methodology we have presented.

Combination Therapies: Number of Drugs Needed for Treatment

We start our discussion by considering combination treatments. We would like to determine the number of drugs, m, needed for successful treatment (Komarova and Wodarz 2005, 2009). An important assumption we make in this section is the absence of cross-resistance. In other words, a single mutation event is assumed to create resistance to one drug only, and no single mutation is enough to make a cell resistant to two or more additional drugs. Later in this chapter we will address the question of cross-resistance in some detail.

In our framework, the absence of cross-resistance imposes certain constraints on mutation networks. To explain this, let us use the Manhattan distance to measure distances between nodes in mutation networks (the Manhattan distance is the number of positions where the two binary indices differ). The absence of cross-resistance means that only the nodes of distance 1 can be connected by a mutation arrow.

We will fix parameters l and d, and assume for simplicity that $h_i = h$, and $u_i = u$ for all the drugs used (these assumptions will be removed later on). We will vary the mutation rate u (with which resistant mutants are generated) and the number of tumor cells at which treatment is started, N. For each pair (u,N), we will find the minimum number of drugs that gives a probability of treatment success greater than $1-\delta$, where δ is the maximum tolerated failure rate. The result for a particular choice of parameters is presented in Figure 15.2. We can see that for smaller mutation rates and smaller tumor cell numbers, two drugs are enough to treat successfully. For larger tumors or mutation rates, the number of drugs grows significantly.

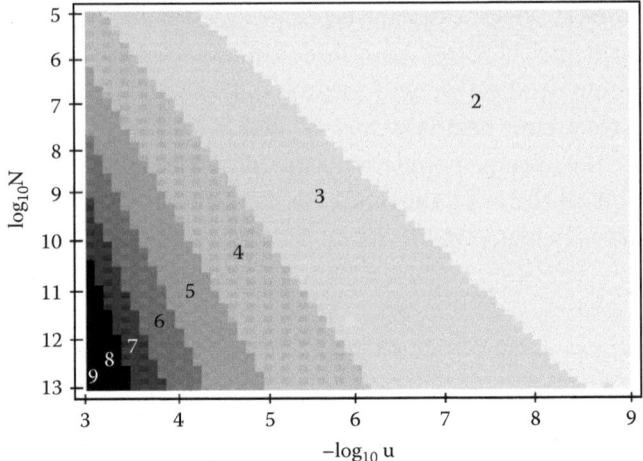

FIGURE 15.2   The number of drugs needed for the probability of treatment success $1-\delta$ with $\delta = 0.01$, as a function of u and N. Other parameters are the following: $l = 1$, $d = 0$, $h = 2$, $M_0 = 100$. The jagged appearance of the boundaries between the domains is due to a discrete method used to calculate it and can be smoothed by using a simulation with a higher resolution.

Can We Reduce the Number of Drugs in the Course of Treatment?

In the diagram of Figure 15.2 we notice that smaller tumors require fewer drugs in combination to achieve the given probability of treatment success. This leads us to ask the question, as the cancer population decreases during therapy, is it possible to reduce the number of drugs used for treatment without reducing the likelihood of treatment success (Komarova and Wodarz 2009)? In Figure 15.3 we calculate the number of tumor cells at which one drug can be taken away such that the probability of treatment success does not decrease by more than δ. The figure presents the probability of treatment success as a function of the quantity N_{off}, the number of tumor cells at which treatment with three drugs is replaced by treatment with only two of those three drugs. This is plotted for different values of N, the number of tumor cells at which treatment is started. We observe the following interesting trends.

The probability of treatment success is flat for small values of N_{off}, and it is equal to the probability of treatment success for $m = 3$ drugs. In other words, if we remove one of the drugs when the number of cancerous cells is very low, it does not change the probability of treatment success. Note that in the particular example of Figure 15.3, the probability of treatment

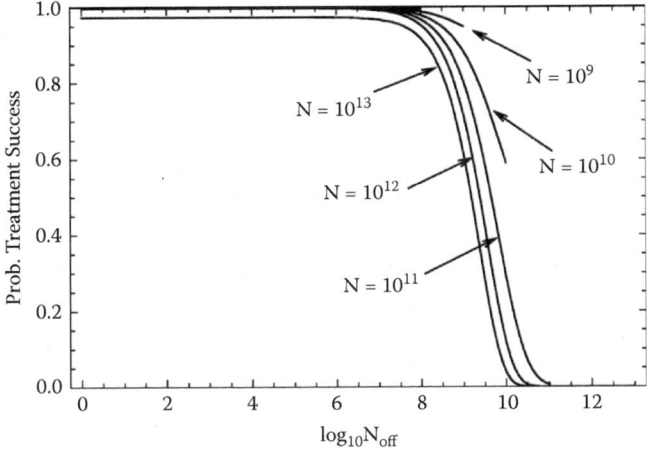

FIGURE 15.3   Probability of treatment success as a function of N_{off}, the number of tumor cells at which the number of drugs is reduced by one. Different curves correspond to different numbers of tumor cells at start of treatment, N. The initial number of drugs is 3, $u = 10^{-6}$, and the rest of the parameters are as in Figure 15.2.

success for low N_{off} is very close (closer than δ) to 1 for all curves except for the one with $N = 10^{13}$. If $N = 10^{13}$ and $u = 10^{-6}$, we can see from Figure 15.2 that the number of drugs necessary for a success rate greater than $1 - \delta$ is $m = 4$, that is, if $m = 3$, the probability of success is smaller than $1 - \delta$. This can be seen in the limiting value of the probability of success rate as $N_{off} \to 0$ for $N = 10^{13}$ in Figure 15.3.

On the other hand, if we take away one drug too early (N_{off} near N), then the probability of treatment success is lower than that with m drugs throughout the course of therapy. There is a relatively sharp transition between maximum possible probability of treatment success and a low probability of treatment success.

The main finding is that if we wait sufficiently long, the number of drugs used in combination treatment can be safely removed by one. Using similar methods, we can show that for some parameter combinations, there are conditions when more than one drug can be removed from treatment without compromising the ability of treatment to suppress the tumor.

Which Drug Should Be Removed?

The next step is to remove the symmetry assumptions, $h_i = h$, and $u_i = u$ for all the drugs, and discuss drugs with different characteristics (Komarova and Wodarz 2009). In particular, we will concentrate on the activity spectrum and potency of the drugs, which are defined as follows.

We will say that a drug is characterized by a broad activity spectrum if it is effective against a large spectrum of mutant cells. On the other hand, a drug with a narrow activity spectrum is a more specific agent, which is active against a relatively small number of cell variants. In terms of our modeling approach, the drug with a broader activity spectrum will be characterized by a smaller mutation rate, u, with which mutants resistant to the drug are generated. The more specific, or narrow, drug is characterized by a larger mutation rate associated with the generation of mutants resistant to the drug. By potency, we mean how effectively a drug kills cells that are susceptible to the drug; this is reflected in the drug-induced cell death rate characteristic of each drug, where higher potency is correlated with higher values of h.

Consider a two-drug treatment with drugs 1 and 2. Suppose that drug 2 is characterized by a broader activity compared with drug 1. Let us envisage a schedule whereby the tumor is first treated with a combination of drugs 1 and 2, and then the number of drugs is reduced by one. Which drug should remain and which one should be taken off?

The results are presented in Figure 15.4. The top two graphs assume that the potency of both drugs is the same, that is, $h_1 = h_2$. The rightmost dashed thick line in both graphs corresponds to a combination therapy. The other lines show the probability of treatment success when a drug is discontinued; the drug is removed once the number of tumor cells is reduced by a factor R, which appears above each line. The goal is to keep the probability of treatment success as high as possible (the thick dashed line). We can see that in both cases, if the reduction happens sufficiently late in treatment (low values of R), the probability of treatment success will be unchanged.

Comparing (a) and (b) in Figure 15.4, we can see that if the drug with a broader activity spectrum is removed then we need to wait until the number of tumor cells is reduced by a factor of $R = 10^{-8}$; otherwise, the probability of treatment success is significantly lower compared with that of continuous combination therapy. On the other hand, if the drug with a narrower activity spectrum is removed, then we can withdraw the drug earlier in treatment (more precisely, when the number of tumor cells is reduced by a factor of $R = 10^{-7}$). This suggests that a more effective strategy is the one when the drug with a narrower activity spectrum is removed.

The bottom two graphs in Figure 15.4 correspond to the cases where the two drugs also differ by their potency. In Figure 15.4c, the drug with a narrower activity has a higher potency, while the drug with a broader activity has a lower potency. We can see that in order to maximize treatment success in this scenario, the drug with a narrower activity and higher potency should be removed. In Figure 15.4d, the drug with narrower activity has

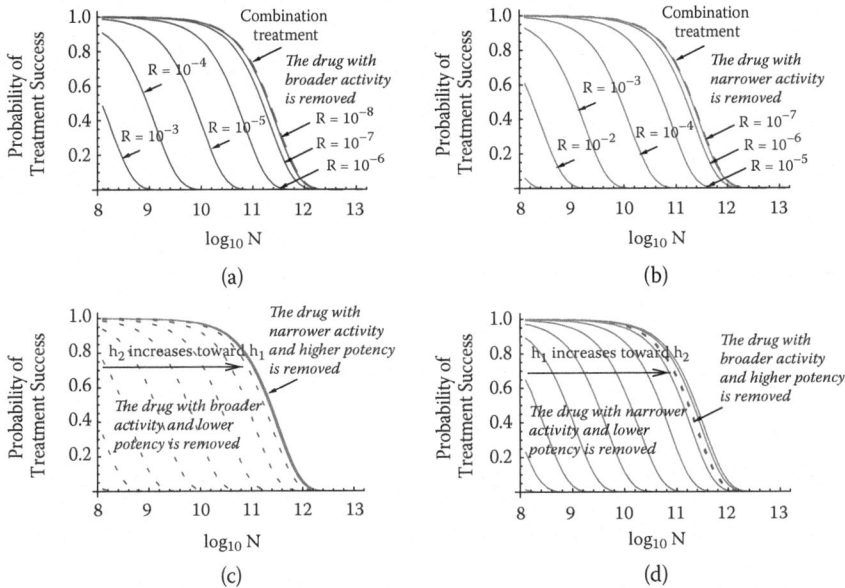

FIGURE 15.4 **(See color insert following page 40)** The probability of treatment success as a function of N, the number of tumor cells, for different treatment schedules. *Top,* the two drugs have equal potency ($h/l = 5$). The dashed thick line corresponds to the combination treatment of two drugs. The other lines correspond to treatment schedules where one of the drugs is removed once the number of tumor cells is reduced by a factor R, indicated in the figure. (a) The drug with broader activity is removed. (b) The drug with narrower activity is removed. *Bottom,* the two drugs have different potency. The dotted lines correspond to the strategy whereby after a combination treatment only the drug with broader activity is removed. Solid lines correspond to the strategy whereby the drug with narrower activity is removed. *(c)* The drug with a narrower activity has a larger potency; different lines correspond to values of $h_2/l \in \{5, 10, ..., 50\}$ such that $h_2 \le h_1 = 50l$. Note that multiple solid lines are on top of each other, as the success of the corresponding strategy does not depend crucially on h_2. (d) The drug with narrower activity has a lower potency; different lines correspond to values of $h_2/l \in \{5, 10, ..., 50\}$, such that $h_1 \le h_2 = 50l$. The parameters are $d/l = 0.2$, $u_1 = 10^{-7}$, $u_2 = 10^{-6}$, $M_0 = 10^2$.

a lower potency. In this case, if the potencies of the drugs are significantly different, it is the broader activity and higher potency drug that should be removed. Results presented here follow from stochastic simulations. Biological explanations for these outcome are presented in Komarova and Wodarz (2009) and based on the rather complex interplay between the number of resistant mutants produced and the magnitude of the drug-induced killing rate.

We conclude that if drugs with differential activity spectra are used, then the drugs with the broader activity should be continued, and the drug(s) with the narrowest activity can be removed. On the other hand, if the broader-activity drug is also considerably more potent, the less potent and narrower drug should remain and the more potent and broader one can be removed.

Cross-Resistance and the Utility of Drug Combinations

In this subsection, we will include the possibility of drug cross-resistance and study how this can affect the probability of treatment success, and the choice of optimal treatment strategies (Komarova et al. 2009). Formally speaking, in the presence of cross-resistance any node in the mutation network can be connected to another node. The corresponding mutation rate reflects mutation events that comprise resistance to more than one drug at the same time. We will use the short-hand notation "cross-resistant mutations" to refer to such events. In the presence of two drugs, only two situations are logically possible: the drugs are cross-resistant or non-cross-resistant. With a higher number of drugs, the situation quickly becomes more complex. For example, for $m = 3$ drugs, there are five possibilities (see Figure 15.5, top panel: (1) no cross resistance, (2) partial cross-resistance when only one of the three pairs of drugs has cross-resistant mutations, (3) partial cross-resistance when two of three pairs of drugs have (different sets of) cross-resistant mutations, (4) complete pairwise cross-resistance when there are three (nonintersecting) sets of cross-resistant mutations, corresponding to each of the three pairs of drugs, and (5) triple cross-resistance when there are mutations conferring resistance to all the three drugs simultaneously.

We will start our exploration of cross-resistance by considering a two-drug case. Figure 15.5 shows the probability of treatment success for one and two drugs, assuming the absence of cross-resistance. This is compared to the cross-resistance scenario. While the probability of treatment success is lower in the presence than in the absence of cross-resistance, combining

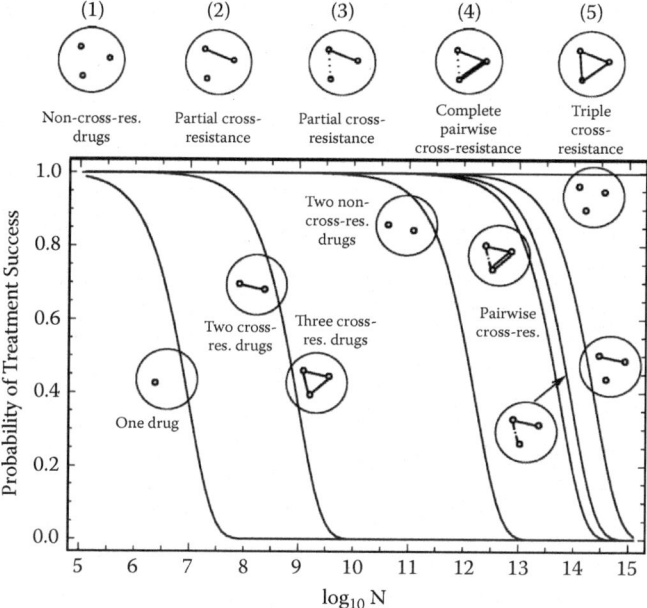

FIGURE 15.5 The role of cross-resistance in drug treatment. *Top*, the cross-resistance networks are presented for *m* = 3 drugs, by using connected and disconnected nodes. The number of nodes corresponds to the number of drugs used. Connected nodes correspond to the existence of a cross-resistant mutation. Identical connecting lines indicate that the same mutation confers cross-resistance to all connected drugs. Different (single, double, dashed) lines correspond to different mutations. *Bottom*, the probability of treatment success is plotted as a function of the colony size, *N*. Different curves correspond to different combination treatments, with one, two, and three drugs. Simulation parameters are as follows: $u = 10^{-7}$, the cross-resistance mutation rate is 10^{-9}, $M_0 = 100$, $d/l = 0.5$, $h/l = 3$.

two cross-resistant drugs clearly improves the probability of treatment success relative to the use of only one drug. This pattern was observed as long as the probability of cross-resistant mutations was not too high (see Komarova et al. [2009] for details). In this case, it is more likely to acquire a mutation that confers resistance against only one drug than to acquire a cross-resistance mutation. Hence, for most mutations, combination therapy will not be challenged by cross-resistance.

Next, consider the combination of three drugs, and assume the existence of triply-cross-resistant mutants (scenario 5 in Figure 15 5, top panel). The striking result is that in this case, combining three drugs will not lead to any further advantage compared to the combination of two drugs (see the graphs in Figure 15.5). For a triple combination therapy to be advantageous, most resistant cells must harbor mutations that render them resistant against two of the drugs (but not the third one). Accumulating two separate resistance mutations, however, is a relatively rare event. It is much more likely that a cell acquires the single cross-resistance mutation. Hence, a triple combination therapy (with triply-cross-resistant drugs) does not improve the probability of treatment success compared to double combination therapy.

Finally, the existence of partial or complete pairwise cross-resistance, while decreasing the chances of treatment success compared to non-cross-resistant three-drug therapy, still provides a significant advantage over non-cross-resistant two-drug treatments. Therefore, adding a third (even cross-resistant!) drug to two non-cross-resistant drugs improves the chances for treatment success. Clinical applicability of our results is briefly discussed in the next section.

DISCUSSION

Chronic myeloid leukemia (CML) is a cancer of the hematopoietic system and is initiated and driven by the product of the BCR-ABL fusion gene (Calabretta and Perrotti 2004; Melo et al. 2003; Shet et al. 2002; Yoshida and Melo 2004). It proceeds in three stages: the chronic phase in which the number of cells is relatively low and the degree of cellular differentiation is relatively high; the accelerated phase during which the number of cells starts to rise to higher levels and the degree of differentiation declines; and blast crisis, which is characterized by explosive cell growth and a low degree of differentiation. Small molecules that specifically target the BCR-ABL gene product provide a successful treatment approach that can lead to a reduction of BCR-ABL+ cells below detectable levels, at least during the early stages of the disease. The drug Imatinib has been mostly used in this respect (Daley 2003; Deininger and Druker 2003; Gorre et al. 2001; O'Dwyer et al. 2002; Shah et al. 2004; Yoshida and Melo 2004). As the disease advances, however, the chances of treatment failure rise due to the presence of drug-resistant mutants that are generated mostly through point mutations, but also through gene duplications (Daub et al. 2004; Druker 2003, 2004; Gambacorti-Passerini et al. 2003; Nardi et al. 2004; Shannon 2002). Drug resistance can potentially be overcome by the combination of multiple drugs, where a mutation that confers resistance against one drug does not confer resistance against any of the other drugs in use. In addition to Imatinib, the drugs Dasatinib and Nilotinib are alternative inhibitors of the BCR-ABL gene product. The reality, however, is that these three drugs are cross-resistant because of one mutation (T315I) that confers resistance against all those drugs (Bradeen et al. 2006; Deininger 2007; Talpaz et al. 2006; Weisberg et al. 2007). In addition, there are more than 50 mutations that confer resistance against only one of the three drugs and not against the others.

The mathematical framework presented in this chapter naturally applies to the evolution of drug-resistant cancer cells in CML treated with small-molecule inhibitors. This framework suggests that a combination of three or four different non-cross-resistant small-molecule inhibitors needs to be administered to avoid treatment failure as a result of drug resistance. We can generate charts (such as that presented in Figure 15.2) that map the number of drugs needed for various numbers of tumor cells and mutation rates.

We further investigated the effect of withdrawing one or more drugs from the combination therapy regimen once the number of cancer cells has declined to lower levels. On the most basic level, the modeling tells us that the number of drugs can be reduced for long-term therapy without significantly reducing the chances of tumor suppression. This is important in the face of side effects that are likely to become problematic if treatment is continued in the long term. Using our computational approach, we calculated the threshold number of cancer cells at which the number of drugs can be reduced without significantly altering the chances of tumor suppression.

Our method also helps determine which drugs should be removed and which ones continued. For example, in two-drug treatments, if one of the drugs has a narrower activity spectrum, and it is equally potent or more potent than the second drug, then it is the drug with the narrower activity that should be discontinued. On the other hand, if the narrower drug is significantly less potent than the broader activity drug, then it is the broader drug that should be removed from treatment. Experimental and clinical studies performing measurements of the rate at which resistance mutations are generated and the drug-induced death rates will be vital to make clinically relevant predictions. Cancer kinetic parameters (such as the division and death rate of the cancer cells) are essential to predict the time course of treatment, such as treatment length and the duration of time until the number of drugs can be reduced without compromising continued tumor suppression.

Finally, we addressed the question of cross-resistance and studied whether drug combinations can still improve treatment outcomes if the drugs possess a degree of cross-resistance. Our results are relevant to the treatment of CML with the three currently available drugs: Imatinib, Dasatinib, and Nilotinib. There is one point mutation, T315I, which confers resistance against all of the three drugs, thus making them triply-cross-resistant. Therefore, in this context, our calculations suggest that combining two of the three currently available drugs can provide an advantage over using one drug alone. Recent experimental data support this notion (Bradeen et al. 2006). However, combining all the three drugs does not improve the chances of treatment success beyond that of a double combination therapy. Hence, at present, the two most effective drugs should be given simultaneously to treat CML. Once a drug effective against T315I mutants becomes available, the most effective treatment strategy will be to combine that drug with two of the three presently existing drugs.

ACKNOWLEDGMENTS

The author gratefully acknowledges the support of NIH grant R01 CA129286, and the Sloan Fellowship.

REFERENCES

Axelrod, D. E., Baggerly, K. A., and Kimmel, M. 1994. Gene amplification by unequal sister chromatid exchange: probabilistic modeling and analysis of drug resistance data. *J Theor Biol* 168, 151–9.

Bradeen, H. A., Eide, C. A., O'Hare, T. et al. 2006. Comparison of Imatinib mesylate, Dasatinib (BMS-354825), and Nilotinib (AMN107) in an N-ethyl-N-nitrosourea (ENU)-based mutagenesis screen: high efficacy of drug combinations. *Blood* 108, 2332–8.

Byrne, H. M. and Chaplain, M. A. 1995. Growth of nonnecrotic tumors in the presence and absence of inhibitors. *Math Biosci* 130, 151–81.

Calabretta, B. and Perrotti, D. 2004. The biology of CML blast crisis. *Blood* 103, 4010–22.

Cojocaru, L. and Agur, Z. 1992. A theoretical analysis of interval drug dosing for cell-cycle-phase-specific drugs. *Math Biosci* 109, 85–97.

Daley, G. Q. 2003. Towards combination target-directed chemotherapy for chronic myeloid leukemia: role of farnesyl transferase inhibitors. *Semin Hematol* 40, 11–4.

Daub, H., Specht, K., and Ullrich, A. 2004. Strategies to overcome resistance to targeted protein kinase inhibitors. *Nat Rev Drug Discov* 3, 1001–10.

Deininger, M. W. 2007. Optimizing therapy of chronic myeloid leukemia. *Exp Hematol* 35, 144–54.

Deininger, M. W. and Druker, B. J. 2003. Specific targeted therapy of chronic myelogenous leukemia with Imatinib. *Pharmacol Rev* 55, 401–23.

Druker, B. J. 2003. Overcoming resistance to Imatinib by combining targeted agents. *Mol Cancer Ther* 2, 225–6.

Druker, B. J. 2004. Imatinib as a paradigm of targeted therapies. *Adv Cancer Res* 91, 1–30.

Gaffney, E. A. 2004. The application of mathematical modelling to aspects of adjuvant chemotherapy scheduling. *J Math Biol* 48, 375–422.

Gaffney, E. A. 2005. The mathematical modelling of adjuvant chemotherapy scheduling: incorporating the effects of protocol rest phases and pharmacokinetics. *Bull Math Biol* 67, 563–611.

Gambacorti-Passerini, C. B., Gunby, R. H., Piazza, R. et al. 2003. Molecular mechanisms of resistance to Imatinib in Philadelphia-chromosome-positive leukaemias. *Lancet Oncol* 4, 75–85.

Gardner, S. N. and Fernandes, M. 2003. New tools for cancer chemotherapy: computational assistance for tailoring treatments. *Mol Cancer Ther* 2, 1079–84.

Glauche, I., Moore, K., Thielecke, L. et al. 2009. Stem cell proliferation and quiescence—two sides of the same coin. *PLoS Comput Biol* 5, e1000447.

Goldie, J. H. and Coldman, A. J. 1979. A mathematic model for relating the drug sensitivity of tumors to their spontaneous mutation rate. *Cancer Treat Rep* 63, 1727–33.

Goldie, J. H. and Coldman, A. J. 1983. Quantitative model for multiple levels of drug resistance in clinical tumors. *Cancer Treat Rep* 67, 923–31.

Goldie, J. H. and Coldman, A. J. 1985a. Genetic instability in the development of drug resistance. *Semin Oncol* 12, 222–30.

Goldie, J. H. and Coldman, A. J. 1985b. A model for tumor response to chemotherapy: an integration of the stem cell and somatic mutation hypotheses. *Cancer Invest* 3, 553–64.

Goldie, J. H. and Coldman, A. J. 1986. Application of theoretical models to chemotherapy protocol design. *Cancer Treat Rep* 70, 127–31.

Gorre, M. E., Mohammed, M., Ellwood, K. et al. 2001. Clinical resistance to STI-571 cancer therapy caused by BCR-ABL gene mutation or amplification. *Science* 293, 876–80.

Harnevo, L. E. and Agur, Z. 1991. The dynamics of gene amplification described as a multitype compartmental model and as a branching process. *Math Biosci* 103, 115–38.

Harnevo, L. E. and Agur, Z. 1993. Use of mathematical models for understanding the dynamics of gene amplification. *Mutat Res* 292, 17–24.

Jackson, T. L. and Byrne, H. M. 2000. A mathematical model to study the effects of drug resistance and vasculature on the response of solid tumors to chemotherapy. *Math Biosci* 164, 17–38.

Jaffrezou, J. P., Chen, G., Duran, G. E., Kuhl, J. S., and Sikic, B. I. 1994. Mutation rates and mechanisms of resistance to etoposide determined from fluctuation analysis. *J Natl Cancer Inst* 86, 1152–8.

Kendal, W. S. and Frost, P. 1988. Pitfalls and practice of Luria-Delbruck fluctuation analysis: a review. *Cancer Res* 48, 1060–5.

Kimmel, M. and Axelrod, D. E. 1990. Mathematical models of gene amplification with applications to cellular drug resistance and tumorigenicity. *Genetics* 125, 633–44.

Kimmel, M., Chakraborty, R., King, J. P. et al. 1998. Signatures of population expansion in microsatellite repeat data. *Genetics* 148, 1921–30.

Kimmel, M. and Stivers, D. N. 1994. Time-continuous branching walk models of unstable gene amplification. *Bull Math Biol* 56, 337–57.

Komarova, N. 2006. Stochastic modeling of drug resistance in cancer. *J Theor Biol* 239, 351–66.

Komarova, N. L., Katouli, A. A., and Wodarz, D. 2009. Combination of two but not three current targeted drugs can improve therapy of chronic myeloid leukemia. *PLoS One* 4, e4423.

Komarova, N. L. and Wodarz, D. 2005. Drug resistance in cancer: principles of emergence and prevention. *Proc Natl Acad Sci U S A* 102, 9714–9.

Komarova, N. L. and Wodarz, D. 2007a. Effect of cellular quiescence on the success of targeted CML therapy. *PLoS ONE* 2, e990.

Komarova, N. L. and Wodarz, D. 2007b. Stochastic modeling of cellular colonies with quiescence: an application to drug resistance in cancer. *Theor Popul Biol* 72, 523–38.

Komarova, N. L. and Wodarz, D. 2009. Combination therapies against chronic myeloid leukemia: short-term versus long-term strategies. *Cancer Res.*

Komarova, N. L., Wu, L., and Baldi, P. 2007. The fixed-size Luria-Delbruck model with a nonzero death rate. *Math Biosci* 210, 253–90.

Melo, J. V., Hughes, T. P., and Apperley, J. F. 2003. Chronic myeloid leukemia. *Hematology (Am Soc Hematol Educ Program)*, 132–52.

Michor, F., Nowak, M. A., and Iwasa, Y. 2006. Evolution of resistance to cancer therapy. *Curr Pharm Des* 12, 261–71.

Nardi, V., Azam, M., and Daley, G. Q. 2004. Mechanisms and implications of Imatinib resistance mutations in BCR-ABL. *Curr Opin Hematol* 11, 35–43.

O'Dwyer, M. E., Mauro, M. J., and Druker, B. J. 2002. Recent advancements in the treatment of chronic myelogenous leukemia. *Annu Rev Med* 53, 369–81.

Roeder, I. and Glauche, I. 2008. Pathogenesis, treatment effects, and resistance dynamics in chronic myeloid leukemia—insights from mathematical model analyses. *J Mol Med* 86, 17–27.

Shah, N. P., Tran, C., Lee, F. Y. et al. 2004. Overriding Imatinib resistance with a novel ABL kinase inhibitor. *Science* 305, 399–401.

Shannon, K. M. 2002. Resistance in the land of molecular cancer therapeutics. *Cancer Cell* 2, 99–102.

Shet, A. S., Jahagirdar, B. N., and Verfaillie, C. M. 2002. Chronic myelogenous leukemia: mechanisms underlying disease progression. *Leukemia* 16, 1402–11.

Talpaz, M., Shah, N. P., Kantarjian, H. et al. 2006. Dasatinib in Imatinib-resistant Philadelphia chromosome-positive leukemias. *N Engl J Med* 354, 2531–41.

Weisberg, E., Manley, P. W., Cowan-Jacob, S. W., Hochhaus, A., and Griffin, J. D. 2007. Second generation inhibitors of BCR-ABL for the treatment of Imatinib-resistant chronic myeloid leukaemia. *Nat Rev Cancer* 7, 345–56.

Yoshida, C. and Melo, J. V. 2004. Biology of chronic myeloid leukemia and possible therapeutic approaches to Imatinib-resistant disease. *Int J Hematol* 79, 420–33.

CHAPTER 16

Bridging from Multiscale Modeling to Practical Clinical Applications in the Study of Human Gliomas

Gargi Chakraborty, Rita Sodt,
Susan Massey, Stanley Gu, Russell Rockne,
Ellsworth C. Alvord, Jr., and Kristin R. Swanson

CONTENTS

INTRODUCTION

Gliomas are malignant lesions of the brain characterized by their propensity to proliferate and invade the normal-appearing tissue. These brain tumors show diverse anatomic and metabolic traits that manifest differently across patients, ranging in aggressiveness from low-grade gliomas (LGGs) to high-grade gliomas (HGGs). Approximately 50% of glioma patients are diagnosed with glioblastoma multiforme (GBM), a World Health Organization grade IV glioma, which is the most aggressive type of brain tumor in adults (Louis et al. 2007). Difficulty in treating gliomas arises primarily from the invasiveness of the cell populations comprising the tumor. Despite extensive treatments including surgical resection, radiotherapy, and chemotherapy, GBMs have a propensity to recur and prove to be fatal with a median survival of approximately 1 year (Burnet et al. 2007; Welsh et al. 2009).

Modern radiology offers imaging techniques such as magnetic resonance imaging (MRI) and positron emission tomography (PET) that allow noninvasive detection of brain tumor anatomy and aberrant molecular activity, respectively. Clinically, routine MRIs comprise a variety of modalities, including T1-weighted gadolinium-enhanced (T1Gd), which images the highly vascularized portion of the tumor and T2-weighted and fluid attenuation inversion recovery (FLAIR) MRI, which image the bulk mass as well as surrounding edema with isolated invading tumor cells. However, because of the diffuse invasion of gliomas cells into the normal appearing brain, even current advances in radiology do not allow detection of the glioma cells that lie beyond the imageable tumor periphery (Harpold et al. 2007).

Mathematical modeling utilizing experimental and/or human data provides a novel tool for understanding the spatial and temporal growth of brain tumors by allowing us to create patient-specific virtual tumors that show the distribution of glioma cells that exist beyond the imageable boundary (Harpold et al. 2007; Rockne et al. 2008a; Swanson et al. 2003; Hatzikirou et al. 2005). These virtual controls serve as novel tools for assessing and predicting response to therapy as well as for designing new therapy. In this chapter, we will consider recent contributions to the mathematical modeling of gliomas, focusing on work bridging the gap between predictive multiscale mathematical modeling and current clinical care of gliomas.

QUANTIFYING IN VIVO GLIOMA KINETICS NONINVASIVELY

Recent models of tumor growth look at both the spatial motility and proliferation of tumor cells, and can be divided into two general modeling approaches: discrete and continuous. Cellular automata (CA) models are discrete in their approach (Hatzikirou and Deutsch 2008; Huang et al. 2008) and model individual tumor cells at discrete time-points along with cell–cell and cell–microenvironment interactions at the microscopic level. CA model predictions apply mostly to cell cultures because of the computational limitations of large numbers of cells typically seen in tumors of clinically detectable size. Continuum models assess the growth and spread of tumor cell density macroscopically, and have applications in predicting tumor growth on the scale of clinical imaging and gross tissue histology but with limited applicability at the cellular level. Both modeling approaches are necessary to understand glioma growth and connect cell or tissue culture experiments to clinical reality, which has led to proposals for multiscale models. A recent study by Tanaka et al. shows that hybridizing discrete continuum and compartment modeling allows assessment of tumor size as detected clinically, and individual cell behavior as detected in a cell culture (Tanaka et al. 2009). Deroulers et al. focus on multiscale migration study of gliomas and compare their results to in vitro data (Deroulers et al. 2009). These studies utilize mathematical tools to understand biological processes underlying glioma behavior; however, their current results cannot be paralleled to clinically available data such as medical imaging or gross tissue histology. For gliomas in particular, an ideal mathematical model would predict invasive tumor growth beyond the imageable abnormality and be easily amenable to input from the most commonly used (clinical) tools to assess glioma growth and response to therapy, namely clinical imaging and gross histology.

Professor J.D. Murray and colleagues, as reviewed in Murray (2003) and Harpold et al. (2007), proposed one of the first glioma models with prospects of clinical application. This is a reaction-diffusion model that captures both the diffuse dispersal and proliferation of glioma cells, hence denoted as the proliferation-invasion (PI) model. In other words, this equation says that the rate of change of glioma cell density is equal to the net diffusion of the glioma cells plus the net proliferation of the glioma cells. Translated to math, we have

$$\frac{\partial c}{\partial t} = \nabla \bullet (D(x)\nabla c) + \rho c \left(1 - \frac{c}{K} \right)$$

$$D(x) = \begin{cases} D_g & x \in \text{Grey} \\ D_w & x \in \text{White} \end{cases} \tag{16.1}$$

where c(x,t) defines the concentration of glioma cells over space (x) and time (t) and is spatially limited by the boundaries of the skull and ventricles, which are anatomical structures of the brain that serve as impenetrable walls for glioma cells. This model utilizes two main terms: *net dispersal/diffusion* (D) of tumor cells differentiated spatially in three dimensions, and *net proliferation* (ρ) of tumor cells with a logistic growth term, where K is the carrying capacity or maximum cell density allowed per cubic centimeter of tissue (about 10^8 cells/cc, assuming a typical cell with diameter of 10 μm). Simulations of this model are performed in the three-dimensional, anatomically-accurate brain domain, approximated by the BrainWeb atlas (Collins et al. 1998; Cocosco et al. 1997), to capture tumor growth in the complex architecture of the human brain (Swanson et al. 2002b; Swanson 1999).

In order to incorporate heterogeneous motility of tumor cells throughout the brain yet keep the model simple, the diffusion coefficient, D, is spatially defined according to the grey and white matter composition of the brain such that $D(x) = D_g$ in gray matter, which consists of densely packed neuronal cell bodies, and $D(x) = D_w$ in white matter, which comprises axon tracts originating from, and leading to, neuronal cell bodies in the gray matter (Swanson 1999). The density of glial cells is highest in white matter and organized as myelin surrounding axons. Gliomas preferentially migrate along white matter tracts with diffusion fastest parallel and slowest perpendicular to these fiber tracts (Stadlbauer et al. 2009). The variation in the net rate of diffusion in white matter (D_w) can be 5 to 100 times faster than in gray matter (D_g) (Swanson et al. 2000, 2002b; Harpold et al. 2007).

One result of the PI model is that of linear radial growth. That is, the radial growth pattern asymptotically approaches Fisher's approximation for a constant velocity of expansion $v = 2\sqrt{D\rho}$. An analysis of velocity for a set of patients with LGGs (one example shown in Figure 16.2) suggested that the rate of change of radial expansion is constant in untreated

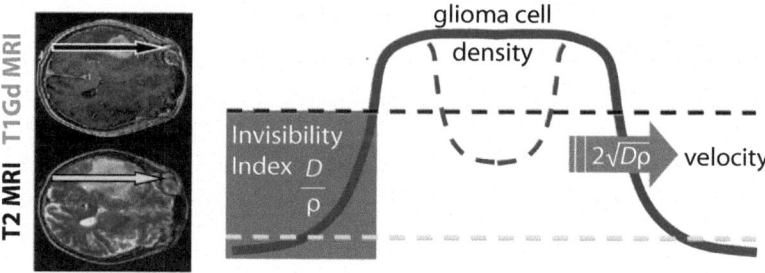

FIGURE 16.1 The left side shows a 2D slice through MRI scans (T1Gd on the top and T2 on the bottom). If we noted the diameter of the tumor abnormality through both images, we would obtain a theoretical 1D profile as shown on the right where glioma cell density is represented on the vertical axis and space is represented on the horizontal axis. The black dotted line is the threshold of detection for the T1Gd MRI (approximately 80% of total cell density), and the white dotted line is the threshold of detection for the T2 MRI (approximately 16% of total cell density). Velocity of the tumor growth asymptotically approaches Fisher's approximation and the invisibility index is given by the ratio of the diffusion and proliferation parameters.

gliomas (Swanson et al. 2004; Mandonnet et al. 2003, 2008) and that this constant velocity served as a prognostic factor for LGGs (Pallud et al. 2006). Swanson showed that the two parameters in Equation 16.1, D and ρ, could be derived from volumetric measurements of tumor abnormality as imaged with T1Gd and T2 MRI scans in humans (Harpold et al. 2007). Using this set of data collected at two time points prior to treatment or operation allows for the calculation of radial velocity of tumor growth (Figure 16.2). These results demonstrated that glioma growth prior to treatment could be predictively modeled in vivo in humans.

Many LGGs do not contrast-enhance, that is, these tumors are visible on T2 and FLAIR modalities, but not on T1Gd. This is because LGGs have little angiogenesis or new blood vessel formation, which allow leakage of the gadolinium-enhancement tracer. However, many LGGs progress to higher grades, although the time to progression (TTP) is quite variable (Chaichana et al. 2009; Michotte et al. 2004). In the study of LGGs, velocity of tumor growth predicted time to contrast enhancement on T1Gd MRI, thereby velocity could be utilized to predict TTP (Alvord and Swanson 2007). Prior to this result, malignant progression was considered an unpredictable event.

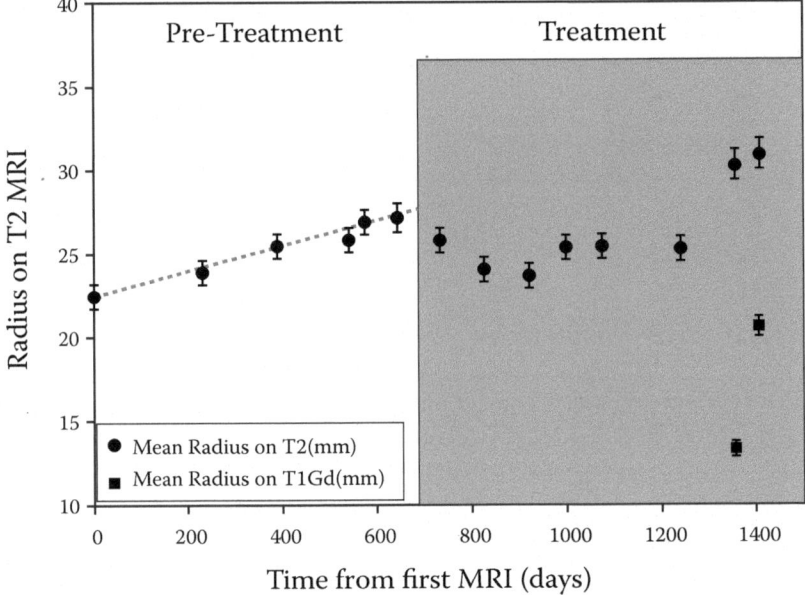

FIGURE 16.2 A patient with an LGG assessed serially on T1Gd and T2. The left part of the graph indicates measurements of mean tumor radius prior to treatment and shows that the velocity of radial expansion on T2 MRI is constant (change in radius is linear as displayed by the line of best fit—dotted gray line). The right part of the graph indicates tumor dynamics after treatment with radiation therapy and subtotal resection. The increase in T2 tumor radius along with contrast-enhancement on T1Gd seen after day 1200 is indicative of tumor progression. The drop in both T1Gd and T2 radius thereafter is due to the surgical removal of tumor.

ANISOTROPIC TUMOR CELL MOTILITY IN HETEROGENEOUS BRAIN TISSUE

Recently, the simulations of glioma invasion was improved by incorporating anisotropy using information from diffusion tensor imaging (DTI), a type of MRI that detects the directionality of water diffusion in the brain (Jbabdi et al. 2005). DTI scans provide a tensor at each spatial location (voxel) in the brain indicating the magnitude and direction that glioma cells would presumably prefer to migrate. It was shown that the shape and kinetic evolution of the tumors were better simulated with anisotropic rather than isotropic diffusion as expected because the tensors provided a patient-specific direction schema for the motility of glioma cells (Jbabdi et al. 2005). Implementation

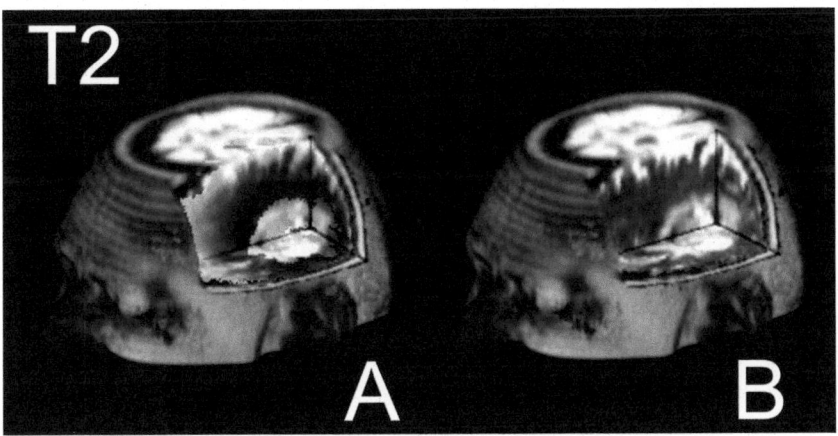

FIGURE 16.3 **(See color insert following page 40)** Panel A shows a patient-specific 3D simulation of the PI model including differential motility in the grey and white matter, while panel B shows the actual tumor detected by T2 MRI.

of the PI model in three dimensions by the Swanson group showed that the architecture of the brain segregated into gray and white matter dictating the tumor's spatial morphology as compared with T2-MRI, visualized in three dimensions (Figure 16.3). Including input from DTI scans to further dictate the motility of glioma cells, simulations of the PI model looking at radial growth of tumor over time indicates that there is indeed a difference in the predicted tumor radius in T1Gd and T2 MRIs with isotropy versus anisotropy (Sodt et al. 2009).

Other groups have considered extensions of the PI model incorporating DTI information as well as a biomechanical component of GBM growth that exerts a mass effect on surrounding brain structures such as skull and ventricles (Bondiau et al. 2008; Clatz et al. 2005). Mass effect is considered to be responsible for the mechanical impact of the tumor on surrounding brain tissue. However, for glioma models incorporating mass effect, the problem of parameter estimation from patient data such as imaging was unaddressed until a recent preliminary study by Hogea et al., which compares optimized parameter estimates for one patient case (Hogea et al. 2008).

ASSESSING GLIOMA ABLATION AND REGROWTH POSTTHERAPY WITH MATHEMATICAL PARAMETERS

Although glioma growth kinetics differ across patients, the standard of care is surprisingly homogeneous and consists of resection or biopsy, followed by radiation therapy along with adjuvant and concurrent chemotherapy. Mathematical modeling tools may be used to understand the extent to which certain therapies may improve survival for any individual patient and to develop a control method by which to assess whether certain patients will respond better to particular therapies.

Does More Treatment Mean Better Survival?

With the continual introduction and investigation of new and more aggressive therapies, assessing the benefits of treatment on decreasing tumor burden and improving a patient's quality of life can be a challenge for physicians. Recurrence of gliomas is common and likely arises from the fraction of cells that cannot be imaged and thus targeted for therapy. Although the diffuse extent of malignant glioma cell invasion cannot be imaged, it can be modeled in a virtual glioma with three-dimensional simulations of the PI model. As presented in Figure 16.1, 80% of the total tumor cell concentration is imageable by the T1Gd MRI and 16% (a fivefold increase) is imageable by the T2 MRI (Swanson et al. 2002b), from which a tumor cell gradient is inferred. The spatial distribution of invisible cells has been shown to lead to a better assessment of treatment planning and detect potential sites of tumor recurrence, which may aid physicians in designing an optimal time course for a glioma treatment-regimen that can be tailored to the patient (Swanson et al. 2004).

A survival analysis of 32 GBM patients indicated that regardless of therapy, the model-defined parameters correlated with patient prognosis (Wang et al. 2009). This result maintained statistical significance even while controlling for standard clinical parameters such as histology, age, and Karnofsky Performance Status (KPS) (Wang et al. 2009), showing the result to be robust. In particular, this pioneering work showed that in a univariate survival analysis, the net proliferation rate, and the ratio ρ/D, which is related to the shape of the gradient of invading glioma cells outside the T1Gd imageable region, strongly correlated with survival. This significant modeling result showed that patient-specific PI model parameters allow assessment of survival independent of standard clinical approaches.

Resection

It is not clear whether glioma patients in whom a larger mass of the tumor is removed tend to do better prognostically than those patients receiving a biopsy (McGirt et al. 2008, 2009; Laws et al. 2003; Devaux et al. 1993). Further, predicting survival based solely on the extent of resection (EOR) is ambiguous since it is difficult to quantitatively determine the percent of tumor removed due to the diffuse nature of gliomas, as not all tumor cells can be visualized through current imaging or histology (Harpold et al. 2007). The tumor's location in the brain also plays a significant role in determining how much of the tumor is resected. Large resections are avoided in cases where removal of more tumor region in critical parts of the brain may lead to a worse quality of life. For example, for tumors straddling the precentral gyrus, which is the cortical area involved in controlling most body motor functions, large resection would result in impaired movement control and execution. Thus, to quantitatively assess whether the EOR is related to survival requires other tools such as mathematical models.

Woodward et al. (1996) considered resection in the context of the spatially homogeneous PI model showing that even with varying extents of surgical resection, gliomas could not be completely "cured" due to the diffuse infiltration of malignant tumor cells throughout the brain. Building upon this idea, Swanson et al. (2008) considered glioma patients who had undergone either partial or near-total resection of the tumor and applied the patient-specific growth kinetic parameters D and ρ from the PI model to model and predict survival based on patient-specific virtual controls. Interestingly, survival correlated better with D and ρ than with the extent of resection. Thus, tumors with a higher propensity to diffuse and proliferate led to shorter survival regardless of the extent of resection (Swanson et al. 2008). Further assessment of the net diffusion and net proliferation parameters combined with the spatial location of gliomas suggests that these parameters are independent of tumor location (Szeto et al. 2009). These studies lead to a dominating hypothesis in glioma that the kinetic parameters of growth dictate patterns of tumor growth irrespective of location and may allow better prediction of survival than location-dependent surgical removal of the tumor.

Radiation Therapy

After surgery, external beam radiation therapy (XRT) is typically the second-line of treatment in the clinical management of gliomas. XRT targets the tumor mass not removed by surgery and has been shown to improve survival over other postoperative therapies (Erpolat et al. 2009; Bloor et al. 1962). However, XRT does not affect all patients equally: some patients respond better to XRT than others, and this phenomenon cannot be predicted clinically prior to treatment. Rockne et al 2009, combined classic linear-quadratic radiobiological mathematical models of the delivery of and response to XRT to equation (1). When applying this model to patient-specific tumor evolution seen on MR imaging, Rockne et al. (2008b) found that the model may help predict which patients will respond better to XRT and to what degree each will respond. Further, this sheds light on the patient-specific tailoring of XRT dose to actual disease distribution as opposed to the one-size-fits-all philosophy currently applied clinically in XRT dose design for each patient's tumor.

Dionysiou et al. (2004) introduced a four-dimensional model that considers the effects of XRT on the growth of GBMs. This model incorporated sensitivity to radiation, genetic profile of tumors, and dose fractionation. Updated work on the model explored the effect of XRT on tumor and normal tissue to simulate fractionation doses (Stamatakos et al. 2007). While the model utilizes tumor size from patient MRI to apply radiation fields, it does not yet have the predictive capacity to suggest patient-specific tailoring of therapy, although because of the large number of parameters to be estimated a method of optimizing treatment is provided (Dionysiou et al. 2008).

Swanson et al. (2008) provided a means of distinguishing glioma patients that are sensitive or resistant to XRT by using PI model-defined, patient-specific virtual controls. Studying eight patients with GBM, Swanson et al. (2008) developed virtual controls using the PI model that predicted untreated patient-specific glioma growth. Patients received XRT as per the standard of care, whereas virtually, patient-specific glioma growth was simulated without treatment. Determining time to reach fatal tumor burden as defined by the T1Gd radius of 35 mm allowed assessment of survival if the patient received no treatment (virtual control). Swanson et al. (2008) compared actual survival after treatment with XRT to the survival predicted by the PI model. Results indicated that approximately half the population survived as predicted by the PI model suggesting that even with XRT, survival was not improved. The other half of the population survived twice as long as predicted by the PI model suggesting that this population had radio-sensitive tumors.

Advancing the modeling methodology, Rockne et al. (2009; Rockne et al. 2008b) proposed an extension of the PI model to incorporate the delivery and effect of XRT as quantified by the classic linear-quadratic model (Hall 1994) for radiation efficacy. This study presents the effect of changing dose delivery from the current clinical standard on a virtual case study. Results suggest that a low-frequency and high-dose delivery scheme would be optimal in reducing the targeted tumor volume (Rockne et al. 2009). Further, for the extended model Rockne et al. show that the parameter of radiation sensitivity, α, can be predicted from ρ (patient-specific parameter of net proliferation in the PI model) prior to treatment, and can be applied to study XRT in virtual glioma patients that can be paralleled to clinical cases (Rockne 2008b). Powathil et al. (2007) described a similar reaction-diffusion model, however, their model did not estimate patient-specific parameters.

Other Therapies

Beyond surgery and radiation therapy, chemotherapy targeting aberrant cell-cycle activity or factors controlling neoangiogenesis among others is utilized to minimize tumor recurrence. Chemotherapeutic agents such as Temozolomide and Bevacizumab have been shown to improve patient survival (Erpolat et al. 2009; Nishikawa 2009; Zhang et al. 2009). However, survival for glioma patients still remains dismal and the incidence of drug-induced toxicity in patients is increased (Calabrese and Schlegel 2009). Mathematical modeling may shed light on the design of optimal chemotherapy dosing such that drug toxicity can be minimized and the attack on tumor cells maximized. Challenges include modeling the spatial and temporal drug delivery, along with simulating how drugs cross the blood–brain barrier (BBB).

Tracqui et al. (1995) first applied a continuous model to study the effects of chemotherapy spatially and temporally in brain tumors. This model extended Equation 16.1 to incorporate a loss term for cells that died from chemotherapy and investigated the effect of treatment on cancer cells resistant and sensitive to therapy. Results from these studies indicated that tumor cell subpopulations would react differently to treatment doses, and that this could explain the failure of the standard management of gliomas (Tracqui et al. 1995).

Similar work by Eikenberry et al. (2009) using a reaction-diffusion model provides an initial framework of how mathematical modeling can be utilized to monitor effects of multiple treatments in virtual GBM patients, but only qualitatively ties model simulations to clinical data. A model developed by Tian et al. (2009) on the other hand provides a purely theoretical perspective on assessing the effects of therapy on glioma growth. Powathil et al. (2007) come closer to describing effects of chemotherapy by modifying the base model studied by Murray's group, by appending death terms to a log-kill model. Modeling radiotherapy and chemotherapy simultaneously, this investigation reports the clinical result that chemotherapy with Temozolomide following XRT results in less neurotoxicity, and considers improving the targeting of tumor cells with concurrent Temozolomide and XRT, which is the current clinical practice (Powathil et al. 2007). This result, though in an early phase, promises further study and validation in the clinic.

Swanson et al. (Swanson et al. 2002a) compare a clinical case study to homogenous and heterogeneous models of drug delivery, which suggests that drug delivery is necessarily a function of the varying vascular density in brain tissue and thus accounts for the clinical observation of tumor reduction in some locations and continued growth in other regions of the tumor surroundings. A recent model (Hinow et al. 2009) based on the spatial dynamics of tumor–host interaction, incorporates the tumor microenvironment along with angiogenesis and suggests that the time course and spatial delivery of drugs needs to be considered, and supports Swanson's result from 2002. In addition, model simulations by Hinow et al. (2009) propose that chemotherapy may either heavily reduce the tumor mass or increase tumor growth substantially depending on how the drug interacts with both tumor and nontumor cells (Hinow et al. 2009). These results provide support for clinicians to plan out both the spatial and temporal effects of chemotherapeutic agents, again with the use of virtual controls and careful consideration of the spatial element of glioma disease distribution and response to therapy.

CAPTURING INFLUENCE OF TUMOR MICROENVIRONMENT ON GLIOMA GROWTH BY PARALLEL COMPARISON OF MRI AND PET

While clinically routine MRIs conducted serially provide information about changes to gross tumor anatomy over the scale of months, PET imaging provides a dynamic picture of information about the tumor on the molecular level, conferred through radiolabeled tracer activity monitored over minutes to hours (Ullrich et al. 2008). Two common PET radio tracers used in assessing brain tumors are: 18F-Fluoromisonidazole (FMISO), which detects hypoxia or regions of low oxygenation (Rajendran et al., 2006), and 3′-Deoxy-3′-[18F]-Fluorothymidine (FLT), which detects cell proliferation (Hatakeyama et al. 2008). Another common PET tracer used in the clinical management of gliomas is [18F]-Fluorodeoxyglucose (FDG), which is a marker of hyperglycolytic activity (Spence et al. 2004). Studies focusing on the clinical use of PET and MRI show that functional imaging tends to aid diagnosis and treatment planning of brain tumors (Krohn et al. 2007; Spence et al. 2008) because these scans capture regions of aberrant activity in the tumor microenvironment.

Hypoxia Assessed with FMISO-PET

Hypoxia not only decreases the radiation sensitivity of tumor cells, but also contributes to tumor progression and drives selection pressure for more aggressive cancer phenotypes (Lunt et al. 2009; Sullivan and Graham 2007). Hypoxic burden assessed in vivo by FMISO-PET (Figure 16.4c) correlates significantly with decreased survival and shorter time to tumor progression (Swanson et al. 2009; Spence et al. 2008). Comparison of hypoxic burden with tumor abnormality observed in spatially registered T1Gd MRI suggests a spatial link between bulk tumor mass and hypoxia (Swanson et al. 2009).

Due to the spatial link between MRI and FMISO-PET, the Swanson group hypothesized that metrics of glioma growth as quantified by the PI model along with MRI-based parameters might shed light upon the aggressiveness determined by the presence of hypoxic burden. This was in fact the case, and they confirmed the link between hypoxia and biological aggressiveness quantified by the ratio ρ/D, where higher ρ/D corresponded to a more aggressive tumor growth (Szeto et al. 2009). While hypoxia is a well-known hallmark of aggressive behavior, this work displayed the first quantitative connection between hypoxia imaged by PET and anatomical tumor data imaged by MRI as well as the profound implication that a mathematical model-based metric could link the two in a patient-specific manner.

While the PI model has been validated and shown to be accurate for assessing anatomical tumor growth and survival, it does not account for the role of the tumor microenvironment. Incorporating nutrient availability and deprivation into glioma kinetics necessitates an extension of the PI model that considers hypoxic burden and its connection to normoxic (tumor cells sufficiently supplied with oxygen) and necrotic (tumor cells that have died due to anoxia) cells. The link between the tumor cell populations relates the presence of vasculature, development of new vessels (angiogenesis) as needed to continue growth, and the communication between tumor cells and vessels through angiogenic growth factors. Incorporating a heterogeneous cell population and capturing the dynamics of tumor cells through the angiogenic cascade, an extended model, denoted the proliferation-invasion-hypoxia-necrosis-angiogenesis

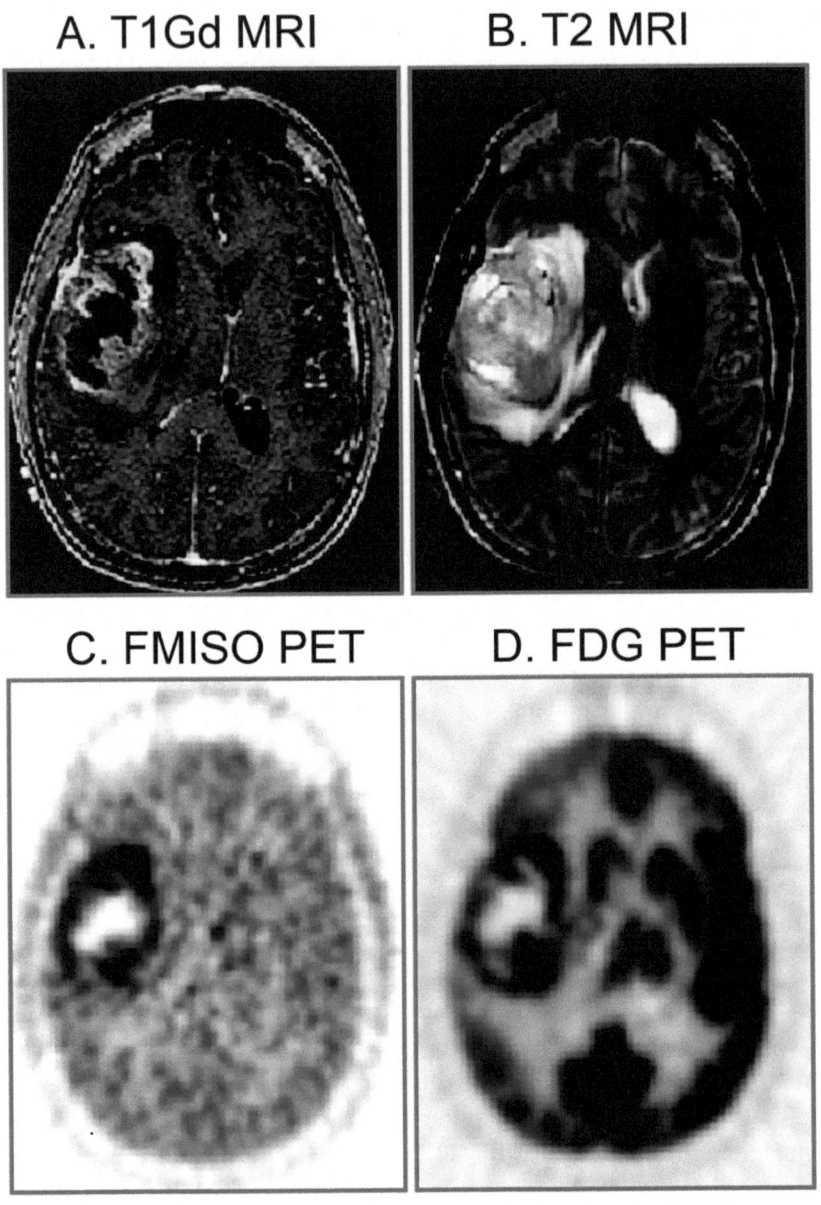

FIGURE 16.4 Panels A through D show a glioblastoma imaged in the axial plane on T1Gd-MRI, T2-MRI, FMISO-PET, and FDG-PET, respectively. The bright regions on the MRIs and the dark regions on the PETs represent tumor abnormality. In panel D, note that for an unprocessed FDG-PET, the gray matter contralateral to the glioma is isointense with the tumor abnormality making tumor identification challenging.

(PIHNA), is currently being investigated that builds upon the success of the PI model and is specifically designed to parallel MRI and PET imaging data (Swanson 2009; Harpold 2006b).

Hyperglycolysis Assessed with FDG-PET

FDG-PET, unlike FMISO-PET and FLT-PET is a clinically routine PET imaging modality, and indicates regions of hyperglycolysis (Figure 16.4d). Gray matter of the brain, however, demonstrates elevated levels of glycolysis relative to most normal body tissue, and even relative to other portions of the brain. This presents a challenge for differentiating regions of increased FDG signal intensity between the normal grey matter and that due to the tumor abnormality. Spence et al. (2004) investigated how to delineate tumor abnormality in FDG-PET using an ordinary differential equation based compartmental model for FDG metabolism kinetics. This study suggested that hyperactive tumor growth could be segmented visually and quantitatively from normal activity of gray matter by increasing the time interval between tracer delivery and PET acquisition (Spence et al. 2004). Additional studies considered patient-specific kinetic parameters derived from FDG-PET proved that different kinetics of glucose metabolism could aid in differentiating gliomas from other brain abnormalities such as lymphomas (Kimura et al. 2009). Further assessment of the abnormality seen on FDG-PET showed that this imaging modality served as a prognostic marker and could be utilized to evaluate response to therapy (Mankoff et al. 2007).

Since FDG-PET relays glucose metabolic activity in vivo, it can serve as a tool to assess the microenvironmental influence on tumor behavior. Harsh tumor microenvironment results from acidosis and hypoxia (Fang et al. 2008). In such conditions, better survival is conferred to those cancer cells that switch over to anaerobic respiration (glycolysis) from aerobic respiration (Gatenby et al. 2006). A recently proposed model of glioma growth applies game theory to capture the interaction between three distinct tumor phenotypes: autonomous growth, glycolysis, and invasion (Basanta et al. 2008). Results of this study implicate a switch to glycolysis in tumor cells to lead to more invasive tumor phenotypes, which agrees with biological hypothesis that cancerous cells display a switch from aerobic respiration to anaerobic respiration (Semenza et al. 2001; Gatenby et al. 2006). However, this model does not consider the spatial distribution of tumor cells and the interaction of the heterogeneous tumor cell population. Cellular automaton models exist for explaining tumor growth spatially and agree with the game theory model results that harsh microenvironments select for more invasive phenotypes (Anderson et al. 2006). These models would suggest that reoxygenating tumor cells as a clinical therapeutic approach may prevent selection of highly invasive phenotypes. However, cellular automata or game theory models cannot be directly translated to monitor clinical tumor growth yet.

Tumor Proliferation Assessed with FLT-PET

The PI model currently utilizes volumetric measurements of tumor from two MRIs taken prior to any treatment to determine the parameters, D and ρ. Although, MRIs are clinically routine, in many instances two observations are not available pretreatment. FLT-PET (Figure 16.5C) conducted in a single day can capture dynamic hyperproliferative activity of brain tumors (Hatakeyama et al. 2008). This dynamic information in addition to one pretreat MRI showing anatomical tumor abnormality may overcome the need for another pretreat MRI in determining D and ρ for the PI model. A preliminary investigation by Harpold et al. looked at how the cellular proliferation rate in tumor versus nontumor cells could be separated. Voxel data from FLT-PET was utilized as input to generate parametric maps that translated to parameters for a compartmental kinetic model of FLT metabolism (Harpold et al. 2006a). This research is still in its early phase since the population of patients receiving FLT-PET, an image modality still in clinical trials, is a fraction of those receiving MRIs.

A. T1Gd MRI   B. T2 MRI   C. FLT PET

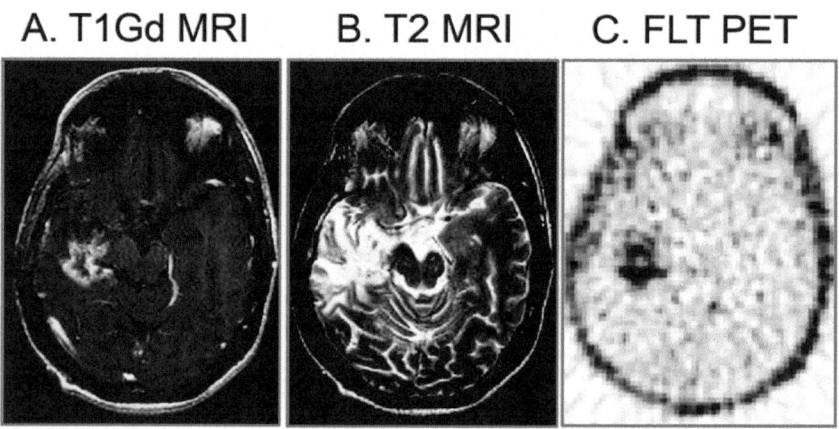

FIGURE 16.5   Panels A through C show a glioblastoma imaged in the axial plane on T1Gd-MRI, T2-MRI and FLT-PET, respectively. Note that the FLT-PET is a summed image of FLT activity recorded dynamically.

However, future studies are necessary to assess if D and ρ from combined FLT-PET and MRI at one time-point can be successfully applied to the PI model.

DEVELOPING MULTISCALE MODELS OF GLIOMAS FROM ANIMAL EXPERIMENTAL MODELS

New approaches aimed at resolving the challenges of conventional therapy to treat GBMs lie in understanding how these tumors originate. One hypothesis in tumorigenesis suggests that progenitor cells, which are precursors to glial cells, may be recruited to the heterogeneous population of brain tumor cells via platelet-derived growth factor (PDGF), a signaling protein with increasing evidence of involvement in brain tumors (Assanah et al. 2006; Shih et al. 2004). This pioneering work showed that injecting fluorescently-labeled retrovirus expressing human PDGF into the white matter of genetically normal adult rats resulted in the rats rapidly developing tumors, which appeared to be identical on histology and on MRI to human GBMs (Assanah et al. 2006, 2009). Assessing tumor cell populations through immunofluorescence analysis 14 days postinjection (DPI) of the retrovirus showed that less than 20% of the malignant mass consisted of the retrovirus-infected cells, and the tumor primarily comprised glial progenitors, which exhibited similar growth kinetics as retrovirus-infected tumor cells (Assanah et al. 2006). This experimental research suggested that simple overexpression of PDGF resulted in growth of a GBM-appearing abnormality in a previously disease-free brain. Moreover, the results pointed to a novel hypothesis for tumorigenesis: normal progenitor cells in the tumor periphery are recruited to the tumor mass through paracrine signaling of PDGF.

Results from in vivo experiments on rats highlighted a new avenue of glioma modeling. Capturing the microscopic interaction between tumor cells and normal progenitor cells via PDGF, while still modeling macroscopic diffusive and proliferative growth of the tumor, necessitated the development of a multiscale model. Like the PIHNA model, this model builds upon the success of the PI model and uses a reaction-diffusion continuum approach and incorporates two cell populations—the retrovirus infected tumor cells and uninfected glial progenitor cells—and looks at the molecular interaction between these populations via PDGF as illustrated in a biological schematic in Figure 16.6. Currently, this model is being studied to develop virtual controls that can provide simulations of both tissue histology and tumor appearance on MRIs (Massey et al. 2009).

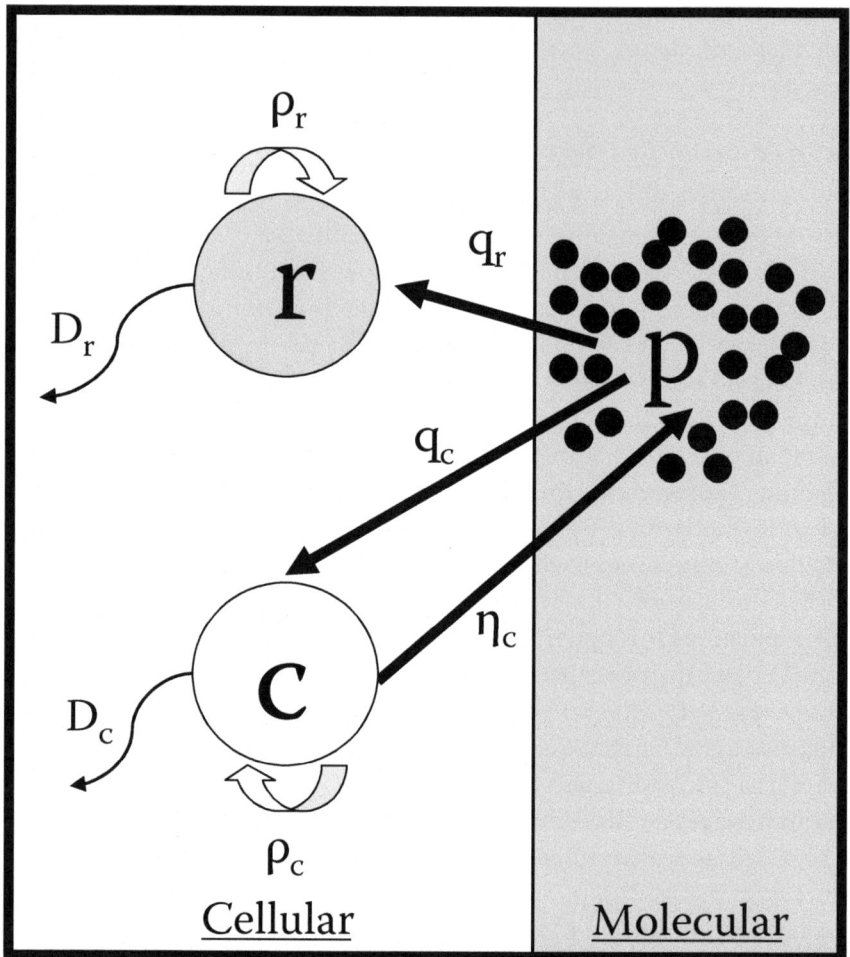

FIGURE 16.6 This schematic shows the hypothesized connection between recruited or uninfected glial progenitor cells (r) and infected tumor cells (c) that serves as a base for the multiscale model of tumorigenesis. Infected cells (c) secrete PDGF (p) at rate η_c in the extracellular space compartment. Via paracrine signaling, these ligands stimulate both the r and c cell populations to proliferate and migrate in a dose-dependent manner. The consumption of p by r and c in the process occurs at rates q_r and q_c, respectively, resulting in migration rates D_r and D_c, and proliferation rates ρ_r and ρ_c.

Such simulations can be compared to histological and MRI data on rats and can provide a channel to connect experimental animal models to in vivo human imaging data. Further, such multiscale models based in animal experimentation provide novel tools for bridging experimental results to human data by providing some assessment of the reasonability of the biological mechanism described.

DISCUSSION

The past few years have seen a surge of investigations in the mathematical modeling of gliomas. In this chapter, we have provided a glance at how mathematical modeling of gliomas relates to the clinical reality of brain tumor assessment. Multiple approaches exist to model glioma growth from the microscopic to the macroscopic scales. We specifically focus on the reaction-diffusion continuum models (PI, PIHNA, PDGF) as they come the closest to informing clinical reality because they can (a) be made specific to patients, thereby providing virtual controls of glioma growth, and (b) parallel clinical tools are utilized to assess gliomas such as MRI and PET imaging, and therefore show the highest promise of being clinically informative. However, quantitatively assessing the biological basis of glioma behavior necessitates both discrete and continuous modeling approaches, which can inform new areas of experimentation and modeling with the goal of generating tools for clinical use. Future directions of glioma modeling look towards the development of mathematical models that can be applied in optimizing glioma therapy, that can utilize PET imaging to extract biochemical parameters for reproducing the tumor microenvironment and parallel patient-specific PET scans, and that can connect analysis of animal-based experimentation to the clinical scale.

ACKNOWLEDGMENTS

We gratefully acknowledge the McDonnell Foundation, the Academic Pathology Fund, and the NIH (R01 NS060752, P01 CA42045) for their generous support.

REFERENCES

Alvord, E. C., Jr., and K. R. Swanson. 2007. Using mathematical modeling to predict survival of low-grade gliomas. *Ann Neurol* 61 (5):496; author reply 496–7.

Anderson, A. R., A. M. Weaver, P. T. Cummings, and V. Quaranta. 2006. Tumor morphology and phenotypic evolution driven by selective pressure from the microenvironment. *Cell* 127:905–15.

Assanah, M. C., J. N. Bruce, S. O. Suzuki, A. Chen, J. E. Goldman, and P. Canoll. 2009. PDGF stimulates the massive expansion of glial progenitors in the neonatal forebrain. *Glia* 57:1835–47.

Assanah, M., R. Lochhead, A. Ogden, J. Bruce, J. Goldman, and P. Canoll. 2006. Glial progenitors in adult white matter are driven to form malignant gliomas by platelet-derived growth factor-expressing retroviruses. *J Neurosci* 26 (25):6781–90.

Basanta, D., M. Simon, H. Hatzikirou, and A. Deutsch. 2008. Evolutionary game theory elucidates the role of glycolysis in glioma progression and invasion. *Cell Prolif* 41 (6):980–7.

Bloor, R. J., A. W. Templeton, and R. S. Quick. 1962. Radiation therapy in the treatment of intracranial tumors. *Am J Roentgenol Radium Ther Nucl Med* 87:463–72.

Bondiau, P., O. Clatz, M. Sermesant, P. Marcy, H. Delingette, M. Frenay, and N. Ayache. 2008. Biocomputing: numerical simulation of glioblastoma growth using diffusion tensor imaging. *Phys Med Biol* 53:879–93.

Burnet, N. G., A. G. Lynch, S. J. Jefferies, S. J. Price, P. H. Jones, N. M. Antoun, J. H. Xuereb, and U. Pohl. 2007. High grade glioma: imaging combined with pathological grade defines management and predicts prognosis. *Radiother Oncol* 85 (3):371–8.

Calabrese, P., and U. Schlegel. 2009. Neurotoxicity of treatment. *Recent Results Cancer Res* 171:165–74.

Chaichana, K. L., M. J. McGirt, A. Niranjan, A. Olivi, P. C. Burger, and A. Quinones-Hinojosa. 2009. Prognostic significance of contrast-enhancing low-grade gliomas in adults and a review of the literature. *Neurol Res.* 31:931–9.

Clatz, O., M. Sermesant, P. Y. Bondiau, H. Delingette, S. K. Warfield, G. Malandain, and N. Ayache. 2005. Realistic simulation of the 3-D growth of brain tumors in MR images coupling diffusion with biomechanical deformation. *IEEE Trans Med Imaging* 24 (10):1334–46.

Collins, D. L., A. P. Zijdenbos, V. Kollokian, J. G. Sled, N. J. Kabani, C. J. Holmes, and A. C. Evans. 1998. Design and construction of a realistic digital brain phantom. *IEEE Trans Med Imaging* 17 (3):463–8.

Cocosco, C. A., V. Kollokian, Kwan R. K.-S., and A. C. Evans. 1997. Brainweb: online interface to a 3D simulated brain database. *Neuroimage* 5:S425.

Deroulers, C., M. Aubert, M. Badoual, and B. Grammaticos. 2009. Modeling tumor cell migration: From microscopic to macroscopic models. *Phys Rev E Stat Nonlin Soft Matter Phys* 79 (3 Pt 1):031917.

Devaux, B. C., J. R. O'Fallon, and P. J. Kelly. 1993. Resection, biopsy, and survival in malignant glial neoplasms: a retrospective study of clinical parameters, therapy, and outcome. *J Neurosurg* 78 (5):767–75.

Dionysiou, D. D., G. S. Stamatakos, D. Gintides, N. Uzunoglu, and K. Kyriaki. 2008. Critical parameters determining standard radiotherapy treatment outcome for glioblastoma multiforme: a computer simulation. *Open Biomed Eng J* 2:43–51.

Dionysiou, D. D., G. S. Stamatakos, N. K. Uzunoglu, K. S. Nikita, and A. Marioli. 2004. A four-dimensional simulation model of tumour response to radiotherapy in vivo: parametric validation considering radiosensitivity, genetic profile and fractionation. *J Theor Biol* 230 (1):1–20.

Eikenberry, S. E., T. Sankar, M. C. Preul, E. J. Kostelich, C. J. Thalhauser, and Y. Kuang. 2009. Virtual glioblastoma: growth, migration and treatment in a three-dimensional mathematical model. *Cell Prolif* 42 (4):511–28.

Erpolat, O. P., M. Akmansu, F. Goksel, H. Bora, E. Yaman, and S. Buyukberber. 2009. Outcome of newly diagnosed glioblastoma patients treated by radiotherapy plus concomitant and adjuvant temozolomide: a long-term analysis. *Tumori* 95 (2):191–7.

Fang, J. S., R. D. Gillies, and R. A. Gatenby. 2008. Adaptation to hypoxia and acidosis in carcinogenesis and tumor progression. *Semin Cancer Biol* 18 (5):330–7.

Gatenby, R. A., E. T. Gawlinski, A. F. Gmitro, B. Kaylor, and R. J. Gillies. 2006. Acid-mediated tumor invasion: a multidisciplinary study. *Cancer Res* 66 (10):5216–23.

Hall, E. 1994. *Radiobiology for the Radiologist*. 4th ed. Philadelphia: J.B. Lippincott Company.

Harpold, H. L., E. C. Alvord, Jr., and K. R. Swanson. 2007. The evolution of mathematical modeling of glioma proliferation and invasion. *J Neuropathol Exp Neurol* 66 (1):1–9.

Harpold, H. L. P., P. Vicini, and K. R. Swanson. 2006a. Kinetic modeling of FLT-PET to generate parametric maps of proliferation. *J Undergraduate Res Bioeng* 6 (1):49–68.

Harpold, H.L.P., Anderson, A.R.A., Alvord, E.C., Jr., and Swanson, K.R. 2006b. Simulating low and high-grade human gliomas: an in silico model integrating the angiogenic cascade. Paper read at Society for Neuro-Oncology.

Hatakeyama, T., N. Kawai, Y. Nishiyama, Y. Yamamoto, Y. Sasakawa, T. Ichikawa, and T. Tamiya. 2008. 11C-methionine (MET) and 18F-fluorothymidine (FLT) PET in patients with newly diagnosed glioma. *Eur J Nucl Med Mol Imaging* 35 (11):2009–17.

Hatzikirou, H., A. Deutsch, C. Schaller, M. Simon, and K. R. Swanson. 2005. Mathematical modelling of glioblastoma tumour development: a review. *Math Models Methods Appl Sci*, 11:1779–1794.

Hatzikirou, H., and A. Deutsch. 2008. Cellular automata as microscopic models of cell migration in heterogeneous environments. *Curr Top Dev Biol* 81:401–34.

Hinow, P., P. Gerlee, L. J. McCawley, V. Quaranta, M. Ciobanu, S. Wang, J. M. Graham, B. P. Ayati, J. Claridge, K. R. Swanson, M. Loveless, and A. R. Anderson. 2009. A spatial model of tumor-host interaction: application of chemotherapy. *Math Biosci Eng* 6 (3):521–46.

Hogea, C., C. Davatzikos, and G. Biros. 2008. An image-driven parameter estimation problem for a reaction-diffusion glioma growth model with mass effects. *J Math Biol* 56 (6):793–825.

Huang, S., D. Vader, Z. Wang, A. Stemmer-Rachamimov, D. A. Weitz, G. Dai, B. R. Rosen, and T. S. Deisboeck. 2008. Using magnetic resonance microscopy to study the growth dynamics of a glioma spheroid in collagen I: a case study. *BMC Med Imaging* 8:3.

Jbabdi, S., E. Mandonnet, H. Duffau, L. Capelle, K. R. Swanson, M. Pelegrini-Issac, R. Guillevin, and H. Benali. 2005. Simulation of anisotropic growth of low-grade gliomas using diffusion tensor imaging. *Magn Reson Med* 54 (3):616–24.

Kimura, N., Y. Yamamoto, R. Kameyama, T. Hatakeyama, N. Kawai, and Y. Nishiyama. 2009. Diagnostic value of kinetic analysis using dynamic 18F-FDG-PET in patients with malignant primary brain tumor. *Nucl Med Commun* 30 (8):602–9.

Krohn, K. A., F. O'Sullivan, J. Crowley, J. F. Eary, H. M. Linden, J. M. Link, D. A. Mankoff, M. Muzi, J. G. Rajendran, A. M. Spence, and K. R. Swanson. 2007. Challenges in clinical studies with multiple imaging probes. *Nucl Med Biol* 34 (7):879–85.

Laws, E. R., I. F. Parney, W. Huang, F. Anderson, A. M. Morris, A. Asher, K. O. Lillehei, M. Bernstein, H. Brem, A. Sloan, M. S. Berger, and S. Chang. 2003. Survival following surgery and prognostic factors for recently diagnosed malignant glioma: data from the Glioma Outcomes Project. *J Neurosurg* 99 (3):467–73.

Louis, D.N., H. Ohgaki, O. D. Wiestler, and W. K. Cavenee. 2007. *WHO Classification of Tumours of the Central Nervous System*. Edited by D. N. Louis, H. Ohgaki, O. D. Wiestler and W. K. Cavenee. 4th ed. Geneva, Switzerland: Renouf Publishing Co. Ltd.

Lunt, S. J., N. Chaudary, and R. P. Hill. 2009. The tumor microenvironment and metastatic disease. *Clin Exp Metastasis* 26 (1):19–34.

Mandonnet, E., J. Y. Delattre, M. L. Tanguy, K. R. Swanson, A. F. Carpentier, H. Duffau, P. Cornu, R. Van Effenterre, E. C. Alvord, Jr., and L. Capelle. 2003. Continuous growth of mean tumor diameter in a subset of grade II gliomas. *Ann Neurol* 53 (4):524–8.

Mandonnet, E., J. Pallud, O. Clatz, L. Taillandier, E. Konukoglu, H. Duffau, and L. Capelle. 2008. Computational modeling of the WHO grade II glioma dynamics: principles and applications to management paradigm. *Neurosurg Rev* 31 (3):263–9.

Mankoff, D. A., J. F. Eary, J. M. Link, M. Muzi, J. G. Rajendran, A. M. Spence, and K. A. Krohn. 2007. Tumor-specific positron emission tomography imaging in patients: [18F] fluorodeoxyglucose and beyond. *Clin Cancer Res* 13 (12):3460–9.

Massey, S., Canoll, P., and Swanson, K. R. 2009. Paracrine PDGF signaling and progression in experimental gliomas. Paper read at the International Conference on Mathematical Biology and Annual Meeting of the Society for Mathematical Biology., at the University of British Columbia.

McGirt, M. J., K. L. Chaichana, F. J. Attenello, J. D. Weingart, K. Than, P. C. Burger, A. Olivi, H. Brem, and A. Quinones-Hinojosa. 2008. Extent of surgical resection is independently associated with survival in patients with hemispheric infiltrating low-grade gliomas. *Neurosurgery* 63 (4):700–7; author reply 707–8.

McGirt, M. J., K. L. Chaichana, M. Gathinji, F. J. Attenello, K. Than, A. Olivi, J. D. Weingart, H. Brem, and A. R. Quinones-Hinojosa. 2009. Independent association of extent of resection with survival in patients with malignant brain astrocytoma. *J Neurosurg* 110 (1):156–62.

Michotte, A., B. Neyns, C. Chaskis, J. Sadones, and P. In 't Veld. 2004. Neuropathological and molecular aspects of low-grade and high-grade gliomas. *Acta Neurol Belg* 104 (4):148–53.

Murray, J. D. 2003. *Mathematical Biology II. Spatial Models and Biological Applications.* 3 ed. 2 vols. Vol. 2. New York: Springer-Verlag.

Nishikawa, R. 2009. Treatment of glioma with temozolomide. *Brain Nerve* 61 (7):849–54.

Pallud, J., E. Mandonnet, H. Duffau, M. Kujas, R. Guillevin, D. Galanaud, L. Taillandier, and L. Capelle. 2006. Prognostic value of initial magnetic resonance imaging growth rates for World Health Organization grade II gliomas. *Ann Neurol* 60 (3):380–3.

Powathil, G., M. Kohandel, S. Sivaloganathan, A. Oza, and M. Milosevic. 2007. Mathematical modeling of brain tumors: effects of radiotherapy and chemotherapy. *Phys Med Biol* 52 (11):3291–306.

Rajendran, J. G., Hendrickson, K. R., Spence, A. M., Muzi, M., Krohn, K. A., and Mankoff, D. A. 2006. Hypoxia imaging–directed radiation treatment planning, *Eur J Nucl Mol Imaging* 33 Suppl 13: 44–53.

Rockne, R., E. C. Alvord, Jr., J. K. Rockhill, and K. R. Swanson. 2009. A mathematical model for brain tumor response to radiation therapy. *J Math Biol* 58 (4–5):561–78.

Rockne, R., E. C. Alvord, P. Reed, and K. R. Swanson. 2008a. Modeling the growth and invasion of gliomas, from simple to complex: the Goldie Locks paradigm. Paper read at BIOMAT, at Rio de Janeiro.

Rockne, R., J. K. Rockhill, M. Mrugala, A. M. Spence, I. Kalet, K. Hendrickson, A. Lai, T. Cloughesy, E. C. Alvord, Jr., and K. R. Swanson. 2010. Predicting the efficacy of radiotheraphy in individual glioblastoma patients *in vivo*: a mathematical modeling approach *Phys Med Biol* 55: 3271–85.

Semenza, G. L., D. Artemov, A. Bedi, Z. Bhujwalla, K. Chiles, D. Feldser, E. Laughner, R. Ravi, J. Simons, P. Taghavi, and H. Zhong. 2001. "The metabolism of tumours": 70 years later. *Novartis Found Symp* 240:251–60; discussion 260–4.

Shih, A. H., C. Dai, X. Hu, M. K. Rosenblum, J. A. Koutcher, and E. C. Holland. 2004. Dose-dependent effects of platelet-derived growth factor-B on glial tumorigenesis. *Cancer Res* 64 (14):4783–9.

Sodt, R., R. Rockne, K. R. Swanson, I. Kalet. 2009. Simulation of anisotropic growth of gliomas using diffusion tensor imaging. International Conference on Mathematical Biology and Annual Meeting of the Society for Mathematical Biology, July 2009.

Spence, A. M., M. Muzi, D. A. Mankoff, S. F. O'Sullivan, J. M. Link, T. K. Lewellen, B. Lewellen, P. Pham, S. Minoshima, K. Swanson, and K. A. Krohn. 2004. 18F-FDG PET of gliomas at delayed intervals: improved distinction between tumor and normal gray matter. *J Nucl Med* 45 (10):1653–9.

Spence, A. M., M. Muzi, K. R. Swanson, F. O'Sullivan, J. K. Rockhill, J. G. Rajendran, T. C. H. Adamsen, J. M. Link, P. E. Swanson, K. J. Yagle, R. C. Rostomily,

D. L. Silbergeld, and K. Krohn. 2008. Regional hypoxia in glioblastoma multiforme quantified with [18F]fluoromisonidazole positron emission tomography before radiotherapy: correlation with time to progression and survival. *Clin Cancer Res* 14 (9):2623–30.

Stadlbauer, A., E. Polking, O. Prante, C. Nimsky, M. Buchfelder, T. Kuwert, R. Linke, M. Doelken, and O. Ganslandt. 2009. Detection of tumour invasion into the pyramidal tract in glioma patients with sensorimotor deficits by correlation of (18)F-fluoroethyl-L: -tyrosine PET and magnetic resonance diffusion tensor imaging. *Acta Neurochir (Wien)* 151 (9):1061–9.

Stamatakos, G., V. P. Antipas, and N. K. Ozunoglu. 2007. A patient-specific in vivo tumor and normal tissue model for prediction of the response to radiotherapy. *Methods Inf Med* 46 (3):367–75.

Sullivan, R., and C. H. Graham. 2007. Hypoxia-driven selection of the metastatic phenotype. *Cancer Metastasis Rev* 26 (2):319–31.

Swanson, K. R. 1999. Mathematical Modeling of the Growth and Control of Tumors. PhD Thesis, University of Washington, Seattle, WA.

———. 2008. Quantifying glioma cell growth and invasion in vitro. *Math Computer Modeling* 47:638–48.

Swanson, K. R., E. C. Alvord, Jr., and J. D. Murray. 2000. A quantitative model for differential motility of gliomas in grey and white matter. *Cell Prolif* 33 (5):317–29.

———. 2002a. Quantifying efficacy of chemotherapy of brain tumors with homogeneous and heterogeneous drug delivery. *Acta Biotheoretica* 50 (4):223–37.

———. 2002b. Virtual brain tumours (gliomas) enhance the reality of medical imaging and highlight inadequacies of current therapy. *Br J Cancer* 86 (1):14–18.

Swanson, K. R., C. Bridge, J. D. Murray, and E. C. Alvord, Jr. 2003. Virtual and real brain tumors: using mathematical modeling to quantify glioma growth and invasion. *J Neurol Sci* 216 (1):1–10.

Swanson, K. R., G. Chakraborty, C. H. Wang, R. Rockne, H. L. Harpold, M. Muzi, T. C. Adamsen, K. A. Krohn, and A. M. Spence. 2009. Complementary but distinct roles for MRI and 18F-fluoromisonidazole PET in the assessment of human glioblastomas. *J Nucl Med* 50 (1):36–44.

Swanson, K. R., H. L. Harpold, D. L. Peacock, R. Rockne, C. Pennington, L. Kilbride, R. Grant, J. M. Wardlaw, and E. C. Alvord, Jr. 2008a. Velocity of radial expansion of contrast-enhancing gliomas and the effectiveness of radiotherapy in individual patients: a proof of principle. *Clin Oncol (R Coll Radiol)* 20 (4):301–8.

Swanson, K. R., R. C. Rostomily, and E. C. Alvord, Jr. 2008b. A mathematical modelling tool for predicting survival of individual patients following resection of glioblastoma: a proof of principle. *Br J Cancer* 98 (1):113–9.

Swanson, K. R. 2009. Bridging from Anatomic Imaging to Molecular Imaging through Multi-scale Models for Brain Tumor Growth and Invasion. Paper read at International Conference on Mathematical Biology and Annual Meeting of The Society for Mathematical Biology at University of British Columbia.

Swanson, K. R., E. C. Alvord, Jr., and J. D. Murray. 2004. Dynamics of a model for brain tumors reveals a small window for therapeutic intervention. *Discrete Cont Dyn—B* 4 (1):289–95.

Szeto, M. D., G. Chakraborty, J. Hadley, R. Rockne, M. Muzi, E. C. Alvord, Jr., K. A. Krohn, A. M. Spence, and K. R. Swanson. 2009a. Quantitative metrics of net proliferation and invasion link biological aggressiveness assessed by MRI with hypoxia assessed by FMISO-PET in newly diagnosed glioblastomas. *Cancer Res* 69 (10):4502–9.

Szeto, M., Rockne, R., and Swanson, K. R. 2009b. Anatomic Variation in Quantitative Measures of Glioma Aggressiveness. Paper read at International Conference on Mathematical Biology and Annual Meeting of the Society of Mathematical Biology., at University of British Columbia.

Tanaka, M. L., W. Debinski, and I. K. Puri. 2009. Hybrid mathematical model of glioma progression. *Cell Prolif* 42 (5):637–46.

Tian, J. P., A. Friedman, J. Wang, and E. A. Chiocca. 2009. Modeling the effects of resection, radiation and chemotherapy in glioblastoma. *J Neurooncol* 91 (3):287–93.

Tracqui, P., G. C. Cruywagen, D. E. Woodward, G. T. Bartoo, J. D. Murray, and E. C. Alvord, Jr. 1995. A mathematical model of glioma growth: the effect of chemotherapy on spatio-temporal growth. *Cell Prolif* 28 (1):17–31.

Ullrich, R. T., L. W. Kracht, and A. H. Jacobs. 2008. Neuroimaging in patients with gliomas. *Semin Neurol* 28 (4):484–94.

Wang, C. H., J. K. Rockhill, M. Mrugala, D. L. Peacock, A. Lai, K. Jusenius, J. M. Wardlaw, T. Cloughesy, A. M. Spence, R. Rockne, E. C. Jr. Alvord, and K. R. Swanson. 2009. Prognostic significance of growth kinetics in newly diagnosed glioblastomas revealed by combining serial imaging with a novel bio-mathematical model. *Cancer Res.* 69:9133–40.

Welsh, J. W., R. K. Ellsworth, R. Kumar, K. Fjerstad, J. Martinez, R. B. Nagel, J. Eschbacher, and B. Stea. 2009. Rad51 protein expression and survival in patients with glioblastoma multiforme. *Int J Radiat Oncol Biol Phys* 74 (4):1251–5.

Woodward, D. E., J. Cook, P. Tracqui, G. C. Cruywagen, J. D. Murray, and E. C. Alvord, Jr. 1996. A mathematical model of glioma growth: the effect of extent of surgical resection. *Cell Prolif* 29 (6):269–88.

Zhang, W., X. G. Qiu, B. S. Chen, S. W. Li, Y. Cui, H. Ren, and T. Jiang. 2009. Antiangiogenic therapy with Bevacizumab in recurrent malignant gliomas: analysis of the response and core pathway aberrations. *Chin Med J (Engl)* 122 (11):1250–4.

CHAPTER 17

Personalization of Reaction-Diffusion Tumor Growth Models in MR Images

Application to Brain Gliomas Characterization and Radiotherapy Planning

Ender Konukoglu, Olivier Clatz,
Hervé Delingette, and Nicholas Ayache

CONTENTS

INTRODUCTION

Gliomas are neoplasms of glial cells that support and nourish the brain. These tumors have varying histopathological features and biological behavior showing different aggressiveness levels, from benign-grade I to malignant-grade IV (glioblastoma multiforme). There has been a vast amount of research in mathematical modeling to describe the growth dynamics of these tumors (Byrne et al. 2006; Cristini et al. 2003; Frieboes et al. 2007; Patel et al. 2001; Stamatakos et al. 2006; Zhang et al. 2007). Lately, specific type of macroscopic models, the reaction-diffusion models, received considerable attention from the literature in the attempt to link glioma growth models to medical images (Tracqui et al. 1995; Swanson et al. 2000; Clatz et al. 2005; Jbabdi et al. 2005; Hogea et al. 2007; Mandonnet et al. 2008). These recent models integrate information coming from medical images, specifically through anatomical and diffusion images, in their formulation. This integration is crucial for the transfer of mathematical models to the clinical applications since medical images are conventionally used for diagnosis and patient follow-up in the clinical routine. One of the biggest challenges in this transfer is the automatic adaptation of mathematical models to the patient, based on images. In this chapter, first we address the problem of adapting the recent reaction-diffusion models to specific patient cases using the time series of medical (magnetic resonance: MR) images (Konukoglu et al. 2009a). Following this, we address the question of retrieving relevant information for radiotherapy planning from the personalized reaction-diffusion models (Konukoglu 2009).

Reaction-Diffusion Models

Reaction-diffusion models describe the evolution of gliomas via proliferation of tumor cells and infiltration of the surrounding healthy tissue.

The building block of these models is the reaction-diffusion type partial differential equations of this form:

$$\frac{\partial u}{\partial t} = \nabla \cdot \left(\mathbf{D}(\mathbf{x})\nabla u\right) + \rho u\left(1-u\right), \; \mathbf{D}\nabla u \cdot \mathbf{n}_{\partial\Omega}, \tag{17.1}$$

where u represents the tumor cell density, $\mathbf{D}$ is the local diffusion tensor (i.e., symmetric positive definite 3×3 matrix), ρ is the proliferation rate, Ω is the brain domain, and $\partial\Omega$ represents the boundaries of the brain. The two terms on the right hand side correspond to the two phenomena described by the model: the diffusion term $\nabla \cdot (\mathbf{D}\nabla u)$ models the migration of tumor cells within the brain tissue and the reaction term $\rho u(1-u)$ models the proliferation of tumor cells.

Different models proposed in the literature mostly differ by the construction of the $\mathbf{D}$ tensor and the form of the proliferation term (Tracqui et al. 1995; Swanson et al. 2000; Giese 1996; Clatz et al. 2005; Jbabdi et al. 2005; Hogea et al. 2007). In this chapter we concentrate on the personalization of the growth models proposed in Clatz et al. (2005) and Jbabdi et al. (2005). However, the methods explained in this chapter are independent from the exact construction of the tensor and can be applied to both of these formulations, as well as to other reaction-diffusion models. The two aforementioned models are based on Equation 16.1, and they construct the diffusion tensor $\mathbf{D}$ as follows:

$$\mathbf{D}(\mathbf{x}) = \begin{cases} d_g\mathbf{I}, \; \mathbf{x} \in \text{gray matter} \\ f\left(d_w, \mathbf{D}_{\text{water}}\right), \; \mathbf{x} \in \text{white matter} \end{cases} \tag{17.2}$$

The choice of the function f is defined differently in the two models:

$$f = \begin{cases} d_w\mathbf{D}_{\text{water}}, \; \text{(Clatz et al. 2005)} \\ \mathbf{V}(\mathbf{x})\left[\text{diag}\left(e_1(\mathbf{x})d_w, d_g, d_g\right)\right]\mathbf{V}(\mathbf{x})^{\text{T}}, \; \text{(Jbabdi et al. 2005)} \end{cases}$$

where d_w is the diffusion rate in the white matter (Clatz et al. 2005) or along the white matter fiber tracts (Jbabdi et al. 2005), d_g is the diffusion rate in the gray matter, I is the unit tensor and $\mathbf{D}_{\text{water}} = \mathbf{V}\Lambda\mathbf{V}^{\text{T}}$ is the diffusion tensor of water molecules at point $\mathbf{x}$ with Λ representing the eigenvalue matrix and e_1 being its highest eigenvalue.

Image-Guided Model Personalization

Once the mathematical model describing the general dynamics of the tumor growth process is created, the personalization is defined as the estimation of the model parameters based on the observations. The image-guided personalization focuses on determining the reaction-diffusion parameters, the diffusion tensor $\mathbf{D}$ (d_w and d_g), and the proliferation rate ρ, based on the observed evolution of the tumor in the time series medical images. The difficulty in this estimation is due to the sparsity of the available information. The reaction-diffusion models describe the temporal evolution of *tumor cell densities,* whereas in the images we only observe the evolution of the *tumor delineation,* which is assumed to correspond to an iso-density contour (Burger 1988) as shown in Figure 17.1. Therefore, reaction-diffusion models are not directly applicable for the estimation problem.

The problem of parameter estimation in the context of tumor growth models is a rather unexplored problem. A first attempt was made by Tracqui et al. (1995) where they optimized the parameters of their model by comparing the area of the tumor observed in CT images at different times and the area of the simulated tumor. The drawback of this approach was the use of tumor cell densities requiring an initialization of the density distribution throughout the brain although these densities are not observable in the images (again, we observe tumor delineations). More recently, in (Hogea et al. 2008), Hogea et al. have optimized their parameters using two different methods: by comparing locations of some manually-placed

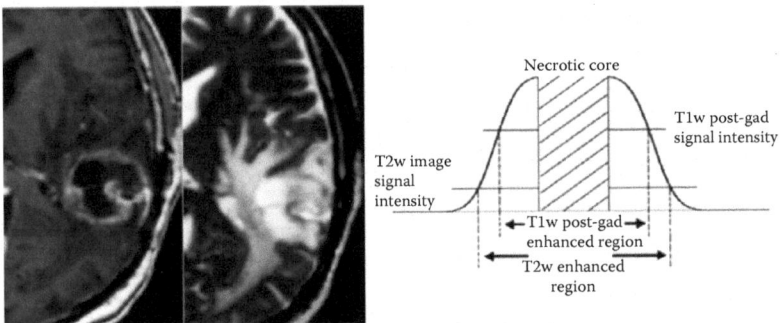

FIGURE 17.1   (Left) T2-weighted and (middle) T1-weighted post-gad MR images showing a high grade glioma. The bulk tumor and the infiltrated edema are the enhanced regions in the left and the middle images respectively. (Right) Hypothetical distribution of tumor cell density is a smooth transition starting from the bulk tumor extending beyond the edema.

landmarks with the model-generated ones and by comparing tumor cell densities extracted from the images and generated by the model. In addition to the parameters of the reaction-diffusion model, they optimize the parameters of their mechanical model as well. The use of tumor cell densities in the optimization process has the same problems as the previously mentioned method; they are not observable. The landmarks on the other hand, seems to be a promising approach; however, the parameter values obtained in this approach depend on the assumed coupling between the mechanical and the pathophysiological model. The resulting parameters are not purely inherent to the tumor. Recently, Swanson (2008) proposed a parameter estimation method for the diffusion process in petri dish experiments, which is consistent with the observed information in the images as it uses the tumor boundaries rather than tumor cell densities. They have derived analytical approximations for the evolution of the tumor delineation for two-dimensional circular growth. Using the formulation for the tumor delineation, they have estimated the diffusion coefficient for the petri dish experiments. The difficulty one would encounter if one wants to apply this method to medical images is that the method assumes radial symmetric growth, which is not the case in the brain (in vivo). Moreover, the existence of a reaction term results in a different evolution than pure diffusion.

In this chapter, we explain and analyze a parameter estimation method for reaction-diffusion based tumor growth models using the time series of MR images. The method is based on the evolution of the tumor delineation rather than tumor cell densities, and in this respect it is consistent with the observations in the images. It also takes into account tissue inhomogeneities, fiber structures, and the real geometry of both the patient's brain and the tumor while remaining consistent with the image information. We also present preliminary promising results on two sets of patient data.

Infiltration Extent of Gliomas and Model-Based Irradiation Margins

For the diagnosis and the therapy of gliomas, clinicians rely on medical images, such as magnetic resonance (MR) and computed tomography (CT) images, which show the mass part of the tumor. Current imaging techniques are not able to expose the low-density infiltration (Tovi et al. 1994; Tracqui et al. 1995; Swanson et al. 2004) posing a problem for the experts in outlining the whole tumor and in delineating its extent (see Figure 17.1). In radiotherapy, this problem of visualizing low density infiltration is addressed by outlining a constant margin of 2 cm around the visible tumor boundary for irradiation and assuming the whole tumor infiltration is contained within this region (Seither 1995; Kantor et al. 2001). This approach does not take into account the infiltration dynamics of gliomas, particularly the higher motility of tumor cells in white matter compared to gray matter (Giese 1996). As a result, the irradiation region ignoring these dynamics might not reach the full extent of the tumor infiltration in white matter and irradiate healthy gray matter. The importance of this problem was shown by Swanson et al. (2002) where they compare the visible part of the tumor with the extent of the invisible infiltration for virtual tumors grown by reaction-diffusion models. Personalized tumor growth models can offer solutions to this problem by integrating clinical information and theoretical knowledge about tumor cell dynamics. In this chapter we describe a new formulation that aims to solve the problem of estimating tumor cell density distribution beyond the visible boundary of gliomas in the images. Starting from the tumor delineation either found by a segmentation algorithm or manually drawn by an expert, it produces a map of possible tumor infiltration. It uses the anatomical MR images and diffusion tensor images (DTI) to suggest irradiation margins, taking into account the growth dynamics. We then use the extrapolated infiltration extents to create variable irradiation margins. The potential benefits of such margins in targeting tumor cells are also described.

METHODS

Parameter Estimation and Model Personalization

The reaction-diffusion model given in Equations 17.1 and 17.2 describes the temporal evolution of local tumor cell densities. In order to solve the parameter estimation problem we need a formulation consistent with the image: it should model the evolution of the tumor delineation rather than the evolution of the cell densities. In this section, first we present the traveling time formulation for tumor delineation which captures the same growth dynamics as the reaction-diffusion models while remaining consistent with the image information. Then we formulate the parameter estimation problem as an optimization problem based on the MR images, using the traveling time formulation.

Traveling Time Formulation for Tumor Delineation
The asymptotic properties of the reaction-diffusion equations under certain conditions allow us to construct a formulation for the tumor

delineation as described above, the traveling time formulation. Reaction-diffusion equations and their asymptotic properties have been well studied in the literature (Aronson and Weinberger 1978; Sarlos and Ebert 2000). The property on which we base our derivation is the existence of *traveling wave solutions* of the reaction-diffusion equations and the convergence of different initial conditions to these solutions in time.

The constant coefficient case of Equation 17.1 admits a traveling wave solution in the infinite cylinder given as $u(x,t) = u(\mathbf{n} \cdot \mathbf{x} - vt) = u(\xi)$, where v is the speed of the wave (front of the u distribution), $\mathbf{n}$ is the direction of motion of the wave, and $\xi = \mathbf{n} \cdot \mathbf{x} - vt$ is the moving frame of the traveling wave. There are two important characteristics of the reaction-diffusion equations and the traveling wave solutions that are very useful for our derivation:

- For the traveling wave solution, all iso-density contours of the distribution u move with the same speed v.

- Any initial condition $u(\mathbf{x},0)$ with compact support (1) converges to the traveling wave solution in time (see Figure 17.2). Therefore, the traveling wave solution serves as a good approximation for a certain class of solutions of the reaction-diffusion equation.

Based on these two characteristics we can model the evolution of the tumor delineation through the speed of the traveling wave.

The asymptotic speed v of the traveling wave solution is given as a function of the model parameters

$$v = 2\sqrt{\rho \mathbf{n}' \mathbf{D} \mathbf{n}}, \tag{17.3}$$

where (.)′ is the transpose operator. This speed is defined in the infinite cylinder under the constant coefficient assumption. In the case of the tumor modeling the shape of the tumor is arbitrary and the model coefficients can be spatially and temporally varying. Therefore, the above assumptions should be relaxed. Under the assumptions that the tumor delineation is planar and coefficients are constant within each image voxel, we can write a preliminary traveling time formulation for the tumor delineation using v:

$$2\sqrt{\rho}\sqrt{\nabla T' \mathbf{D} \nabla T} = 1, \tag{17.4}$$

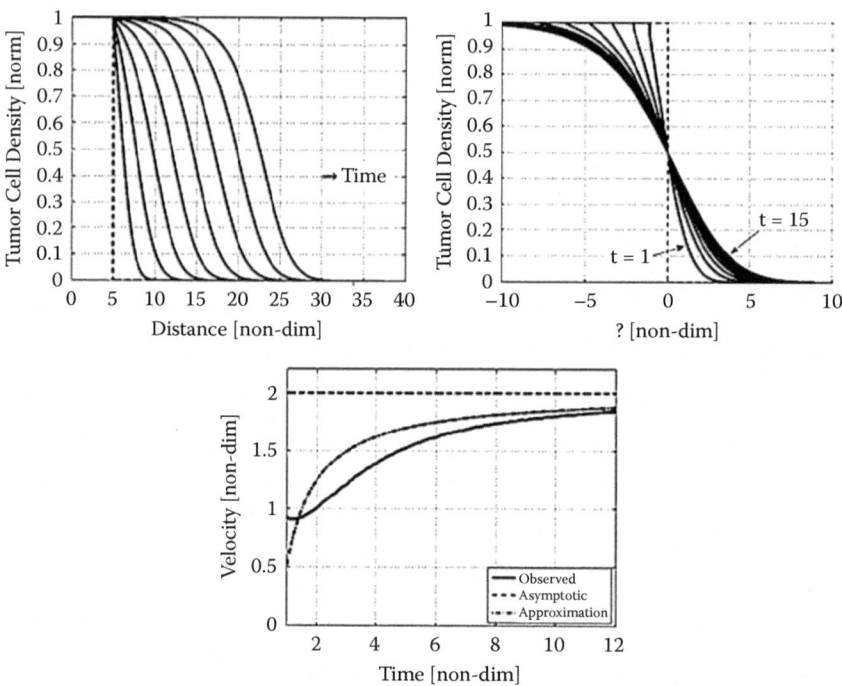

FIGURE 17.2 Any initial condition with compact support when evolved with reaction-diffusion equations (in the infinite cylinder and constant coefficients) converges to the traveling wave solution. (Left) A Heaviside initialization is evolved in time. (Middle) We plot the distribution at different times on the moving frame. (Right) The traveling wave solution has an asymptotic speed plot as dashed curve; however, when we observe the speed of the tumor front (u = 0.5 iso-density contour) we notice the low rate of convergence (solid curve). We can get a better approximation for the speed of tumor front by including the effect of time convergence and curvature (point-dashed curve).

$$T(\mathbf{x}) = T_0 \; \forall \mathbf{x} \in \Gamma, \tag{17.5}$$

where $T(\mathbf{x})$ is an implicit time function representing the time when the tumor delineation passes through the point $\mathbf{x}$. Equation 17.5 represents the Dirichlet-type boundary condition, stating that the tumor delineation reaches the surface Γ in T_0 time since its emergence. The value of T_0 is the absolute time value, and it is not known from the images. However, we do not need to know this value in order to evolve the tumor delineation with the model given in Equation 17.4. The formulation given by Equations 17.4

and 17.5 is, in a sense, a first-order approximation. It does not take into account the convergence rate of the initial distribution u to the traveling wave solution (see Figure 17.2) and the effect of the curvature of the tumor delineation on its speed. These effects can be included, leading to the final traveling time formulation for the tumor delineation as follows (Konukoglu 2009a)

$$\left\{ \frac{4\rho T - 3}{2T\sqrt{\rho}} - 0.3\sqrt{\rho}\left(1 - e^{-|\kappa_{\text{eff}}|/(0.3\sqrt{\rho})}\right) \right\} \sqrt{\nabla T' \mathbf{D} \nabla T} = 1 \qquad (17.6)$$

$$\kappa_{\text{eff}} = \nabla \cdot \frac{\mathbf{D}\nabla T}{\sqrt{\nabla T' \mathbf{D} \nabla T}}, \ T(\mathbf{x}) = T_0 \ \forall \mathbf{x} \in \Gamma,$$

where the term $(4\rho T - 3)/(2T\sqrt{\rho})$ is the effect of the tumor rate of convergence, and the term with κ_{eff} is the effect of the tumor delineation's curvature on its evolution. The traveling time formulation given in Equation 17.6 describes the evolution of the tumor delineation in the MR images based on the same growth dynamics as the reaction-diffusion models. In the formulation given in Equation 17.6 we notice the T dependence of the equation. This means that the value of T_0 becomes important for the simulation. As we have noted, this value would not be known in the clinical routine. However, we solve this problem by treating this value as another model parameter to be optimized for.

In Figure 17.3 we show an example evolution simulated using the traveling time formulation to show that it captures the same growth dynamics as the reaction-diffusion model given in Equations 17.1 and 17.2. We compare the evolution of a synthetic tumor delineation simulated by the reaction-diffusion model (in white) and by the traveling time formulation (in black). In the case of the reaction-diffusion model, the synthetic delineation is obtained by thresholding the tumor cell density distribution, which would not be available in patient cases.

The Parameter Estimation Formulation
Since we have linked the reaction-diffusion model and the evolution of the tumor delineation through the traveling time formulation, we can formulate the parameter estimation problem using this link. In the

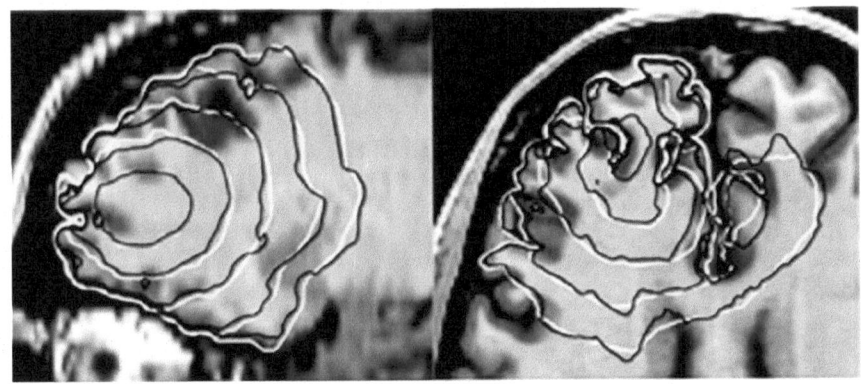

FIGURE 17.3 Comparison between the reaction-diffusion model and the traveling time formulation: The temporal evolution of the iso-density contour is demonstrated for a synthetic tumor. Contours are shown for days 400, 600, 800, 1000, and 1200 from the innermost to outermost respectively. The synthetic tumor is virtually grown using the reaction-diffusion model. White contours are obtained by thresholding the tumor cell densities at u = 0.4 for the respective day values (400-600-800-1000-1200). Then in order to simulate the evolution of the iso-density contour (assumed to correspond to tumor delineation in real images) starting from day = 400, without the knowledge of the tumor cell density distribution we use the traveling time formulation. Black curves are the contours we obtain at days 600 (2nd innermost) to 1200 (outermost). The tumors were grown in the images of a healthy subject for whom we also have the DT-MRIs. Parameters: (dw = 0.25 mm²/day, dg = 0.01 mm²/day, ρ = 0.012 day–1). The figure is Copyright © 2009 IEEE.

reaction-diffusion model given in Equations 17.1 and 17.2, there are three different parameters: d_w, d_g, and ρ. In addition to these, in the previous section we added another parameter T_0 in the traveling time formulation as a result of integrating the convergence characteristics of the reaction-diffusion solutions. This results in four parameters to estimate for. Our aim is to optimize these parameters such that the evolution simulated using the traveling time formulation best matches the real evolution observed in the time series of MR images.

At the first step we define a discrepancy measure between the simulated tumor evolution and the observed one for a given set of parameters. Minimizing this discrepancy then would provide us with the optimum model parameters. Our strategy in defining this discrepancy is to use the symmetric surface distances between the simulated and the real delineations and also add a criterion regarding the initial size of the tumor in the first acquired image. The resulting criterion becomes a quadratic measure as follows:

$$C\left(d_w, d_g, \rho, T_0\right) = \sum_{1}^{N-1} \left\{ \mathrm{dist}(\Gamma_i, \hat{\Gamma}_i) \right\} + v_{\min} \left| T_{\min} - T_0 \right|^2 \qquad (17.7)$$

$$\hat{\Gamma}_i = \left\{ \mathbf{x} \mid T(\mathbf{x}) = T_0 + \Delta t_i \right\} \text{ with } T(\mathbf{x}) = T_0 \; \forall \mathbf{x} \in \Gamma_0 \qquad (17.8)$$

$$v_{\min} = 0.1 \sqrt{\rho \mathbf{n}_{\max}' \mathbf{D}(\mathbf{x}_{\min}) \mathbf{n}_{\max}}, \; T_{\min} = T(\mathbf{x}_{\min}) \qquad (17.9)$$

where N is the number of images, Γ_i is the delineation enclosing the tumor in the i-th image, dist() is the symmetric distance between two surfaces, Δt_i is the time difference between the i-th, and the first MR image, $\mathbf{x}_{\min}$ is the point where the simulated T reaches its minimum and $\mathbf{n}_{\max}$ is the principal eigenvector of $\mathbf{D}(\mathbf{x}_{\min})$. In Equation 17.7 the summation measures the distance between the simulated and real observations. This is obtained by running the traveling time formulation outwards, starting from the delineation of the tumor obtained in the first image and comparing the corresponding simulated and observed delineations for the other images. The other term in the same equation, $v_{\min} |T_{\min} - T_0|^2$, takes into account the size of the visible tumor in the first image and quantifies the coherency of this size with the model parameters. We compute this value by simulating the traveling time evolution within the tumor delineation in the first image and compare the minimum T value obtained with the T_0 parameter. If all the parameters of the model are coherent with the observations, then this term should be equal to zero.

We define the parameter estimation problem as the minimization of the error measure defined in Equation 17.7. Different optimization algorithms can be used for this purpose. Since the gradients of C are not trivial to find analytically, we prefer to use a general algorithm that approximates the gradient directions. In this work we use the algorithm proposed by Powell in (Powell 2001) that finds the gradient directions by fitting quadratic surfaces to the underlying minimization surface. Computation times depend on the size of the tumor and the number of images. As an example for a high grade glioma with a time series of 4 images of size ($256 \times 256 \times 53$) the minimization takes around 50 min on a 2.4 GHz Intel Pentium machine

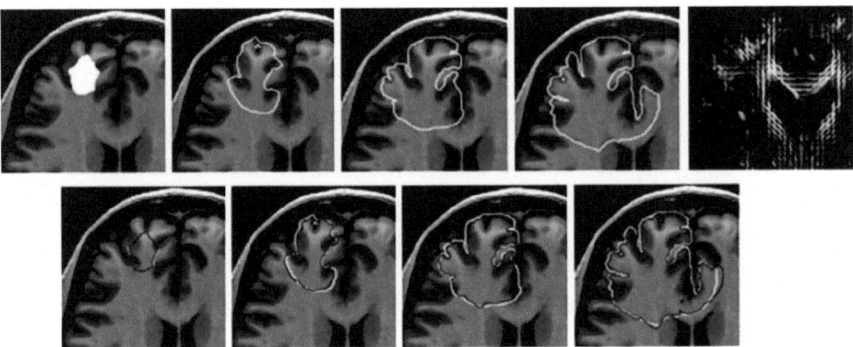

FIGURE 17.4 **(See color insert following page 40)** The inputs and the outputs of the parameter estimation method for a virtual tumor grown by reaction-diffusion models are shown. Top row shows the tumor delineations as observed in medical images and also the DT-MRI image of the subject. Bottom row shows the outputs of the method. The white contours are the tumor delineations observed in the images, the red contours are the evolution of the tumor delineation simulated by the traveling time formulation with optimized parameters.

with 1 GB of RAM. In Figure 17.4 we provide an example showing the inputs and the outputs of the parameter estimation method.

Extrapolating Infiltration Extent and Adaptive Irradiation Margins

In the previous section we presented a method to personalize the reaction-diffusion-type tumor growth models using the MR images of the patient. Here, we assume that we have the personalized model and we present a method to extrapolate the cell density distribution of the tumor beyond its delineation in the image. This method produces a map of possible tumor infiltration that is not visible in the image. Then, using this map, we describe a method for constructing variable irradiation margins, taking into account the growth dynamics of the tumor.

Extrapolating Invisible Infiltration Extents of Gliomas

In order to formulate the extrapolation method, first we need an imaging model for the gliomas. We assume that this imaging process can be modeled using a Heaviside function as done in (Tracqui et al. 1995) and in (Swanson et al. 2000). The imaging function Im is given as

$$Im(u(\mathbf{x},t)) = \begin{cases} 1 \text{ if } u \geq u_0 \\ 0 \text{ if } u < u_0 \end{cases};$$ (17.10)

we define 1 as enhanced, 0 as nonenhanced regions in the image, and u_0 is the tumor cell density threshold of the imaging modality. This simplistic model assumes that the delineations in the images correspond to an iso-density contour of the tumor cell density distribution. A more general imaging function can be used, and similar ideas presented below would be applicable. Based on this definition, the problem of extrapolating cell density distribution of a tumor beyond its delineation in the image is defined as

$$\tilde{u}(\mathbf{x}) \approx u(\mathbf{x}, T_0) \ \forall \mathbf{x} \in \left\{ \mathbf{x} \mid \mathrm{Im}(\mathbf{x}) = 0 \right\}, \tag{17.11}$$

where $\tilde{u}$ approximates the actual tumor distribution at a time instant T_0. Unlike the forward modeling of the tumor growth, the construction of the approximation $\tilde{u}$ is a static problem, and *it does not involve the time evolution* of the tumor. Moreover, in the clinical situations, the value of T_0, which indicates the time elapsed between the emergence of the tumor and the *image acquisition* is not available.

The ability to personalize reaction-diffusion models give us the opportunity to construct $\tilde{u}$ for patient images. The asymptotic properties of the reaction-diffusion equations explained in section titled "Parameter Estimation and Model Personalization" allow us to write the following formulation

$$\frac{\sqrt{\nabla \tilde{u} \cdot \left(\mathbf{D} \nabla \tilde{u} \right)}}{\sqrt{\rho \tilde{u} \left(1 - \tilde{u} \right)}} = 1, \tilde{u}\left(\Gamma \right) = u_0. \tag{17.12}$$

In this equation Γ is the tumor delineation in the image either found by a segmentation algorithm or drawn manually. For the derivation of this formulation we ask the reader to refer to (Konukoglu 2009). Equation 17.12 provides us a gradient relationship for $\tilde{u}$. In order to solve this equation we start from the enhanced part of the tumor in the image and sweep the brain tissue outwards, computing possible tumor cell density values at every point. Since the equation is an anisotropic Eikonal equation, it can be rapidly solved as described using fast-marching methods suitable to the exact form of the equation (Konukoglu et al. 2007). In Figures 17.5a–17.5c, for a virtual tumor grown by the reaction-diffusion model, we demonstrate the visible part of the tumor, the invisible infiltration, and the reconstructed tumor cell density based only on the visible part of the

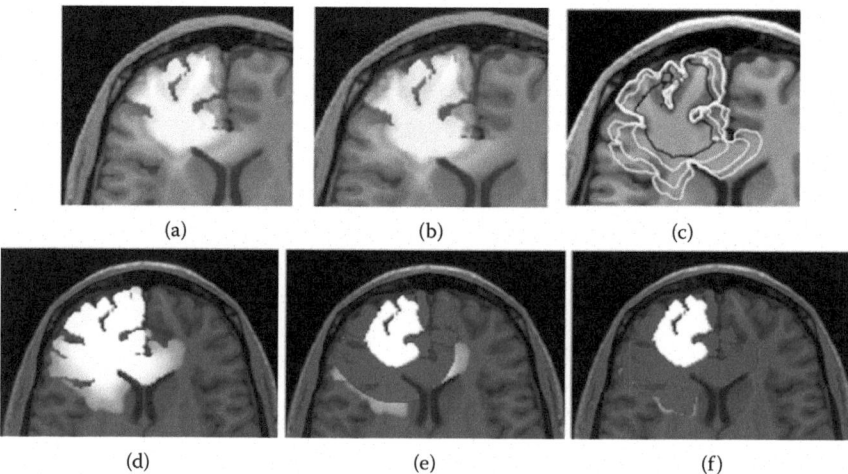

(a) (b) (c)

(d) (e) (f)

FIGURE 17.5 **(See color insert following page 40)** (a) to (c): Example of an extrapolated image for a synthetic tumor. (a) The image of a synthetic tumor is shown, where the white region is the visible part. The not-imageable infiltration region is also shown in color (from white = high density to green = low density). (b) The low density infiltration extrapolated by the proposed method starting from the visible part. (c) Iso-density contours of the actual distribution (red solid) and the corresponding ones of the extrapolated distribution (white solid). The black contour is the tumor delineation visible in the image. We observe the high global resemblance. (d) to (f): The proposed variable irradiation region construction takes into account the growth dynamics of the tumor. Figures show the two irradiation margin construction approaches and the synthetic tumor cell distribution they aim to target. (d) Cell distribution of the synthetic tumor. The white region is the visible part while the colored region is the infiltration not visible in the image. Figures (e) and (f) show constant and variable irradiation regions overlaid on the tumor distribution respectively. Blue regions represent the areas to be irradiated. For the synthetic tumor the variable margin better covers the extent of the infiltration therefore provides a better targeting.

tumor and personalized parameters. We observe the similarity between the distributions.

Adaptive Irradiation Margins

In conventional radiotherapy, the irradiation margins are constructed based on the tumor geometry visible in the medical images. The margin takes into account the enhanced area of the tumor in the image, plus a constant margin around the delineation to deal with the low cell density infiltration (Kantor et al. 2001). This approach assumes that the tumor cells would diffuse within the brain tissue homogeneously around the visible tumor. It does not take into account the growth dynamics of the tumor, especially the differential motility (Giese 1996).

The extrapolation method presented in the previous section gives us a map of possible tumor infiltration that is not visible in the images. We can use this to construct irradiation margins adapted to this map. Our strategy is to first construct the conventional constant margin irradiation region M_c and then create the adapted margins M_v by molding M_c such that M_v takes into account the infiltration extent constructed as explained in the previous section. We refer the reader to (Konukoglu 2009) for the details of this construction. In Figures 17.5d–17.5f we demonstrate the two margins, the constant and the variable ones. For a virtually grown tumor we show the visible part of the tumor, its invisible infiltration and the constant and the variable irradiation margins constructed based on its visible part.

RESULTS

Parameter Estimation and Model Personalization

The evaluation of parameter estimation for tumor growth models using real patient images is not easy because we do not have access to the real values of the parameters. The real values could be found using microscopic in vivo analysis; however, up to the best of our knowledge such a study has not been performed yet. In this work we perform an indirect evaluation for the proposed parameter estimation method using patient images. For a given patient dataset, we estimate the parameters using all but the image taken at the last time point. Then, using the estimated parameters we simulate the evolution of the tumor delineation for the time elapsed between the last two images. We then compare the evolution predicted, using the estimated parameters and the traveling time formulation with the one observed in the last image. The correlation between the prediction and the observed delineation provides us with a qualitative evaluation of the estimated parameters.

In the parameter estimation process it is shown in Konukoglu et al. (2009a) that estimating all the parameters of the model independently using the evolution of the tumor delineations in the images results in non-unique solutions. Therefore, in this study we fix the proliferation rates ρ of the tumors to the values suggested in the literature. We only estimate for the diffusion rates and the initial time estimates. In fixing the value of ρ we assume that this value can be estimated based on the biopsy results and microscopic analysis using the relation between the mitotic index (MI) and the labeling index (LI) (Johannessen and Torp 2006).

As a preliminary step we use two patient datasets that include anatomical and diffusion tensor MR images. The dataset for the first patient, who suffers from a high-grade glioma (glioblastoma multiforme), includes T1-post gadolinium MR images (with the resolution of $0.5 \times 0.5 \times 6.5$ mm^3) at three successive different time points and diffusion tensor MR image (with the resolution of $2.5 \times 2.5 \times 2.5$ mm^3) taken at the second time point. The second patient suffers from a low-grade glioma (second grade astrocytoma) and the dataset for this patient includes T2 flair MR images (with the resolution of $0.5 \times 0.5 \times 6.5$ mm^3) at five successive time points and the DT-MRI image (with the resolution of $2.5 \times 2.5 \times 2.5$ mm^3) taken at the first time point. For both cases, the tumor boundaries were manually delineated by an expert in each image separately.

The images used to estimate parameters, the estimated parameters and the predicted evolution of the tumor delineations along with the real delineations are given in Figures 17.6 and 17.7. In the images in both figures, first we show the anatomical images at the time of detection and the intermediate images used in the parameter estimation. On the intermediate images we also plot the manual delineations for the underlying image (white contour) and the simulated evolution of the tumor delineation with the estimated parameters (dark contour) obtained in the course of estimation. Following this, we start from the last image (in time) used in the parameter estimation and predict the evolution of the tumor delineation until the acquisition of the final image (which was not used in the estimation). In the corresponding images we show the anatomical MR image taken at the last time point showing the final state of the tumor along the tumor delineation predicted, using the estimated parameters drawn as the dark contour. In the accompanying tables we provide the values of the estimated parameters.

We observe in Figure 17.6c that the prediction of the tumor delineation is in very good agreement with the final state of the tumor. In the case of the low-grade tumor shown in Figure 17.7, the correlation between the predicted tumor delineation and the final state of the tumor is in line with our previous arguments. We observe that the slow evolution of the tumor is well captured by the estimated parameters. For the proliferation rate

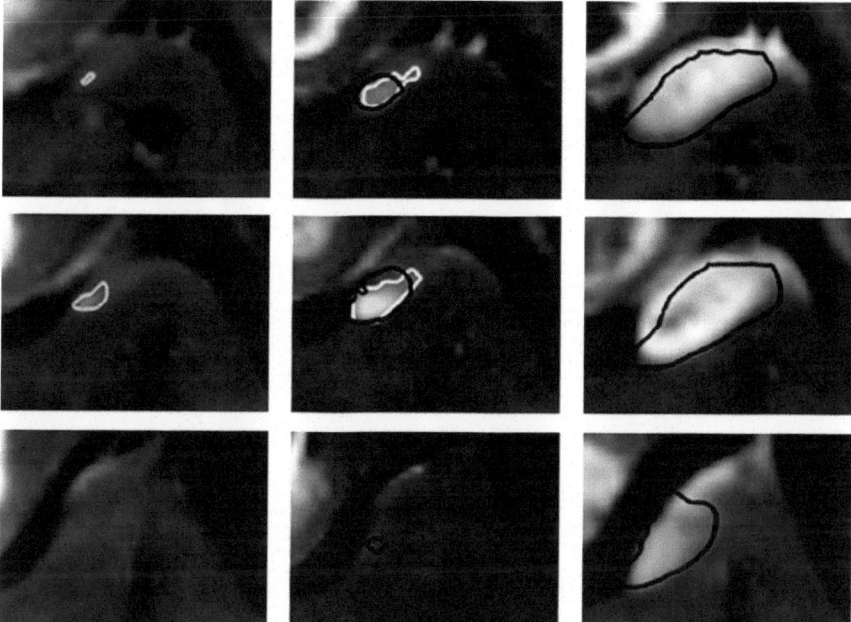

FIGURE 17.6 The parameter estimation method is applied to the images of a real patient suffering from high grade glioma. Images in columns (left column) and (middle column) shows different slices of the T1-post gadolinium images which are used to estimate the parameters of the growth model as: ρ = 0.05/day (set), dw = 0.66 mm²/day, dg = 0.0013 mm²/day. In (middle column) we also show the manual delineation of the tumor (in white) used in parameter estimation along with the optimum simulation obtained by the estimated parameters (in black) (only white contour is shown in (left column) since it is the same as the black one). (right column) The final image shows the final state of the tumor and the evolution of the delineation predicted by the estimated parameters (the black contour). The figure is Copyright © 2009 IEEE.

we pick a lower value than the one in the previous case since it is a lower grade tumor.

Extrapolating Infiltration Extent and Adaptive Irradiation Margins

In this section we assess the quality of the extrapolation method and the method for constructing the variable, adaptive irradiation margins. In order to understand the potential benefits of extrapolating the infiltration extents of gliomas and constructing irradiation margins adapted to these results, we perform experiments on synthetic images. We create a

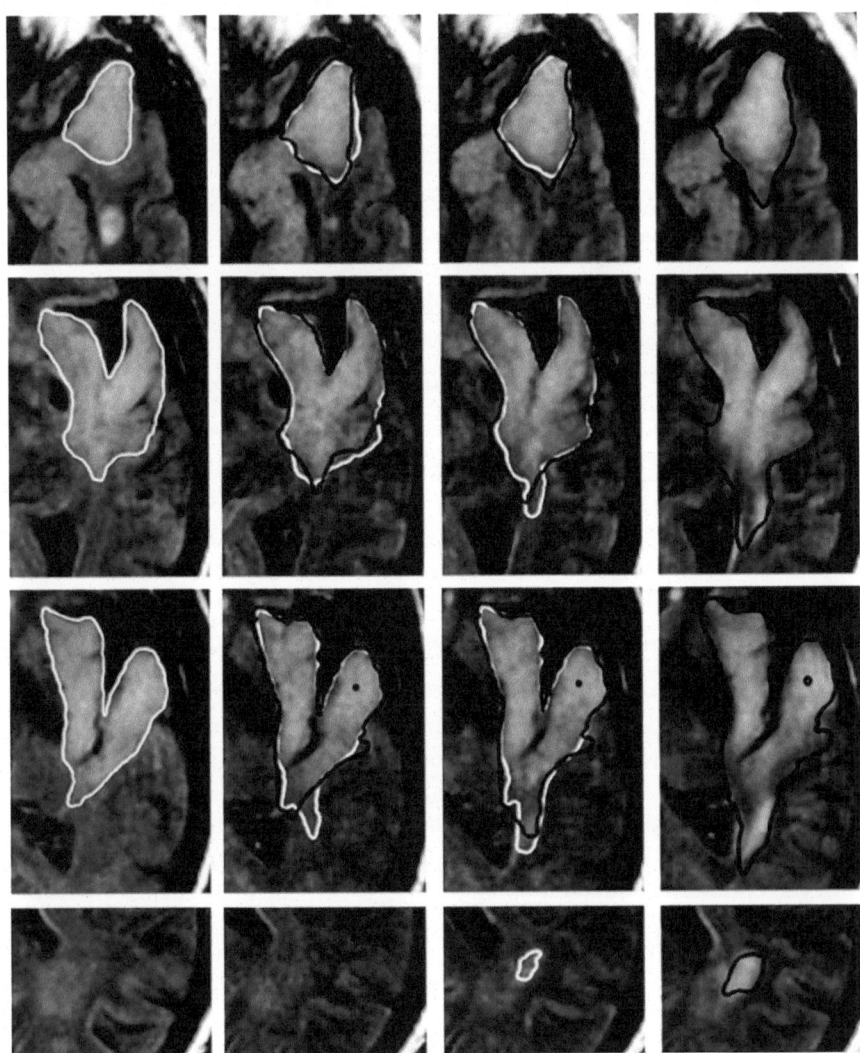

FIGURE 17.7 As a second case we applied our methodology to the images of a patient suffering from a low-grade tumor. Images (first three columns) show different slices of the T2 flair images and the manual delineations (in white) which are used to estimate the parameters of the growth model as: $\rho = 0.008$/day (set), dw = 0.20 mm²/day, dg =7 × 10-4 mm²/day. Also in these images we show the simulated evolution of the tumor delineation obtained by the estimated parameters in black contours. The simulated evolution starts from the white contour in the image (left column). Images (right column) are the slices of the final image showing the final state of the tumor and the delineation predicted by the estimated parameters as the black contour. The figure is Copyright © 2009 IEEE.

dataset of virtually grown tumors using the reaction-diffusion models where the tumors are grown in the MR images of a healthy subject ($1 \times 1 \times 2.6$ mm^3 resolution). Different tumors in the dataset are created using different model parameters and different seed points in the brain. For each tumor we construct synthetic images using the Im() function given in Equation 17.10 with $u_0 = 0.4$, value consistent with (Tracqui et al. 1995). We assume that the detection and the first image acquisition take place when the average diameter of the visible tumor reaches 1.5 cm. After the detection for each tumor we construct a synthetic image every 50 days for 1 year (8 images in total). An example of the created synthetic image with the synthetic tumor is shown in Figures 17.5d–17.5f. For the tumors in the dataset we construct the constant and the variable irradiation margins based on their visible parts in the images. Since for the virtual tumors, the cell density at every location is known—even if the image does not show a tumor at that point—we carry out a quantitative comparison. We geometrically compare these margins based on two criteria:

1. *R*: number of tumor cells not targeted

2. *Vol*: volume of healthy tissue targeted by the irradiation margin

In order to compute these values we follow the values given in (Tracqui et al. 1995) we assume that a voxel of $1 \times 1 \times 2.6$ mm^3 can hold a maximum of 9.1×10^4 tumor cells. At the time of detection of high grade gliomas, isolated tumor cells can be found in any region in the brain. Therefore, there is no completely healthy brain. In order to compute the *Vol* value we need to define "healthy tissue." In this work we define a voxel to be healthy if there are on the average less than 1 tumor cell in it. In Figure 17.8 for a synthetic tumor we show the comparison between the constant and the variable irradiation margins. We plot the *R* versus time and *Vol* versus time graphs. We observe that the variable irradiation margins adapted to the extrapolated infiltration extents targets more tumor cells and less healthy tissue.

The experiments presented above use synthetic images and virtually grown tumors. Validating the presented methods on real images would require the knowledge of tumor cell density distributions throughout the brain. This could be obtained with postmortem analysis or with animal models. This would be the topic of a future study.

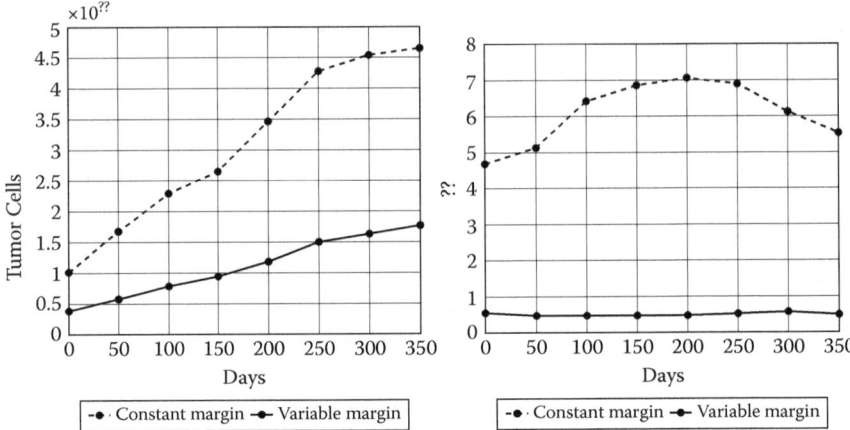

FIGURE 17.8 For an example virtual tumor we plot the R vs. time (left) and Vol versus time plots (right). Each time point corresponds to an image taken in the study as explained in the section titled "Extrapolating Infiltration Extent and Adaptive Irradiation Margins." We see that potentially variable irradiation margins can improve the accuracy of irradiation margins.

CONCLUSION

In this chapter we presented methods to personalize reaction-diffusion type tumor growth models in MR images. Our aim was to bridge the gap between the mathematical tumor growth models and the clinical applications. In this spirit, first we focused on a method for personalizing the generic reaction-diffusion growth models based on time series of MR images. The method described the evolution of the tumor delineation in the images based on the growth dynamics captured by the reaction-diffusion models. Based on this formulation, it optimized for the model parameters such that the resulting simulations best matches the observed evolution of the tumor boundaries in the MR images. The results shown in section titled "Results" demonstrated promising preliminary findings.

As a second part, we presented an application of personalized tumor growth models for radiotherapy to demonstrate clinical relevance of the personalization process. We constructed the possible tumor infiltration map of gliomas starting from their visible delineation in the MR images. Using this map we proposed variable irradiation margins that are adapted to the modeled growth dynamics of the observed tumor. We also showed potential benefits of such variable irradiation margins compared to constant ones.

Although a preliminary validation was performed a thorough validation remains to be done on animal models and postmortem analysis.

ACKNOWLEDGMENTS

We would like to thank Bjoern H. Menze, Emmanuel Mandonnet, Bram Stieltjes Marc-André Weber, and the German Cancer Research Institute (DKFZ) for their support both for providing the data used in this work and also for fruitful discussions. We would also like to acknowledge our funding agencies: INRIA, PanEuropean Health-e-Child project, CompuTumor project, and Microsoft Research at Cambridge.

REFERENCES

Aronson, D. and Weinberger, H. 1978. Multidimensional nonlinear diffusion arising in population genetics. *Adv. Math.* 30:33–76.

Burger, P.C., Heinz, E.R., Shibata, T., and Kleihues, P. 1988. Topographic anatomy and CT correlations in the untreated glioblastomamultiforme. *J. Neurosurg.* 68:698–704.

Byrne, H.M., Owen, M.R., Alarcon, T., Murphy, J., and Maini, P.K., 2006. Modelling the response of vascular tumours to chemotherapy: a multiscale approach. *Math. Models Methods Appl. Sci.*, 16:1219–41.

Clatz, O., Sermesant, M., Bondiau, P-Y. et al. 2005. Realistic simulation of the 3D growth of brain tumors in MR images coupling diffusion with biomechanical deformation. *IEEE TMI* 24.

Cristini, V., Lowengrub, J., and Nie, Q. 2003. Nonlinear simulation of tumor growth. *J. Math. Biol.* 46:191–224.

Ebert, U., and Saarloos, W.V. 2000. Front propagation into unstable states: universal algebriac convergence toward uniformly translating pulled fronts. *Physica D: Nonlinear Phenomena* 146:1–99.

Frieboes, H., Lowengrub, J., Wise, S. et al. 2007a. Computer simulation of glioma growth and morphology. *NeuroImage* 37:S59–S70.

Hogea, C., Davatzikos, C., and Biros, G. 2007a. Modeling glioma growth and mass effect in 3D MR images of the brain. *LNCS Proc. MICCAI* 4791:642–650.

Hogea, C., Davatzikos, C., and Biros, G. 2007b. An image-driven parameter estimation problem for a reaction-diffusion glioma growth model with mass effects. *J. Math. Biol.* 56:793–825.

Giese, A., Kluwe, L. Laube, B., Meissner, H., Berens, M.E., and Westaphal, M. 1996. Migration of human glioma cells on megelino: *J. Neurosurgery*, 38:755–764.

Jbabdi, S., Mandonnet, E., Duffau, H. et al. 2005. Simulation of anisotropic growth of low-grade gliomas using diffusion tensor imaging. *Magn. Reson. Med.* 54.

Johannessen, A.L. and Torp, S.H. 2006. The clinical value of Ki-67/MIB-1 labeling index in human astrocytomas. *Pathol. Oncol. Res.* 12:143–47.

Kantor, G., Loiseau, H., Vital, A., and Mazeron, J., 2001. Descriptions of GTV and CTV for radiation therapy of adult glioma. *Cancer Radiother.* 5:571–80.

Konukoglu, E. 2009. Modeling Glioma Growth and Personalizing Growth Models in Medical Images, Ph.D. thesis, Université de Nice, Sophia Antipolis.

Konukoglu, E., Sermesant, M., Clatz, O. et al. 2007. A recursive anisotropic fast marching approach to reaction diffusion equation: application to tumor growth modeling. *LNCS Proc. IPMI.* 4584.

Konukoglu, E., Clatz, O., Menze, B.H. et al. 2010a. Image guided personalization of reaction-diffusion type tumor growth models using modified anisotropic eikonal equations. *IEEE TMI* 29:77–95.

Konukoglu, E., Clatz, O., Bondiau, P.-Y., Delingette, H., and Ayache, N., 2010b. Extrapolating glioma invasion margin in brain magnetic resonance images: suggesting new irradiation margins, *Medical Image Analysis* 14:111–125.

Mandonnet, E., Pallud, J., Clatz, O. et al. 2008. Computational modeling of the WHO grade II glioma dynamics: principles and applications to management paradigm. *Neurosurgery Rev.* 31:263–69.

Patel, A., Gawlinski, E., Lemieux, S., and Gatenby, R. 2001. A cellular automaton model of early tumor growth and invasion. *J. Theor. Biol.* 213.

Powell, M., 2001. UOBYQA: Unconstrained optimization by quadratic approximation. *Math. Prog. Ser. B* 92.

Saarlos, W.V. and Ebert, U., 2000. Front propagation into unstable states: universal algebraic convergence towards uniformly translating pulled fronts. *Physica D: Nonlinear Phenomena* 146.

Seither, R.B., Jose, B., Paris, K.J., Lindberg, R.D., and Spanos, W.J. 1995. Results of irradiation in patients with high-grade gliomas evaluated by magnetic resonance imaging. *Am. J. Clin. Oncol.*, 18:297–299.

Stamatakos, G., Antipas, V., and Uzunoglu, N. 2006. A spatiotemporal, patient individualized simulation model of solid tumor response to chemotherapy in vivo. *IEEE Trans. Biomed. Eng.* 53:1467–77.

Swanson, K. 2008. Quantifying glioma cell growth and invasion in vitro. *Math. Comp. Model* 47:638–48.

Swanson, K.R., Alvord, E.C., and Murray, J.D. 2000. The modelling of diffusive tumours. *J. Biol. Syst.* 3.

Swanson, K.R., Alvord, E.C., and Murray, J.D. 2002. Virtual brain tumours (gliomas) enhance the reality of medical imaging and highlight inadequacies of current therapy. *Br. J. Cancer* 86:14–18.

Swanson, K.R., Alvord, E.C., and Murray, J.D. 2004. Dynamics of a model for brain tumors reveals a small window for therapeutic intervention. *Discrete and Continuous Dynamical Systems—Series B* 4:289–95.

Tovi., M., Hartman, M., Lilja, A., and Ericsson, A. 1994. MR imaging of cerebral gliomas: tissue component analysis in correlation with histopathology of whole-brain specimens. *Acta Radiologica* 35:495–505.

Tracqui, P., Cruywagen, G., Woodward, D. et al. 1995. A mathematical model of glioma growth: the effect of chemotherapy on spatio-temporal growth. *Cell Proliferation* 28:17–31.

Zhang, L., Athale, C.A., and Deisboeck, T.S. 2007. Development of a three-dimensional multiscale agent-based tumor model: simulating gene-protein interaction profiles, cell phenotypes and multicellular patterns in brain cancer. *J. Theor. Biol.* 244:96–1077.

CHAPTER 18

In Silico Oncology

Part I—Clinically Oriented Cancer Multilevel Modeling Based on Discrete Event Simulation

Georgios S. Stamatakos

CONTENTS

INTRODUCTION

Most cancer modeling techniques developed up to now adopt the straightforward "bottom-up" approach, focusing on the better understanding and quantification of rather microscopic tumor dynamics mechanisms and the investigation of crucial biological entity interdependences including tumor response to treatment in the generic investigational context. To this end several combinations of mathematical concepts, entities, and techniques have been developed and/or recruited and appropriately adapted. They include population dynamics models (Guiot et al. 2006), diffusion-related continuous and finite mathematics treatments (Murray 2003; Swanson et al. 2002; Breward et al. 2003; Cristini et al. 2005; Frieboes et al. 2006; Enderling et al. 2007; Ramis-Conde et al. 2008), cellular automata and hybrid techniques (Duechting and Vogelsaenger 1981; Duechting et al. 1992; Ginsberg et al. 1993; Kansal et al. 2000; Stamatakos et al. 2001a, 2001b; Zacharaki et al. 2004), agent-based techniques (Mansury and Deisboeck 2003), etc. Additionally, a number of bulky clinical tumor models focusing mainly on invasion and tumor growth morphology rather than on tumor response to concrete therapeutic schemes as administered in the clinical setting have appeared. Finite difference and finite element-based solutions of the diffusion and classical mechanics equations constitute the core working tools of the corresponding techniques (Murray 2003; Swanson et al. 2002; Clatz et al. 2005).

However, a number of concrete and pragmatic clinical questions of importance cannot be dealt with neither by the bottom-up approach nor by the morphology-oriented bulky tumor growth models in a direct and efficient way. Two examples of such questions are the following (Graf and Hoppe 2006): Can the response of the local tumor and the metastases to a given treatment be predicted in size and shape over time?, What is the best treatment schedule for a patient regarding drugs, surgery, irradiation and their combination, dosage, time schedule, and duration? A promising modeling method designed with the primary aim of answering such questions is the Discrete Event-Based Cancer Simulation Technique (DEBCaST) (Stamatakos et al. 2001c, 2002, 2006a, 2006b, 2006c, 2007a, 2007b, 2009; Dionysiou et al. 2007, 2008; Stamatakos and Uzunoglu 2006; Dionysiou et al. 2004, 2006a, 2006b; Antipas et al. 2004, 2007; Stamatakos and Dionysiou 2009). DEBCaST is basically a "top-down" biomodeling approach, in the sense that macroscopic data, including anatomic and metabolic tomographic images of the tumor, provide the framework for the integration of available and clinically trusted biological information pertaining to lower and lower biocomplexity levels such as clinically approved histological and molecular markers. However, DEBCaST does also provide a powerful framework for the investigation of multiscale tumor biology in the generic investigational context.

From the mathematical standpoint, DEBCaST is primarily a discrete mathematics method, although continuous mathematics (continuous functions, differential equations) are used in order to tackle specific aspects of the models, such as pharmacokinetics and cell survival probabilities based on pharmacodynamical and radiobiological models. Adoption of the discrete approach as the core mathematical strategy of DEBCaST has been dictated by the obvious fact that from the cancer treatment perspective it is the discrete (i.e., the integer) number of the usually few tumor cells surviving treatment and their discrete mitotic potential categorization (stem cells, progenitor cells of various mitotic potential levels, and differentiated cells) that really matters. These discrete entities and quantities in conjunction with their complex interdependences may give rise to tumor relapse or to ensure tumor control over a given time interval following completion of the treatment course. Cell-cycle phases have a clearly discrete character, too. Moreover, the properties of the different cell phases may vary immensely from the clinical significance perspective. A classic example is the lack of effect of cell-cycle-specific drugs on living tumor cells residing in the quiescent G0 phase.

It is noted that complex interdependencies of microscopic factors in the surrounding milieu of the cells such as oxygenation, nutrient supply, and molecular signals emitted by other cells play a critical role in the mitotic fate of tumor cells. Their effect is taken into account in DEBCaST through the local mean values of the corresponding model parameters. To this end, imaging, histological, and molecular data are exploited, as will be described later.

Due to the numerical character of the method, a careful and realistically thorough numerical analysis concerning consistency, convergence, and sensitivity/stability issues is absolutely necessary before any application is envisaged. A discussion of this critical issue is included in the section titled "Discussion and Future Perspectives."

Tumor neovascularization is taken into account in an indirect yet pragmatic way by exploiting grey level and/or color information contained within slices of tomographic imaging modalities sensitive to blood perfusion and/or the metabolic status of the tumor. (Stamatakos et al. 2001a, 2002, 2006a; Dionysiou et al. 2004, 2007; Marias et al. 2007). The reason for adopting the above-mentioned strategy rather than developing or integrating detailed tumor angiogenesis models is that no microscopic information regarding the exact mesh of the neovascularization capillaries throughout the tumor can be currently extracted from clinically utilized imaging modalities. Nevertheless, the microscopic functional capillary density distribution over the tumor can be grossly estimated, based on various imaging modalities such as T1-gadolinium-enhanced MRI in the case of glioblastoma multiforme (GBM) and arterial spin labeling (ASL) MRI.

Precursors of DEBCaST can be traced in the well-established and clinically applicable disciplines of pharmacology and radiobiology. Integration of molecular biology in DEBCaST may be viewed as the introduction of a perturbator or adaptor of the cellular and higher biocomplexity level parameters. In such a way, the *in vivo* measurable clinical manifestation of tumor dynamics is placed in the foreground. This is one of the reasons why DEBCaST is gaining wider and wider acceptance within the clinical and the industrial environment, including the emergent domain of *in silico* oncology (Stamatakos et al. 2002, 2007b, 2009; Stamatakos 2006, 2008; Graf and Hoppe 2006; Graf et al. 2007, 2009). Both the large scale European Commission (EC) and Japan-funded research and development (R&D) project ACGT [ACGT: Advancing Clinicogenomic Trials on Cancer: Open Grid Services for Improving Medical Knowledge Discovery, FP6-2005-IST-026996, http://eu-acgt.org/acgt-for-you/researchers/in-silico-oncology/oncosimulator.html and http://www.eu-acgt.org/] and the EC-funded R&D project ContraCancrum [ContraCancrum: Clinically Oriented Translational Cancer Multilevel Modeling FP7-ICT-2007-2-223979, www.contracancrum.eu] have adopted DEBCaST as their core cancer simulation method. It is worth noting that in both projects the role of clinicians is prominent. A biomedical engineering concept and construct tightly associated with DEBCaST, the Oncosimulator, which is currently under clinical adaptation, optimization, and validation, is also sketched.

In order to convey the core philosophy of the method to the reader in a concise way, a symbolic mathematical formulation of DEBCaST in terms of a hypermatrix and discrete operators is introduced. Two specific models of tumor response to chemotherapeutic and radiotherapeutic schemes are briefly outlined so as to exemplify DEBCaST's application potential. The chapter concludes with a discussion of several critical aspects including numerical analysis, massive parallel code execution, associated technologies, extensions, and validation within the framework of clinico-genomic trials and future challenges and perspectives.

An encouraging fact as far as industrial and eventually clinical translation of the method is concerned is that both DEBCaST and the Oncosimulator have been selected and endorsed by a worldwide leading medical technology company and now constitute modules of their research and development line (ContraCancrum project). One of the envisaged final products of this endeavor is a radiotherapy treatment planning system based on both physical and multiscale biological optimization of the spatiotemporal dose administration scheme. A clinical trial-based validation process for the system is currently at the final stage of its detailed formulation.

THE ONCOSIMULATOR

The Oncosimulator is, at the same time, a concept of multilevel integrative cancer biology, a complex algorithmic construct, a biomedical engineering system, and eventually a clinical tool that primarily aims at supporting the clinician in the process of optimizing cancer treatment in the patient-individualized context through conducting experiments *in silico* (i.e., on the computer). Additionally, it is a platform for simulating, investigating, better understanding, and exploring the natural phenomenon of cancer, supporting the design and interpretation of clinicogenomic trials and finally training doctors, researchers, and interested patients alike (Stamatakos and Uzunoglu 2006; Stamatakos et al. 2007a; Graf et al. 2009).

A synoptic outline of the clinical utilization of a specific version of the Oncosimulator, as envisaged to take place following an eventually successful completion of its clinical adaptation, optimization, and validation process, is provided in the form of the following seven steps (Figure 18.1):

First step: Obtain patient's individual multiscale and inhomogeneous data. Data sets to be collected for each patient include: clinical data (age, sex, weight, etc.), possible previous antitumor treatment history, imaging data (e.g., MRI, CT, PET, etc., images), histopathological data (e.g., detailed identification of the tumor type, grade and stage, histopathology slide images whenever biopsy is allowed and feasible, etc.), molecular data (DNA array data, selected molecular marker values or statuses, serum markers, etc.). It is noted that the last two data categories are extracted from biopsy material and/or body fluids.

Second step: Preprocess patient's data. The data collected are preprocessed in order to take an adequate form, allowing its introduction into the Tumor and Normal Tissue Response Simulation Module of the Oncosimulator. For example, the imaging data are segmented, interpolated, and eventually fused; subsequently, the anatomic entities of interest are three-dimensionally reconstructed.

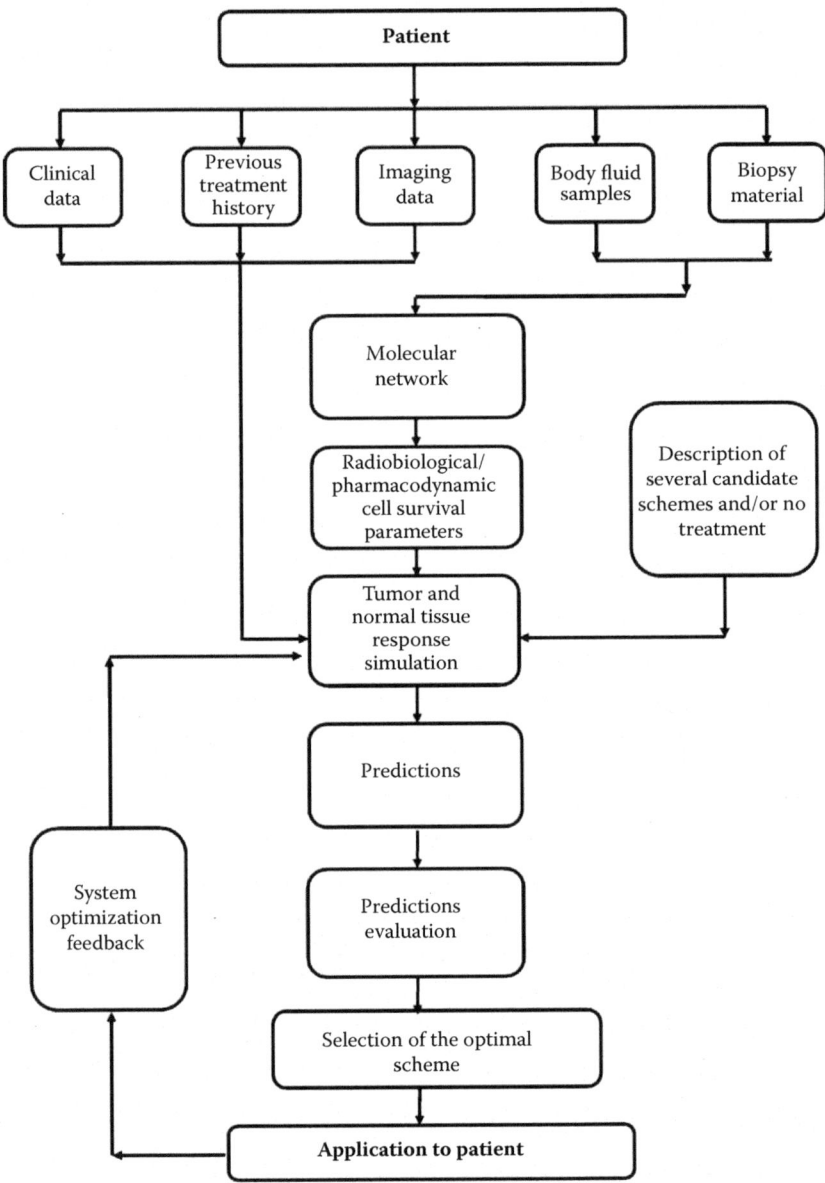

FIGURE 18.1 Oncosimulator: a synoptic workflow diagram of one of the system versions.

This reconstruction will provide the framework for the integration of the rest of data and the execution of the simulation. In parallel the molecular data is processed via molecular interaction networks so as to perturb and individualize the average pharmacodynamic or radiobiological cell survival parameters.

Third step: Describe one or more candidate therapeutic schemes and/or schedules. The clinician describes a number of candidate therapeutic schemes and/or schedules or no treatment (obviously leading to free, i.e., noninhibited tumor growth), to be simulated *in silico*, that is, on the computer.

Fourth step: Run the simulation. The computer code of tumor growth and treatment response is massively executed on distributed grid or cluster computing resources so that several candidate treatment schemes and schedules are simulated for numerous combinations of possible tumor parameter values *in parallel* (see the section titled "Discussion and Future Perspectives" for detailed justification). Predictions concerning the toxicological compatibility of each candidate treatment scheme are also produced.

Fifth step: Visualize the predictions. The expected reaction of the tumor as well as toxicologically relevant side effect estimates for all scenarios simulated are visualized using several techniques ranging from simple graph plotting to four-dimensional virtual reality rendering.

Sixth step: Evaluate the predictions and decide on the optimal scheme or schedule to be administered to the patient. The Oncosimulator's predictions are carefully evaluated by the clinician by making use of their logic, medical education, and even qualitative experience. If no serious discrepancies are detected, the predictions support the clinician in taking their final and expectedly optimal decision regarding the actual treatment to be administered to the patient.

Seventh step: Apply the theoretically optimal therapeutic scheme or schedule and further optimize the Oncosimulator. The expectedly optimal therapeutic scheme or schedule is administered to the patient. Subsequently, the predictions regarding the finally adopted and applied scheme or schedule are compared with the actual tumor course and a *negative feedback signal* is generated and used in order to optimize the Oncosimulator.

MODEL—A SYMBOLIC MATHEMATICAL FORMULATION OF DEBCaST

Hypermatrix of the Anatomical Region of Interest

A special operator notation is introduced in order to concisely describe DEBCaST as applied to the modeling of malignant tumor response to therapeutic schemes in the clinically oriented setting. In order to convey the basic philosophy of the method, a deliberately simplified version of DEBCaST is considered throughout the chapter. The anatomical region of interest, primarily including the tumor and possibly adjacent normal tissues and edema, in conjunction with its biological, physical, and chemical dynamics is represented by hypermatrix $\bar{\mathbf{a}}$. A *hypermatrix* is a

Matrix of (Matrices of (Matrices...of (Scalars or Vectors or Matrices)...)).

The hypermatrix $\bar{\mathbf{a}}$ is created by a cubic discretization mesh that is virtually superimposed upon the anatomical region of interest. Biological cells residing within each geometrical cell of the mesh are conceptually clustered into mathematical *equivalence classes*. Equivalence classes primarily correspond to the various phases within or out of the cell cycle in which a biological cell of the tumor resides. Since a tumor cell at any given instant also belongs to a mitotic potential category (stem, progenitor, terminally_differentiated), the latter acts as a further partitioner of the biological cells into equivalence classes. One of the reasons, though *not* the single most important, for clustering biological cells into equivalence classes within each geometrical cell of the discretization mesh, is computing resource limitations. Complex computational treatment of each single cell of a large clinical tumor undergoing therapeutic treatment as a separate entity is still not achievable within acceptable resource and time limits.

Discrete time represents a further dimension of the hypermatrix. An important discretization aspect of the method is the mean time spent in the phase of an equivalence class by the biological cells belonging to the equivalence class (Stamatakos et al. 2001c, 2002, 2006a, 2006c; Dionysiou et al. 2004, 2006; Stamatakos and Dionysiou 2009). In order to allow for spatiotemporal perturbations of critical parameter values throughout the tumor and also avoid artificial cell synchronizations due to discretization, use of pseudorandom numbers is extensively made (Monte Carlo technique).

Practical Considerations Regarding the Construction of the Discretization Mesh

Collection of the appropriate monomodality or, far better, multimodality tomographic data of the patient such as slices of T1-weighted contrast-enhanced MRI, T2-weighted MRI, CT, PET, or other modalities, image segmentation, slice interpolation, three-dimensional reconstruction of the anatomical entities of interest centered at the tumor, and eventually fusion of more than one modality images constitute the initial steps for the creation of the discretization mesh. The latter discretizes and covers the anatomical region of interest. Processed microscopic data (histological, molecular) are then utilized in order to enhance the patient individualization of the hypermatrix.

Hypermatrix and Operator Formulation of DEBCaST

The following mathematical entities are introduced:

$\bar{\bar{a}}$ denotes the hypermatrix corresponding to the anatomical region of interest that includes the tumor and possibly parts of the surrounding normal tissue and edema. $\bar{\bar{a}}$ describes explicitly or implicitly the biological, physical, and chemical dynamics of the anatomical region of interest. For a given geometrical cell of the discretization mesh, each vector element of the hypermatrix $\bar{\bar{a}}$ can be written as follows:

$$\bar{\bar{a}}\left(x_i, y_j, z_k, t_l, c_m, p_n\right) = \left(g^{ijklmn}, N^{ijklmn}, X^{ijklmn}, h^{ijklmn}, \tilde{h}^{ijklmn}\right) \quad (18.1)$$

$$\bar{\bar{a}}(t_0) = \bar{\bar{a}}_0 \quad (18.2)$$

The following symbols are used—x_i, y_j, and z_k—to denote the spatial coordinates of the discrete points of the discretization mesh with spatial indices i, j, and k, respectively. Each discrete spatial point lies at the center of a geometrical cell of the mesh. The temporal coordinate of the discrete time point is denoted as t_l with temporal index l. The mitotic potential category (i.e., stem, progenitor, or terminally_differentiated) of the biological cells is represented as c_m with mitotic potential category index m. It is noted that the term tumor *progenitor cell* denotes a tumor cell with limited mitotic potential (i.e., number of possible mitoses); p_n denotes the cell phase (within or out of the cell cycle) of the biological cells with cell phase index n.

For the biological cells belonging to the mitotic potential category c_m *and* residing in cell phase p_n *and* being accommodated within the geometrical cell whose center lies at the spatial point (x_i, y_j, z_k) *and* being considered at the time point t_l—in other words, for the biological cells clustered in the same equivalence class denoted by ijklmn—the following state parameters are provided by the corresponding hypermatrix element $\bar{a}(x_i, y_j, z_k, t_l, c_m, p_n)$: g^{ijklmn}: local oxygen and nutrient provision level; N^{ijklmn}: number of biological cells; X^{ijklmn}: average time spent by the biological cells in phase n; h^{ijklmn}: number of biological cells hit by treatment; $\tilde{h}^{ijklmn}$: number of biological cells *not* hit by treatment. The symbol $\bar{a}_0$ represents the initial biological, physical, and chemical state of the entire anatomical region of interest or equivalently of the hypermatrix under consideration. This state corresponds to the instant just before the start of the treatment course to be simulated.

The previously mentioned discrete variables can take values from proper or improper subintervals of the discrete intervals appearing below or from proper or improper subsets of the discrete sets appearing also below.

$$x_i \in [x_{min}, x_{max}] \tag{18.3}$$

where x_{min} and x_{max} denote the minimum and the maximum value, respectively, of the parameter x_i within the discretization mesh,

$$y_j \in [y_{min}, y_{max}] \tag{18.4}$$

where y_{min} and y_{max} denote the minimum and the maximum value, respectively, of the parameter y_j within the discretization mesh,

$$z_k \in [z_{min}, z_{max}] \tag{18.5}$$

where z_{min} and z_{max} denote the minimum and the maximum value, respectively, of the parameter z_k within the discretization mesh,

$$t_l \in [t_{min}, t_{max}] \tag{18.6}$$

where t_{min} and t_{max} denote the initial and the final simulation time point, respectively

$$c_m \in \{\text{stem, progenitor, terminally\_differentiated}\} \tag{18.7}$$

$$p_n \in \{G_1, S, G_2, M, G_0, A, N, D\} \tag{18.8}$$

where G_1 denotes the G_1 cell cycle phase; S denotes the DNA synthesis phase; G_2 denotes the G_2 cell cycle phase; M denotes mitosis; G_0 denotes the quiescent (dormant) G0 phase; A denotes the apoptotic phase; N denotes the necrotic phase; and D denotes the remnants of dead cells.

$$g^{ijklmn} \in \{s, \tilde{s}\} \tag{18.9}$$

where s stands for "oxygen and nutrient provision level sufficient for tumor cell proliferation"; $\tilde{s}$ stands for "oxygen and nutrient provision level insufficient for tumor cell proliferation." Obviously, the binary character of the oxygen and nutrient provision level is to be considered only a first simplifying approximation. More elaborate descriptions have been proposed and applied (Stamataos et al. 2002, 2006a, 2006b; Dionysiou et al. 2004, 2006a; Antipas et al. 2004).

$$N^{ijklmn} \in \{0, \dots, N_{max}\} \tag{18.10}$$

where N_{max} denotes the maximum number of biological cells irrespectively of mitotic potential category or cell phase that can be accommodated in any geometrical cell under either normal or transient conditions.

$$X^{ijklmn} \in [0, X_{max}] \tag{18.11}$$

where X_{max} denotes the maximum average time that can be spent by biological cells in phase n

$$h^{ijklmn} \in \{0, N_{max}\} \tag{18.12}$$

$$\tilde{h}^{ijklmn} \in \{0, N_{max}\} \tag{18.13}$$

In the six dimensional *discrete abstract space* of tumor dynamics delineated above, three discrete dimensions represent space, one discrete dimension represents time, another one represents mitotic potential, and the last one represents the cell phase. Combinations of all possible values from these six discrete dimensions produce all the biological cell equivalence classes that have been introduced so far.

The entire simulation can be viewed as the periodic and sequential application of a number of discrete algorithmic operators on the hypermatrix $\bar{\bar{a}}$ of the anatomic region of interest. The *operator application period* is equal to the time separating two consecutive *complete scans* of the discretization mesh. A complete scan includes mesh scans performed by all operators for any given time point. The operator application period is usually taken 1 h since this is approximately the duration of mitosis, the shortest of the cell-cycle phases. It should be noted that although the parameter values exported by the simulation execution at any desired instant for visualization and analysis purposes have a discrete character, certain parameters are handled by the computer internally and temporarily as real numbers (even with enhanced precision) in order to minimize discretization error propagation, in particular when dealing with small numbers of discrete entities in the stochastic context. By no means, however, does this technicality affect the fundamentally discrete character of DEBCaST.

The various processes or modules of the complex (not in the sense of complex numbers) algorithmic manipulations applied on the hypermatrix can be thought of as corresponding to discrete operators acting on the discrete hypermatrix of the anatomical region of interest in analogy to the action of continuous operators on a continuous wave function in quantum mechanics (Schiff 1981). The great importance of introducing abstract (vector) spaces and operators has been established in virtually all fields of the essentially nonliving matter physics (Morse and Feshbach 1953). Compactness, clarity, amenability to in-depth mathematical analysis and propensity to provide stimulating and eventually highly creative analogies with other scientific fields are but a few of the advantages of the particular formalism. Therefore, extending such a strategy to multi-scale living matter physics (i.e., biology) seems to be a straightforward step forward.

In order to proceed to the operator application on the hypermatrix the following symbols are introduced:

f stands for the composite discrete operator (i.e., the operator formed by the synthesis of all partial operators that are sequentially applied on the hypermatrix at each discrete time step. Therefore, the updated hypermatrix at the time point t_{l+1} is given by

$$f\left(\bar{\bar{a}}(t_l)\right) = \bar{\bar{a}}(t_{l+1}) \tag{18.14}$$

The composite operator can be written as

$$f = f^U \, f^E \, f^C \, f^H \, f^O \, f^T \tag{18.15}$$

where

$$f^J, J \in \{U, E, C, H, O, T\} \text{ stands for a } \textit{partial operator} \tag{18.16}$$

T stands for time updating (i.e., increasing time by a time unit [e.g., 1h]). Action of operator T on the hypermatrix should not be confused with the updated state of the entire hypermatrix $\bar{a}$ at time t_1. O stands for estimating the local oxygen and nutrient provision level. H stands for the effect of treatment (therapy) referring mainly to cell hit by treatment, cell kill, and cell survival. C stands for cell cycling, possibly perturbed by treatment. Transition between mitotic potential cell categories such as transition of the offspring of a finally divided progenitor cell into the terminally differentiated cell category is also tackled by this operator. E stands for differential expansion or shrinkage or more generally for geometry and mechanics handling. U stands for the updating of the local oxygen and nutrient provision level following application of the rest of the operators at each time step.

It is noted parenthetically that the outcome of appropriate processing of the molecular and/or histopathological data via, for example, molecular networks and signaling pathways is used as a perturbator of the cell survival probabilities included in operator H, so as to considerably enhance patient individualization of the simulation. A realistic estimate of the extent of such perturbations for a given tumor-type subclass in the framework of a clinico-genomic trial is achieved in a stepwise way. Initial rough modifications of the cell survival probabilities based on the baseline-pretreatment data, pertinent literature information, and logic are subsequently corrected through utilization of the corresponding posttreatment data via a process of parameter fitting.

Therefore, Equation 18.14 becomes:

$$f^U\left(f^E\left(f^C\left(f^H\left(f^O\left(f^T\bar{a}(t_1)\right)\right)\right)\right)\right) = \bar{a}(t_{1+1}), \quad 1 = 0,1,2,\dots \tag{18.17}$$

or in a more compact form:

$$f^U \, f^E \, f^C \, f^H \, f^O \, f^T \bar{a}(t_1) = \bar{a}(t_{1+1}), \quad 1 = 0,1,2,\dots \tag{18.18}$$

where the application of operators takes place from the right to the left.

It is obvious that the above mentioned concepts and symbols cannot convey all the details needed for the simulation to run. Their role is, rather, to identify and decompose the major conceptual mathematical and computational steps than to list all modeling details. The interested reader is referred to the Web site of the *In Silico* Oncology Group, Institute of Communications and Computer Systems, National Technical University of Athens (www.in-silico-oncology.iccs.ntua.gr) where they may find lists of pertinent publications providing detailed descriptions of several DEBCaST models including assumptions, mathematical treatment, numerical aspects such as convergence and discretization error minimization, sensitivity analysis, validation, applications, and suggested extensions.

It is worth noting that discrete simulation under certain constraints can efficiently replace analytical solutions to a wide range of mathematical problems which, although being formulated in terms of continuous mathematics—usually including symbolically formulated differential equations—refer in fact to *discrete* physical quantities such as biological cells and cell state transition rates. Moreover, in many cases the continuous symbolic formulation of mathematical operators, such as the well-known differential operator, when acting on discrete physical quantities can be readily replaced by a conceptually more straightforward *algorithmic formulation*. Several techniques leading to the minimization of error propagation for those cases where small numbers of discrete entities are dealt with by stochastic processes are available. The above generic policy has been extensively adopted in DEBCaST models. For extensions of DEBCaST currently under implementation see section titled "Discussion and Future Perspectives."

RESULTS

In this section two indicative models denoted by Model A and Model B are briefly outlined so as to exemplify the application potential of DEBCaST in the clinical context.

Model A: Tumor Response to Chemotherapeutic Schemes

Model A is a four-dimensional, patient-specific DEBCaST simulation model of solid tumor response to chemotherapeutic treatment *in vivo*. The special case of imageable glioblastoma multiforme (GBM) treated by temozolomide (TMZ) has been selected as a simulation paradigm. Nevertheless, a considerable number of the involved algorithms are quite generic. The model is based on the patient's imaging, and histopathologic and genetic data. For a given drug administration schedule lying within acceptable toxicity boundaries, the concentration of the prodrug and its metabolites within the tumor is calculated as a function of time based on the drug pharmacokinetics and is conceptually included in operator H. A discretization mesh is superimposed upon the anatomical region of interest and within each geometrical cell of the mesh the basic biological, physical, and chemical "laws" such as the rules concerning oxygen and nutrient provision (operators O and U), cell cycling (operator C) (Salmon and Sartorelli 2001), mechanical deformation (operator E), and so forth are applied at each discrete time point (operator T). The biological cell fates are predicted based on the drug pharmacodynamics, constituting part of operator H (Perry 2001; Katzung 2001; FDA 1999; Newlands et al. 1992; Bobola et al. 1996; Stupp et al. 2001). The outcome of the simulation is a prediction of the spatiotemporal activity of the entire tumor and is virtual reality visualized. A good qualitative agreement of the model's predictions with clinical experience (Stamatakos et al. 2006b, 2006c) supports the applicability of the approach. Model A has provided a basic platform for performing patient individualized *in silico* experiments as a means of chemotherapeutic treatment optimization in the theoretical context. A few indicative aspects of the model are described below. Since the complexity of the analysis is high, the interested reader is referred to (Stamatakos et al. 2006b, 2006c) for a detailed description of the model. The work has also provided the basis for the development of the chemotherapy treatment response models of the ACGT and the ContraCancrum projects.

Figure 18.2 depicts the simplified cytokinetic model of a tumor cell that has been proposed and adopted in Model A. The cytotoxicity produced by TMZ is primarily modeled by a delay in the S phase compartment (TDS), which is denoted by "Delay due to the effect of chemotherapy" in the diagram of Figure 18.2 and by subsequent apoptosis. Further details are provided in the caption of Figure 18.2.

Figure 18.3 provides a three-dimensional visualization of the simulated response of a clinical GBM tumor to one cycle of the TMZ chemotherapeutic scheme: 150 mg/m^2 orally once daily for 5 consecutive days per 28-day treatment cycle (Stamatakos et al. 2006b). Panel (a) shows the external surface of the tumor before the beginning of chemotherapy. Panel (b) shows the internal structure of the tumor before the beginning of chemotherapy. Panel (c) shows the predicted external surface of the tumor 20 days after the beginning of chemotherapy. Panel (d) shows the predicted internal

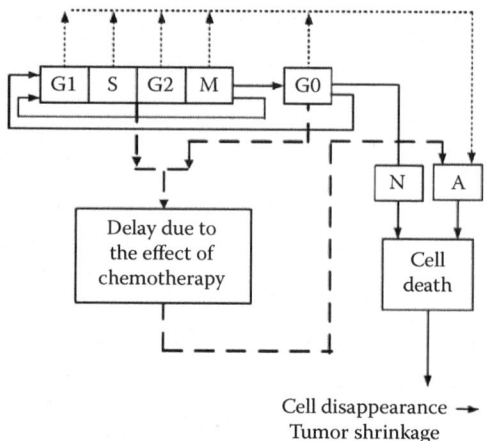

FIGURE 18.2 Simplified cytokinetic model of a tumor cell proposed and adopted in model A. Symbols: G1: G1 phase; S: DNA synthesis phase; G2: G2 phase; G0: G0 phase; N: necrosis; A: apoptosis. The cytotoxicity produced by TMZ is primarily modeled by a delay in the S-phase compartment (TDS) ("Delay due to the effect of chemotherapy" in the diagram) and subsequent apoptosis. The delay box simply represents the time corresponding to, at most, two cell divisions that are required before the emergence of temozolomide cytotoxicity. It is not a time interval additional to the times represented by the cell cycle phase boxes. (From Stamatakos G.S. et al. 2006b. *IEEE Trans Biomed Eng* 53: 1467–1477. Reprinted with permission from the Institute of Electrical and Electronics Engineers [IEEE], © 2006 IEEE.)

structure of the tumor 20 days after the beginning of chemotherapy. The pseudocoloring criterion proposed and utilized is described in the caption of Figure 18.3.

Model B: Tumor Response to Radiotherapeutic Schemes

Model B is a spatiotemporal simulation model of in vivo tumor growth and response to radiotherapy exemplified by the special case of imageable GBM treated with radiation. The main constitutive processes of the model can be summarized as follows. A discretizing cubic mesh is superimposed upon a three-dimensional virtual reconstruction of the tumor including its necrotic region and the surrounding anatomical features based on imaging data. In a way analogous to Model A, within each geometrical cell of the mesh a number of biological cell equivalence classes are defined based, inter alia, on the biological cell distribution over the various phases within or out of the cell cycle for the various mitotic potential

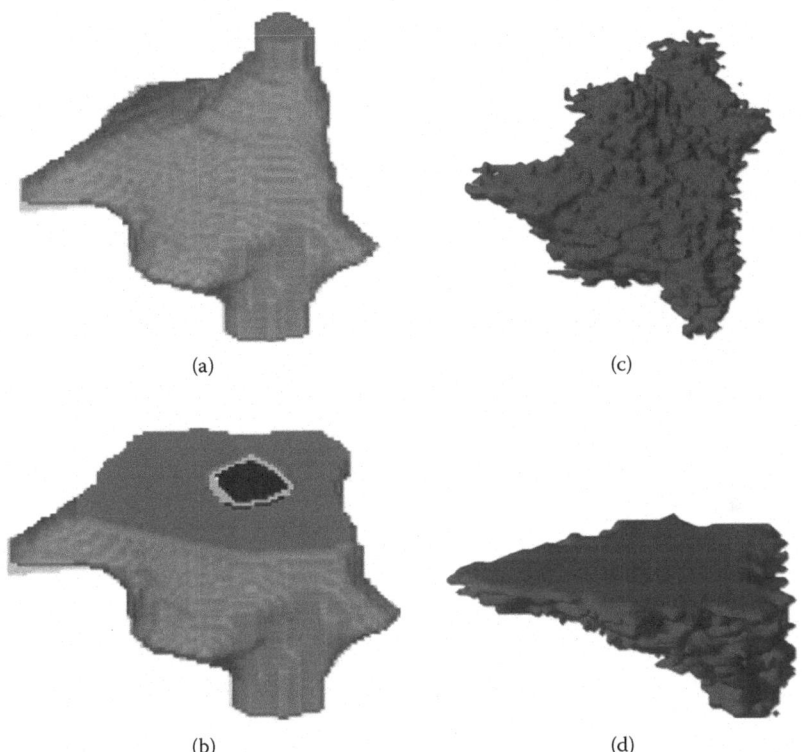

FIGURE 18.3 **(See color insert following page 40)** A three-dimensional visualization of the simulated response of a clinical GBM tumor to one cycle of the TMZ chemotherapeutic scheme: 150 mg/m² orally once daily for five consecutive days per 28-day treatment cycle. (a) External surface of the tumor before the beginning of chemotherapy, (b) internal structure of the tumor before the beginning of chemotherapy, (c) predicted external surface of the tumor 20 days after the beginning of chemotherapy (d) predicted internal structure of the tumor 20 days after the beginning of chemotherapy. The following pseudocolor code has been applied: red: proliferating cell layer; green: dormant cell layer (G0); blue: dead cell layer. The "99.8%" pseudocoloring criterion has been devised and applied as follows (pseudocode version): "For a geometrical cell of the discretizing mesh, if the percentage of dead cells within it is lower than 99.8% then {if percentage of proliferating cells > percentage of G0 cells, then paint the geometrical cell red (proliferating cell layer) else paint the geometrical cell green (G0 cell layer)}else paint the geometrical cell blue (dead cell layer)." (From Stamatakos G.S. et al. 2006b. *IEEE Trans Biomed Eng* 53(8): 1467–1477. Reprinted with permission from the Institute of Electrical and Electronics Engineers [IEEE], © 2006 IEEE.)

categories. Sufficient registers are used in order to store the current state of each equivalence class such as the average time spent by clustered biological cells in phase G1, etc. The mesh is scanned every one hour (operator T). The basic biological, physical, and chemical "laws" including the metabolic activity dynamics [operators O and U], cell cycling [operator C], mechanical and geometrical aspects [operator E], cell survival probability following irradiation with dose D [operator H] (Perez and Brady 1998; Steel 2002) are applied on each geometrical cell at each complete scan. A spatial and functional restructuring of the tumor takes place during each discrete time point since new biological cells are eventually produced, leading to differential tumor growth, or existing cells eventually die and subsequently disappear through specific molecular and cellular event cascades, thus leading to differential tumor shrinkage. Simulation predictions can be two- or three-dimensionally visualized at any simulated instant of interest. In the particular model special attention has been paid to the influence of oxygenation on radiosensitivity in conjunction with the introduction of a refined imaging-based description of the neovasculature density distribution. In order to validate the model two identical (except for the status of the p53 gene) virtual GMB tumors of large size, complex shape, and complex geometry of their internal necrotic region were considered. The first one possessed a wild-type p53 gene whereas the second one was characterized by a mutated p53 (Stamatakos et al. 2006a; Dionysiou et al. 2004). The values of the α and β parameters of the standard linear quadratic radiobiological model for cell survival (Steel 2002, Perez and Brady 1998) have been determined experimentally for the two cell lines considered (Haas-Kogan et al. 1995). Simulation predictions agree at least semiquantitatively with clinical experience and in particular with the outcome of the Radiation Therapy Oncology Group RTOG Study 83 02 (Werner-Wasik et al. 1996). The model allows for a quantitative study of the interrelationship between the competing influences in a complex, dynamic tumor environment. Therefore, the model is already useful as an educational tool with which to theoretically study, understand, and demonstrate the role of various parameters on tumor growth and response to irradiation. A long-term quantitative clinical adaptation and validation of a considerably extended version of the model is in progress within the framework of the ContraCancrum project. The long-term goal is integration into the clinical treatment planning procedure.

Figure 18.4 provides the simulation outcomes corresponding to several branches of the clinical study RTOG Study 83 02. Panel (a) shows the total

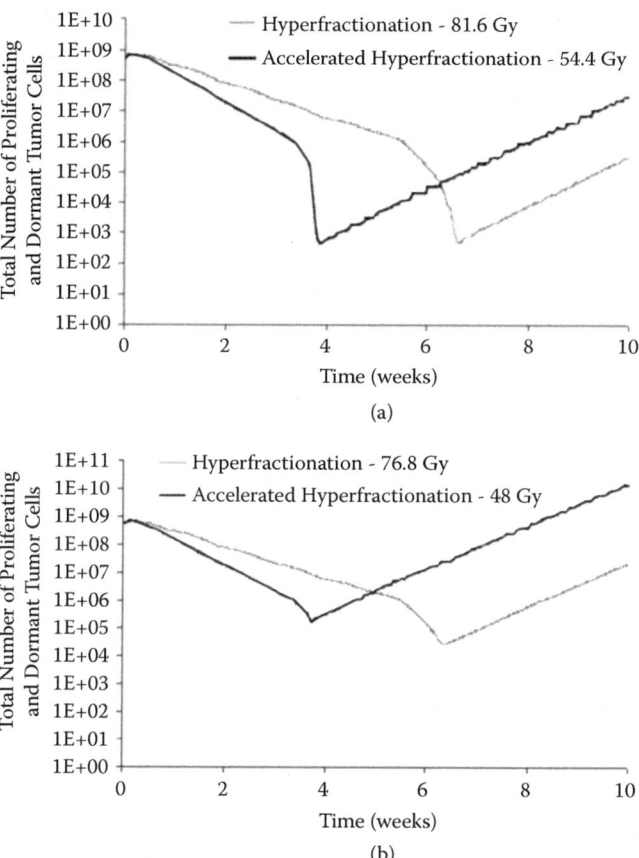

FIGURE 18.4 Simulation predictions corresponding to several branches of the RTOG Study 83 02. (a) total predicted number of proliferating and dormant tumor cells as a function of time for the hyperfractionated (1.2 Gy twice daily to the dose of 81.6 Gy, "HF-81.6") and accelerated hyperfractionated (1.6 Gy twice daily to the dose of 54.4 Gy, "AHF-54.4") radiotherapy schedules. All schemes start on the first day of the radiotherapy course. HF-81.6 is completed on day 46 after initiation of treatment, whereas AHF-54.4 is completed on day 23. (b) total number of proliferating and dormant tumor cells as a function of time for the hyperfractionated (1.2 Gy twice daily to the dose of 76.8 Gy, "HF-76.8") and accelerated hyperfractionated (1.6 Gy twice daily to the dose of 48 Gy, "AHF-48") radiotherapy schedules. Both irradiation schedules start on the first day of the first week of treatment. HF-76.8 is completed on day 44 after initiation of treatment whereas AHF-48 is completed on day 19. (From Stamatakos G.S. et al. 2006a. *Br J Radiol* 79: 389–400. Reprinted with permission from the British Institute of Radiology, © 2006 BIR.)

predicted number of proliferating and dormant tumor cells as a function of time for the hyperfractionated (1.2 Gy twice daily to the dose of 81.6 Gy, "HF-81.6") and accelerated hyperfractionated (1.6 Gy twice daily to the dose of 54.4 Gy, "AHF-54.4") radiotherapy schedules. All schemes start on the first day of the radiotherapy course. HF-81.6 is completed on day 46 after initiation of treatment whereas AHF-54.4 is completed on day 23. Figure 18.4 (b) depicts the total number of proliferating and dormant tumor cells as a function of time for the hyperfractionated (1.2 Gy twice daily to the dose of 76.8 Gy, "HF-76.8") and accelerated hyperfractionated (1.6 Gy twice daily to the dose of 48 Gy, "AHF-48") radiotherapy schedules. Both irradiation schedules start on the first day of the first week of treatment. HF-76.8 is completed on day 44 after initiation of treatment whereas AHF-48 is completed on day 19. According to the graphs, before completion of the AHF course cell kill due to AHF irradiation is more pronounced than cell kill induced by the HF scheme. This can be explained by the fact that a higher total dose has been administered to the tumor by the AHF scheme, whereas for the period under consideration both schemes are characterized by the same time intervals between consecutive sessions. If not all living cells have been killed by AHF irradiation, tumor repopulation is considerable so that by the time the HF scheme is completed living tumor cells and their progeny that have escaped AHF irradiation outnumber tumor cells that have escaped HF irradiation. Improved tumor control following the HF irradiation scheme in comparison with tumor control following the AHF scheme is in agreement with the conclusions of the clinical trial RTOG-83-02.

DISCUSSION AND FUTURE PERSPECTIVES

Since DEBCaST is a numerical method, a thorough convergence and sensitivity/stability analysis that includes the study of multiple parameter interdependences is imperative before any application is envisaged. Numerical analysis should satisfactorily cover at least those regions of the abstract parameter space that correspond to the envisaged applications. It is noted that of particular importance is the creation of the baseline tumor constitution by exploiting the relevant multiscale data available. Convergence of the tumor initialization has to be ensured. All of the above issues have been successfully addressed for specific tumor treatment cases such as breast cancer treated with epirubicin and nephroblastoma treated with vincristine and dactinomycin. The numerical behavior of the corresponding models has been checked through massive numerical experimentation.

Concrete applicability intervals, restrictions, and limitations have been identified (Internal ACGT project reports and deliverables; Kolokotroni et al. 2008; Georgiadi et al. 2008). Since the entire parameter space of DEBCaST models is rather large, numerical studies covering regions that correspond to further applications are in progress. Special attention is paid to the inherent relative biological instability of the cancer system itself when the model's stability is investigated.

It is well known that the values of critical parameters determining treatment outcome can vary considerably around what is assumed to be their population-based average values. Even after incorporation of patient-specific multiscale data into the simulation model, several critical model parameters cannot be accurately evaluated. Moreover, as already mentioned, a tumor may behave as a relatively unstable system. Therefore, in order to compare candidate treatment schemes and/or schedules *in silico*, several possible combinations of parameter values lying around their apparently most probable estimates have to be constructed so as to cover the abstract parameter space as best as possible. Code executions have to be performed for all these selected parameter combinations. If, for example, the clinical question addressed is "Which one of the two candidate treatment schedules denoted by I and II is the most promising for a given patient?" simulations have to be run for both schedules I and II and for all parameter value combinations selected in the way briefly delineated above. If based on the simulation predictions schedule I outperforms schedule II for a sufficiently large percentage of the total parameter combinations considered, say 90%, then there is ground to suggest adoption of schedule I. Candidate scheme/schedule selection criteria are currently under formulation in tight collaboration with specialist clinicians within the framework of the ACGT and the ContraCancrum projects. Obviously, the above-drafted treatment optimization strategy dictates the need for a large number of parallel code executions on either cluster or grid platforms. This necessity has been addressed by specific actions of the previously mentioned projects.

Critical constraints imposed by toxicological limits of the treatment affected normal tissues should also be taken into account in order to judge whether or not a candidate scheme could be toxicologically acceptable. This issue may be addressed by exploiting the outcome of eventually relevant clinical trials and in particular of their phase I results. Ideally, direct multiscale spatiotemporal simulation of the effects of a given candidate scheme on specific normal tissues would provide quantitatively refined predictions. However, due to the extremely high complexity of the homeostatic mechanisms governing normal tissue dynamics, the large number of functional aspects of normal tissue and the potential induction of serious late effects by treatment such as radiotherapy, clinical translation of the second scenario seems to be a long-term enterprise (Antipas et al. 2007).

Since many solid tumors are microscopically inhomogeneous in space, the applications presented so far essentially make use of the *mean values* of certain biological parameters over each imaging-based, segmented subregion of the tumor (Stamatakos et al. 2002, 2006a, 2006b, 2006c, 2007b; Stamatakos and Uzunoglu 2006; Dionysiou et al. 2004, 2006a, 2006b; 2007; 2008; Antipas et al. 2004, 2007). Small perturbations around these values are nevertheless implemented across each region through Monte Carlo simulation by DEBCaST. In the paradigmal case of MRI T1 gadolinium-enhanced imaging modality, strong gray-level fluctuations over a tomographic slice can lead to an approximate delineation of the internal necrotic and the well-neovascularized region of the tumor. Despite the fact that different values of certain parameters may be assigned to these two regions, subimaging-scale inhomogeneities may still create spatial fluctuations of certain parameter values. In order to theoretically investigate the role of such biological inhomogeneities, pertaining for example to the genotypic and/or phenotypic tumor constitution, as well as the role of biochemical inhomogeneities of the extra tumoral environment such as acidity, necrosis exudate concentration, etc., the DEBCaST basic platform can be still used, provided that specific adaptations have taken place. Furthermore, tumor cell–tumor cell, tumor cell–host cell, and tumor cell–local environment interactions in the microscopic setting can, in principle, be studied. In order to implement the above scenarios, the density of the discretization mesh should considerably increase, a deeper level partitioning into more equivalence (sub-)classes has to be introduced into the hypermatrix of the anatomical region of interest, and the operators should be extended accordingly. However, such an approach dictates a sharp increase in computing memory and time demands, and therefore the tumor size must be kept small if restrictions in these resources apply, as is usually the case.

Following appropriate adaptation of specific modeling modules or equivalently operators such as operator H, DEBCaST is, in principle, able to simulate tumors of any shape, size, geometry, macroscopic distribution of the metabolic or neovascularization field, differentiation grade, spatial inhomogeneities, molecular profile, and treatment scheme/schedule such as radiotherapeutic, chemotherapeutic, combined, and new treatment modalities. However, great care should be taken so that the model parameter values are estimated as accurately as possible, based on real multiscale data. If such data is not available, use of at least population-based average parameter value estimates or qualitative experience-based plausible values may be utilized only for generic exploratory reasons.

Hybridization of DEBCaST with continuous and finite mathematics approaches such as diffusion-based tumor growth modeling and detailed biomechanics is currently under implementation (ContraCancrum project). The aim of the task is to integrate into a DEBCaST GBM model the microscopic tumor invasion process that conceptually constitutes part of operator E. The detailed biomechanics of the system calculated via a finite element module and constituting also part of operator E is also being integrated. Such a hybrid model is expected to be able to reproduce in relative detail both physical and biological aspects of tumor dynamics within the generic investigational framework. It should be noted that the nonimageable diffusive component of GBM does play an important role in the development of the disease and therefore merits an in-depth theoretical investigation. However, since the nonimageable boundaries of GBM cannot be defined and monitored in a sufficiently objective way, that is, based on observational data such as clinically obtainable tomographic images, direct handling of the nonimageable component by treatment planning systems in the patient-individualized treatment context seems not to be a fully mature scenario as yet. Furthermore, by focusing on the imageable component within the treatment optimization context, one may argue that if for the imageable component a candidate treatment scheme denoted by scheme I outperforms another candidate scheme denoted by scheme II *in silico*, the same would be true for the nonimageable component of the tumor as well. Furthermore, by focusing on the imageable component within the treatment optimization context, one may expect that if for the imageable component a candidate treatment scheme denoted by scheme I outperforms another candidate scheme denoted by scheme II *in silico*, the same would be true for the nonimageable component of the tumor as well. The latter can be substantiated as follows. Although there may be differing gene activity, blood–tumor barrier status, and perhaps other microscopic characteristics between the imageable (predominantly proliferative) and the nonimageable (predominantly invasive) component of a GBM tumor, the above assumption reflects the current clinical practice. The latter is based on the absence of clinically tested and exploitable information supporting the opposite hypothesis (i.e., an uncorrelated treatment sensitivity behavior of the imageable and the nonimageable component). Such an approach is in line with the principle of parsimony (Occam's razor). As a consequence, up to now no differing standard treatments have been developed for the two tumor components. Nevertheless, in case that clinically reliable information supporting an uncorrelated treatment response behavior of the two major GBM components becomes available in the future, adaptations in the simulation exploitation policy can readily be made.

From the treatment perspective, again, the main advantage of focusing on the imageable component, although this may represent even less than half of the total number of all viable tumor cells, is that this very component is amenable to relatively objective measurement *in vivo* and not only postmortem. Therefore, glioma dynamics models based on the imageable component are amenable to validation, at least in part, in vivo. Besides, incorporation of the immune system response to the tumor (D'Onofrio 2005) and simulation of the effects of antiangiogenetic drugs on the tumor are two further scenarios, currently under investigation in the ACGT Oncosimulator extension context.

Referring to the molecular level from the generic investigational standpoint, a large number of mechanisms that can be informed by available molecular data, such as pathways leading to apoptosis or survival, can be readily integrated into DEBCaST models. The latter can be achieved by applying the *summarize* and *jump* strategy of biodata and bioknowledge integration across biocomplexity scales (Stamatakos et al. 2009). This is, in fact, one of the actions currently taking place within the framework of the ContraCancrum project. However, if the same biocomplexity level is viewed from the clinical perspective, care has to be taken so that only those characteristics and mechanisms whose predictive potential has been proved and established in the clinical setting—normally through clinical trials—may be incorporated into the models.

Regarding the anticipated clinical translation of DEBCaST-based models and systems, including the Oncosimulator, a *sine qua non* prerequisite is a systematic, formal, and strict clinical validation. Designing the models so as to mimic actual clinical or—far better—clinicogenomic trials seems to be the optimal way to achieve this goal (Stamatakos et al. 2006a, 2007a; Graf and Hoppe 2006; Graf et al. 2007, 2008, 2009). Therefore, involvement of clinicians in the model and system design and validation process should start at the very beginning of the endeavor (Graf et al. 2007, 2008,

2009). Real clinicogenomic trials can provide invaluable multiscale data (imaging, histological, molecular, clinical, treatment) before, during, and after a treatment course so as to best adapt and optimize the models and subsequently validate them. This is one of the core tasks of both the ACGT and the ContraCancrum projects. Nephroblastoma and breast cancer are the tumor types addressed by ACGT, whereas gliomas and lung cancer are the ones addressed by ContraCancrum.

A further important challenge is to develop reliable, efficient, highly versatile and user-friendly technological platforms which, following clinical adaptation, optimization, and validation of the models would facilitate translation of Oncosimulators into the clinical practice so as to efficiently support, enhance, and accelerate patient-individualized treatment optimization. Advanced image processing, visualization, and parallel-code execution modules are but a few of the components necessary to achieve this goal.

In summary, having taken into account the hypercomplex and heavily multiscale character of the natural phenomenon of cancer, as well as the constantly expanding accumulation of experimental and clinical knowledge pertaining to the disease, both DEBCaST and the Oncosimulator have been designed so as to be readily optimizable, extensible, and adaptable to varying clinical, biological, and research contexts. In other words both entities, being primarily biomedical-engineering-geared, have a clearly pragmatic and evolutionary character. This has been largely achieved through the extensive application of the principles of discrete event simulation and system modularity.

ACKNOWLEDGMENTS

This work has been supported in part by the European Commission under the projects "ACGT: Advancing Clinicogenomic Trials on Cancer" (FP6-2005-IST-026996) and ContraCancrum: Clinically Oriented Translational Cancer Multilevel Modelling" (FP7-ICT-2007-2- 223979).

REFERENCES

Antipas, V., G.S. Stamatakos, N. Uzunoglu, D. Dionysiou, and R. Dale. 2004. A spatiotemporal simulation model of the response of solid tumors to radiotherapy *in vivo*: parametric validation concerning oxygen enhancement ratio and cell cycle duration. *Phys Med Biol* 49: 1–20.

Antipas, V.P., G.S. Stamatakos, and N.K. Uzunoglu. 2007. A patient-specific *in vivo* tumor and normal tissue model for prediction of the response to radiotherapy: a computer simulation approach. *Meth Inf Med* 46: 367–375.

Bobola, M.S., S. H. Tseng, A. Blank, M.S. Berger, and J.R. Silber. 1996. Role of O6-methylguanine-DNA methyltransferase in resistance of human brain tumor cell lines to the clinically relevant methylating agents temozolomide and streptozotocin. *Clin Cancer Res* 2: 735–741.

Breward, C.J., Byrne H.M., and C.E. Lewis. 2003. A multiphase model describing vascular tumour growth. *Bull Math Biol* 65(4): 609–640.

Clatz, O., Sermesant M., Bondiau P.Y., Delingette H., Warfield S.K., Malandain G., and Ayache N. 2005. Realistic simulation of the 3-D growth of brain tumors in MR images coupling diffusion with biomechanical deformation. *IEEE Trans Med Imaging* 24(10): 1334–1346.

Cristini, V., H.B. Frieboes, R. Gatenby, S. Caserta, M.Ferrari, and J.P. Sinek. 2005. Morphological instability and cancer invasion. *Clin Cancer Res* 11: 6772–6779.

Dionysiou, D.D., G.S. Stamatakos, N.K. Uzunoglu, K.S. Nikita, and A. Marioli, 2004. A four-dimensional simulation model of tumour response to radio-therapy *in vivo*: parametric validation considering radiosensitivity, genetic profile and fractionation. *J Theor Biol* 230: 1–20.

Dionysiou, D.D. and G.S. Stamatakos. 2006. Applying a 4D multiscale in vivo tumor growth model to the exploration of radiotherapy scheduling: the effects of weekend treatment gaps and p53 gene status on the response of fast growing solid tumors. *Cancer Informatics* 2: 113–121.

Dionysiou, D.D., G.S. Stamatakos, N.K., Uzunoglu, and K.S. Nikita. 2006a. A computer simulation of in vivo tumour growth and response to radiotherapy: new algorithms and parametric results. *Comput Biol Med* 36: 448–464.

Dionysiou, D.D., G.S. Stamatakos, N.K. Uzunoglu, and K.S. Nikita. 2006b. A computer simulation of *in vivo* tumour growth and response to radiotherapy: new algorithms and parametric results. *Comput Biol Med* 36: 448–464.

Dionysiou, D.D., G.S.Stamatakos, and K. Marias. 2007. Simulating cancer radio-therapy on a multi-level basis: biology, oncology and image processing. *Lect Notes Comput Sci* 4561: 569–575.

Dionysiou, D.D., G.S. Stamatakos, D. Gintides, N. Uzunoglu, and K. Kyriaki. 2008. Critical parameters determining standard radiotherapy treatment outcome for glioblastoma multiforme: a computer simulation. *Open Biomed Eng J* 2: 43–51.

D'Onofrio, A. 2005. A general framework for modeling tumor-immune system competition and immunotherapy: analysis and medical inferences. *Physica D* 208: 220–235.

Duechting, W. and T. Vogelsaenger. 1981. Three-dimensional pattern generation applied to spheroidal tumor growth in a nutrient medium. *Int J Biomed Comput* 12 (5): 377–392.

Duechting, W., W. Ulmer, R. Lehrig, T. Ginsberg, and E. Dedeleit. 1992. Computer simulation and modeling of tumor spheroid growth and their relevance for optimization of fractionated radiotherapy. *Strahlenther Onkol* 168(6): 354–360.

Enderling, H., M.A.J. Chaplain, A.R.A Anderson, and J.S. Vaidya. 2007. A mathematical model of breast cancer development, local treatment and recurrence. *J Theor Biol* 246: 245–259.

FDA, Center for Drug Evaluation and Research. 1999. Application Number: 21029, Drug Name: Temozolomide (Temodal), New Drug Application, Clinical Pharmacology and Biopharmaceutics Reviews (Revision date: 2/2/1999), pp. 19–20.

Frieboes, H.B., X. Zheng, C.H. Sun, B. Tromberg, R. Gatenby, and V. Cristini. 2006. An integrated computational/experimental model of tumor invasion. *Cancer Res* 66(3): 1597–604.

Georgiadi, E. et al. 2008. Multilevel cancer modeling in the clinical environment: simulating the behavior of Wilms tumor in the context of the SIOP 2001/GPOH clinical trial and the ACGT project. *Proc. 8th IEEE International Conference on Bioinformatics and Bioengineering (BIBE 2008)*, Athens, Greece, October 8–10, 2008. IEEE Catalog Number: CFP08266, ISBN: 978-1-4244-2845-8, Library of Congress: 2008907441, Paper No. BE-2.1.2.

Ginsberg, T., W. Ulmer, and W. Duechting. 1993. Computer simulation of fractionated radiotherapy: further results and their relevance to percutaneous irradiation and brachytherapy. *Strahlenther Onkol* 169: 304–310.

Graf, N. and Hoppe, A. 2006. What are the expectations of a clinician from *in silico* oncology? *Proc. 2nd International Advanced Research Workshop on In Silico Oncology*, Kolympari, Chania, Greece, September 25–26, 2006. Edited by K. Marias and G. Stamatakos. pp. 36–38. [www.ics.forth.gr/bmi/2nd-iarwiso/pdf/conf_proceedings_final.pdf].

Graf, N., C. Desmedt, A. Hoppe, M. Tsiknakis, D. Dionysiou, and G. Stamatakos. 2007. Clinical requirements of "*in silico* oncology" as part of the integrated project ACGT (Advancing Clinico-Genomic Trials on Cancer). *Eur J Cancer Suppl*. 5(4): 83.

Graf, N., C. Desmedt, F. Buffa, D. Kafetzopoulos, N. Forgo, R. Kollek, A. Hoppe, G. Stamatakos, and M. Tsiknakis. 2008. Post-genomic clinical trials—the perspective of ACGT. *Ecancermedicalscience* 2.

Graf, N., A. Hoppe, E. Georgiadi, R. Belleman, C. Desmedt, D. Dionysiou, M. Erdt, J. Jacques, E. Kolokotroni, A. Lunzer, M. Tsiknakis, and G. Stamatakos. 2009. "*In silico* oncology" for clinical decision making in the context of nephroblastoma. *Klin Paediatr* 221: 141–149.

Guiot, C, P.P. Delsanto, A. Carpinteri, N. Pugno, Y. Mansury, and T.S. Deisboeck. 2006. The dynamic evolution of the power exponent in a universal growth model of tumors. *J Theor Biol* 240(3): 459–463.

Haas-Kogan, D.A., G. Yount, M. Haas, D. Levi, S.S. Kogan, L. Hu et al. 1995. p53-dependent G1 arrest and p53 independent apoptosis influence the radiobiologic response of glioblastoma. *Int J Radiat Oncol Biol Phys* 36: 95–103.

Kansal, A.R, S. Torquato, G.R. Harsh, E.A. Chiocca, and T.S. Deisboeck. 2000. Simulated brain tumour growth dynamics using a three-dimensional cellular automaton. *J Theor Biol* 203: 367–382.

Katzung, B.G. Ed. 2001. *Basic and Clinical Pharmacology*, 8th ed. Lange Medical Books, McGraw-Hill, New York.

Kolokotroni, E.A., G.S. Stamatakos, D.D. Dionysiou, E.Ch. Georgiadi, Ch. Desmedt, and N.M. Graf. 2008. Translating multiscale cancer models into clinical trials: simulating breast cancer tumor dynamics within the framework of the "Trial of Principle" clinical trial and the ACGT project, *Proc. 8th IEEE International Conference on Bioinformatics and Bioengineering (BIBE 2008)*, Athens, Greece, October 8–10, 2008. IEEE Catalog Number: CFP08266, ISBN: 978-1-4244-2845-8, Library of Congress: 2008907441, Paper No. BE-2.1.1.

Mansury, Y. and T.S. Deisboeck. 2003. The impact of "search precision" in an agent-based tumor model. *J Theor Biol* 224(3): 325–337.

Marias, K., D. Dionysiou, G.S. Stamatakos, F. Zacharopoulou, E. Georgiadi, T.G. Maris, and I. Tollis. 2007. Multi-level analysis and information extraction considerations for validating 4D models of human function. *Lect Notes Comput Sci* 4561: 703–709.

Morse, P.M. and H. Feshbach. 1953. *Methods of Theoretical Physics*. Part I, McGraw-Hill, New York. pp. 76–92.

Murray, J.D. 2003. *Mathematical Biology II. Spatial Models and Biomedical Applications*, 3rd ed. Springer-Verlag, Heidelberg, pp. 543–546.

Newlands, E.S., G.R. P. Blackledge, J.A. Slack et al. 1992. Phase I trial of temozolomide (CCRG 81045: M&B 39831:NSC 362856), *Br. J. Cancer* 65: 287–291.

Perez, C. and L. Brady, Eds., 1998. *Principles and Practice of Radiation Oncology*, 3rd ed. Lippincott-Raven, Philadelphia.

Perry, M.C. Ed. 2001. *The Chemotherapy Source Book*, 3rd ed. Lippincott Williams and Wilkins, Philadelphia.

Ramis-Conde, I., M.A.J. Chaplain, and A.R.A. Anderson. 2008. Mathematical modelling of cancer cell invasion of tissue. *Math Comput Model* 47: 533–545.

Salmon, S.E. and A.C. Sartorelli. 2001. Cancer chemotherapy. In *Basic and Clinical Pharmacology*. B.G. Katzung, Ed. Lange Medical Books/McGraw-Hill, International edition, pp. 923–1044.

Schiff, L.I. 1981. *Quantum Mechanics*, 3rd ed. McGraw Hill Kogakusha, Tokyo. pp. 148–186.

Stamatakos, G., E. Zacharaki, M. Makropoulou, N. Mouravliansky, A. Marsh, K. Nikita, and N. Uzunoglu. 2001a. Modeling tumor growth and irradiation response in vitro—a combination of high-performance computing and web based technologies including VRML visualization. *IEEE Trans Inform Technol Biomed* 5(4): 279–289.

Stamatakos, G.S, E.I. Zacharaki, N.K. Uzunoglu, and K.S. Nikita. 2001b. Tumor growth and response to irradiation in vitro: a technologically advanced simulation model. *Int J Radiat Oncol Biol Phys* 51(3) Suppl. 1: 240–241.

Stamatakos, G.S., D. Dionysiou, K. Nikita, N. Zamboglou, D. Baltas, G. Pissakas and N. Uzunoglu. 2001c. *In vivo* tumour growth and response to radiation therapy: a novel algorithmic description. *Int J Radiat Oncol Biol Phys* 51(3) Suppl. 1: 240.

Stamatakos, G.S., D.D. Dionysiou, E.I. Zacharaki, N.A. Mouravliansky, K.S. Nikita, and N.K. Uzunoglu 2002. *In silico* radiation oncology: combining novel simulation algorithms with current visualization techniques. *Proc. IEEE. Special Issue on Bioinformatics: Adv. Challenges* 90(11): 1764–1777.

Stamatakos, G. 2006. Spotlight on cancer informatics. *Cancer Informatics* 2: 83–86.

Stamatakos, G.S. and Uzunoglu, N. 2006. Computer simulation of tumour response to therapy. In S. Nagl, Ed. *Cancer Bioinformatics: From Therapy Design to Treatment.* John Wiley & Sons, Chichester, U.K. pp. 109–125.

Stamatakos, G.S., V.P. Antipas, N.K. Uzunoglu, and R.G. Dale. 2006a. A four dimensional computer simulation model of the in vivo response to radiotherapy of glioblastoma multiforme: studies on the effect of clonogenic cell density. *Br J Radiol* 79: 389–400.

Stamatakos, G.S., V.P. Antipas, and N.K. Uzunoglu. 2006b. A spatiotemporal, patient individualized simulation model of solid tumor response to chemotherapy *in vivo*: the paradigm of glioblastoma multiforme treated by temozolomide. *IEEE Trans Biomed Eng* 53: 1467–1477.

Stamatakos, G.S., V.P. Antipas, N.K. Uzunoglu. 2006c. Simulating chemotherapeutic schemes in the individualized treatment context: the paradigm of glioblastoma multiforme treated by temozolomide *in vivo*. *Comput Biol Med.* 36(11): 1216–1234.

Stamatakos, G.S., D.D. Dionysiou, N.M. Graf, N.A. Sofra, C. Desmedt, A. Hoppe, N. Uzunoglu and M. Tsiknakis. 2007a. The Oncosimulator: a multilevel, clinically oriented simulation system of tumor growth and organism response to therapeutic schemes. Towards the clinical evaluation of *in silico* oncology. *Proc. 29th Annual Intern Conf IEEE EMBS.* Cite Internationale, Lyon, France, August 23–26. SuB07.1: 6628–6631.

Stamatakos, G.S., D.D. Dionysiou, and N.K. Uzunoglu. 2007b. *In silico* radiation oncology: a platform for understanding cancer behavior and optimizing radiation therapy treatment. In M. Akay, Ed. *Genomics and Proteomics Engineering in Medicine and Biology.* Wiley-IEEE Press, Hoboken, NJ. pp. 131–156.

Stamatakos, G.S., 2008. *In silico* oncology: a paradigm for clinically oriented living matter engineering. *Proc. 3rd International Advanced Research Workshop on In Silico Oncology,* Istanbul, Turkey, September 23–24, 2008. Edited by G. Stamatakos and D. Dionysiou. pp. 7–9 [www.3rd-iarwiso.iccs.ntua.gr/procs.pdf].

Stamatakos, G.S., E. Kolokotroni, D. Dionysiou, E. Georgiadi, and S. Giatili. 2009. *In silico* oncology: a top-down multiscale simulator of cancer dynamics. Studying the effect of symmetric stem cell division on the cellular constitution of a tumour. In O. Dössel and W.C. Schlegel (Eds.): WC 2009, *IFMBE Proceedings 25/IV,* pp. 1830–1833, 2009.

Stamatakos, G. and D. Dionysiou. 2009. Introduction of hypermatrix and operator notation into a discrete mathematics simulation model of malignant tumour response to therapeutic schemes *in vivo*. Some operator properties. *Cancer Informatics* 2009:7 239–251.

Steel, G. Ed. *Basic Clinical Radiobiology,* 3rd ed., 2002. Oxford University Press, Oxford.

Stupp, R., M. Gander, S. Leyvraz, and E. Newlands. 2001. Current and future developments in the use of temozolomide for the treatment of brain tumours. *Lancet Oncol Rev* 2(9): 552–560.

Swanson, K.R., E.C. Alvord, and J.D. Murray. 2002. Virtual brain tumours (gliomas) enhance the reality of medical imaging and highlight inadequacies of current therapy. *Br J Cancer* 86: 14–18.

Werner-Wasik, M., C.B. Scott, D.F. Nelson, L.E. Gaspar, K.J. Murray, J.A. Fischbach, J.S. Nelson, A.S. Weinstein, and W.J. Jr. Curran. 1996. Final report of a phase I/II trial of hyperfractionated and accelerated hyperfractionated radiation therapy with carmustine for adults with supratentorial malignant gliomas. Radiation Oncology Therapy Group Study 83-02. *Cancer* 77(8): 1535–1543.

Zacharaki, E.I., G.S. Stamatakos, K.S. Nikita, and N.K. Uzunoglu. 2004. Simulating growth dynamics and radiation response of avascular tumour spheroid model validation in the case of an EMT6/Ro multicellular spheroid. *Comput Methods Programs Biomed* 76: 193–206.

CHAPTER 19

In Silico Oncology

Part II—Clinical Requirements Regarding In Silico *Oncology*

Norbert Graf

CONTENTS

INTRODUCTION

This chapter deals with the clinical potential and some clinical translation prerequisites of multiscale cancer modeling as perceived from the standpoint of the clinician engaged in clinical trial research. Therefore, the aim is to propose clinical translation guidelines for clinically oriented cancer modeling, rather than to provide the description of any particular mathematical model.

Cancer as a leading cause of death accounted for 7.9 million or 13% of all deaths worldwide in 2007 [1]. The main types of cancer leading to overall mortality are lung, stomach, liver, colon, and breast cancer in adults. To reduce the burden of cancer, evidence-based strategies are needed for its prevention, early detection, treatment, and care. In 2007 the WHO initiated an action plan against cancer with the following four goals:

- Prevent what is preventable.

- Cure what is curable.

- Provide palliative care for all cancer patients.

- Manage and monitor results.

In clinical oncology, today's treatment is still based on surgery, radio-therapy, chemotherapy, and supportive care, including psychosocial support, pain relievers, and palliation. Individualization of treatment is just starting by introducing small molecules as targeted therapies for specific diseases. But there is a significant amount of further knowledge needed, and to be translated to the clinics, before personalized medicine becomes reality. This knowledge is mainly coming from molecular biology (genomics, proteomics, metabolomics, etc.) and clinico-genomic trials. A third source for the individualization of treatment is based on system-biology-driven concepts in biomedicine, which are based on mathematical cancer models, and that belong to *in silico* oncology. Such mathematical models are just at the beginning. But it can be anticipated that they will play an important role as clinical decision makers in the near future. The demand for better treatments based on individual risk factors is obvious as most patients do respond to treatment but fewer will be cured.

ENROLLMENT IN CLINICAL TRIALS

An increase in the number of patients that are enrolled in clinical trials is needed. This can be achieved by simplifying and supporting the clinical trial process—from obtaining the idea, writing of the protocol, and setting up the trial, to collecting and analyzing the data. Today, only 5–10% of adult cancer patients are enrolled in prospective and randomized clinical trials [2] (see Figure 19.1). The fact that a patient is enrolled in a clinical trial increases his chance to be cured per se [3]. In pediatric oncology, more than 90% of patients are enrolled in prospective clinical trials [4]. This is one of the reasons, why survival rates in childhood cancer have risen to 80% today. The need for clinical trials is also based on the fact that for translational research, as well as for mathematical modeling of

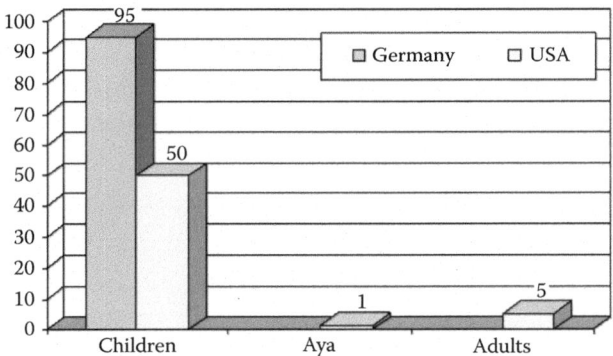

FIGURE 19.1 Enrollment of patients in clinical trials in percentage of patients. (AYA: adolescents and young adults). (From Couzin, J. *Science* 317:1160–1162, 2007.)

cancer, reliable clinical data are needed. These data can only be generated in prospective clinico-genomic trials. Only with such trials can new risk factors for outcome be found, which will allow a better stratification of patients according to their individual risk factors. This will help to spare toxic chemotherapy (CT) in patients who do not need them and increase dose intensity of CT or change of treatment options for those patients with a poor prognosis. Such a treatment plan is given in Figure 19.2 for the prospective randomized trial SIOP 2001/GPOH for the treatment of nephroblastoma in children, adolescents, and young adults in Europe.

NEED OF BIO-BANKS

In every clinical trial there is a demand for bio-banking and analyzing tumor samples in molecular genetic laboratories to gain more insights into the biology of cancer. An example of such a bio-bank is given in Figure 19.3 for the SIOP nephroblastoma trials. A variety of different analytical techniques and methods are available today. This will increase our knowledge with regard to the molecular biology of cancer. As a result, new therapeutics can be awaited that will lead to higher cure rates. Such therapeutics will be small molecules and used as targeted therapies. Drivers for such molecular therapeutics are the following subjects:

- Human Genome Project

- Bioinformatics

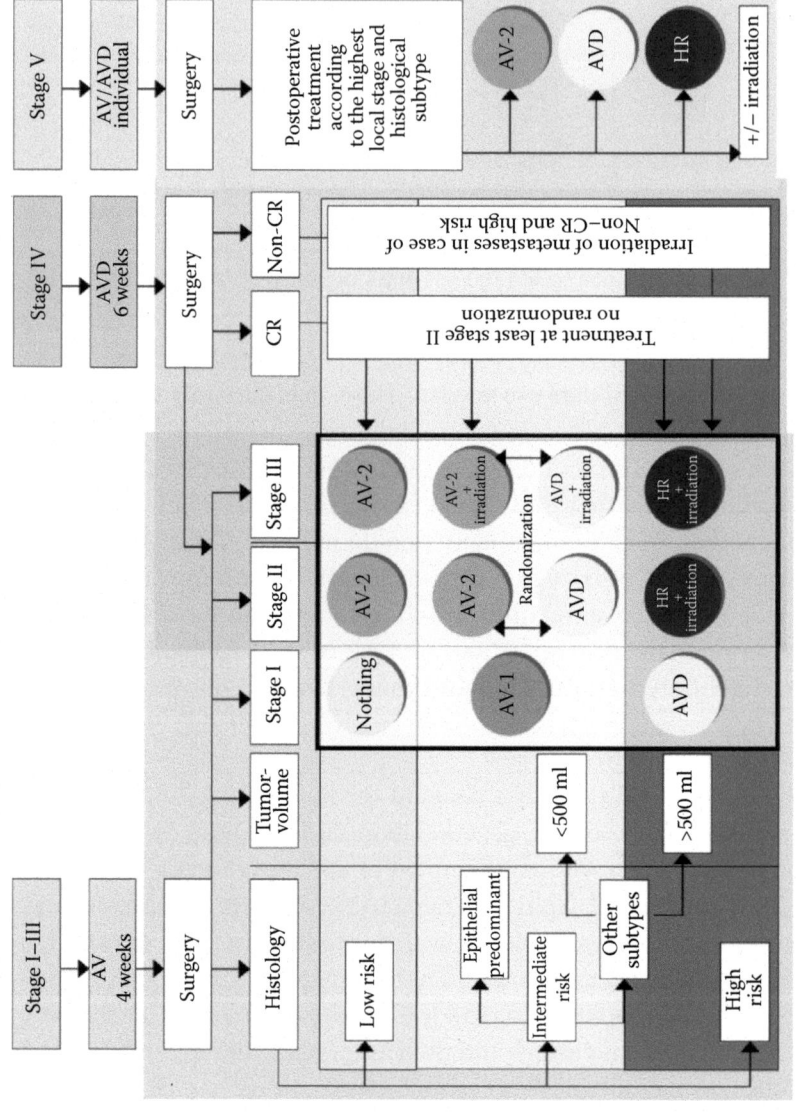

FIGURE 19.2 **(See color insert following page 40)** Treatment schema for nephroblastoma. SIOP 2001/GPOH trial for children, adolescents, and young adults.

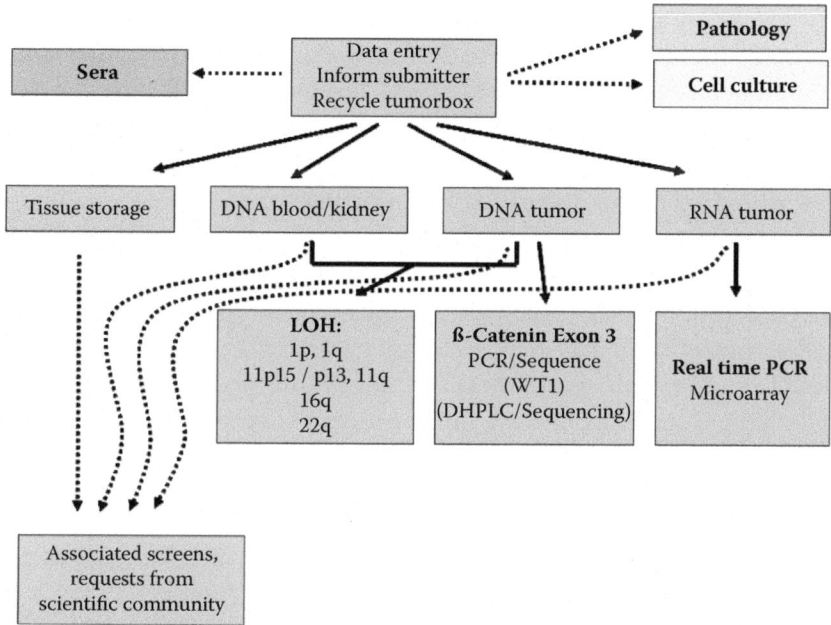

FIGURE 19.3   Schema for a bio-bank as used in the SIOP nephroblastoma trials.

- Robotic high-throughput screening

- Computer-aided drug design

- Expression vectors for novel protein target production

SEAMLESS INTEGRATION OF HETEROGENEOUS DATA

Besides clinical data and molecular biological data, imaging data, pathological data, and other data will be collected in prospective clinical trials. Such heterogeneous data need to be seamlessly integrated in the individual patient's clinical record. An integrated IT platform as it is developed in ACGT (Advancing Clinico Genomic Trials, a European Integrated Project of Framework Package 6) is very helpful for this purpose. Pseudonymized or anonymized data can then be used for translational research in accordance to legal, ethical, and regulatory requirements. Risk-adapted treatment in the future is only possible by fostering translational research [5,6]. Statisticians and bioinformatics can then analyze these data to gain new knowledge in an effort to get closer to personalized medicine.

IN SILICO ONCOLOGY

The aim of *in silico* oncology is to develop patient-specific computer simulation models of malignant tumors and normal tissues in order to optimize the planning of various therapeutic schemes. Ultimately, the goal is to aid in the process of effectively treating cancer and to contribute to the understanding of the disease at the molecular, cellular, organ, and body level.

Only with the help of powerful integrative platforms for the seamless integration of molecular biological, clinical, imaging, and other data, as well as public databases mathematical models of cancer can be generated. To run *in silico* experiments successfully, the characterization of the underlying disease, its biology, and treatment is needed, as well as computational mathematical methods and techniques. From system biology, physiological and pharmacokinetic models have to be developed for usage in such experiments.

From a clinical point of view it is expected that cancer growth and response to different treatments can be simulated. In simulating response to treatment in a given cancer, this knowledge is significant for assessing better methods for treatment efficacy as the RECIST criteria [7,8] provide. It might be time to improve traditional measures of clinical response as trial end points and to evaluate the activity on rare and resistant cancer cells. If *in silico* experiments are to be of more help for a clinician than providing a prediction of changes in tumor volume and shape, the response of treatment of the small fraction of resistant cancer cells will be of utmost importance in the future [9].

There are two possibilities for *in silico* experiments, the top-down and the bottom-up approach. The top-down approach uses clinical observations and the knowledge about the behavior of a cancer as a whole [10]. This approach tries to identify subsystems based on physiological and biological findings that are required to build a reproducible model of a specific cancer [11]. An iterative process continues to find the highest granularity of the system in making the *in silico* model as accurate as possible in reflecting reality. In doing so, the model is kept under surveillance by the overall behavior of the entire system [12,13]. In contrast, a bottom-up approach assembles all known parts of a system, starting with genes and proteins, and brings them into a formal structure until a model of the system is attained [12,14]. The disadvantage of the bottom-up approach is the fact that the discovery of each new component requires a reconfiguration of the whole model [12]. For further information, available biosimulation software is summarized by Ho and Bartsell 2004 [15] and Deisboeck et al., 2009 [16], who gives an excellent overview of *in silico* cancer modeling by reviewing selected studies on modeling the progression and therapy of highly malignant brain tumors.

CLINICAL PRECONDITIONS OF IN VIVO CANCER MODELING

Such *in silico* experiments might help clinicians in the future to find the best way of treating an individual patient by simulating different treatments in the computer before starting the treatment in reality. Two preconditions are of utmost importance before one can rely on *in silico* oncology models [17]:

1. Every *in silico* method should be part of a clinico-genomic trial.

2. Every prediction of an *in silico* method has to be compared with the reality.

After establishing the *in silico* model it is necessary to define the needed data in a first step, including data from the tumor (molecular biology, pathology, imaging), from the patient (clinical data), and from the possible treatment (pharmacokinetics of drugs that will be used, the treatment schema), as well as from the literature and open-source databases. To make the simulation predictions precise and realistic, it is crucial to get as much information as possible from each of the different categories. The amount of data will be restricted by the availability of tumor material, imaging data, and clinical data. Therefore, *in silico* oncology should always be integrated into or be part of a clinico-genomic trial, where data management including data security and anonymization or pseudonymization of data, along with tumor banking, are well established. In addition, the trial is always reviewed by an ethical committee and fulfills all other GCP criteria to get approval by regulatory authorities [18].

The simulation prediction of each *in silico* model must always be compared with the reality, in other words, the actual treatment outcome. The actual outcome provides feedback for tuning the *in silico* model to get better predictions. Such a control loop, executed automatically, must be a component in any *in silico* model whose predictions are to be used for treatment. In that way *in silico* experiments should be considered and established as learning systems. Only if there are no or minimal deviations between the prediction and the reality can the *in silico* method be allowed to be used in a clinical setting. The clinician has to define what can be accepted as a maximal deviation between prediction and reality in a trial. This definition should always be included in the biometrics part of a clinico-genomic trial protocol. For the safety of patients, a stopping rule has to be defined if clinical decisions are based on *in silico* experiments.

CLINICAL QUESTIONS TO BE ADDRESSED

For a clinician it is important that the *in silico* experiments can address and answer precisely for each patient the following questions [10,17]:

1. What is the natural course of the tumor growth over time, in size and shape?

2. When and to which site(s) is the tumor metastasizing?

3. Can the response of the local tumor and the metastases to a given treatment be predicted over time, in size and shape?

4. What is the best treatment schedule regarding medication, surgery, irradiation, and their combination, dosage, time schedule, and duration to achieve a cure?

5. Is it possible to predict severe adverse events (SAE) of a treatment and to propose alternatives to avoid them without deteriorating the outcome?

6. Is it possible to predict a cancer before it becomes clinically manifest, and to recommend treatment options to prevent the occurrence or a recurrence?

Which question will be addressed is decided by the clinician and will influence the model. It has to be accentuated as stated before, that every *in silico* experiment should be part of a prospective clinical trial.

CONCLUSIONS

The question "What is the best treatment for a given tumor in an individual patient?" leads to a high level of individualization. To attain this goal it is of utmost importance that the result of the *in silico* experiment is available in a short timeframe after diagnosis. This implies that all data that are necessary for running the *in silico* experiment have to be available in a timely manner. This is especially important for molecular biologists, radiologists, and clinicians, who have to produce reliable data very fast [10]. Besides this, a high acceptance rate of *in silico* oncology experiments by clinicians is needed in the future. For this purpose, curricula of medical schools need to be changed and physicians need to be trained to use *in silico* experiments as a support system for clinical decision making in the daily clinical care of patients.

REFERENCES

1. WHO: http://www.who.int/mediacentre/factsheets/fs297/en/index.html (last accessed 07/19/2009).
2. Couzin, J. 2007. Survival in young adults with Cancer. *Science* 317:1160–1162.
3. Pritchard-Jones, K. et al. 2008. Improving recruitment to clinical trials for cancer in childhood. *The Lancet Oncol* 9:392–399.
4. Graf, N. and Göbel, U. 2004. Therapieoptimierungsstudien der Gesellschaft für Pädiatrische Onkologie und Hämatologie (GPOH) und 12. Novelle des Arzneimittelgesetzes zur Umsetzung der EU-Richtlinie. *Klin Pädiatr* 216:129–131.
5. Göbel, U. and Jürgens, H. 2003. Translation der kliniknahen Grundlagenforschung in der pädiatrischen Onkologie. *Klin Pädiatr* 215:289–290.
6. Graf, N. et al. 2008. Post-genomic clinical trials: The perspective of ACGT. *eCancerMedicalScience J*, (online journal) 1, Article Num. 66, DOI:10.3332/eCMS.2007.66.
7. Bogaerts, J. et al. 2009. Individual patient data analysis to assess modifications to the RECIST criteria. *EJC* 45:248–260.
8. Therasse, P. et al. 2000. New guidelines to evaluate the response to treatment in solid tumors. *J Natl Cancer Inst* 92:205–216.
9. Huff, C.A. et al. 2006. The paradox of response and survival in cancer therapeutics. *Blood* 107:431–434.
10. Tsiknakis, M. et al. 2008. A semantic grid infrastructure enabling integrated access and analysis of multilevel biomedical data in support of post-genomic clinical trials on Cancer. *IEEE Trans Information Technology in Biomedicine, Special issue on Bio-Grids* 12:191–204.
11. Stamatakos, G. et al. 2002. In silico radiation oncology: combining novel simulation algorithms with current visualization techniques, *Proc. IEEE, Special Issue on Bioinformatics: Advances and Challenges* 90:1764–1777.
12. Friedrich, C.M. and Paterson, T.S. 2004. In silico predictions of target clinical efficacy. *Drug Disc Today* 3:216–222.
13. Michelson, S. 2006. In silico prediction of clinical efficacy. *Curr Opin Biotechnol* 17:666–670.
14. Noble, D. 2002. Modeling the heart—from genes to cells to the whole organ. *Science* 295:1678–1682.
15. Ho, R.L. and Bartsell, L.T. 2004. Biosimulation software is changing research. *Biotechnol Annu Rev* 10:297–302.

16. Deisboeck, T.S. et al. 2009. In silico cancer modeling: is it ready for prime time? *Nat Clin Pract Oncol* 6:34–42.

17. Graf, N. and Hoppe, A. 2006. What are the expectations of a clinician from in silico oncology? In Marias K, Stamatakos G (Hrsg). *Proc. 2nd Int Adv Res Workshop on In Silico Oncology*, Kolympari, Chania, Greece, September 25–26, 2006, pp. 36–38 (http://www.ics.forth.gr/bmi/2nd-iarwiso/; last accessed 07/19/2009).

18. Graf, N. et al. 2009. "In Silico" oncology for clinical decision-making in the context of nephroblastoma. *Klin Pädiatr* 221:141–149.

Abbreviations

A: Apoptosis
ABM: Agent-based model
ACGT: Advancing Clinico-Genomic Trials on Cancer
AHF: Accelerated Hyper-Fractionation
BIR: British Institute of Radiology
Cpav: Average plasma concentration
CT: Computerized tomography
DEBCaST: Discrete Event Based Cancer Simulation Technique
EC: European Commission
ECM: Extracellular matrix
EGF: Epidermal growth factor
EGFR: *EGF* receptor
EMT: Epithelial–mesenchymal transition
ERK: Extracellular signal-regulated kinase
GBM: Glioblastoma multiforme
G_0 or G0: Dormant phase
G_1 or G1: Gap 1 phase of the cell cycle
G_2 or G2: Gap 2 phase
HF: Hyper-Fractionation
HMM: Heterogeneous multiscale method
IBM: Individual-based model
IEEE: Institute of Electrical and Electronics Engineers
M: Mitosis
MRI: Magnetic resonance imaging
N: Necrosis
NSCLC: Non-small cell lung cancer
ODE: Ordinary differential equation
PDE: Partial differential equation
PET: Positron emission tomography
PLCγ: Phospholipase Cγ

R&D: Research and Development
RTOG: Radiation Therapy Oncology Group
S: DNA synthesis phase
TDS: Time Delay in the S phase compartment
TGFβ: Transforming growth factor β
TMZ: Temozolomide
2D: Two-dimensional
3D: Three-dimensional

Index